Theory of Ground Vehicles

Theory of Ground Vehicles

Fifth Edition

J.Y. Wong, Ph.D., D.Sc.
Professor Emeritus
Department of Mechanical and Aerospace Engineering
Carleton University
Ottawa, Canada

This edition first published 2022
© 2022 John Wiley & Sons, Inc.

Edition history
John Wiley & Sons, Inc (1e 1978; 2e 1993; 3e 2001; 4e 2008)

Registered Office
John Wiley & Sons, Inc., 111 River Street, Hoboken, NJ 07030, USA

Editorial Office
111 River Street, Hoboken, NJ 07030, USA

For details of our global editorial offices, customer services, and more information about Wiley products visit us at www.wiley.com.

Wiley also publishes its books in a variety of electronic formats and by print-on-demand. Some content that appears in standard print versions of this book may not be available in other formats.

Library of Congress Cataloging-in-Publication Data

Names: Wong, J. Y. (Jo Yung), author.
Title: Theory of ground vehicles / J.Y. Wong, Ph.D., D.Sc., Professor
  Emeritus, Department of Mechanical and Aerospace Engineering, Carleton
  University, Ottawa, Canada.
Description: Fifth edition. | Hoboken, N.J., USA : John Wiley & Sons, Inc.,
  2022. | Includes bibliographical references and index.
Identifiers: LCCN 2021050924 (print) | LCCN 2021050925 (ebook) | ISBN
  9781119719700 (hardback) | ISBN 9781119719915 (adobe pdf) | ISBN
  9781119719939 (epub)
Subjects: LCSH: Motor vehicles–Design and construction. | Motor
  vehicles–Dynamics.
Classification: LCC TL240 .W66 2022 (print) | LCC TL240 (ebook) | DDC
  629.2/31–dc23/eng/20211110
LC record available at https://lccn.loc.gov/2021050924
LC ebook record available at https://lccn.loc.gov/2021050925

Cover Design: Wiley
Cover Image: © narvikk/Getty Images

Set in 9.5/12.5pt STIXTwoText by Straive, Pondicherry, India

SKY10035128_071222

To May
  Chak
  Anna
  Amanda
  Ben
  Jing
  Kay
  Leo
  Sang
  Loretta
  San
  Nicholas
  the memory of my parents and
  the glory of the Almighty

Contents

Author Biography

Dr. Jo Yung Wong is Professor Emeritus, Department of Mechanical and Aerospace Engineering, Carleton University, Ottawa, Canada. He received the degrees of PhD and DSc from the University of Newcastle-upon-Tyne, England. He has been engaged in research and development of ground vehicle technologies for decades. Many of his research findings have found wide applications in industry. In recognition of his significant contributions to the field, Dr. Wong has been presented with many awards by learned societies, including the George Stephenson Prize and Starley Premium Award (twice) of the Institution of Mechanical Engineers, United Kingdom.

He has published extensively with numerous research papers and two books. In addition to *Theory of Ground Vehicles*, he is the author of another book, *Terramechanics and Off-Road Vehicle Engineering,* currently in its second edition, published by Elsevier. He is on the editorial/advisory boards of several international scientific journals, including *Vehicle System Dynamics*, *Journal of Terramechanics*, *International Journal of Heavy Vehicle Systems*, and *International Journal of Vehicle Performance*. He has also been a member of the editorial board of the *Journal of Automobile Engineering*, Part D of the Proceedings of the Institution of Mechanical Engineers. Dr. Wong is a former President of the International Society for Terrain-Vehicle Systems. He is Fellow of the American Society of Mechanical Engineers, Institution of Mechanical Engineers, Canadian Society for Mechanical Engineering, and International Society for Terrain–Vehicle Systems.

Dr. Wong has been a consultant to vehicle industry and governmental agencies in North America, Europe, Asia, and Africa. He has lectured on ground vehicle technologies in many countries around the world. At the invitations of Glenn Research Center of the U.S. National Aeronautics and Space Administration, European Space Research and Technology Centre, and Canadian Space Agency, he has presented special professional development programs on extraterrestrial rover mobility to their staff members.

Preface to the Fifth Edition

Reducing greenhouse gas emissions is a central issue for curbing climate change which is of great concern to the global community. As a significant portion of the emissions is associated with transportation, the use of the electric drive, hybrid electric drive, and fuel cells for ground vehicles, to facilitate replacing the internal combustion engine powered by fossil fuels, has received intense worldwide attention. To enhance safety, traffic flow, and operational efficiency of road transport, automated driving systems and cooperative driving automation have been under active development. With the growing interest shown by an increasing number of countries in the exploration of extraterrestrial bodies, such as the moon, Mars, and beyond, research on the mobility of rovers for extraterrestrial surface exploration has attracted considerable interest. Studies of the application of terramechanics to modeling and evaluating rover mobility have been intensified. The discussions of these and other topics of current and future interest in ground vehicle technologies are included in this edition.

While in this edition new topics are introduced and discussions of certain topics covered in previous editions are expanded, emphasis continues being placed on elucidating the physical nature and the mechanics of ground vehicle–environment interactions as in previous editions. Notable features of this edition are highlighted below.

In Chapter 1, definitions of tire slip associated with the application of a driving torque and of tire skid associated with a braking torque are reviewed and updated. Identifications of tire design features, such as load carrying capacity, operating speed range, quality (treadwear), and traction, are included. Discussions of tire/road noise are expanded. In Chapter 2, a method for characterizing terrain behavior pertinent to vehicle mobility using the Bekker–Wong terrain parameters is described. Comparisons of physics-based models with empirically based models for predicting the cross-country performance of off-road vehicles are presented. Approaches to the development of next-generation mobility models are indicated. Discussions on the physical nature of slip sinkage, which may lead to vehicle immobilization on weak terrain as well as soft regolith on extraterrestrial surfaces, and methods for characterizing the relationship between sinkage and slip are included. Applications of terramechanics to the study of extraterrestrial rover mobility are outlined. Methods for predicting the performance of the rover and/or its running gear on extraterrestrial surfaces based on test data obtained on earth under earth gravity are explored. Applications of the discrete (distinct) element method to the study of vehicle–terrain interaction are updated. In Chapter 3, discussions of using the electric drive, hybrid electric drive, and fuel cell to eliminate or reduce greenhouse gas emissions are expanded. Various configurations for the hybrid electric drive are evaluated. Energy consumption characteristics of battery electric vehicles, hybrid electric vehicles, plug-in hybrid electric vehicles, and fuel cell vehicles are presented. To provide a common basis for comparing the performance characteristics of internal combustion engines, methods for converting engine power measured under test atmospheric conditions to that under reference atmospheric conditions are updated. Based on test data, comparisons of fuel consumption characteristics of passenger vehicles with all-wheel drive and those with two-wheel drive, as well as that of vehicles with automatic transmissions and those with manual transmissions, are also included. In Chapter 4, discussions on the necessary and sufficient conditions for achieving the optimal tractive efficiency of all-wheel drive off-road vehicles are further elucidated. In Chapter 5, an introduction to the automated driving system and cooperative driving automation is presented. The classification of the levels of driving automation and the associated enabling technologies are outlined. In Chapter 7, the evaluation of human exposure to whole-body vibration is updated, in accordance with ISO 2631-1, Amendment 1 : 2010. The characterization of vertical road surface profiles by displacement power spectral density for vehicle ride assessment, in accordance with ISO 8608 : 2016, is presented. The use of the International Roughness Index for classifying longitudinal profile of traveled surfaces is introduced. While commercial services with passenger-carrying air-cushion vehicles across stretches of water have declined since the early 2000s, the air-cushion vehicle still plays an active role in defense and coast guard operations, in search and rescue missions, and in recreational activities, because of its unique capability of being able to travel across a variety of surfaces. Consequently, Chapter 8 on the introduction to air-cushion vehicles is retained in this edition.

Since the publication of the first edition of this book in 1978, three subsequent editions with many printings each have been continually published over a period of more than four decades and they have been widely cited. The translations of the first edition into Russian and Chinese languages were published in Moscow and Beijing, respectively. The translation of the fourth edition into Chinese language was published in Beijing in 2018. It is gratifying to see that the previous editions have served the needs of the global educational and professional communities in the field of ground vehicle technology. It is anticipated that this fifth edition will continue serving their needs in the years to come.

I wish to express my appreciation to many collaborators in the automotive industry, research institutions, and universities for their contributions to, as well as industrial organizations and governmental agencies for their support for, our research enterprise over the years. The technical assistance provided by Jon Preston-Thomas and Wei Huang for the preparation of this fifth edition is appreciated.

Jo Yung Wong
Toronto, Canada

Preface to the Fourth Edition

Since the publication of the third edition of this book in 2001, it has again been well received by both the educational and professional communities and has gone through more than six printings. A growing number of universities and colleges in North America, Europe, the Middle East, Africa, and Asia have adopted it either as a text or as a reference book for undergraduate or postgraduate courses in the subject area. Professionals in industry and in research institutions have continued using it as a reference in the design, development, and research of automotive vehicles, as well as extraterrestrial vehicles. All of these have encouraged the author to publish an updated edition of this book. While the general objective, contents, and format of the third edition have been retained, new material has been introduced to this edition to reflect recent advances in ground transportation technology.

With the continuing interest in the use of computer simulation models in vehicle design and development, the latest examples of applications of the simulation model for tracked vehicles, developed by the author and his associates, have been included in Chapter 2. An outline of the basic features of a computer-aided method for performance and design evaluation of off-road wheeled vehicles has been added. These computer simulation models not only can be used in guiding the development of a new generation of terrestrial off-road vehicles but also have potential applications to the development of mobility subsystems of manned or unmanned robotic rovers for the exploration of the moon, Mars, and beyond. A brief discussion on the applications of finite element and discrete (distinct) element methods to the analysis of vehicle–terrain interaction has also been included in Chapter 2. With the growing concerns over global climate change and future oil supply, "clean-vehicle" technology and alternative power sources have attracted a great deal of attention. A brief introduction to exhaust emissions of internal combustion engines, electric drives, hybrid drives, and fuel cells has been added to Chapter 3. Discussions on the optimization of the tractive performance of four-wheel drive off-road vehicles have been expanded in Chapter 4. Guidelines for the selection of a suitable vehicle configuration for off-road operations, including the issue of "wheeled vehicles vs. tracked vehicles," have been further discussed. In view of the introduction of the Federal Motor Vehicle Safety Standard (FMVSS) No. 126 in the United States, discussions of vehicle stability control systems have been expanded in Chapter 5. In relation to the assessment of vehicle ride, an introduction to the International Standard ISO 2631-1 : 1997 on the evaluation of human exposure to whole-body vibration has been included in Chapter 7. In addition, technical data, such as fuel economy ratings of passenger cars, including hybrid electric vehicles, have been updated.

I would like to express my appreciation to many collaborators in industry, research institutions, and universities, particularly Jon Preston-Thomas and Wei Huang, for their contributions to our research, some of which is presented in this book. Appreciation is due also to Mike Galway, Changhong Liu, and Jiming Zhou for their technical assistance in the preparation of this new edition.

Jo Yung Wong
Ottawa, Canada

Preface to the Third Edition

More than two decades have elapsed since the first publication of this book in the United States in 1978. During this period, the first edition went through 10 printings, and the second edition, which first appeared in 1993, went through more than 7 printings. An increasing number of universities in North America, Europe, Asia, and elsewhere have adopted it as a text for courses in automotive engineering, vehicle dynamics, off-road vehicle engineering, or terramechanics. Many professionals in the vehicle industry around the world have also used it as a reference. It is gratifying indeed to see that the book has achieved such wide acceptance.

As we enter a new millennium, the automotive industry is facing greater challenges than ever before in providing safer, more environmentally friendly, and more energy-efficient products to meet increasingly stringent demands of society. As a result, new technologies have continually been developed and introduced into its products. Accordingly, to better serve the changing needs of the educational and professional communities related to ground transportation technology, this third edition has been prepared.

To improve competitiveness, shortening the product development cycle is of critical importance to vehicle manufacturers. Virtual prototyping is therefore widely adopted in the industry. To implement this process effectively, however, the development of reliable computer simulation models for vehicle performance evaluation is essential. For a realistic simulation of the handling behavior of road vehicles, a method referred to as the Magic Formula for characterizing tire behavior from test data is gaining increasingly wide acceptance. A discussion of the basic features of the Magic Formula is included in Chapter 1 of this edition. For performance and design evaluation of off-road vehicles, particularly with respect to their soft ground mobility, a variety of computer simulation models have emerged, including those developed by myself along with my associates. It is encouraging that our models have since played a significant role in assisting vehicle manufacturers in the development of a new generation of high-mobility off-road vehicles as well as governmental agencies in evaluating vehicle candidates in North America, Europe, Asia, Africa, and elsewhere. In recognition of our contributions to the development of these simulation models, we have been presented with a number of awards by learned societies. These include the George Stephenson Prize, the Starley Premium Award (twice), and the Engineering Applied to Agriculture Award presented by the Institution of Mechanical Engineers. The major features and practical applications of these simulation models are described in Chapter 2. New experimental data on the optimization of the tractive performance of four-wheel drive off-road vehicles based on our own investigations are presented in Chapter 4.

To further enhance the active safety of road vehicles, systems known as "vehicle stability control" or "vehicle dynamics control" have been introduced in recent years. The operating principles of these systems are described in Chapter 5. A new theory developed by us for skid-steering of tracked vehicles on firm ground is presented in Chapter 6. It is shown that this new theory offers considerable improvement over existing theories and provides a unified approach to the study of skid-steering of tracked vehicles. Experimental data, obtained from our own research, on the performance of an electrorheological damper in improving the ride comfort of ground vehicles are presented in Chapter 7.

While new topics are introduced and new data are presented in this third edition, the general objective, contents, and format remain similar to those of previous editions. The fundamental engineering principles underlying the rational development and design of road vehicles, off-road vehicles, and air-cushion vehicles are emphasized.

To a certain extent, this book summarizes some of my experience of more than three decades in teaching, research, and consulting in the field of ground transportation technology. I would like to take this opportunity once again to record my appreciation to my colleagues and collaborators in industry, research institutions, and universities for inspiration and cooperation, particularly Dr. Alan R. Reece, Professor Leonard Segel, and the late Dr. M. Gregory Bekker. I wish also to express my appreciation to staff members of Transport Technology Research Laboratory, Carleton University, and Vehicle Systems Development Corporation, Nepean, Ontario, and to my postdoctoral fellows and postgraduate students, former and present, for their contributions and assistance. Thanks are also due to governmental agencies and vehicle manufacturers for supporting our research effort over the years.

Jo Yung Wong
Ottawa, Canada

Preface to the Second Edition

Since the first edition of this book was published in 1978, it has gone through 10 printings. A number of engineering schools in North America, Europe, Asia, and elsewhere have adopted it as a text for courses in automotive engineering, vehicle dynamics, off-road vehicle engineering, agricultural engineering, etc. It was translated into Russian and published in Moscow, Russia in 1982, and into Chinese and published in Beijing, China in 1985. Meanwhile, significant technological developments in the field have taken place. To reflect these new developments and to serve the changing needs of the educational and professional communities, the time is ripe for the second edition of this book.

With the growing emphasis being placed by society on energy conservation, environmental protection, and safety, transportation technology is under greater challenge than ever before. To improve fuel economy and to reduce undesirable exhaust emissions, in addition to improvements in power plant design, measures such as improving vehicle aerodynamic performance, better matching of transmission with engine, and optimizing power requirements have received intense attention. To improve driving safety, antilock brake systems and traction control systems have been introduced. To provide better ride comfort while maintaining good roadholding capability, active and semi-active suspension systems have attracted considerable interest. To expedite the development of new products, computer-aided methods for vehicle performance and design optimization have been developed. Discussions of these and other technological developments in the field have been included in this second edition. Furthermore, data on various topics have been updated.

As with the first edition, this second edition of *Theory of Ground Vehicles* is written with the same philosophy of emphasizing the fundamental engineering principles underlying the rational development and design of nonguided ground vehicles, including road vehicles, off-road vehicles, and air-cushion vehicles. Analysis and evaluation of performance characteristics, handling behavior, and ride comfort of these vehicles are covered. A unified method of approach to the analysis of the characteristics of various types of ground vehicle is again stressed. This book is intended primarily to introduce senior undergraduate and beginning graduate students to the study of ground vehicle engineering. However, it should also be of interest to engineers and researchers in the vehicle industry.

Similar to the first edition, this second edition consists of eight chapters. Chapter 1 discusses the mechanics of pneumatic tires. Practical methods for predicting the behavior of tires subject to longitudinal or side force, as well as under their combined action, are included. New experimental data on tire performance are added. Chapter 2 examines the mechanics of vehicle–terrain interaction, which has become known as "terramechanics." Computer-aided methods for the design and performance evaluation of off-road vehicles are included. Experimental data on the mechanical properties of various types of terrain are updated. Chapter 3 deals with the analysis and prediction of road vehicle performance. Included is updated information on the aerodynamic performance of passenger cars and articulated heavy commercial vehicles. Procedures for matching transmission with engine to achieve improved fuel economy while maintaining adequate performance are outlined. Characteristics of continuously variable transmissions and their effects on fuel economy and performance are examined. The operating principles of antilock brake systems and traction control systems and their effects on performance and handling are presented in some detail. The performance of off-road vehicles is the subject of Chapter 4. Discussions on the optimization of the performance of all-wheel drive off-road vehicles are expanded. In addition, various criteria for evaluating military vehicles are included. Chapter 5 examines the handling behavior of road vehicles. In addition to discussions of the steady-state and transient handling behavior of passenger cars, the handling characteristics of tractor–semitrailers are examined. The handling diagram for evaluating directional response is included. The steering of tracked vehicles is the topic of Chapter 6. In addition to skid-steering, articulated steering for tracked vehicles is examined. Chapter 7 deals with vehicle ride comfort. Human tolerance to vibration, vehicle ride models, and applications of the random vibration theory to the evaluation of ride comfort are covered. Furthermore, the effects of suspension spring stiffness, damping, and unsprung mass on vibration isolation characteristics, roadholding, and suspension travel are examined. The principles of active and semi-active suspensions are also discussed. In addition to conventional road vehicles and off-road vehicles, air-cushion vehicles have found applications in ground transportation. The basic principles of air-cushion systems and the unique characteristics of air-cushion vehicles for overland and overwater operations are treated in Chapter 8. New data on the mechanics of skirt–terrain interaction are included.

The material included in this book has been used in the undergraduate and graduate courses in ground transportation technology that I have been teaching at Carleton for some years. It has also been presented, in part, at seminars and in professional development programs in Canada, China, Finland, Germany, Italy, Singapore, Spain, Sweden, Taiwan, the United Kingdom, and the United States.

In preparing the second edition of this book, I have drawn much on my experience acquired from collaboration with many of my colleagues in industry, research organizations, and universities in North America, Europe, Asia, and elsewhere. The encouragement, inspiration, suggestions, and comments that I have received from Dr. A. R. Reece, formerly of the University of Newcastle-upon-Tyne, and currently Managing Director, Soil Machine Dynamics Limited, England; Professor L. Segel, Professor Emeritus, University of Michigan; and Professor E. H. Law, Clemson University, are particularly appreciated. I would also like to record my gratitude to the late Dr. M. G. Bekker, with whom I had the good fortune to collaborate in research projects and in joint offerings of professional development programs, upon which some of the material included in this book was developed.

The typing of the manuscript by D. Dodds and the preparation of additional illustrations by J. Brzezina for this second edition are appreciated.

JO YUNG WONG
Ottawa, Canada

Preface to the First Edition

Society's growing demand for better and safer transportation, environmental protection, and energy conservation has stimulated new interest in the development of the technology for transportation. Transport technology has now become an academic discipline in both graduate and undergraduate programs at an increasing number of engineering schools in North America and elsewhere. While preparing lecture notes for my two courses at Carleton on ground transportation technology, I found that although there was a wealth of information in research reports and in journals of learned societies, there was yet no comprehensive account suitable as a text for university students. I hope this book will fill this gap.

Although this book is intended mainly to introduce senior undergraduate and beginning graduate students to the study of ground vehicles, it should also interest engineers and researchers in the vehicle industry. This book deals with the theory and engineering principles of nonguided ground vehicles, including road, off-road, and air-cushion vehicles. Analysis and evaluation of performance characteristics, handling behavior, and ride qualities are covered. The presentation emphasizes the fundamental principles underlying rational development and design of vehicle systems. A unified method of approach to the analysis of the characteristics of various types of ground vehicle is also stressed.

This book consists of eight chapters. Chapter 1 discusses the mechanics of pneumatic tires and provides a basis for the study of road vehicle characteristics. Chapter 2 examines the vehicle running gear–terrain interaction, which is essential to the evaluation of off-road vehicle performance. Understanding the interaction between the vehicle and the ground is important to the study of vehicle performance, handling, and ride, because, aside from aerodynamic inputs, almost all other forces and moments affecting the motion of a ground vehicle are applied through the running gear–ground contact. Chapter 3 deals with analysis and prediction of the performance of road vehicles. Included in the discussion are vehicle power plant and transmission characteristics, performance limits, acceleration characteristics, braking performance, and fuel economy. The performance of off-road vehicles is the subject of Chapter 4. Drawbar performance, tractive efficiency, operating fuel economy, transport productivity and efficiency, mobility map, and mobility profile are discussed. Chapter 5 examines handling behavior of road vehicles, including steady-state and transient responses, and directional stability. The steering of tracked vehicles is the subject of Chapter 6. Included in the discussion are the mechanics of skid-steering, steerability of tracked vehicles, and steering by articulation. Chapter 7 examines vehicle ride qualities. Human response to vibration, vehicle ride models, and the application of random process theory to the analysis of vehicle vibration are covered. In addition to conventional road and off-road vehicles, air-cushion vehicles have found applications in ground transport. The basic engineering principles of air-cushion systems and the unique features and characteristics of air-cushion vehicles are treated in Chapter 8.

A book of this scope limits detail. Since it is primarily intended for students, some topics have been given a simpler treatment than the latest developments would allow. Nevertheless, this book should provide the reader with a comprehensive background on the theory of ground vehicles.

I have used part of the material included in this book in my two engineering courses in ground transport technology at Carleton. It has also been used in two special professional programs. One is "Terrain–Vehicle Systems Analysis," given in Canada and Sweden jointly with Dr. M. G. Bekker, formerly with AC Electronics-Defense Research Laboratories, General Motors Corporation, Santa Barbara, California. The other is "Braking and Handling of Heavy Commercial Vehicles" given at Carleton jointly with Professor J. R. Ellis, School of Automotive Studies, Cranfield Institute of Technology, England, and Dr. R. R. Guntur, Transport Technology Research Laboratory, Carleton University.

In writing this book, I have drawn much on the knowledge and experience acquired from collaboration with many colleagues in industry, research organizations, and universities. I wish to express my deep appreciation to them. I am especially indebted to Dr. A. R. Reece, University of Newcastle-upon-Tyne, England, Dr. M. G. Bekker, and Professor J. R. Ellis for stimulation and encouragement.

I also acknowledge with gratitude the information and inspiration derived from the references listed at the end of the chapters and express my appreciation to many organizations and individuals for permission to reproduce illustrations and other copyrighted material.

Appreciation is due to Dr. R. R. Guntur for reviewing part of the manuscript and to Dean M. C. de Malherbe, Faculty of Engineering, Professor H. I. H. Saravanamuttoo, Chairman, Department of Mechanical and Aeronautical Engineering, and many colleagues at Carleton University for encouragement.

Jo Yung Wong
Ottawa, Canada
July 1978

Conversion Factors

Quantity	US customary unit	SI equivalent
Acceleration	ft/s^2	0.3048 m/s^2
Area	ft^2	0.0929 m^2
	$in.^2$	645.2 mm^2
Energy	ft·lb	1.356 J
Force	lb	4.448 N
Length	ft	0.3048 m
	in.	25.4 mm
	mile	1.609 km
Mass	slug	14.59 kg
	ton	907.2 kg
Moment of a force	lb·ft	1.356 N·m
Power	hp	745.7 W
Pressure or stress	lb/ft^2	47.88 Pa
	$lb/in.^2$ (psi)	6.895 kPa
Speed	ft/s	0.3048 m/s
	mph	1.609 km/h
Volume	ft^3	0.02832 m^3
	$in.^3$	16.39 cm^3
	gal (liquids)	3.785 liter

Abbreviations and Acronyms

ANSI	American National Standards Institute
ASABE	American Society of Agricultural and Biological Engineers
ASAE	American Society of Agricultural Engineers (name changed to ASABE in 2005)
ASTM	American Society for Testing and Materials
DEM	discrete (distinct) element method
DOT	U.S. Department of Transportation
EPA	U.S. Environmental Protection Agency
EU	European Union
FEM	finite element method
FFT	fast Fourier transform
FMVSS	U.S. Federal Motor Vehicle Safety Standards
IRI	International Roughness Index
ISO	International Organization for Standardization
ISTVS	International Society for Terrain–Vehicle Systems
NASA	U.S. National Aeronautics and Space Administration
NATO	North Atlantic Treaty Organization
NG-NRMM	Next Generation NATO Reference Mobility Model
NHTSA	U.S. National Highway Traffic Safety Administration
NRMM	NATO Reference Mobility Model
NTVPM	Nepean Tracked Vehicle Performance Model
NWVPM	Nepean Wheeled Vehicle Performance Model
RTVPM	Rigid-Link Tracked Vehicle Performance Model
SAE	Society of Automotive Engineers International
USCS	Unified Soil Classification System
WES	U.S. Army Corps of Engineers Waterways Experiment Station

List of Symbols

A	area
A_c	cushion area
A_f	frontal area of a vehicle
A_u	parameter characterizing terrain response to repetitive normal loading
a	acceleration
a_x	acceleration component along the x axis
a_y	acceleration component along the y axis
a_z	acceleration component along the z axis
B	axle or vehicle tread (track) (i.e., transverse distance between left- and right-side wheels on an axle, or transverse distance between left- and right-side tracks of a tracked vehicle)
B_m	working width of machinery
b	width
C, CI	cone index
C_D	aerodynamic resistance (drag) coefficient
C_f	ratio of braking effort to normal load of vehicle front axle
C_i	longitudinal stiffness of tire subject to a driving torque
C_L	aerodynamic lift coefficient
C_{ld}	lift/drag ratio
C_M	aerodynamic pitching moment coefficient
C_r	ratio of braking effort to normal load of vehicle rear axle
C_{ro}	restoring moment coefficient
C_s	longitudinal stiffness of tire subject to a braking torque
C_{se}	ratio of braking effort to normal load of semitrailer axle
C_{sk}	coefficient of skirt contact drag
C_{sp}	coefficient of power spectral density function
C_{sr}	speed ratio of torque converter
C_{tr}	torque ratio of torque converter
C_α	cornering stiffness of tire
$C_{\alpha f}$	cornering stiffness of front tire
$C_{\alpha r}$	cornering stiffness of rear tire
$C_{\alpha s}$	cornering stiffness of semitrailer tire

C_γ	camber stiffness of tire
c	cohesion
c_a	adhesion
c_{eq}	equivalent damping coefficient
c_{sh}	damping coefficient of shock absorber
c_t	damping coefficient of tire
D	diameter
D_c	discharge coefficient
D_h	hydraulic diameter
D_r	relative density of soil
E	energy
E_d	pulling energy available at vehicle drawbar hitch
F	force, thrust, or tractive effort
F_b	braking force
F_{bf}	braking force of vehicle front axle
F_{br}	braking force of vehicle rear axle
F_{bs}	braking force of semitrailer axle
F_{cu}	lift generated by air cushion
F_d	drawbar pull
F_{emax}	maximum tractive effort on earth surface
F_{exmax}	maximum tractive effort on extraterrestrial surface
F_f	thrust of vehicle front axle
F_h	hydrodynamic force acting on a tire over flooded surfaces
F_{hi}	horizontal force acting at the hitch point of a tractor–semitrailer
F_i	thrust of the inside track of a tracked vehicle in a turn
F_l	lift generated by the change of momentum of an air jet
F_{net}	net thrust
F_o	thrust of the outside track of a tracked vehicle in a turn
F_p	resultant force due to passive earth pressure
F_{pn}	normal component of the resultant force due to passive earth pressure
F_r	thrust of vehicle rear axle
F_s	side force
F_x	force component along the x axis
F_y	force component along the y axis
F_{yf}	cornering force of front tire
F_{yr}	cornering force of rear tire
$F_{y\alpha}$	cornering force of tire
$F_{y\gamma}$	camber thrust of tire
F_z	force component along the z axis
f	frequency
f_a	atmospheric factor for power correction of the compression-ignition engine
f_c	center frequency

f_{eq}	equivalent coefficient of motion resistance of a vehicle wholly supported by air cushion
f_m	engine factor for power correction of the compression-ignition engine
f_{n-s}	natural frequency of sprung mass
f_{n-us}	natural frequency of unsprung mass
f_r	coefficient of rolling resistance
G	grade
G_{acc}	lateral acceleration gain
G_s	sand penetration resistance gradient
G_{yaw}	yaw velocity gain
g	acceleration due to gravity
g_e	gravity on earth surface
g_{ex}	gravity on extraterrestrial surface
h	height of center of gravity of a vehicle
h_a	height of the point of application of aerodynamic resistance above ground level
h_b	depth
h_c	clearance height
h_d	height of drawbar hitch
I	mass moment of inertia
I_w	mass moment of inertia of a wheel
I_y	mass moment of inertia of a vehicle about the y axis
I_z	mass moment of inertia of a vehicle about the z axis
i	slip
i_f	slip of front tire
i_i	slip of the inside track of a tracked vehicle in a turn
i_o	slip of the outside track of a tracked vehicle in a turn
i_r	slip of rear tire
i_s	skid
J_j	momentum flux of an air jet
j	shear displacement
K	shear deformation parameter
K_a	augmentation factor
K_{bf}	proportion of total braking force placed on vehicle front axle
K_{br}	proportion of total braking force placed on vehicle rear axle
K_{bs}	proportion of total braking force placed on semitrailer axle
K_d	coefficient of thrust distribution
K_{di}	gear ratio of a controlled differential
K_e	engine capacity factor
K_p	passive earth pressure coefficient
K_{por}	coefficient considering the effect of ground porosity on the air flow and power requirement of an air-cushion vehicle
K_r	ratio of residual shear stress to the maximum shear stress

K_s	ratio of the angular speed of the outside track sprocket to that of the inside track sprocket in a turn
K_{st}	stiffness modulus of terrain
K_{tc}	torque converter capacity factor
K_{us}	understeer coefficient
$K_{us.s}$	understeer coefficient of semitrailer
$K_{us.t}$	understeer coefficient of tractor
K_v	ratio of the theoretical speed of the front tire to that of the rear tire
K_w	shear displacement where the maximum shear stress occurs for shear curves with a hump
K_{we}	weight utilization factor
k_c	pressure–sinkage parameter
k_f	front suspension spring stiffness
k_p	stiffness of underlying peat for organic terrain (muskeg)
k_r	rear suspension spring stiffness
k_s	stiffness of suspension spring
k_{tr}	equivalent spring stiffness of tire
k_u	parameter characterizing terrain response to repetitive normal loading
k_ϕ	pressure–sinkage parameter
k_0	parameter characterizing terrain response to repetitive normal loading
L	wheelbase or liter
L_c	characteristic length
L_s	wheelbase of semitrailer
L_t	wheelbase of tractor
l	length
l_{cu}	cushion perimeter
l_j	nozzle perimeter
l_o	distance between oscillation center and center of gravity of a vehicle
l_t	contact length
l_1	distance between front axle and center of gravity of a vehicle
l_2	distance between rear axle and center of gravity of a vehicle
M_a	aerodynamic pitching moment
M_b	braking torque
M_e	engine output torque
M_r	moment of turning resistance
M_{ro}	restoring moment in roll
M_{tc}	torque converter output torque
M_w	wheel torque
M_x	moment about the x axis
M_y	moment about the y axis
M_z	moment about the z axis
MI	mobility index
MMP	mean maximum pressure

m	vehicle mass
m_m	pressure–sinkage parameter for organic terrain (muskeg)
m_s	sprung mass
m_{us}	unsprung mass
N	exponent of power spectral density function
N_c, N_q, N_γ	bearing capacity factors
N_{st}	exponent of terrain deformation considering effects of slip and other factors
N_ϕ	flow value for soils
n	exponent of terrain deformation
n_e	engine speed
n_g	number of speeds of a gearbox
n_r	number of teeth of ring gear of a planetary gearset
n_s	number of teeth of sun gear of a planetary gearset
n_{tc}	torque converter output speed
P	engine power
P_a	power required to sustain the air cushion
P_{br}	engine brake power under reference atmospheric conditions
P_{bt}	engine brake power under test atmospheric conditions
P_d	drawbar power
P_{ft}	engine friction power under test atmospheric conditions
P_{ir}	engine indicated power under reference atmospheric conditions
P_{it}	engine indicated power under test atmospheric conditions
P_m	power required to overcome momentum drag
P_{re}	engine power under reference atmospheric conditions
P_{st}	power consumption of a tracked vehicle in straight line motion
P_t	power consumption of tracked vehicle during a turn
P_{te}	engine power under test atmospheric conditions
p	pressure
p_c	pressure exerted by tire carcass
p_{cr}	critical pressure
p_{cu}	cushion pressure
p_d	dynamic pressure
p_g	ground pressure at the lowest point of contact
p_{gcr}	critical ground pressure
p_i	inflation pressure
p_j	total jet pressure
p_t	dry atmospheric pressure under test conditions
Q	volume flow
q	surcharge
R	turning radius
R_a	aerodynamic resistance (drag)
R_c	motion resistance due to terrain compaction
R_{ce}	motion resistance due to terrain compaction on earth surface
R_{cex}	motion resistance due to terrain compaction on extraterrestrial surface

R_d	drawbar load
R_g	grade resistance
R_h	motion resistance of tire due to hysteresis and other internal losses
R_i	motion resistance of the inside track of a tracked vehicle in a turn
R_{in}	internal resistance of track system
R_L	aerodynamic lift
R_l	lateral resistance of track
R_m	momentum drag
R_o	motion resistance of the outside track of a tracked vehicle in a turn
R_r	rolling resistance
R_{rf}	rolling resistance of front tire
R_{rr}	rolling resistance of rear tire
R_{rs}	rolling resistance of semitrailer tire
R_{sk}	skirt contact drag
R_{tot}	total motion resistance
R_w	wave-making drag
R_{wave}	drag due to wave
R_{wet}	wetting drag
RCI	rating cone index
RCI_x	excess soil strength (i.e., difference between rating cone index and vehicle cone index, $RCI - VCI$)
r	radius of wheel or sprocket
r_e	effective rolling radius of tire
r_p	pitch radius of pinon of a planetary gearset
r_r	pitch radius of ring gear of a planetary gearset
r_s	pitch radius of sun gear of a planetary gearset
r_y	radius of gyration of a vehicle about the y axis
S	distance
S_b	braking distance
$S_g(f)$	power spectral density function of terrain profile (temporal frequency)
$S_g(\Omega)$	power spectral density function of terrain profile (spatial frequency)
S_s	stopping distance
$S_v(f)$	power spectral density function of vehicle response (temporal frequency)
s	displacement
T_b	braking torque on a tire
T_t	temperature under test conditions
t	time
t_j	thickness of air jet
t_p	pneumatic trail of tire
t_t	track pitch
U	energy dissipation
u_a	fuel consumed for work performed per unit area
u_e	energy available at the drawbar hitch per unit volume of fuel consumed
u_h	fuel consumed per hour

u_s	specific fuel consumption
u_t	fuel consumed during time t
u_{tr}	fuel consumed per unit payload per unit distance
V	speed
V_a	speed of wind relative to vehicle
V_c	speed of air escaping from cushion
V_{char}	characteristic speed of an understeer vehicle
V_{crit}	critical speed of an oversteer vehicle
V_i	speed of the inside track of a tracked vehicle in a turn
V_j	slip speed
V_{jc}	jet speed
V_m	average operating speed
V_o	speed of the outside track of a tracked vehicle in a turn
V_p	hydroplaning speed of tire
V_t	theoretical speed
V_{tf}	theoretical speed of front tire
V_{tr}	theoretical speed of rear tire
VCI	vehicle cone index
W	normal load or weight
W_a	load supported by air cushion
W_c	critical load
W_d	proportion of vehicle weight applied to driven wheels
W_e	normal load on rover wheel on earth surface
W_{ex}	normal load on rover wheel on extraterrestrial surface
W_f	normal load on vehicle front axle
W_{hi}	normal load at the hitch point of a tractor-semitrailer
W_p	payload
W_r	normal load on vehicle rear axle
W_s	normal load on semitrailer axle
z	depth, penetration, or sinkage
z_{cr}	critical sinkage
z_e	rover wheel sinkage on earth surface
z_{ex}	rover wheel sinkage on extraterrestrial surface
z_j	sinkage of vehicle running gear due to slip
z_s	sinkage of vehicle running gear due to static normal load
z_ω	pressure–sinkage parameter for snow cover
α	angle or slip angle of tire
α_a	angle of attack,
α_{ae}	power correction factor for the spark-ignition engine (for ISO 1585:1997-04)
α_{an}	angular acceleration
α_b	inclination angle
α_{be}	power correction factor for the spark-ignition engine (for SAE J1349 SEP 2011)
α_{ce}	power correction factor for the compression-ignition engine (for ISO 1585:1997-04)
α_f	slip angle of front tire

α_r	slip angle of rear tire
α_s	slip angle of semitrailer tire
β	vehicle sideslip angle
β_b	inclination angle of blade
Γ	articulation angle
γ	camber angle of tire
γ_m	vehicle mass factor
γ_{mt}	mass density of terrain
γ_s	specific weight of terrain
δ	angle of interface friction, angle of inclination
δ_f	steer angle of front tire
δ_i	steer angle of inside front tire
δ_o	steer angle of outside front tire
δ_t	tire deflection
ϵ	strain
ζ	damping ratio
η_b	braking efficiency
η_c	torque converter efficiency
η_{cu}	cushion intake efficiency
η_d	drawbar (tractive) efficiency (i.e., ratio of the drawbar power to power at driven wheels or sprockets)
η_{do}	overall drawbar (tractive) efficiency (i.e., ratio of the drawbar power to engine output power)
η_m	efficiency of motion
η_p	propulsive efficiency
η_s	slip efficiency
η_{st}	structural efficiency
η_t	transmission efficiency
η_{tr}	transport efficiency
θ_j	nozzle angle
θ_s	slope angle
θ_t	trim angle
μ	coefficient of road adhesion
μ_p	peak value of coefficient of road adhesion
μ_s	sliding value of coefficient of road adhesion
μ_t	coefficient of lateral resistance
μ_{tr}	coefficient of traction
υ	concentration factor
ξ	gear ratio
ξ_o	overall reduction ratio
ξ_s	steering gear ratio
ρ	air density
ρ_f	density of fluid

ρ_w	density of water
σ	normal stress
σ_a	active earth pressure
σ_p	passive earth pressure
σ_r	radial stress
σ_z	vertical stress
τ	shear stress
$\tau_{\max}$	maximum shear stress
τ_r	residual shear stress
ϕ	angle of internal shearing resistance, or angle of friction
Ω	spatial frequency
Ω_x	angular speed about the x axis
Ω_y	angular speed about the y axis
Ω_z	angular speed about the z axis
ω	angular speed
ω_c	angular speed of carrier of a planetary gearset
ω_i	angular speed of the sprocket of the inside track of a tracked vehicle in a turn
ω_n	circular natural frequency
ω_o	angular speed of the sprocket of the outside track of a tracked vehicle in a turn
ω_r	angular speed of ring gear of a planetary gearset
ω_s	angular speed of sun gear of a planetary gearset

Introduction

Ground vehicles are those vehicles that are supported by the ground, in contrast with aircraft and marine craft, which in operation are supported by air and water, respectively. Ground vehicles may be broadly classified as guided and nonguided. Guided ground vehicles are constrained to move along a fixed path (guideway), such as railway vehicles and tracked levitated vehicles. Nonguided ground vehicles, such as road and off-road vehicles, can move in various directions on the ground, in response to the command of the driver or the automated driving system. The mechanics of nonguided ground vehicles is the subject of this book.

The prime objective of the study of the mechanics of ground vehicles is to establish guiding principles for the rational development, design, and selection of vehicles to meet various operational requirements. In general, the characteristics of a ground vehicle may be described in terms of its performance, handling, and ride. Performance refers to the ability of the vehicle to accelerate, to develop drawbar pull, to overcome obstacles, and to decelerate. Handling is concerned with the response of the vehicle to steering inputs and its ability to stabilize its motion against external disturbances. Ride is related to vibrations of the vehicle excited by surface irregularities and their effects on passengers and goods. The theory of ground vehicles is concerned with the study of the performance, handling, and ride and their relationships with the design of ground vehicles under various operating conditions.

The behavior of a ground vehicle represents the results of the interactions of the driver or the automated driving system, the vehicle, and the driving environment, as illustrated in Figure 1.

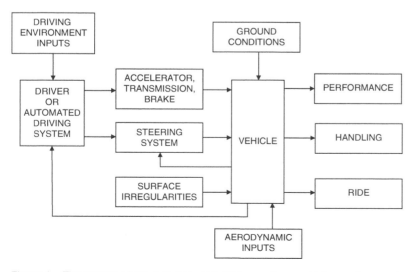

Figure 1 The system of driver/automated driving system–vehicle–environment.

An understanding of the behavior of the human driver (or the automated driving system), the characteristics of the vehicle, and the engineering and geometric properties of the ground is essential to the design and evaluation of ground vehicle systems.

About the Companion Website

Theory of Ground Vehicles, Fifth Edition, is accompanied by a companion website:

www.wiley.com/go/wong/TGV5e

The website includes:

- Instructor's Manual

1

Mechanics of Pneumatic Tires

Aside from aerodynamic and gravitational forces, all other major forces and moments affecting the motion of a ground vehicle are applied through the running gear–ground contact. An understanding of the basic characteristics of the interaction between the running gear and the ground is, therefore, essential to the study of performance characteristics, handling behavior, and ride quality of ground vehicles. The running gear of a ground vehicle is generally required to fulfill the following functions:

- To support the weight of the vehicle
- To cushion the vehicle over surface irregularities
- To provide sufficient traction for driving and braking
- To provide adequate steering control and direction stability

Pneumatic tires can perform these functions effectively and efficiently; thus, they are universally used in road vehicles, and are also widely used in off-road vehicles. The study of the mechanics of pneumatic tires is of fundamental importance to the understanding of the performance and behavior of ground vehicles. Two basic types of problem in the mechanics of tires are of special interest to vehicle engineers. One is the mechanics of tires on paved roads, which is essential to the study of the characteristics of road vehicles. The other is the mechanics of tires on deformable surfaces (unprepared terrain), which is of prime importance to the study of off-road wheeled vehicle behavior.

The mechanics of tires on paved roads is discussed in this chapter, whereas the behavior of tires over unprepared terrain is discussed in Chapter 2.

A pneumatic tire is a flexible structure of the shape of a toroid filled with compressed air. The most important structural element of the tire is the carcass. It is made up of several layers of flexible cords of high modulus of elasticity encased in a matrix of low modulus rubber compounds, as shown in Figure 1.1. The cords are made of fabrics of natural, synthetic, or metallic composition, and are anchored around the beads made of high tensile strength steel wires. The beads serve as the foundations for the carcass and provide adequate seating of the tire on the rim. The ingredients of the rubber compounds are selected to provide the tire with specific properties. The rubber compounds for the sidewall are generally required to be highly resistant to fatigue and scuffing, and styrene–butadiene compounds are widely used (French 1989).[1] The rubber compounds for the tread vary with the type of tire. For instance, for heavy truck tires, the high load intensities necessitate the use of tread compounds with high resistance to abrasion, tearing, and crack growth, and with low hysteresis to reduce internal heat generation and rolling resistance. Consequently, natural rubber compounds are used for truck tires, although they intrinsically provide lower values of the

_____
1 References cited are listed at the end of the chapter.

(a)

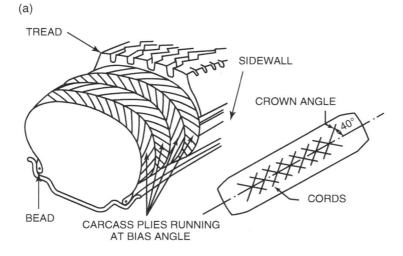

(b)

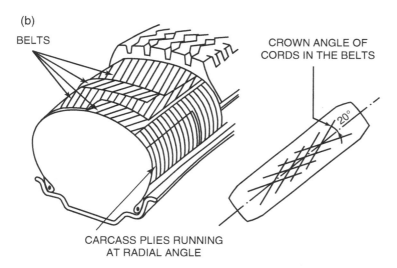

Figure 1.1 Tire construction: (a) bias-ply tire; (b) radial-ply tire.

coefficient of road adhesion, particularly on wet surfaces, than various synthetic rubber compounds universally used for passenger car and racing car tires (French 1989). For tubeless tires, which have become dominant, a thin layer of rubber with high impermeability to air (such as butyl rubber compounds) is attached to the inner surface of the carcass.

The load transmission of a pneumatic tire is analogous to that of a bicycle wheel, where the hub hangs on the spokes from the upper part of the rim, which in turn is supported at its lower part by the ground. For an inflated pneumatic tire, the inflation pressure causes tension to be developed in the cords comprising the carcass. The load applied through the rim of the wheel hangs primarily on the cords in the sidewalls through the beads.

The design and construction of the carcass determine, to a great extent, the characteristics of the tire. Among the various design parameters, the geometric dispositions of layers of rubber-coated cords (plies), particularly their directions, play a significant role in the behavior of the tire. The direction of the cords is usually defined by the crown angle, which is the angle between the cord and the circumferential centerline of the tire, as shown in Figure 1.1. When the cords have a low crown angle, the tire will have good cornering characteristics, but a harsh ride. On the other hand, if the cords are at right angle to the centerline of the tread, the tire will be capable of providing a comfortable ride, but poor handling performance.

A compromise is adopted in a bias-ply tire, in which the cords extend diagonally across the carcass from bead to bead with a crown angle of approximately 40°, as shown in Figure 1.1(a). A bias-ply tire has two plies (for light-load tires) or more (up to 20 plies for heavy-load tires). The cords in adjacent plies run in opposite directions. Thus, the cords overlap in a diamond-shaped (crisscross) pattern. In operation, the diagonal plies flex and rub, thus elongating the diamond-shaped elements and the rubber filler. This flexing action produces a wiping motion between the tread and the road, which is one of the main causes of tire wear and high rolling resistance (Gough 1971; Moore 1975).

The radial-ply tire, on the other hand, is constructed very differently from the bias-ply tire. It was first introduced by Michelin in 1948 and has now become dominant for passenger cars and trucks and increasingly for heavy-duty earth-moving machinery. The radial-ply tire has one or more layers of cords in the carcass extending radially from bead to bead, resulting in a crown angle of 90°, as shown in Figure 1.1(b). A belt of several layers of cords of high modulus of elasticity (usually steel or other high-strength materials) is fitted under the tread, as shown in Figure 1.1(b). The cords in the belt are laid at a low crown angle of approximately 20°. The belt is essential to the proper functioning of the radial-ply tire. Without it, a radial-ply carcass can become unstable since the tire periphery may develop into a series of buckles due to the irregularities in cord spacing when inflated. For passenger car tires, usually there are two radial plies in the carcass made of synthetic material, such as rayon or polyester, and two plies of steel cords and two plies of cords made of synthetic material, such as nylon, in the belt. For truck tires, usually there is one radial steel ply in the carcass and four steel plies in the belt. For the radial-ply tire, flexing of the carcass involves very little relative movement of the cords forming the belt. In the absence of a wiping motion between the tire and the road, the power dissipation of the radial-ply tire could be as low as 60% of that of the bias-ply tire under similar operating conditions, and the life of the radial-ply tire could be as long as twice that of the equivalent bias-ply tire (Moore 1975). For a radial-ply tire, there is a relatively uniform ground pressure over the entire contact area. In contrast, the ground pressure for a bias-ply tire varies greatly from point to point as tread elements passing through the contact area undergo complex localized wiping motion.

There are also tires built with belts in the tread on bias-ply construction. This type of tire is usually called the bias-belted tire. The cords in the belt are of materials with a higher modulus of elasticity than those in the bias plies. The belt provides high rigidity to the tread against distortion and reduces tread wear and rolling resistance in comparison with the conventional bias-ply tire. Generally, the bias-belted tire has characteristics midway between those of the bias-ply and the radial-ply tire.

In the United States, it is required that tire manufacturers use the tire code to identify tire dimensions and ratings on the sidewall of every tire. The tire identification requirements are given in 49 CFR (Code of Federal Regulations) V – National Highway Traffic Safety Administration (NHTSA), Department of Transportation (DOT). This tire identification code is used worldwide. For instance, for a tire identified as P185/70 R14 87S:

- P is for passenger car tires; LT for light truck tires; ST for special trailer tires; and T for temporary usage tires (restricted for "space-saver" spare tires).
- 185 is the nominal width of the cross-section in millimeters.
- 70 is the aspect ratio, which is the ratio of the height of the sidewall to the cross-sectional width.
- R is for radial-ply tires; B for bias-belt tires; and D for diagonal (bias-ply) tires.

Table 1.1 Load Index (LI) for tires typically used on passenger cars and light trucks.

LI	Load		LI	Load	
	kg	lb		kg	lb
70	335	739	99	775	1709
71	345	761	100	800	1764
72	355	783	101	825	1819
73	365	805	102	850	1874
74	375	827	103	875	1929
75	387	853	104	900	1984
76	400	882	105	925	2039
77	412	908	106	950	2094
78	425	937	107	975	2149
79	437	963	108	1000	2205
80	450	992	109	1030	2271
81	462	1019	110	1060	2337
82	475	1047	111	1090	2403
83	487	1074	112	1120	2469
84	500	1102	113	1150	2535
85	515	1135	114	1180	2601
86	530	1168	115	1215	2679
87	545	1201	116	1250	2756
88	560	1235	117	1285	2833
89	580	1279	118	1320	2910
90	600	1323	119	1360	2998
91	615	1356	120	1400	3086
92	630	1389	121	1450	3197
93	650	1433	122	1500	3307
94	670	1477	123	1550	3417
95	690	1521	124	1600	3527
96	710	1565	125	1650	3638
97	730	1609	126	1700	3748
98	750	1653			

Note: For this table, kg is used as the unit for force. A force of 1 kg is equal to a force of 2.2046 lb.

- 14 is the rim diameter in inches.
- 87 is a load index (LI) and, according to the code for LI shown in Table 1.1 (for tires typically used on passenger cars and light trucks), it identifies that the maximum load for the tire is 545 kg (1201 lb).
- S is a speed index (SI) and, according to the code for SI shown in Table 1.2, it identifies that the maximum speed that the tire can sustain without failure is 180 km/h (112 mph).

Table 1.2 Speed Index (SI) for tires typically used on passenger cars and light trucks.

SI	Speed		Applications
	km/h	mph	
L	up to 120	up to 75	Light truck tires
M	up to 130	up to 81	Temporary spare tires
N	up to 140	up to 87	
P	up to 150	up to 93	
Q	up to 160	up to 99	Winter tires
R	up to 170	up to 106	Heavy-duty light truck tires
S	up to 180	up to 112	Family sedans and vans
T	up to 190	up to 118	Family sedans and vans
U	up to 200	up to 124	
H	up to 210	up to 130	Sport sedans and coupes
V	up to 240	up to 149	Sport sedans, coupes, and sports cars

In addition, NHTSA of the United States has established the Uniform Tire Quality Grading Standards (49 CFR 575.104) for rating tread wear, traction on wet pavement (asphalt and concrete), and temperature tolerance.

- Tread wear index is an indication of expected tire life. It is graded against a reference tire, referred to as the Course Monitoring Tire, with an index of 100. For instance, a tread wear rating of 420 means that the tire should last 4.2 times as long as the reference tire.
- Traction grades, AA, A, B, and C, are based on the coefficient of sliding friction (or locked-wheel traction coefficient) of the tire on wet asphalt and concrete, as shown in Table 1.3.
- Temperature grades, A, B, and C, are based on the tire's tolerance to heat generated at speeds. Grade A indicates that the tire can effectively dissipate heat up to a maximum speed greater than 115 mph (185 km/h). B rates at a maximum speed between 100 and 115 mph (161–185 km/h), while C rates at a maximum speed between 85 and 100 mph (137–161 km/h). Tires that cannot be graded up to C are not allowed to be sold in the United States.

Although the design and construction of pneumatic tires differ from one type to another, the basic issues involved are not dissimilar. In the following sections, the mechanics fundamental to all types of tire are discussed. The characteristics particular to a specific type of tire are also described.

Table 1.3 NHTSA tire traction grades.

Grade	Coefficient of sliding friction on wet asphalt	Coefficient of sliding friction on wet concrete
AA	Above 0.54	0.38
A	Above 0.47	0.35
B	Above 0.38	0.26
C	Less than 0.38	0.26

1.1 Tire Forces and Moments

To describe the characteristics of a tire and the forces and moments acting on it, it is necessary to define a tire axis system that serves as a reference for the definitions of various parameters. Previously, in the Society of Automotive Engineers (SAE) Surface Vehicle Recommended Practice – SAE J670e (July 1976), an axis system based on aeronautical practice was recommended, with positive X forward, positive Y to the right, and positive Z down. This axis system is different from that recommended in the earlier version of the International Organization for Standardization Standard ISO 8855 : 1991 (Road vehicles – Vehicle dynamics and road-holding ability – Vocabulary), as well as the current versions ISO 8855 : 2011 (in German) and ISO 8855 : 2013–11 (English translation). For both the previous and current versions of the ISO Standard, it utilizes an axis system with positive X forward, positive Y to the left, and positive Z up. In the revised SAE J670 JAN2008, which supersedes J670e, it embraces both axis orientations: one with positive X forward, positive Y to the left, and positive Z up; the other with positive X forward, positive Y to the right, and positive Z down. These two axis systems (axis orientations) are equally acceptable, and the selection of an appropriate orientation would be based on the requirements of the analysis or test being performed.

One of the two tire axis systems, with positive X forward, positive Y to the right, and positive Z down, recommended in SAE J670 JAN2008 is shown in Figure 1.2. The origin of the axis system is the center of the tire contact. The X axis is the intersection of the wheel plane and the ground plane with a positive direction forward. The Z axis is perpendicular to the ground plane with a positive direction downward. The Y axis is in the ground plane, and its positive direction is to the right to form a right-handed orthogonal axis system.

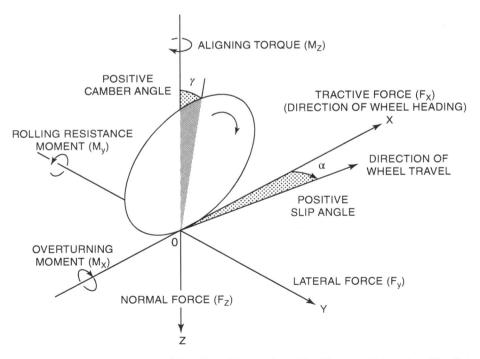

Figure 1.2 A tire axis system with positive X forward, positive Y to the right, and positive Z down.

There are three forces and three moments acting on the tire from the ground. Tractive force (or longitudinal force) F_x is the component in the X direction of the resultant force exerted on the tire by the road. Lateral force F_y is the component in the Y direction of the resultant force, and normal force F_z is the component in the Z direction of the resultant force. Overturning moment M_x is the moment about the X axis exerted on the tire by the road. Rolling resistance moment M_y is the moment about the Y axis and aligning torque (moment) M_z is the moment about the Z axis.

With this axis system, many performance parameters of the tire can be conveniently defined. For instance, the longitudinal shift of the center of normal pressure is determined by the ratio of the rolling resistance moment to the normal load. The lateral shift of the center of normal pressure is defined by the ratio of the overturning moment to the normal load. The longitudinal shift of the lateral force is determined by the ratio of the aligning torque to the lateral force. The integration of longitudinal shear stresses over the entire contact patch represents the tractive or braking force. A driving torque about the axis of rotation of the tire produces a force (tractive effort) for accelerating the vehicle, and a braking torque produces a force (braking effort) for decelerating the vehicle.

There are two important angles associated with a rolling tire: the slip angle and the camber angle. Slip angle α shown in Figure 1.2 is the angle formed between the direction of wheel travel and the line of intersection of the wheel plane with the road surface. Camber angle γ shown in Figure 1.2 is the angle formed between the XZ plane and the wheel plane. The lateral force on the tire–ground contact patch is a function of both the slip angle and the camber angle.

1.2 Rolling Resistance of Tires

The rolling resistance of tires on paved roads is primarily caused by the hysteresis in tire materials due to the deflection of the carcass while rolling. Friction between the tire and the road caused by sliding, the resistance due to air circulating inside the tire, and the fan effect of the rotating tire on the surrounding air also contribute to the rolling resistance of the tire, but they are of secondary importance. Available experimental data give a breakdown of tire losses in the speed range 128–152 km/h (80–95 mph) as 90–95% due to internal hysteresis losses in the tire, 2–10% due to friction between the tire and the ground, and 1.5–3.5% due to air resistance (French 1959; van Eldik Thieme and Pacejka 1971). Of the total energy losses within the tire structure, it is found that for a radial truck tire, hysteresis in the tread region, including the belt, contributes 73%, the sidewall 13%, the region between the tread and the sidewall, commonly known as the shoulder region, 12%, and the beads 2%.

When a tire is rolling, the carcass is deflected in the ground contact area. As a result of tire distortion, the normal pressure in the leading half of the contact patch is higher than that in the trailing half. The center of normal pressure is shifted in the direction of rolling. This shift produces a moment about the axis of rotation of the tire, which is the rolling resistance moment. In a free-rolling tire, the applied wheel torque is zero; therefore, a horizontal force at the tire–ground contact patch must exist to maintain equilibrium. This resultant horizontal force is generally known as the rolling resistance. The ratio of the rolling resistance to the normal load on the tire is defined as the coefficient of rolling resistance.

Several factors affect the rolling resistance of a pneumatic tire. They include the structure of the tire (design, construction, and materials) and its operating conditions (surface conditions, inflation pressure, speed, temperature, etc.). Tire design and construction have a significant influence on its rolling resistance. Figure 1.3 shows the average values of the coefficient of rolling resistance at

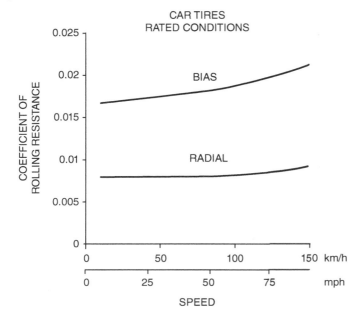

Figure 1.3 Variations of the coefficient of rolling resistance of radial-ply and bias-ply car tires with speed on a smooth, flat road surface under rated load and inflation pressure.

various speeds of a range of bias-ply and radial-ply passenger car tires at rated loads and inflation pressures on a smooth road, based on test data from various sources. It is shown that for the radial-ply passenger car tire, the average value of rolling resistance coefficient is approximately constant at speeds up to 130 km/h (81 mph) and slightly increases beyond that. The average values of the coefficient of rolling resistance at various speeds of a bias-ply and a radial-ply truck tire of the same size under rated conditions is shown in Figure 1.4 (Segel 1984). Thicker treads and sidewalls and an increased number of carcass plies tend to increase the rolling resistance because of greater hysteresis losses. Tires made of synthetic rubber compounds generally have higher rolling resistance

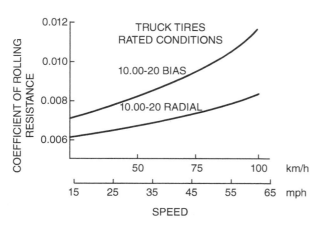

Figure 1.4 Variations of the coefficient of rolling resistance of radial-ply and bias-ply truck tires with speed under rated load and inflation pressure. *Source:* Segel (1984). Reproduced with permission of the University of Michigan Transportation Research Institute.

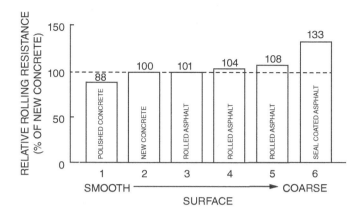

Figure 1.5 Variations of tire rolling resistance with pavement surface texture. Reprinted with permission from SAE Special Publication P-74 © 1977 Society of Automotive Engineers, Inc.

than those made of natural rubber. Tires made of butyl rubber compounds, which are shown to have better traction and road-holding properties, have an even higher rolling resistance than those made of conventional synthetic rubber. It is found that the rolling resistance of tires with tread made of synthetic rubber compounds and that made of butyl rubber compounds are approximately 1.06 and 1.35 times that made of natural rubber compounds, respectively (Hunt et al. 1977).

Surface conditions also affect the rolling resistance. On hard, smooth surfaces, the rolling resistance is considerably lower than that on a rough road. On wet surfaces, a higher rolling resistance than on dry surfaces is usually observed. Figure 1.5 shows a comparison of the rolling resistance of passenger car tires over six road surfaces with different textures, ranging from polished concrete to coarse asphalt (DeRaad 1977). The profiles of these six surfaces are shown in Figure 1.6 (DeRaad 1977). It can be seen that on the asphalt surface with coarse sealcoat (surface no. 6) the

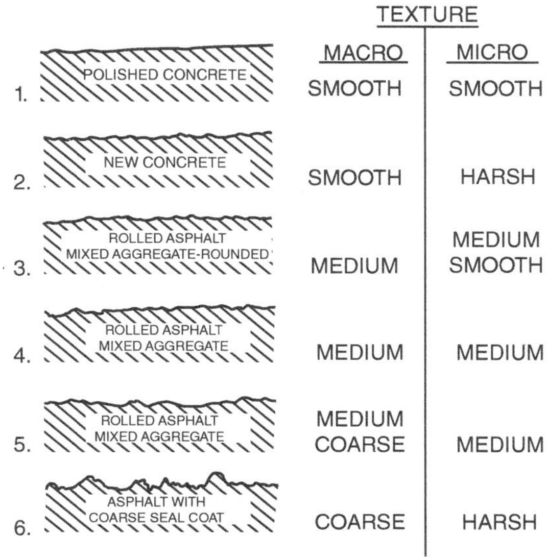

Figure 1.6 Textures of various types of pavement surface. Reprinted with permission from SAE Special Publication P-74 © 1977 Society of Automotive Engineers, Inc.

	TEXTURE	
	MACRO	**MICRO**
1. POLISHED CONCRETE	SMOOTH	SMOOTH
2. NEW CONCRETE	SMOOTH	HARSH
3. ROLLED ASPHALT MIXED AGGREGATE-ROUNDED	MEDIUM	MEDIUM SMOOTH
4. ROLLED ASPHALT MIXED AGGREGATE	MEDIUM	MEDIUM
5. ROLLED ASPHALT MIXED AGGREGATE	MEDIUM COARSE	MEDIUM
6. ASPHALT WITH COARSE SEAL COAT	COARSE	HARSH

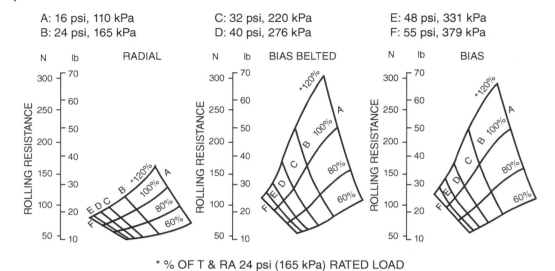

A: 16 psi, 110 kPa   C: 32 psi, 220 kPa   E: 48 psi, 331 kPa
B: 24 psi, 165 kPa   D: 40 psi, 276 kPa   F: 55 psi, 379 kPa

* % OF T & RA 24 psi (165 kPa) RATED LOAD

Figure 1.7 Variations of rolling resistance of radial-ply, bias-belted, and bias-ply car tires with load and inflation pressure. Reprinted with permission from SAE paper 800087 © 1980 Society of Automotive Engineers, Inc.

rolling resistance is 33% higher than that on a new concrete surface (surface no. 2), while on the polished concrete (surface no. 1), it shows a 12% reduction in comparison with that on the new concrete surface.

Inflation pressure affects the flexibility of the tire. Depending on the deformability of the ground, the inflation pressure affects the rolling resistance of the tire in different manners. On paved roads, the rolling resistance generally decreases with the increase in inflation pressure. This is because, with higher inflation pressure, the deflection of the tire decreases, with consequently lower hysteresis losses. Figure 1.7 shows the effects of inflation pressure on the rolling resistance of a radial-ply tire (GR78-15), a bias-ply tire, and a bias-belted tire (both G78-15) under various normal loads, expressed in terms of the percentage of the rated load at an inflation pressure of 165 kPa (24 psi) (Collier and Warchol 1980). The results were obtained with the inflation pressure being regulated; that is, the pressure was maintained at a specific level throughout the tests. It shows that inflation pressure has a much more significant effect on the rolling resistance of the bias and bias-belted tires than the radial-ply tire. On deformable surfaces, such as sand, high inflation pressure results in increased ground penetration work, and therefore higher rolling resistance, as shown in Figure 1.8 (Taborek 1957). Conversely, lower inflation pressure, while decreasing ground penetration, increases the deflection of the tire and hence internal hysteresis losses. Therefore, an optimum inflation pressure exists for a particular tire on a given deformable surface, which minimizes the sum of ground penetration work and internal losses of the tire.

Inflation pressure affects not only the rolling resistance, but also the tread wear of a tire. Figure 1.9 shows the effects of inflation pressure on tread wear of a radial-ply, a bias-ply, and a bias-belted tire (Collier and Warchol 1980). The wear rate at 165 kPa (24 psi) is used as a reference for comparison. It shows that the effects of inflation pressure on tread wear are more significant for the bias-ply and bias-belted tire than the radial-ply tire.

Figure 1.8 Variations of rolling resistance coefficient with inflation pressure of tires on various surfaces. *Source:* Taborek (1957). Reproduced with permission of *Machine Design.*

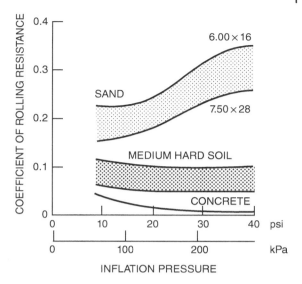

Figure 1.9 Variations of shoulder-crown wear with inflation pressure for radial-ply, bias-ply, and bias-belted car tires. Reprinted with permission from SAE paper 800087 © 1980 Society of Automotive Engineers, Inc.

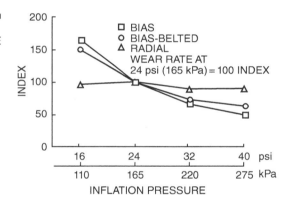

Rolling resistance is also affected by driving speed because of the increase of work in deforming the tire and of vibrations in the tire structure with the increase in speed. The effects of speed on the rolling resistance of bias-ply and radial-ply passenger car and truck tires are illustrated in Figures 1.3 and 1.4, respectively. For a given tire under a particular operating condition, there exists a threshold speed above which the phenomenon popularly known as standing waves will be observed, as shown in Figure 1.10. The approximate value of the threshold speed V_{th} may be determined by the expression $V_{th} = \sqrt{F_t/\rho_t}$, where F_t is the circumferential tension in the tire and ρ_t is the density of tread material per unit area (Hartley and Turner 1956). Standing waves are formed because, owing to high speed, the tire tread does not recover immediately from distortion originating from tire deflection after it leaves the contact surface, and the residual deformation initiates a wave. The amplitude of the wave is greatest immediately on leaving the ground and is damped out in an exponential manner around the circumference of the tire. The formation of the standing wave greatly increases energy losses, which in turn cause considerable heat generation that could lead to tire failure. This places an upper limit on the safe operating speed of tires. The standing wave phenomenon occurs mainly in bias-ply tires.

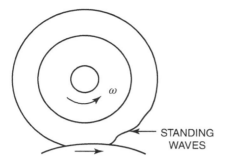

Figure 1.10 Formation of standing waves of a tire at high speeds.

STANDING WAVES

Operating temperature, tire diameter, and tractive force also have effects on the rolling resistance of a tire. Tire temperature affects the rolling resistance in two ways: one is by changing the temperature of the air in the tire cavity, and thereby changing the operating inflation pressure; and the other is by changing the stiffness and hysteresis of the rubber compounds. Figure 1.11 shows the dependence of the rolling resistance on the internal tire temperature for an automobile tire (French 1959). The variation of rolling resistance coefficient with shoulder temperature of a radial-ply passenger car tire is shown in Figure 1.12 (Janssen and Hall 1980). It shows that the rolling resistance at a shoulder temperature of −10°C is approximately 2.3 times that at 60°C for the tire examined. It is also found that the shoulder temperature of the tire, and not the ambient temperature, is a basic determining factor of the tire rolling resistance coefficient. The effect of tire diameter on the coefficient of rolling resistance is shown in Figure 1.13 (Taborek 1957). It shows that the effect of tire diameter is negligible on paved roads but is considerable on deformable or soft ground. Figure 1.14 shows the effects of the braking and tractive effort coefficient on the rolling resistance coefficient (van Eldik Thieme and Pacejka 1971).

When considering the effects of material, construction, and design parameters of tires on rolling resistance, it is necessary to have a proper perspective on the energy losses in the tire and the characteristics of the tire–vehicle system as a whole. Although it is desirable to keep the rolling resistance as low as possible, it should be considered together with other performance parameters, such as tire endurance and life, traction, cornering properties, cushioning effects, and cost.

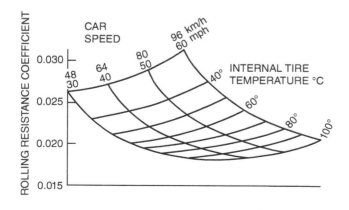

Figure 1.11 Effects of internal temperature on rolling resistance coefficient of a car tire. *Source:* French (1959). Reprinted by permission of the Council of the Institution of Mechanical Engineers.

Figure 1.12 Variations of rolling resistance coefficient with shoulder temperature for a car tire P195/75R14. Reprinted with permission from SAE paper 800090 © 1980 Society of Automotive Engineers, Inc.

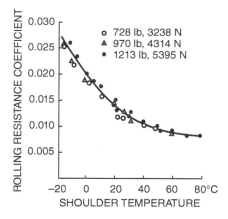

Figure 1.13 Effects of tire diameter on rolling resistance coefficient on various surfaces. *Source:* Taborek 1957. Reproduced with permission of *Machine Design.*

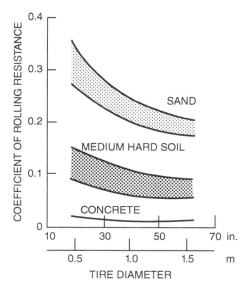

For instance, from the standpoint of rolling resistance, synthetic rubber compounds are less favorable than natural rubber compounds, yet because of significant advantages in cost, tread life, wet-road grip, and tire squeal, they have virtually displaced natural rubber compounds from passenger car tires, particularly for treads. For high-performance vehicles, there may be some advantage for using butyl rubber tires because of the marked gains in traction, road-holding, silence, and comfort, despite their poor hysteresis characteristics (French 1959).

The complex relationships between the design and operational parameters of the tire and its rolling resistance make it extremely difficult, if not impossible, to develop an analytic method for predicting the rolling resistance. The determination of the rolling resistance, therefore, relies almost entirely on experiments. To provide a uniform basis for collecting experimental data, the SAE recommends tire rolling resistance measurement procedures, described in the SAE J2452-JUL2017 (Surface vehicle recommended practice – Stepwise coastdown methodology for measuring tire rolling resistance).

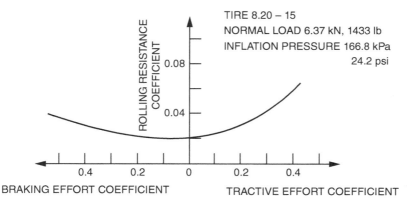

Figure 1.14 Effects of tractive and braking effort on rolling resistance coefficient of a car tire. Reproduced with permission from *Mechanics of Pneumatic Tires*, edited by S.K. Clark, Monograph 122, National Bureau of Standards, 1971.

Based on experimental results, many empirical formulas have been proposed for estimating the rolling resistance of tires on paved roads. For instance, based on the experimental data shown in Figure 1.3, for the radial-ply passenger car tire under rated loads and inflation pressures on a smooth road, the average value of rolling resistance coefficient may be considered approximately constant at speeds up to 130 km/h (81 mph)

$$f_r = 0.008 \tag{1.1}$$

and for the bias-ply passenger car tire at speeds up to 150 km/h (93 mph)

$$f_r = 0.0169 + 0.19 \times 10^{-6}V^2 \tag{1.2}$$

where V is in km/h.

Based on the experimental data shown in Figure 1.4, for the radial-ply truck tire under rated load and inflation pressure, the relationship between the average value of rolling resistance coefficient f_r and speed V (up to 100 km/h or 62 mph) may be described by

$$f_r = 0.006 + 0.23 \times 10^{-6}V^2 \tag{1.3}$$

and for the bias-ply truck tire,

$$f_r = 0.007 + 0.45 \times 10^{-6}V^2 \tag{1.4}$$

where V is in km/h.

The rolling resistance coefficient of truck tires is usually lower than that of passenger car tires on paved roads. This is primarily due to the higher inflation pressure used in truck tires (typically 620–827 kPa or 90–120 psi as opposed to 193–248 kPa or 28–36 psi for passenger car tires).

In preliminary performance estimations, the effect of speed and the type of tire (whether bias-ply or radial-ply) may be ignored. The average values of f_r for the passenger car tire on different surfaces and that for the truck tire on concrete and asphalt are summarized in Table 1.4.

In the European Union, the rolling resistance coefficient f_r, measured in accordance with UNECE Regulation No.117 and its subsequent amendments, is used as the basis to classify energy efficiency of tires. The energy efficiency classes for three types of tires: C1 (passenger car tires), C2 (light utility

Table 1.4 Coefficient of rolling resistance.

Road surface	Coefficient of rolling resistance
Car tires	
Concrete, asphalt	0.008
Rolled gravel	0.02
Tarmacadam	0.025
Unpaved road	0.05
Field	0.1–0.35
Truck tires	
Concrete, asphalt	0.006–0.01

Source: Adapted from Robert Bosch GmbH, *Bosch Automotive Handbook*, 10th Edition (2018). Reproduced with permission of Wiley.

Table 1.5 European Union classification of energy efficiency of tires.

C1 (Passenger car tires)		C2 (Light utility vehicle tires)		C3 (Heavy commercial vehicle tires)	
Energy efficiency class	Rolling resistance coefficient f_r	Energy efficiency class	Rolling resistance coefficient f_r	Energy efficiency class	Rolling resistance coefficient f_r
A	$f_r \leq 0.0065$	A	$f_r \leq 0.0055$	A	$f_r \leq 0.004$
B	$0.0066 \leq f_r \leq 0.0077$	B	$0.0056 \leq f_r \leq 0.0067$	B	$0.0041 \leq f_r \leq 0.0050$
C	$0.0078 \leq f_r \leq 0.0090$	C	$0.0068 \leq f_r \leq 0.0080$	C	$0.0051 \leq f_r \leq 0.0060$
D	Not assigned	D	Not assigned	D	$0.0061 \leq f_r \leq 0.0070$
E	$0.0091 \leq f_r \leq 0.0105$	E	$0.0081 \leq f_r \leq 0.0092$		

Source: European Union Regulation (EC 1222/2009).

vehicle tires), and C3 (heavy commercial vehicle tires), are listed in Table 1.5. The energy efficiency class is required to be shown on the tire label.

1.3 Tractive (Braking) Effort and Longitudinal Slip (Skid)

1.3.1 Tractive Effort and Longitudinal Slip

When a driving torque is applied to a pneumatic tire, a tractive effort (force) is developed at the tire–ground contact patch, as shown in Figure 1.15 (van Eldik Thieme and Pacejka 1971). At the same time, the tire tread in front of and within the contact patch is subjected to compression. A corresponding shear deformation of the sidewall of the tire is also developed. As tread elements

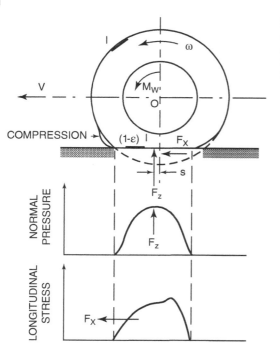

Figure 1.15 Behavior of a tire under the action of a driving torque. Reproduced with permission from *Mechanics of Pneumatic Tires*, edited by S.K. Clark, Monograph 122, National Bureau of Standards, 1971.

are compressed before entering the contact region, the distance that the tire travels when subject to a driving torque will be less than that in free rolling.

The phenomenon of tire rotating without the equivalent translatory progression, when a driving torque is applied, is generally referred to as the longitudinal slip. In the analysis that follows, the longitudinal slip (longitudinal slip ratio or travel reduction) of a tire i, is defined as

$$i = 1 - \frac{V}{r\omega} \tag{1.5}$$

where V is the longitudinal speed of the tire center with the tire moving along a linear path; ω is the angular speed of the tire about its spin axis; and r is the rolling radius of the straight free-rolling tire, which is a loaded, rolling tire moving along a straight path with zero input torque (neither driving nor braking) and with zero slip angle and camber (inclination) angle. The longitudinal speed of the tire center V and the angular speed of the tire ω are determined concurrently under a given set of operating conditions.

For a straight free-rolling tire, its rolling radius r in Eq. (1.5) may be determined as follows:

$$r = S/2\pi n$$

where n is the number of revolutions of the tire for a distance S that the tire center travels along a linear path under a given set of operating conditions.

The definition of longitudinal slip given by Eq. (1.5) is used in this book not only for the analysis of the mechanics of the tire on paved roads, but also for that of the vehicle running gear (wheel, tire, or track) on deformable terrain. For a track, r and ω in Eq. (1.5) are the pitch radius and angular speed of the sprocket, respectively.

In the International Society for Terrain–Vehicle Systems (ISTVS) Standard released in 2020, the definition for the longitudinal slip of the vehicle running gear given by Eq. (1.5) is adopted (He et al. 2020).

When a driving torque is applied, the tire rotates without the equivalent translatory progression (i.e., $r\omega > V$), and a positive value for slip results. If a tire is rotating at an angular speed ω but the longitudinal speed of the tire center V is zero, then in accordance with Eq. 1.5, the longitudinal slip of the tire will be +100%. This phenomenon is often observed on a slippery, icy surface, where the driven tire spins at high angular speeds, while the vehicle does not move forward.

There are other definitions for tire longitudinal slip in the technical literature. For instance, in the SAE Surface Vehicle Recommended Practice – SAE J670 (January 2008), as well as in the International Organization for Standardization Standard ISO 8855 : 2013-11, tire longitudinal slip i' (or longitudinal slip ratio) is defined by

$$i' = \frac{\omega - \omega_0}{\omega_0} \tag{1.6}$$

where i' is tire longitudinal slip defined by Eq.(1.6), to be distinguished from that defined by Eq. (1.5); ω is the wheel-spin velocity or the angular velocity of the wheel about its spin axis; and ω_0 is the wheel-spin velocity of the straight free-rolling tire at a given set of operating conditions. Both ω and ω_0 are determined at the same tire longitudinal velocity, tire load, and hot inflation pressure.

When a driving torque is applied, in Eq. (1.6) ω is greater than ω_0 and the longitudinal slip i' is positive, whereas when a braking torque is applied, ω_0 is greater than ω and the longitudinal slip i' is negative. When the applied braking torque is such that the tire is locked up, the tire no longer rotates, and it slides on the ground. Under these circumstances, wheel-spin velocity ω is zero and, in accordance with Eq. (1.6), longitudinal slip i' is -1 or -100%. This indicates that Eq. (1.6) is suited for characterizing the behavior of a tire subject to a braking torque ranging from zero to that when the tire is locked up. This corresponds to longitudinal slip i' varying from 0 to -100%. Tire behavior when a braking torque is applied will be further discussed later in this section.

Let us consider the behavior of a tire subject to a driving torque when the tire spins with an angular velocity ω while the tire longitudinal velocity is zero. As mentioned previously, this phenomenon is observed when a tire is operating on a slippery, icy road. Under these circumstances, it is uncertain that a unique value of longitudinal slip i' can be determined using Eq. (1.6), as ω could be any value dependent upon the accelerator pedal position (or engine throttle opening). Furthermore, when the tire longitudinal velocity V approaches zero, ω_0 would also approach zero. For any value of ω, the longitudinal slip i' could attain a very high value that may approach infinity, in accordance with Eq. (1.6). To characterize the behavior of a tire over the full range of operations, from driving torque being zero to that when the tire spins with angular velocity ω while the longitudinal velocity V is zero, the scale for the longitudinal slip i' must be extended from zero to a very high value that approaches infinity (i.e., from 0 to ∞). Using a scale from 0 to ∞ for longitudinal slip i' to characterize tire behavior with a driving torque applied would be difficult in practice or would be impractical.

In summary, based on the above-noted analysis, it appears that Eq. (1.6) is suited for defining the longitudinal slip (or longitudinal slip ratio) for characterizing tire behavior with a braking torque applied. It is uncertain, however, that Eq. (1.6) is suited for defining the longitudinal slip for characterizing tire behavior with a driving torque applied over the full range of possible operating conditions, as described above. It should be noted that appropriate characterization of tire behavior with a driving torque applied over the full range of operating conditions is essential for the study of road vehicle dynamics, including traction control. The relationship between tractive effort and longitudinal slip over the full operating range up to wheel spinning is of critical importance to the study of off-road wheeled vehicle mobility and will be discussed in detail in Chapter 2.

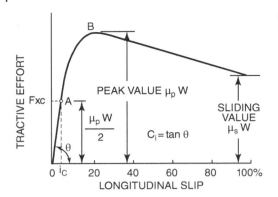

As the tractive force developed by a tire is proportional to the applied wheel torque under steady-state conditions, slip is a function of tractive effort. At first the wheel torque and tractive force increase linearly with slip because, initially, slip is mainly due to elastic deformation of the tire tread. This corresponds to section OA of the curve shown in Figure 1.16. A further increase of wheel torque and tractive force results in part of the tire tread sliding on the ground. Under these circumstances, the relationship between the tractive force and the slip is nonlinear. This corresponds to section AB of the curve shown in Figure 1.16. Based on available experimental data, the maximum tractive force of a pneumatic tire on paved roads is usually reached somewhere between 15% and 20% slip. Any further increase of slip beyond that results in an unstable condition, with the tractive effort falling rapidly from the peak value $\mu_p W$ to the pure sliding value $\mu_s W$, as shown in Figure 1.16, where W is the normal load on the tire; μ_p and μ_s are the peak and sliding values of coefficient of road adhesion, respectively (also known as the static and sliding coefficient of friction, respectively).

A general theory that can accurately predict the relationship between the tractive effort and the longitudinal slip of pneumatic tires on paved roads has yet to be evolved. However, several theories have been proposed that could provide a basic understanding of the physical nature of the processes involved. One of the earliest theoretical treatises on the relationship between the tractive effort and the longitudinal slip of pneumatic tires was presented by Julien (Hadekel 1952).

In Julien's theory, it is assumed that the tire tread can be regarded as an elastic band, and that the contact patch is rectangular, and the normal pressure is uniformly distributed (Hadekel 1952). It is further assumed that the contact patch can be divided into an adhesion region and a sliding region. In the adhesion region, the interacting forces depend on the elastic properties of the tire, whereas in the sliding region, the interacting forces depend upon the adhesive properties of the tire–ground interface. When a driving torque is applied to a tire, in the region in front of the contact patch, the driving torque produces a longitudinal strain ϵ (in compression) in the tread. It remains constant in the adhesion region of the contact patch, where no sliding between the tire tread and the ground takes place. Let e_0 be the longitudinal deformation of the tire tread in front of the contact patch, and let e be the longitudinal deformation of the tread at a point at a distance x behind the front contact point, then

$$e = e_0 + x\epsilon \tag{1.7}$$

Assume that e_0 is proportional to ϵ, and $e_0 = \lambda\epsilon$. Then

$$e = (\lambda + x)\epsilon \tag{1.8}$$

It is further assumed that, within the adhesion region, where no sliding between the tire tread and the ground takes place, the tractive force per unit contact length is proportional to the deformation of the tread. Thus,

$$\frac{dF_x}{dx} = k_t e = k_t(\lambda + x)\epsilon \tag{1.9}$$

where k_t is the tangential stiffness of the tire tread and F_x is the tractive force.

Based on experimental data of a sample of heavy truck tires under rated loads and inflation pressures, it is found that the value of k_t varies in a narrow range from approximately 3930 kN/m^2 (570 lb/in.^{2}) for a radial-ply tire to 4206 kN/m^2 (610 lb/in.^{2}) for a bias-ply tire.

$$F_x = \int_0^x k_t(\lambda + x)\epsilon \, dx = k_t \lambda x \epsilon \left(1 + \frac{x}{2\lambda}\right) \tag{1.10}$$

Let p be the normal pressure, b the width of the contact patch, and μ_p the peak value of the coefficient of road adhesion. Then no sliding will take place between the tread and the ground if the following condition is satisfied:

$$\frac{dF_x}{dx} = k_t(\lambda + x)\epsilon \leq \mu_p p b \tag{1.11}$$

This implies that if a point at a distance of x behind the front contact point is in the adhesion region, then x must be less than a characteristic length l_c, which defines the length of the region where no sliding between the tire tread and the ground takes place; that is,

$$x \leq l_c = \frac{\mu_p p b}{k_t \epsilon} - \lambda = \frac{\mu_p W}{l_t k_t \epsilon} - \lambda \tag{1.12}$$

where W is the normal load on the tire and l_t is the contact length of the tire.

If $l_t \leq l_c$, then the entire contact area is an adhesion region. Letting $x = l_t$ in Eq. (1.10), the tractive force becomes

$$F_x = k_t \lambda l_t \epsilon \left(1 + \frac{l_t}{2\lambda}\right) = K_t \epsilon \tag{1.13}$$

where $K_t = k_t \lambda l_t [1 + l_t/2\lambda]$.

Since the longitudinal strain ϵ is a measure of the longitudinal slip i of the tire, it is concluded that if the entire contact patch is an adhesion region, the relationship between the tractive force F_x and the slip i is linear. This corresponds to the region between points O and A on the tractive effort–slip curve shown in Figure 1.16.

The condition for sliding at the rear edge of the contact area is given by

$$l_t = l_c = \frac{\mu_p W}{l_t k_t i} - \lambda \tag{1.14}$$

This means that, if the slip or tractive force reaches the respective critical value i_c or F_{xc} given below, sliding in the trailing part of the contact patch begins:

$$i_c = \frac{\mu_p W}{l_t k_t (l_t + \lambda)} \tag{1.15}$$

$$F_{xc} = \frac{\mu_p W[1 + (l_t/2\lambda)]}{1 + (l_t/\lambda)} \tag{1.16}$$

A further increase of slip or tractive force beyond the respective critical value results in the spread of the sliding region from the trailing edge toward the leading part of the contact patch. The tractive force F_{xs} developed in the sliding region is given by

$$F_{xs} = \mu_p W (1 - l_c/l_t) \tag{1.17}$$

and the tractive force F_{xa} developed in the adhesion region is given by

$$F_{xa} = k_t \lambda i l_c \left(1 + \frac{l_c}{2\lambda}\right) \tag{1.18}$$

where l_c is determined by Eq. (1.12).

Hence, the relationship between the total tractive force and the slip when part of the tire tread sliding on the ground is expressed by

$$F_x = F_{xs} + F_{xa} = \mu_p W - \frac{\lambda \left(\mu_p W - K'i\right)^2}{2l_t K'i} \tag{1.19}$$

where $K' = l_t k_t \lambda$.

This equation clearly indicates the nonlinear behavior of the tractive effort–longitudinal slip relationship when sliding occurs in part of the contact area. This corresponds to the region beyond point A of the curve shown in Figure 1.16.

When sliding extends over the entire contact patch, the tractive force F_x is equal to $\mu_p W$. Under this condition, the slip i is obtained by setting l_c to zero in Eq. (1.14). The value of the slip i_m where the maximum tractive effort occurs is equal to $\mu_p W/l_t k_t \lambda$ and corresponds to point B shown in Figure 1.16. A further increase of tire slip results in an unstable situation, with the coefficient of road adhesion falling rapidly from the peak value μ_p to the pure sliding value μ_s.

In practice, the normal pressure distribution over the tire–ground contact patch is not uniform. There is a gradual drop of pressure near the edges. It is expected, therefore, that a small sliding region will be developed in the trailing part of the contact area, even at low slips.

Using Julien's theory to define the relationship between tractive effort and longitudinal slip, in addition to the parameters μ_p, W, and l_t, the value of λ, which determines the longitudinal deformation of the tire tread prior to entering the contact patch, must be known. To determine the value of λ for a given tire would require considerable effort and elaborate experiments. In view of this, a simplified theory has been developed in which the effect of λ is neglected. From Eq. (1.9), by neglecting the term λ, the tractive force per unit contact length in the adhesion region at a distance of x from the front contact point is given by

$$\frac{dF_x}{dx} = k_t x \epsilon = k_t x i \tag{1.20}$$

If there is no sliding between the tire tread and the ground for the entire contact patch, the relationship between the tractive force and slip can be expressed by

$$F_x = \int_0^{l_t} k_t i x \, dx = \left(k_t l_t^2/2\right) i \tag{1.21}$$

The term $k_t l_t^2/2$ may be taken as the slope C_i of the tractive effort–slip curve at the origin as shown in Figure 1.16; that is,

$$\frac{k_t l_t^2}{2} = C_i = \tan\theta = \left.\frac{\partial F_x}{\partial i}\right|_{i=0} \tag{1.22}$$

where C_i is usually referred to as the longitudinal stiffness of the tire.

If no sliding takes place on the contact patch, the relationship between the tractive force and the slip will, therefore, be linear:

$$F_x = C_i i \tag{1.23}$$

Equation 1.23 applies to section OA of the curve shown in Figure 1.16.

With the increase of slip beyond point A shown in Figure 1.16, the tractive force per unit contact length at the trailing edge of the contact patch reaches the adhesion limit and sliding between the tread and the ground takes place.

$$\frac{dF_x}{dx} = k_t l_t i = \mu_p p b = \frac{\mu_p W}{l_t} \tag{1.24}$$

This indicates that when the slip or tractive force reaches the respective critical value i_c or F_{xc} given below, sliding in the trailing part of the contact patch begins:

$$i_c = \frac{\mu_p W}{k_t l_t^2} = \frac{\mu_p W}{2C_i} \tag{1.25}$$

$$F_{xc} = C_i i_c = \frac{\mu_p W}{2} \tag{1.26}$$

In other words, if slip $i \le i_c$ or the tractive force $F_x \le F_{xc}$, the relationship between the tractive force and slip is linear, as shown in Figure 1.16. Equation (1.26) indicates that the upper limit for the linear range of the tractive force–slip relationship is identified by the tractive force being equal to one-half of its maximum value ($\mu_p W/2$).

A further increase of slip or tractive force beyond the respective critical value (i.e., $i > i_c$ or $F_x > F_{xc}$) results in the spread of the sliding region from the trailing edge toward the leading part of the contact patch. The tractive force F_{xs} developed in the sliding region is given by

$$F_{xs} = \mu_p W \left(1 - \frac{l_c}{l_t} \right) = \mu_p W \left(1 - \frac{\mu_p W}{2C_i i} \right) \tag{1.27}$$

and the tractive force F_{xa} developed in the adhesion region is expressed by

$$F_{xa} = \frac{1}{2} \frac{\mu_p W l_c}{l_t} = \frac{\mu_p^2 W^2}{4C_i i} \tag{1.28}$$

Hence, the relationship between the total tractive force and the slip when part of the tread is sliding on the ground (i.e., $i > i_c$ or $F_x > F_{xc}$) is given by

$$F_x = F_{xs} + F_{xa} = \mu_p W \left(1 - \frac{\mu_p W}{4C_i i} \right) \tag{1.29}$$

The equation above indicates the nonlinear nature of the tractive effort–longitudinal slip relationship when sliding occurs in part of the contact patch. It is applicable to predicting the tractive effort–slip relation when the tractive effort is lower than its maximum value $\mu_p W$.

In comparison with Julien's theory, the simplified theory described above requires only three parameters, μ_p, W, and C_i, to define the tractive effort–longitudinal slip relationship. As pointed out previously, the value of C_i can easily be identified from the initial slope of the measured tractive effort–slip curve.

1.3.2 Braking Effort and Longitudinal Skid

When a braking torque is applied to a pneumatic tire, a braking effort (force) acting in the opposite direction of tire motion is developed on the tire–ground contact patch. At the same time, stretching of tread elements occurs prior to entering the contact area, as shown in Figure 1.17, in contrast with the compression effect when a driving torque is applied, as discussed previously. The distance that the tire travels when a braking torque is applied is greater than that in free rolling.

The phenomenon of the distance the tire travels when a braking torque is applied being greater than that in free rolling is generally referred to as the longitudinal skid. In the analysis that follows, the longitudinal skid of a tire i_s is defined as

$$i_s = 1 - \frac{r\omega}{V} \qquad (1.30)$$

where r, V, and ω are the same as those in Eq. (1.5).

The definition of longitudinal skid given by Eq. (1.30) is used in this book not only for the analysis of the mechanics of the tire on paved roads, but also for that of the vehicle running gear (wheel, tire, or track) on deformable terrain.

In the ISTVS Standard released in 2020, the definition for the longitudinal skid of the vehicle running gear given by Eq. (1.30) is adopted (He et al. 2020).

For a locked tire, the angular speed ω of the tire is zero, whereas the forward longitudinal speed of the tire center V is not zero. Under this condition, from Eq. (1.30), tire skid is 100%. It should be noted that Eq. (1.30) yields essentially the same result for a locked tire as that from Eq. (1.6), except when using Eq. (1.6) the longitudinal slip i' for a locked tire is denoted as −100%.

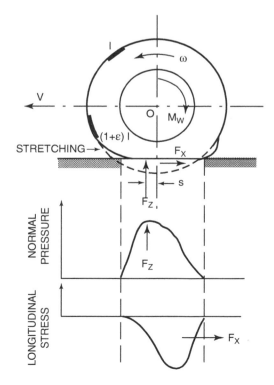

Figure 1.17 Behavior of a tire under the action of a braking torque. Reproduced with permission from *Mechanics of Pneumatic Tires*, edited by S.K. Clark, Monograph 122, National Bureau of Standards, 1971.

If Eq. (1.5) is extended to cover tire behavior during braking, then when the tire is locked up, with angular speed of the tire ω being zero while the tire is sliding on the ground, the value of slip as determined by Eq. (1.5) will approach negative infinity. This will pose a difficulty in practice for characterizing the braking performance of tires, like that using Eq. (1.6) for characterizing the tractive performance of tires.

A simplified theory for the relationship between the braking effort and skid can also be developed, following an approach like that for the relationship between the tractive force and slip described previously.

According to the definitions of slip i and skid i_s given by Eqs. (1.5) and (1.30), respectively, the expressions for slip i and skid i_s are related by

$$|i| = |i_s/(1-i_s)| \tag{1.31}$$

If no sliding takes place on the contact patch, the relationship between the braking effort and the skid can be established by replacing C_i and i in Eq. (1.23) with C_s and $i_s/(1-i_s)$, respectively.

$$F_x = C_s i_s/(1-i_s) \tag{1.32}$$

where F_x is the braking effort acting in the opposite direction of motion of the tire center, and C_s is the slope of the braking effort–skid curve at the origin, and is given by (Segel 1984)

$$C_s = \left.\frac{\partial F_x}{\partial i_s}\right|_{i_s=0} \tag{1.33}$$

C_s is referred to as the longitudinal stiffness of the tire during braking. Like the parameter C_i, the value of C_s can easily be identified from the initial slope of the measured braking effort–skid curve.

It is interesting to note from Eq. (1.32) that, using the definition of skid given by Eq. (1.30), the relationship between braking effort and skid is nonlinear, even at low skids, where no sliding takes place between the tread and the ground.

The critical value of skid i_{sc}, at which sliding between the tread and the ground begins, can be established by replacing C_i and i in Eq. (1.25) with C_s and $i_s/(1-i_s)$, respectively:

$$i_{sc} = \frac{\mu_p W}{2C_s + \mu_p W} \tag{1.34}$$

The corresponding critical value of braking effort F_{xc}, above which sliding between the tread and the ground begins, is given by

$$F_{xc} = \frac{C_s i_{sc}}{1-i_{sc}} = \frac{\mu_p W}{2} \tag{1.35}$$

When sliding takes place in part of the contact patch (i.e., $i_s > i_{sc}$), the relationship between the braking effort and the skid can be established by replacing C_i and i in Eq. (1.29) with C_s and $i_s/(1-i_s)$, respectively.

$$F_x = \mu_p W \left[1 - \frac{\mu_p W(1-i_s)}{4C_s i_s}\right] \tag{1.36}$$

While the theory described above represents a simplified model for the highly complex phenomenon of tire–ground interaction, it has been proven to be useful in representing tire behavior in the simulations of the dynamics of passenger cars (Segel 1984; Fancher and Grote 1971).

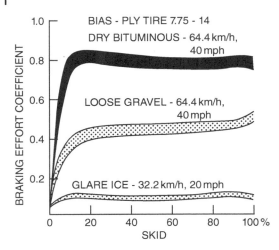

Figure 1.18 Variations of braking effort coefficient with skid of a car tire on various surfaces. Reprinted with permission from SAE paper 690214 © 1969 Society of Automotive Engineers, Inc.

Figure 1.18 shows the variation of the braking effort coefficient, which is the ratio of the braking effort to normal load, with skid for a bias-ply passenger car tire over various surfaces based on test data (Harned et al. 1969). It should be noted that for a locked wheel the skid is denoted by 100%. The peak and sliding values of the coefficient of road adhesion of a bias-ply, a bias-belted, and a radial-ply passenger car tire of the same size with various inflation pressures at a speed of 64 km/h (40 mph) on a dry, aggregate asphalt surface are shown in Figure 1.19 (Collier and Warchol 1980). Sliding coefficient of road adhesion is that when the tire is locked or that when tire skid is 100%. It

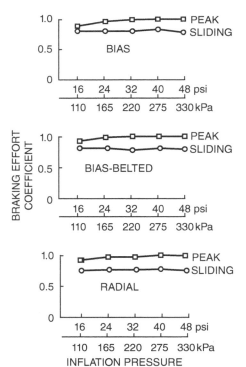

Figure 1.19 Variations of peak and sliding values of braking effort coefficient with inflation pressure for bias-ply, bias-belted, and radial-ply car tires on dry pavement. Reprinted with permission from SAE paper 800087 © Society of Automotive Engineers, Inc.

appears that on a dry surface, the coefficient of road adhesion does not vary significantly with tire type and inflation pressure.

Average values of peak coefficient of road adhesion μ_p (static coefficient of friction) and sliding coefficient of road adhesion under various operating conditions are presented in Table 1.6 (Robert Bosch, *Bosch Automotive Handbook*, 10th Edition, 2018; Taborek 1957).

Table 1.6 Values of peak and sliding coefficient of road adhesion for tires under various operating conditions.

Surface	Driving speed km/h (mph)	Tire condition	Peak value μ_p	Sliding value μ_s
Dry asphalt and concrete	50 (31)	New	0.85	-
		Worn[a]	1	
	90 (56)	New	0.8	
		Worn	0.95	
	130 (81)	New	0.75	
		Worn	0.9	
Wet asphalt and concrete, water level approx. 0.2 mm	50 (31)	New	0.65	-
		Worn	0.5	
	90 (56)	New	0.6	
		Worn	0.2	
	130 (81)	New	0.55	
		Worn	0.2	
Wet asphalt and concrete, heavy rainfall water level approx. 1 mm	50 (31)	New	0.55	-
		Worn	0.4	
	90 (56)	New	0.3	
		Worn	0.1	
	130 (81)	New	0.2	
		Worn	0.1	
Wet asphalt and concrete, puddles water level approx. 2 mm	50 (31)	New	0.5	-
		Worn	0.25	
	90 (56)	New	0.05	
		Worn	0.05	
	130 (81)	New	0	
		Worn	0	
Gravel			0.6	0.55
Earth road (dry)			0.68	0.65
Earth road (wet)			0.55	0.4–0.5
Snow (hard-packed)			0.2	0.15
Ice			0.1	0.07

Sources: Adapted from Robert Bosch GmbH, *Bosch Automotive Handbook*, 10th Edition (2018), with permission of Wiley; and from Taborek (1957), with permission of *Machine Design*.
[a] Worn to 1.6 mm tread depth (legal minimum in Germany).

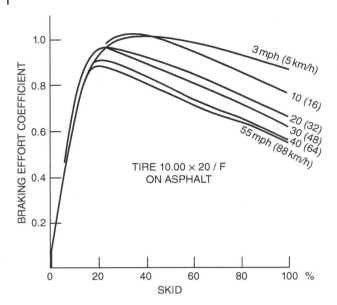

Figure 1.20 Effects of speed on braking performance of a truck tire on asphalt. *Source:* Ervin (1975). Reproduced with permission of the University of Michigan Transportation Research Institute.

Among the operational parameters, speed and normal load have noticeable effects on the tractive (braking) effort–slip (skid) characteristics. Figure 1.20 shows the influence of speed on the braking effort coefficient–skid characteristics of a bias-ply truck tire on a dry asphalt surface (Ervin 1975). As shown in Figure 1.20, speed appears to have a significant effect on the tractive (braking) performance of a tire. Therefore, it has been suggested that to improve the prediction of the relationship between the tractive (braking) effort and the slip (skid), the effect of the sliding speed between the tire tread and the ground should be incorporated into the theories described previously (Segel 1984). Figure 1.21 shows the effect of normal load on the braking performance of a bias-ply truck tire on a dry asphalt surface (Ervin 1975). The value of the longitudinal stiffness C_s increases noticeably with an increase of the normal load. This is because the tire contact length increases with the normal load for a given inflation pressure. According to Eq. (1.21), to develop a given longitudinal force, the longer tire contact length results in lower longitudinal slip (or skid).

A sample of the peak and sliding values of the coefficient of road adhesion μ_p and μ_s for truck tires at 64 km/h (40 mph) on dry and wet concrete pavements is shown in Table 1.7 (Fancher et al. 1986). The pavements were aggressively textured, like those of relatively new paved roads meeting the requirements of the U.S. Federal Interstate Highway System.

It can be seen from Table 1.7 that the ratio of the peak value μ_p to the sliding value μ_s for truck tires on dry concrete pavement is around 1.4, whereas on wet concrete pavement, it ranges from approximately 1.3–1.6. It is also noted that there appears to be no clear distinctions between the tractive (braking) performance of bias-ply and radial-ply truck tires.

The significant difference between the peak values μ_p and the sliding value μ_s of the coefficient of road adhesion indicates the importance of avoiding wheel lockup during braking (skid $i_s = 100\%$) or wheel spinning during acceleration (slip $i = 100\%$). This is one of the impetuses to the development of antilock brake systems and traction control systems for road vehicles, which will be discussed in Chapter 3.

Figure 1.21 Effects of normal load on braking performance of a truck tire on asphalt. *Source:* Ervin (1975). Reproduced with permission of the University of Michigan Transportation Research Institute.

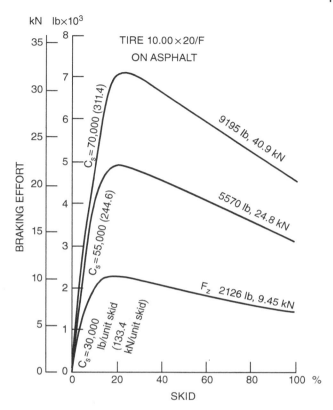

1.4 Cornering Properties of Tires

1.4.1 Slip Angle and Cornering Force

When a pneumatic tire is not subject to any force perpendicular to the wheel plane (i.e., side force), it will move along the wheel plane. If, however, a side force F_s is applied to a tire, a lateral force will be developed on the contact patch, and the tire will move along a path at an angle α with the wheel plane, as shown in Figure 1.22. The angle α is referred to as the slip angle, and the phenomenon of side slip is mainly due to the lateral elasticity of the tire.

The lateral force developed on the tire–ground contact patch is usually called the cornering force $F_{y\alpha}$ when the camber angle of the wheel is zero. The relationship between the cornering force and the slip angle is of fundamental importance to the directional control and stability of road vehicles.

When the tire is moving at a uniform speed in the direction of OA, the side force F_s applied at the wheel center and the cornering force $F_{y\alpha}$ developed in the ground plane are usually not collinear, as shown in Figure 1.22. At small slip angles, the cornering force in the ground plane is normally behind the applied side force, giving rise to a torque (or couple), which tends to align the wheel plane with the direction of motion. This torque is called the aligning or self-aligning torque and is one of the primary restoring moments that help the steered tire return to the original position after negotiating a turn. The distance t_p between the side force and the cornering force is called the pneumatic trail, and the product of the cornering force and the pneumatic trail determines the self-aligning torque.

Table 1.7 Values of coefficient of road adhesion for truck tires on dry and wet concrete pavement at 64 km/h (40 mph).

Tire type	Tire construction	Dry		Wet	
		μ_p	μ_s	μ_p	μ_s
Goodyear Super Hi Miler (rib)	Bias-ply	0.850	0.596	0.673	0.458
General GTX (rib)	Bias-ply	0.826	0.517	0.745	0.530
Firestone Transteel (rib)	Radial-ply	0.809	0.536	0.655	0.477
Firestone Transport 1 (rib)	Bias-ply	0.804	0.557	0.825	0.579
Goodyear Unisteel R-1 (rib)	Radial-ply	0.802	0.506	0.700	0.445
Firestone Transteel Traction (lug)	Radial-ply	0.800	0.545	0.600	0.476
Goodyear Unisteel L-1 (lug)	Radial-ply	0.768	0.555	0.566	0.427
Michelin XZA (rib)	Radial-ply	0.768	0.524	0.573	0.443
Firestone Transport 200 (lug)	Bias-ply	0.748	0.538	0.625	0.476
Uniroyal Fleet Master Super Lug	Bias-ply	0.739	0.553	0.513	0.376
Goodyear Custom Cross Rib	Bias-ply	0.716	0.546	0.600	0.455
Michelin XZZ (rib)	Radial-ply	0.715	0.508	0.614	0.459
Average		0.756	0.540	0.641	0.467

Source: Fancher et al. (1986).

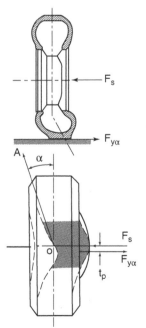

Figure 1.22 Behavior of a tire subject to a side force. Reproduced with permission from *Mechanics of Pneumatic Tires*, edited by S.K. Clark, Monograph 122, National Bureau of Standards, 1971.

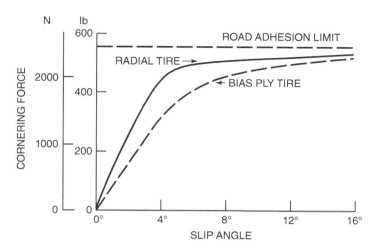

Figure 1.23 Cornering characteristics of a bias-ply and a radial-ply car tire. Reproduced with permission from *Mechanics of Pneumatic Tires*, edited by S.K. Clark, Monograph 122, National Bureau of Standards, 1971.

The relationships between the slip angle and the cornering force of various types of tire under a variety of operating conditions have been investigated extensively. Typical plots of the cornering force as a function of the slip angle for a bias-ply and a radial-ply passenger car tire are shown in Figure 1.23 (van Eldik Thieme and Pacejka 1971). For slip angles below a certain value, such as 4° shown in Figure 1.23, the cornering force is approximately proportional to the slip angle. Beyond that, the cornering force increases at a lower rate with an increase of the slip angle, and it reaches a maximum value where the tire begins sliding laterally. For passenger car tires, the maximum cornering force may occur at a slip angle of about 18°, while for racing car tires, the cornering force may peak at approximately 6°. Figure 1.23 shows that the cornering force of a bias-ply tire increases more slowly with an increase of the slip angle than that of a radial-ply tire. These characteristics are more suited to two-wheeled vehicles, such as motorcycles. A more gradual increase of the cornering force with the slip angle enables the driver to exercise better control over a two-wheeled vehicle. This is one of the reasons why bias-ply tires can still be found in motorcycles (French 1989). Figure 1.24 shows the variations of the ratio of the cornering force to the normal load with slip angle for radial-ply and bias-ply truck tires of size 10.00-20 with different tread designs (ribbed or lugged) (Segel 1984). Like that shown in Figure 1.23 for passenger car tires, the cornering force of radial-ply truck tires increases more rapidly with an increase of the slip angle than that of bias-ply truck tires.

Several factors affect the cornering behavior of pneumatic tires. The normal load on the tire strongly influences the cornering characteristics. Some typical results are shown in Figure 1.25 (van Eldik Thieme and Pacejka 1971). For a given slip angle, the cornering force generally increases with an increase of the normal load. However, the relationship between the cornering force and the normal load is nonlinear. Thus, the transfer of load from the inside to the outside tire during a turning maneuver will reduce the total cornering force that a pair of tires on an axle can develop. Consider a pair of tires on a beam axle, each with normal load F_z, as shown in Figure 1.26. The cornering force per tire with normal load F_z is F_y for a given slip angle. If the

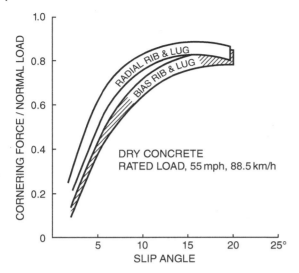

Figure 1.24 Cornering characteristics of bias-ply and radial-ply truck tires on dry concrete. *Source:* Segel (1984). Reproduced with permission of the University of Michigan Transportation Research Institute.

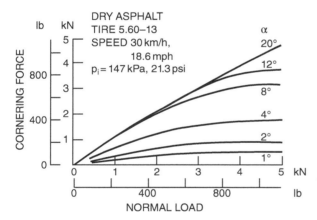

Figure 1.25 Effects of normal load on the cornering characteristics of a car tire. Reproduced with permission from *Mechanics of Pneumatic Tires*, edited by S.K. Clark, Monograph 122, National Bureau of Standards, 1971.

vehicle undergoes a steady-state turn, owing to lateral load transfer, the normal load on the inside tire will be reduced to F_{zi} and that on the outside tire will be increased to F_{zo}. As a result, the total cornering force of the two tires will be the sum of F_{yi} and F_{yo}, which is less than $2F_y$, as shown in Figure 1.26. This implies that for a pair of tires on a beam axle to develop the required amount of cornering force to balance a given centrifugal force during a turn, the lateral load transfer results in an increase in the slip angle of the tires.

To provide a measure for comparing the cornering behavior of different tires, a parameter called cornering stiffness C_α is used. It is defined as the derivative of the cornering force $F_{y\alpha}$ with respect to slip angle α evaluated at zero slip angle:

$$C_\alpha = \left.\frac{\partial F_{y\alpha}}{\partial \alpha}\right|_{\alpha = 0} \tag{1.37}$$

Figure 1.26 Effects of lateral load transfer on the cornering capability of a pair of tires on an axle.

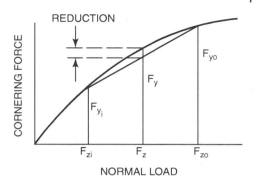

Figure 1.27 shows a comparison of the relationships between the cornering stiffness and the normal load for a sample of passenger car, light truck, and heavy truck tires (Segel 1984). In the figure, RL indicates the rated load for a specific tire. For the three passenger car tires tested, the cornering stiffness reaches a maximum at the rated load and decreases with a further increase in the normal load. However, for the light truck and heavy truck tires shown, the cornering stiffness keeps increasing beyond the rated load, although at a lower rate.

To evaluate the effect of the normal load on the cornering ability of tires, a parameter called the cornering coefficient, which is defined as the cornering stiffness per unit normal load, is often used. Figure 1.28 shows a typical relationship between the cornering coefficient and the normal load of a tire (Taborek 1957). It shows that the cornering coefficient decreases with an increase in normal load.

Inflation pressure usually has a moderate effect on the cornering properties of a tire. In general, the cornering stiffness of tires increases with an increase of the inflation pressure. Figure 1.29 shows

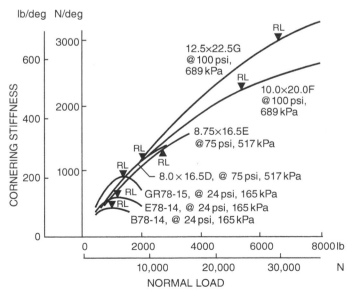

Figure 1.27 Comparison of cornering stiffness of car, light truck, and heavy truck tires. *Source:* Segel (1984). Reproduced with permission of the University of Michigan Transportation Research Institute.

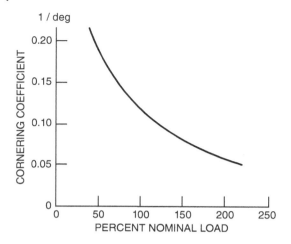

Figure 1.28 Effects of normal load on the cornering coefficient of a tire. *Source:* Taborek (1957). Reproduced with permission of *Machine Design*.

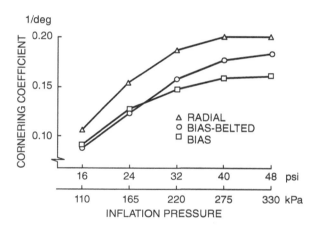

Figure 1.29 Variations of cornering coefficient with inflation pressure for radial-ply, bias-ply, and bias-belted car tires. Reprinted with permission from SAE paper 800087 © 1980 Society of Automotive Engine, Inc.

a comparison of the cornering coefficients at different inflation pressures of a radial-ply, a bias-belted, and a bias-ply passenger car tire (Collier and Warchol 1980). Table 1.8 shows a sample of the values of the cornering coefficient for truck tires at rated loads and inflation pressures (unless specified) (Fancher et al. 1986).

1.4.2 Slip Angle and Aligning Torque

As mentioned in Section 1.4.1, the side force F_s applied at the wheel center and the cornering force $F_{y\alpha}$ developed in the ground plane are usually not collinear, as shown in Figure 1.22. This gives rise to a torque commonly known as the aligning or self-aligning torque (moment). Figure 1.30 shows a plot of the cornering force versus the aligning torque for a passenger car tire at various slip angles and under different normal loads (Gough 1956). Figures 1.31 and 1.32 show the variations of the aligning torque with the slip angle and the normal load for a bias-ply truck tire (10.00-20/*F*) and for a radial-ply truck tire (10.00-20/*G*), respectively (Segel 1984). It is interesting to note that with a

Table 1.8 Cornering coefficients for truck tires at rated loads and inflation pressures (unless specified).

Tire type	Tire construction	Cornering coefficient (deg^{-1})
Michelin Radial XZA (1/3 tread)	Radial-ply	0.1861
Michelin Radial XZA (1/2 tread)	Radial-ply	0.1749
Michelin Pilote XZA	Radial-ply	0.1648
Michelin Radial XZA	Radial-ply	0.1472
Goodyear Unisteel G159, 11R22.5 LRF at 655 kPa (95 psi)	Radial-ply	0.1413
Michelin XZZ	Radial-ply	0.1370
Goodyear Unisteel 11, 10R22.5 LRF at 620 kPa (90 psi)	Radial-ply	0.1350
Goodyear Unisteel G159, 11R22.5 LRG at 792 kPa (115 psi)	Radial-ply	0.1348
Goodyear Unisteel 11, 10R22.5 LRF at 758 kPa (110 psi)	Radial-ply	0.1311
Firestone Transteel	Radial-ply	0.1171
Firestone Transteel Traction	Radial-ply	0.1159
Goodyear Unisteel R-1	Radial-ply	0.1159
Goodyear Unisteel L-1	Radial-ply	0.1121
Firestone Transport 1	Bias-ply	0.1039
General GTX	Bias-ply	0.1017
Goodyear Super Hi Miler	Bias-ply	0.0956
Goodyear Custom Cross Rib	Bias-ply	0.0912
Uniroyal Fleet Master Super Lub	Bias-ply	0.0886
Firestone Transport 200	Bias-ply	0.0789

Source: Fancher et al. (1986).

given normal load, the aligning torque first increases with an increase of the slip angle. It reaches a maximum at a particular slip angle, and then decreases with a further increase of the slip angle. This is mainly caused by the sliding of the tread in the trailing part of the contact patch at high slip angles, which results in shifting the point of application of the cornering force forward. Table 1.9 shows a sample of measured values of pneumatic trail for truck tires at a slip angle of 1° and under rated loads and inflation pressures (unless specified) (Fancher et al. 1986). It is shown that the pneumatic trail for truck tires varies in the range from 4.6 cm (1.8 in.) to 7.1 cm (2.8 in.). A typical value for a new bias-ply truck tire is 5.8 cm (2.3 in.), while that for a new radial-ply tire is 5.3 cm (2.1 in.).

Longitudinal force affects the aligning torque significantly. In general, the effect of a driving torque is to increase the aligning torque for a given slip angle, while a braking torque has the opposite effect. Inflation pressure and normal load also have noticeable effects on the aligning torque because they affect the size of the tire contact patch. Higher normal load and lower inflation pressure result in longer tire contact length, and hence pneumatic trail. This causes an increase in the aligning torque.

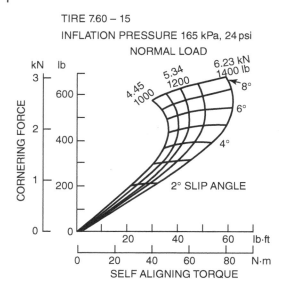

TIRE 7.60 – 15

INFLATION PRESSURE 165 kPa, 24 psi

Figure 1.30 Variations of self-aligning torque with cornering force of a car tire under various normal loads. *Source:* Gough (1956). Reprinted with permission from SAE Transactions, Vol. 64 © 1956 Society of Automotive Engineers, Inc.

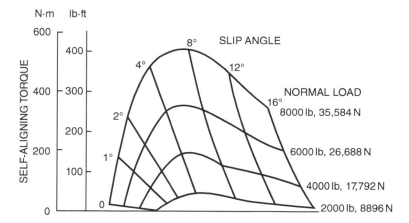

Figure 1.31 Variations of self-aligning torque with normal load and slip angle for a bias-ply truck tire, 10.00–20/F. *Source:* Segel (1984). Reproduced with permission of the University of Michigan Transportation Research Institute.

1.4.3 Camber and Camber Thrust

Camber is the inclination of the wheel plane from a plane perpendicular to the road surface when viewed from the fore and aft directions of the vehicle, as shown in Figure 1.33. Its main purpose is to achieve axial bearing pressure and to decrease the kingpin offset. Camber on passenger cars is between 1/2 and 1°. High camber angles promote excessive tire wear (Taborek 1957).

Camber causes a lateral force developed on the contact patch. This lateral force is usually referred to as camber thrust $F_{y\gamma}$, and the development of this thrust may be explained in the following way. A free-rolling tire with a camber angle would revolve about point O, as shown in Figure 1.33. However, the cambered tire in a vehicle is constrained to move in a straight line.

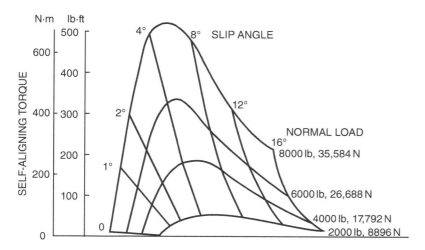

Figure 1.32 Variation of self-aligning torque with normal load and slip angle for a radial-ply truck tire, 10.00–20/G. *Source:* Segel (1984). Reproduced with permission of the University of Michigan Transportation Research Institute.

Table 1.9 Pneumatic trails for truck tires at a slip angle of 1° under rated loads and inflation pressures (unless specified).

Tire type	Tire construction	Pneumatic trails	
		cm	in.
Michelin Radial 11R22.5 XZA (1/3 Tread)	Radial-ply	6.17	2.43
Goodyear Unisteel II, 10R22.5 LRF at 620 kPa (90 psi)	Radial-ply	6.15	2.42
Michelin Radial 11R22.5 XZA (1/2 Tread)	Radial-ply	5.89	2.32
Goodyear Unisteel G159, 11R22.5 LRG at 655 kPa (95 psi)	Radial-ply	5.87	2.31
Michelin Radial 11R22.5 XZA	Radial-ply	5.51	2.17
Goodyear Unisteel G159, 11R22.5 LRG at 792 kPa (115 psi)	Radial-ply	5.46	2.15
Goodyear Unisteel II, 10 R22.5 LRF at 758 kPa (110 psi)	Radial-ply	5.41	2.13
Michelin Radial 11R22.5 XZA	Radial-ply	5.38	2.12
Michelin Pilote 11/80R22.5 XZA	Radial-ply	4.62	1.82
New Unspecified Model 10.00-20/F	Bias-ply	5.89	2.32
Half-Worn Unspecified Model 10.00-20/F	Bias-ply	7.14	2.81
Fully-Worn Unspecified Model 10.00-20/F	Bias-ply	6.55	2.58

Source: Fancher et al. (1986).

A lateral force in the direction of the camber is, therefore, developed in the ground plane. It is interesting to note that the camber thrust acts ahead of the wheel center, and therefore forms a small camber torque. The relationship between the camber thrust and the camber angle (at zero slip angle) for a bias-ply passenger car tire is illustrated in Figure 1.34 (Nordeen and Cortese 1963). It has been shown that the camber thrust is approximately one-fifth the value of the cornering force

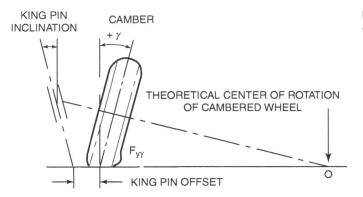

Figure 1.33 Behavior of a cambered tire.

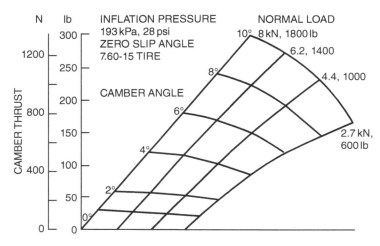

Figure 1.34 Variations of camber thrust with camber angle and normal load for a car tire. Reprinted with permission from SAE paper 713A © 1963 Society of Automotive Engineers, Inc.

obtained from an equivalent slip angle for a bias-ply tire and somewhat less for a radial-ply tire. To provide a measure for comparing the camber characteristics of different tires, a parameter called "camber stiffness" is often used. It is defined as the derivative of the camber thrust with respect to the camber angle evaluated at zero camber angle.

$$C_\gamma = \left.\frac{\partial F_{y\gamma}}{\partial \gamma}\right|_{\gamma = 0} \tag{1.38}$$

Like the cornering stiffness, the normal load and inflation pressure have an influence on the camber stiffness. Figure 1.35 shows the variations of the camber stiffness with normal load for three truck tires at an inflation pressure of 620 kPa (90 psi) (Segel 1984). It is found that for truck tires, the value of the camber stiffness is approximately one-tenth to one-fifth of that of the cornering stiffness under similar operating conditions.

Figure 1.35 Variations of camber stiffness with normal load for heavy truck tires. *Source:* Segel (1984). Reproduced with permission of the University of Michigan Transportation Research Institute.

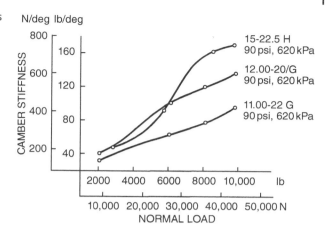

The total lateral force of a cambered tire operating at a slip angle is the sum of the cornering force $F_{y\alpha}$ and the camber thrust $F_{y\gamma}$:

$$F_y = F_{y\alpha} \pm F_{y\gamma} \tag{1.39}$$

If the cornering force and the camber thrust are in the same direction, the positive sign should be used in the above equation. For small slip and camber angles, the relationship between the cornering force and the slip angle and that between the camber thrust and the camber angle are essentially linear; the total lateral force of a cambered tire at a slip angle can, therefore, be determined by

$$F_y = C_\alpha \alpha \pm C_\gamma \gamma \tag{1.40}$$

As discussed previously, the lateral forces due to slip angle and camber angle produce an aligning torque. The aligning torque due to slip angle, however, is usually much greater.

1.4.4 Characterization of Cornering Behavior of Tires

Several attempts have been made to develop mathematical models for the cornering behavior of pneumatic tires. There are two basic types of model. One is based on the assumption that the tread of the tire is equivalent to a stretched string restrained by lateral springs, representative of the sidewall with the wheel rim acting as the base of the springs, as shown in Figure 1.36(a) (Ellis 1969). In the other model, the tread is considered equivalent to an elastic beam with continuous lateral elastic support, as shown in Figure 1.36(b).

In both models, it is assumed that the cornering behavior of a tire can be deduced from the characteristics of the equatorial line of the tire, which is the intersection of the undeformed tire tread with the wheel plane. The portion of the equatorial line in the contact area is called the contact line. One of the major differences in these two basic models is that in the stretched-string model, discontinuities of the slope of the equatorial line are permissible, whereas for the beam model that is not the case. It has been shown that for small slip angles, the stretched-string model can provide a basic understanding of the lateral behavior of a pneumatic tire. In the following, the stretched-string model as proposed by Temple and von Schlippe is discussed (Hadekel 1952).

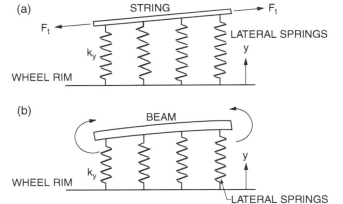

Figure 1.36 Models for cornering behavior of tires: (a) stretched-string model; (b) beam on elastic foundation model. *Source:* Ellis (1969). Reprinted with permission of J.R. Ellis and Business Books.

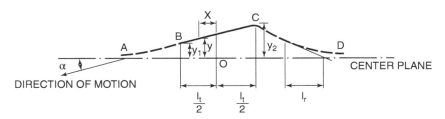

Figure 1.37 Behavior of the equatorial line of a rolling tire subject to a side force.

Consider a tire in a steady-state motion with a fixed slip angle. The shape of the equatorial line *BC* in the contact area shown in Figure 1.37 is the path of the tire, and it is immobile relative to the ground when no sliding takes place. Let the dashed line *AB* in the figure represent the projection of the portion of the equatorial line outside and in front of the contact patch. As the tire rolls forward, points of *AB* becomes points of *BC*. This indicates that *AB* and *BC* must have a common tangent at point *B*. At the rear of the contact patch, such conditions do not hold, and a kink may be present at point *C*. Thus, it can be stated that for a rolling tire the slope of the equatorial line is continuous at the front edge of the contact area, but not necessarily at the rear.

Consider an element of the distorted equatorial line shown in Figure 1.37. Let the lateral deflection from the wheel center plane be *y*, and the distance measured along the undistorted equatorial line be *x*, with the origin at the center of the contact patch. It is assumed that the lateral force applied to the rim by the element due to lateral deflection *y* is given, in differential form, by

$$dF_{y1} = k_y y dx \tag{1.41}$$

where k_y is the lateral stiffness of the tire.

This equation applies at all points of the periphery. Based on experimental data of a sample of bias-ply and radial-ply heavy truck tires under rated loads and inflation pressures, it is found that the value of k_y varies in a narrow range. The average value is approximately 2275 kN/m^2 (330 lb/in.^{2})

In an element of the equatorial line, there is another force component acting in the lateral direction, which is due to the tension in the string. This component is proportional to the curvature of the equatorial line, and for small deflection is given, in differential form, by

$$dF_{y2} = -F_t \frac{d^2y}{dx^2} dx \tag{1.42}$$

where F_t represents the tension in the string. It is usually convenient to write $F_t = k_y l_r^2$, where l_r is termed the "relaxation length," in which the lateral deflection, described by an exponential function, decreases to 1/2.718 of its prior value, as shown in Figure 1.37.

Let l_t be the contact length with the origin for x at the contact center, and let y_1 and y_2 be the deflections of the equatorial line at the front and rear ends of the contact patch, as shown in Figure 1.37. Over the part of the tire not in contact with the ground (i.e., free region) having total length l_h, the tire is not loaded by external means, and therefore from Eqs. (1.41) and (1.42),

$$k_y \left(y - l_r^2 \frac{d^2y}{dx^2} \right) = 0 \tag{1.43}$$

The solution of this differential equation will yield the deflected shape of the equatorial line in the free region, which is given by

$$y = \frac{y_2 \sinh\left[(x - l_t/2)/l_r\right] + y_1 \sinh\left[(l_t/2 + l_h - x)/l_r\right]}{\sinh\left(l_h/l_r\right)} \tag{1.44}$$

If r is the tire radius, under normal conditions l_h lies between $4.5r$ and $6r$, whereas l_r is approximately equal to r (Hadekel 1952). Hence, Eq. (1.44) may be approximated by an exponential function. For the free region near the front of the contact area (i.e., $x > l_t/2$),

$$y = y_1 \exp\left[\frac{-(x - l_t/2)}{l_r}\right] \tag{1.45}$$

For the free region near the rear of the contact area (i.e., $x < l_t/2 + l_h$),

$$y = y_2 \exp\left[\frac{-(l_t/2 + l_h - x)}{l_r}\right] \tag{1.46}$$

Thus, in the free region not in contact with the ground but near either end of the contact patch, the shape of the equatorial line is an exponential curve.

The expressions for the lateral deflection and the lateral forces acting on an element of the tread described above permit the determination of the cornering force and the aligning torque in terms of constants k_y and l_r, and contact length l_t. This can be achieved in two ways:

- Integrating the lateral force exerted on the tire over the contact length and including an infinitesimal length of the equatorial line in the free region at either end, as proposed by Temple (Hadekel 1952).
- Integrating the lateral force exerted on the rim by the tire over the entire circumference, including the contact length, as proposed by von Schlippe (Hadekel 1952).

The essence of these two methods is illustrated in Figure 1.38.

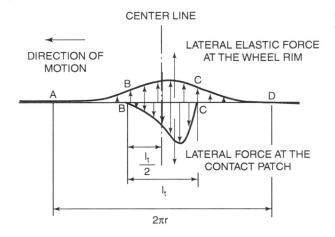

Figure 1.38 Lateral force acting on the wheel rim and the tire–road contact patch.

Following Temple's method, and assuming that the equatorial line in the contact region is a straight line, one can obtain the total lateral force F_y by integration:

$$
\begin{aligned}
F_y &= k_y \int_{-l_t/2}^{l_t/2} \left(y - l_r^2 \frac{d^2 y}{dx^2} \right) dx \\
&= k_y \int_{-l_t/2}^{l_t/2} y\, dx - k_y l_r^2 \left(\frac{dy}{dx} \right) \Bigg]_{-l_t/2}^{l_t/2} \\
&= k_y (y_1 + y_2) l_t / 2 + k_y l_r (y_1 + y_2) \\
&= k_y (y_1 + y_2)(l_r + l_t/2)
\end{aligned}
\tag{1.47}
$$

For a non-rolling tire subject to a pure side force,

$$
y_1 = y_2 = y_0 \text{ and } F_y = 2k_y y_0 (l_r + l_t/2)
\tag{1.48}
$$

The moment of lateral force about a vertical axis through the center of contact (i.e., the aligning torque) is given by

$$
\begin{aligned}
M_z &= k_y \int_{-l_t/2}^{l_t/2} x \left(y - l_r^2 \frac{d^2 y}{dx^2} \right) dx \\
&= k_y \int_{-l_t/2}^{l_t/2} xy\, dx - k_y l_r^2 \left(x \frac{dy}{dx} - y \right) \Bigg]_{-l_t/2}^{l_t/2} \\
&= k_y \frac{(l_t/2)^2}{3} (y_1 - y_2) + k_y l_r \left(l_r + \frac{l_t}{2} \right)(y_1 - y_2) \\
&= k_y (y_1 - y_2) \left[\frac{(l_r/2)^2}{3} + l_r \left(l_r + \frac{l_t}{2} \right) \right]
\end{aligned}
\tag{1.49}
$$

Following von Schlippe's approach, one can obtain the same expressions.

For a tire rolling at a slip angle α, the slope of the equatorial line in the contact area is equal to tan α if the tread in the contact patch is not sliding. Thus,

$$\alpha \approx \tan \alpha = \frac{y_1 - y_2}{l_t} = -\frac{y_1}{l_r} \tag{1.50}$$

Substituting the above expression into Eqs. (1.47) and (1.49), the relationships between the magnitudes of the lateral force and the self-aligning torque and the slip angle become

$$\frac{F_y}{\alpha} = 2k_y \left(l_r + \frac{l_t}{2} \right)^2 \tag{1.51}$$

$$\frac{M_z}{\alpha} = k_y l_t \left[\frac{(l_t/2)^2}{3} + l_r \left(l_r + \frac{l_t}{2} \right) \right] \tag{1.52}$$

The pneumatic trail t_p is given by

$$t_p = \frac{M_z}{F_y} = \frac{(l_t/2)\left[(l_t/2)^2/3 + l_r(l_r + l_t/2)\right]}{(l_r + l_t/2)^2} \tag{1.53}$$

The two basic parameters k_y and l_r, which specify the characteristics of the lateral elasticity of the pneumatic tire, can be measured by suitable tests. It is noted that the ratio of F_y/α to M_z/α is independent of k_y, and therefore l_r can be determined from the measured values of F_y/α and M_z/α (contact length of the tire l_t being known). On the other hand, the ratio of $(F_y/y_0)^2$ for a non-rolling tire to F_y/α is independent of l_r, and therefore k_y can be determined from the measured values of $(F_y/y_0)^2$ and F_y/α. Measurements of k_y and l_r have been carried out by several investigators. For instance, values of l_r for a family of aircraft tires of different sizes but with similar proportion were found by von Schlippe to vary from $0.6r$ to $0.9r$ approximately. Values of k_y measured by von Schlippe were about 90% of the inflation pressure (Hadekel 1952).

Equations (1.51) and (1.52) indicate that, if no sliding between the tread and the ground occurs, the lateral force and the aligning torque increase linearly with the slip angle. This is the case for small slip angles, as shown in Figure 1.23. As the slip angle increases, sliding between the tread and the ground occurs. The assumption that the equatorial line in the contact patch is a straight line is no longer valid. Thus, the theory proposed by Temple and von Schlippe is restricted to small slip angles.

As noted above, using Temple's or von Schlippe's theory to define the relationship between the cornering force and the slip angle, the values of k_y and l_r must be known. Their determination is usually quite an involved process. In view of this, a simplified theory has been proposed (Segel 1984). In the simplified model, it is assumed that if no sliding takes place, the lateral deflection y' of a tread element on the ground at a longitudinal distance of x from the front of the contact patch (along the wheel plane) is proportional to tan α and is given by

$$y' = x \tan \alpha \tag{1.54}$$

where the lateral deflection y' is measured with respect to the front contact point and perpendicular to the wheel plane, and α is the slip angle.

If k'_y is the equivalent lateral stiffness of the tire, then when no lateral sliding between the tire tread and the ground takes place, the lateral force per unit contact length is given by

$$\frac{dF_{y\alpha}}{dx} = k'_y x \tan \alpha \tag{1.55}$$

and the cornering force developed on the entire contact patch is expressed by

$$F_{y\alpha} = \int_0^{l_t} k'_y x \tan\alpha \, dx = \left(k'_y l_t^2/2 \right) \tan\alpha \tag{1.56}$$

where l_t is the contact length of the tire.

The term $k'_y l_t^2/2$ may be taken as the cornering stiffness C_α defined by Eq. (1.37) – that is, the slope of the cornering force–slip angle curve at the origin – which can easily be identified:

$$\frac{k'_y l_t^2}{2} = C_\alpha = \left. \frac{\partial F_{y\alpha}}{\partial \alpha} \right|_{\alpha = 0} \tag{1.57}$$

Therefore, when no lateral sliding takes place on the contact patch, the relationship between the cornering force and the slip angle is expressed by

$$F_{y\alpha} = C_\alpha \tan\alpha \tag{1.58}$$

If the slip angle α is small, $\tan\alpha \approx \alpha$, and Eq. (1.58) may be rewritten as

$$F_{y\alpha} = C_\alpha \alpha \tag{1.59}$$

Following an approach like that for analyzing the relationship between the tractive effort and the longitudinal slip described in Section 1.3, the critical values of the slip angle α_c and the cornering force $F_{y\alpha c}$, at which lateral sliding in the trailing part of the contact patch begins, can be determined. The critical value of α_c is given by

$$\alpha_c = \frac{\mu_p W}{2C_\alpha} \tag{1.60}$$

and the critical value of $F_{y\alpha c}$ is given by

$$F_{y\alpha c} = \frac{\mu_p W}{2} \tag{1.61}$$

Like the relationship between the tractive effort–longitudinal slip described in Section 1.3, Eq. (1.61) indicates that the relationship between the cornering force and the slip angle will be linear and no lateral sliding will take place if the cornering force is less than one-half of its peak value ($\mu_p W/2$).

When lateral sliding between the tire tread and the ground takes place (i.e., $\alpha > \alpha_c$ or $F_{y\alpha} > F_{y\alpha c}$), the relationship between the cornering force and the slip angle, analogous to Eq. (1.29), is expressed by

$$F_{y\alpha} = \mu_p W \left(1 - \frac{\mu_p W}{4C_\alpha \tan\alpha} \right) = \mu_p W \left(1 - \frac{\mu_p W}{4C_\alpha \alpha} \right) \tag{1.62}$$

The above equation indicates the nonlinear nature of the cornering force–slip angle relationship when lateral sliding takes place in part of the contact patch.

While the theories described above provide physical insight into certain aspects of the cornering behavior of the pneumatic tire, they are simplified representations of a highly complex process. In the simulations of the lateral dynamic behavior of road vehicles, to represent tire characteristics more accurately, measured tire data, rather than theoretical relationships, are often used. Measured tire data in tabular form or represented by empirical equations may be entered as input to the simulation models. For instance, the following empirical equation has been proposed to represent the relationship between the cornering force $F_{y\alpha}$ and the slip angle α (Ellis 1994):

$$F_{y\alpha} = c_1\alpha + c_2\alpha^2 + c_3\alpha^3 \tag{1.63}$$

where c_1, c_2, and c_3 are empirical constants derived from fitting Eq. (1.63) to the measured data of a given tire.

As mentioned previously, normal load has a significant influence on the development of cornering force. To take the effects of normal load into account, the coefficients c_1, c_2, and c_3 may be expressed as a quadratic function of normal load (Ellis 1994). This would require an additional curve-fitting exercise.

In the discussion of the cornering behavior of pneumatic tires described above, the effect of the longitudinal force has not been considered. However, quite often both the side force and the longitudinal force are present, such as braking in a turn. In general, tractive (or braking) effort will reduce the cornering force that can be generated for a given slip angle; the cornering force decreases gradually with an increase of the tractive or braking effort. At low values of tractive (or braking) effort, the decrease in the cornering force is mainly caused by the reduction of the cornering stiffness of the tire. A further increase of the tractive (or braking) force results in a pronounced decrease of the cornering force for a given slip angle. This is due to the mobilization of the available local adhesion by the tractive (or braking) effort, which reduces the amount of adhesion available in the lateral direction.

The difference in behavior between a bias-ply and a radial-ply passenger car tire is shown in Figure 1.39 (van Eldik Thieme and Pacejka 1971). It is interesting to note that for a radial-ply tire, the cornering force available at a given slip angle is approximately the same for both braking and driving conditions. For a bias-ply tire, however, at a given slip angle, a higher cornering force is obtained during braking than when the tire is driven. The fact that the presence of the tractive (or braking) effort requires a higher slip angle to generate the same cornering force is also illustrated in Figure 1.39. Figure 1.40 shows the effects of longitudinal force on the development of cornering force for a truck tire at different slip angles (Ford and Charles 1988). Like that shown in Figure 1.39, for a truck tire, the cornering force available at a given slip angle also decreases with an increase of the longitudinal force. Note that if an envelope around each family of curves of Figure 1.39 is drawn, a curve approximately semielliptical in shape may be obtained. This enveloping curve is often referred to as the friction ellipse.

The friction ellipse concept is based on the assumption that the tire may slide on the ground in any direction if the resultant of the longitudinal force (either tractive or braking) and lateral (cornering) force reaches the maximum value defined by the coefficient of road adhesion and the normal load on the tire. However, the longitudinal and lateral force components may not exceed their respective maximum values $F_{x\ max}$ and $F_{y\ max}$, as shown in Figure 1.41. $F_{x\ max}$ and $F_{y\ max}$ can be identified from measured tire data and constitute the major and minor axes of the friction ellipse, respectively, as shown in Figure 1.41.

Based on the experimental observations described above, attempts have been made to formulate an analytical framework for predicting the longitudinal force and cornering force as functions of combined longitudinal slip (or skid) and slip angle.

One of the simplest theories for predicting the cornering force available at a specific slip angle in the presence of a tractive or braking force is based on the friction ellipse concept described above.

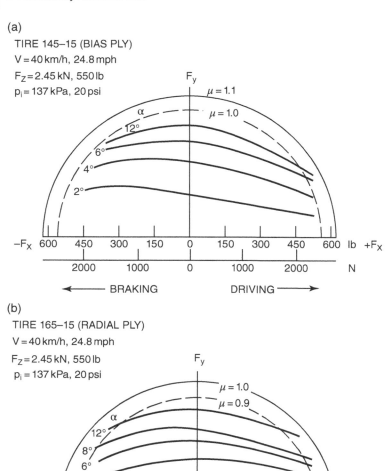

Figure 1.39 Effects of tractive and braking effort on the cornering characteristics of (a) a bias-ply and (b) a radial-ply car tire. Reproduced with permission from *Mechanics of Pneumatic Tires*, edited by S.K. Clark, Monograph 122, National Bureau of Standards, 1971.

The procedure for determining the available cornering force based on this simple theory is outlined below:

- From measured tire data, the relationship between the cornering force and the slip angle under free-rolling conditions (i.e., in the absence of tractive or braking effort) is first plotted, as shown in Figure 1.42(a).

Figure 1.40 Effects of longitudinal force on the cornering characteristics of a truck tire. Reprinted with permission from SAE SP-729 © 1988 Society of Automotive Engineers, Inc.

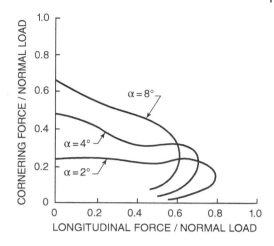

Figure 1.41 The friction ellipse concept relating the maximum cornering force to a given longitudinal force.

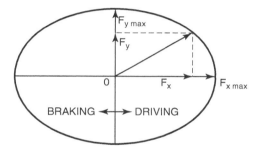

- The cornering forces at various slip angles under free-rolling conditions are then marked on the vertical axis of Figure 1.42(b), as shown. For instance, the cornering force developed at a slip angle of 4° is identified as F_{y4} on the vertical axis, which constitutes the minor axis of an ellipse to be established.
- From measured tire data, the maximum tractive or braking force, $F_{x\,max}$, in the absence of lateral force, is marked on the horizontal axis in Figure 1.42(b) as shown, which constitutes the major axis of the ellipse.
- The available cornering force F_y at a given slip angle, such as the 4° angle shown in Figure 1.42(b), for any given tractive or braking force F_x is then determined from the following equation:

$$\left(F_y/F_{y4}\right)^2 + \left(F_x/F_{x\,max}\right)^2 = 1 \tag{1.64}$$

The above equation describes an ellipse with the measured values of $F_{x\,max}$ and F_{y4} as the major and minor axes, respectively.

Following the procedure outlined above, the available cornering force at any slip angle in the presence of any given tractive or braking force can be determined, and a set of curves illustrating the relationships between the cornering force and the tractive (or braking) force at various slip angles can be plotted, as shown in Figure 1.42(b). It is noted that for a given slip angle, the cornering

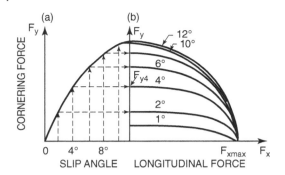

force is reduced as a tractive (or braking) force is applied to the tire. This is consistent with the trends of the measured data shown in Figures 1.39 and 1.40.

Based on the simplified theory for the relationship between the braking force and the longitudinal skid described in Section 1.3 and that between the cornering force and the slip angle described earlier in this section, another semiempirical method for predicting the braking force and cornering force in the presence of both the longitudinal skid and slip angle has been proposed (Segel 1984). In this method, it is assumed that when no sliding takes place, the braking force per unit contact length at a distance of x from the front contact point is given by (see Eqs. (1.20) and (1.31))

$$\frac{dF_x}{dx} = k_t x i_s/(1 - i_s) \tag{1.65}$$

where i_s is the longitudinal skid, as defined by Eq. (1.30).

If, at the same time, the tire develops a slip angle α, then, due to the longitudinal skid, the tread in contact with the ground will be elongated at a rate equal to $1/(1 - i_s)$. As a result, the lateral deflection y' of a point on the tread in contact with the ground is given by (see Eq. (1.54))

$$y' = x \tan \alpha/(1 - i_s) \tag{1.66}$$

The corresponding lateral force per unit contact length is, therefore, expressed by (see Eq. (1.55))

$$\frac{dF_{y\alpha}}{dx} = k'_y x \tan \alpha/(1 - i_s) \tag{1.67}$$

Let p be the uniform normal pressure on the contact patch, b the contact width, and μ the coefficient of road adhesion. Then, based on the concept of friction ellipse described above, no sliding will take place at a point located at a distance of x from the front contact point if the resultant of the braking force and lateral force per unit contact length is less than the minimum value defined by the coefficient of road adhesion μ and the normal pressure p; that is,

$$\sqrt{\left[k_t x i_s/(1 - i_s)\right]^2 + \left[k'_y x \tan \alpha/(1 - i_s)\right]^2} = \mu p b = \frac{\mu W}{l_t} \tag{1.68}$$

where W is the normal load and l_t is the contact length of the tire.

This implies that if a point at a distance x from the front contact point is in the adhesion region, then x must be less than a characteristic length l_c, which defines the length of the adhesion region where no sliding between the tire tread and the ground takes place. The value of l_c in relation to the contact length l_t can be derived from Eq. (1.68), and is given by

$$\frac{l_c}{l_t} = \frac{\mu\,W(1-i_s)}{2\sqrt{\left(k_t l_t^2 i_s/2\right)^2 + \left(k'_y l_t^2 \tan\alpha/2\right)^2}}$$

$$= \frac{\mu\,W(1-i_s)}{2\sqrt{(C_s i_s)^2 + (C_\alpha \tan\alpha)^2}} \tag{1.69}$$

where $k_t l_t^2/2 = C_s$ and $k'_y l_t^2/2 = C_\alpha$, as described by Eqs. (1.33) and (1.57), respectively.

If $l_c/l_t \geq 1$, the entire contact patch is an adhesion region. The braking force is given by

$$F_x = \int_0^{l_t} [k_t x i_s/(1-i_s)]\,dx = k_t l_t^2 i_s/2(1-i_s)$$

$$= C_s i_s/(1-i_s) \tag{1.70}$$

and the cornering force $F_{y\alpha}$ as a function of slip angle α and skid i_s is expressed by

$$F_{y\alpha} = \int_0^{l_t} \left[k'_y x \tan\alpha/(1-i_s)\right]\,dx$$

$$= k'_y l_t^2 \tan\alpha/2(1-i_s) \tag{1.71}$$

$$= C_\alpha \tan\alpha/(1-i_s)$$

If $l_c/l_t < 1$, then sliding between the tread and the ground will take place. The braking force developed on the adhesion region F_{xa} is given by

$$F_{xa} = \int_0^{l_c} [k_t x i_s/(1-i_s)]\,dx$$

$$= \frac{\mu^2 W^2 C_s i_s(1-i_s)}{4\left[(C_s i_s)^2 + (C_\alpha \tan\alpha)^2\right]} \tag{1.72}$$

and the braking force developed on the sliding region F_{xs} is expressed by

$$F_{xs} = \frac{\mu W C_s i_s}{\sqrt{(C_s i_s)^2 + (C_\alpha \tan\alpha)^2}}\left[1 - \frac{\mu W(1-i_s)}{2\sqrt{(C_s i_s)^2 + (C_\alpha \tan\alpha)^2}}\right] \tag{1.73}$$

The total braking force F_x is given by

$$F_x = F_{xa} + F_{xs}$$

$$= \frac{\mu W C_s i_s}{\sqrt{(C_s i_s)^2 + (C_\alpha \tan\alpha)^2}}\left[1 - \frac{\mu W(1-i_s)}{4\sqrt{(C_s i_s)^2 + (C_\alpha \tan\alpha)^2}}\right] \tag{1.74}$$

Similarly, if sliding between the tread and the ground takes place, then the cornering force developed on the adhesion region is given by

$$
\begin{aligned}
F_{yaa} &= \int_0^{l_c} \left[k_y' x \tan \alpha / (1 - i_s) \right] dx \\
&= \frac{\mu^2 W^2 C_\alpha \tan \alpha (1 - i_s)}{4 \left[(C_s i_s)^2 + (C_\alpha \tan \alpha)^2 \right]}
\end{aligned}
\tag{1.75}
$$

and the cornering force developed on the sliding region is expressed by

$$
F_{yas} = \frac{\mu W C_\alpha \tan \alpha}{\sqrt{(C_s i_s)^2 + (C_\alpha \tan \alpha)^2}} \left[1 - \frac{\mu W (1 - i_s)}{2\sqrt{(C_s i_s)^2 + (C_\alpha \tan \alpha)^2}} \right]
\tag{1.76}
$$

The total cornering force F_{ya} is given by

$$
\begin{aligned}
F_{ya} &= F_{yaa} + F_{yas} \\
&= \frac{\mu W C_\alpha \tan \alpha}{\sqrt{(C_s i_s)^2 + (C_\alpha \tan \alpha)^2}} \left[1 - \frac{\mu W (1 - i_s)}{4\sqrt{(C_s i_s)^2 + (C_\alpha \tan \alpha)^2}} \right]
\end{aligned}
\tag{1.77}
$$

The parameters, μ, W, C_s, and C_α may change with operating conditions. For instance, it has been found that on a given surface, the values of μ, C_s, and C_α are functions of the normal load and operating speed of the tire. In a dynamic maneuver involving both braking and steering, the normal load and speed of the tires on a vehicle change as the maneuver proceeds. To achieve more accurate predictions, the effects of normal load and speed on the values of μ, C_s, and C_α, and other tire parameters should be properly taken into account (Segel 1984).

The semiempirical method for modeling tire behavior described above has been incorporated into a computer model for simulating the directional response and braking performance of commercial vehicles (Segel 1984). The method presented above is for predicting the braking force and cornering force of a tire during combined braking and cornering. Following the same approach, however, a method for predicting the tractive force and cornering force as functions of combined longitudinal slip and slip angle can be formulated.

Example 1.1 A truck tire $10 \times 20/F$ with a normal load of 24.15 kN (5430 lb) is traveling on a dry asphalt pavement with a coefficient of road adhesion $\mu = 0.85$. The cornering stiffness of the tire C_α is 133.30 kN/rad (523 lb/deg) and the longitudinal stiffness C_s is 186.82 kN/unit skid (42,000 lb/unit skid).

Estimate the braking force and the cornering force that the tire can develop at a slip angle $\alpha = 4°$ and a longitudinal skid of 10%.

Solution

To determine whether sliding takes place on the tire contact patch under the given operating conditions, the ratio l_c/l_t is calculated using Eq. (1.69):

$$
\begin{aligned}
\frac{l_c}{l_t} &= \frac{\mu W (1 - i_s)}{2\sqrt{(C_s i_s)^2 + (C_\alpha \tan \alpha)^2}} \\
&= \frac{0.85 \times 24.15 \times (1 - 0.1)}{2\sqrt{(186.82 \times 0.1)^2 + (133.30 \times 0.0699)^2}} = 0.442
\end{aligned}
$$

Since $l_c/l_t < 1$, sliding takes place in part of the contact patch.
The braking force can be predicted using Eq. (1.74):

$$F_x = F_{xa} + F_{xs}$$

$$= \frac{\mu W C_s i_s}{\sqrt{(C_s i_s)^2 + (C_\alpha \tan \alpha)^2}} \left[1 - \frac{\mu W (1 - i_s)}{4\sqrt{(C_s i_s)^2 + (C_\alpha \tan \alpha)^2}} \right]$$

$$= \frac{0.85 \times 24.15 \times 186.82 \times 0.1}{\sqrt{(186.82 \times 0.1)^2 + (133.30 \times 0.0699)^2}}$$

$$\cdot \left[1 - \frac{0.85 \times 24.15 \times (1 - 0.1)}{4\sqrt{(186.82 \times 0.1)^2 + (133.30 \times 0.0699)^2}} \right]$$

$$= 14.30 \, \text{kN} \, (3215 \, \text{lb})$$

The cornering force can be predicted using Eq. (1.77):

$$F_{y\alpha} = F_{y\alpha a} + F_{y\alpha s}$$

$$= \frac{\mu W C_\alpha \tan \alpha}{\sqrt{(C_s i_s)^2 + (C_\alpha \tan \alpha)^2}} \left[1 - \frac{\mu W (1 - i_s)}{4\sqrt{(C_s i_s)^2 + (C_\alpha \tan \alpha)^2}} \right]$$

$$= \frac{0.85 \times 24.15 \times 133.30 \times 0.0699}{\sqrt{(186.82 \times 0.1)^2 + (133.30 \times 0.0699)^2}}$$

$$\cdot \left[1 - \frac{0.85 \times 24.15 \times (1 - 0.1)}{4\sqrt{(186.82 \times 0.1)^2 + (133.30 \times 0.0699)^2}} \right]$$

$$= 7.14 \, \text{kN} (1605 \, \text{lb})$$

1.4.5 The Magic Formula

To provide a general method to characterize the relationships between tire forces and moments (such as side force, braking effort, and self-aligning torque) and tire operational parameters (such as slip angle and skid) for vehicle dynamics studies and simulations, the Magic Formula has been developed. It was initially formulated in the 1980s (Bakker et al. 1987; Bakker et al. 1989). Since then, it has evolved (Pacejka and Bakker 1991; Pacejka 1996; Pacejka and Besselink 1997; Pacejka 2012). To provide a general understanding of the approach, the formula in its basic form is outlined below.

The Magic Formula may be used to fit experimental tire data for characterizing the relationships, for instance, between the cornering force and slip angle, self-aligning torque and slip angle, or braking effort and skid. For given values of tire vertical load and camber angle, the general form of the formula is expressed by

$$y(x) = D \sin \{ C \arctan [Bx - E(Bx - \arctan Bx)] \} \tag{1.78}$$

$$Y(X) = y(x) + S_v \tag{1.79}$$
$$x = X + S_h$$

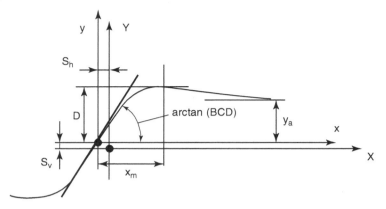

Figure 1.43 Characteristics of the Magic Formula for fitting tire test data. *Source:* Pacejka, H.B. and Besselink, I.J.M. (1997), *Proceedings of the 2nd International Colloquium on Tyre Models for Vehicle Dynamic Analysis*, pp. 234–249, © Swets & Zeitlinger. Used with permission.

where $Y(X)$ represents cornering force, self-aligning torque, or braking effort, and X denotes slip angle or skid. Coefficient B is called the stiffness factor, C the shape factor, D the peak factor, and E the curvature factor. S_h and S_v are horizontal shift and vertical shift, respectively.

The relationships between $Y(X)$ and $y(x)$ and that between X and x are defined by Eq. (1.79), with horizontal shift S_h and vertical shift S_v of the origin of the coordinates of X and Y with respect to that of x and y, respectively, as shown in Figure 1.43 (Pacejka and Besselink 1997). For each of the relationships between the cornering force and slip angle, self-aligning torque and slip angle, or braking effort and skid, coefficients B, C, D, and E and shifts S_h and S_v will have their corresponding values.

Equation (1.78) produces a curve that passes through the origin, $x = y = 0$, and reaches a maximum of y at $x = x_m$, as shown in Figure 1.43. Beyond that y decreases and finally approaches an asymptote y_a. For given values of the coefficients, the curve shows an antisymmetric shape with respect to the origin, $x = y = 0$. To allow the curve to have an offset with respect to the origin, two shifts S_h and S_v are introduced, as shown in Figure 1.43. Consequently, a new set of coordinates, X and Y, representing cornering force, self-aligning torque, or braking effort and slip angle or skid, respectively, is established. This enables the effects of ply-steer, conicity, or rolling resistance on cornering force, self-aligning torque, or braking effort to be considered.

Figure 1.43 illustrates the meaning of some of the coefficients in Eq. (1.78). For instance, if Figure 1.43 represents the cornering force and slip angle relationship of a tire, then coefficient D represents the peak value with respect to x, y coordinates and the product BCD corresponds to the slope of the curve at the origin, representing the cornering stiffness of the tire, as defined by Eq. (1.37).

The Magic Formula can produce characteristics that closely match measured data. Figures 1.44–1.46 show a comparison of the experimental data and fitted curves using Eqs. (1.78) and (1.79) for the relationships of cornering force and slip angle, self-aligning torque and slip angle, and braking effort and skid of a passenger car tire, respectively (Bakker et al. 1989).

As an example, the values of the coefficients in Eqs. (1.78) and (1.79) for predicting cornering force F_y, self-aligning torque, M_z, and braking effort F_x of a car tire are given in Table 1.10. In using the values of the coefficients in the table to predict the cornering force, self-aligning torque, and braking effort, the resulting values are in N, N · m, and N, respectively; the slip angle is in degrees; and skid is defined by Eq. (1.30) and considered to be a negative value.

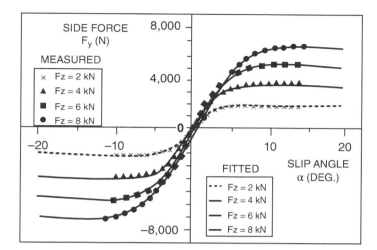

Figure 1.44 Comparison of the measured and fitted relationships between side force and slip angle using the Magic Formula. Reprinted with permission from SAE paper 890087 © 1989 Society of Automotive Engineers, Inc.

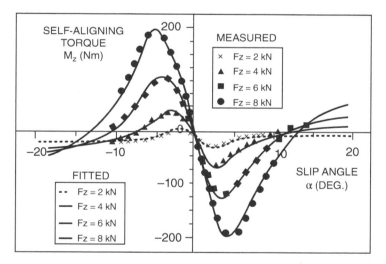

Figure 1.45 Comparison of the measured and fitted relationships between self-aligning torque and slip angle using the Magic Formula. Reprinted with permission from SAE paper 890087 © 1989 Society of Automotive Engineers, Inc.

It is found that some of the coefficients in Eqs. (1.78) and (1.79) are functions of the normal load and/or camber angle of the tire (Bakker et al. 1987). For instance, peak factor D may be expressed as a function of normal load F_z as follows:

$$D = a_1 F_z^2 + a_2 F_z \tag{1.80}$$

where F_z is in kN, and a_1 and a_2 are empirical coefficients.

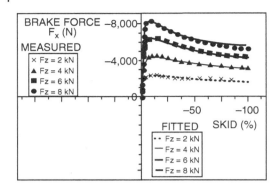

Figure 1.46 Comparison of the measured and fitted relationships between braking force and skid using the Magic Formula. Reprinted with permission from SAE paper 890087 © 1989 Society of Automotive Engineers, Inc.

Table 1.10 Values of the coefficients in the Magic Formula for a car tire (slip angle in degrees and skid in minus %).

	Load, F_z, kN	B	C	D	E	S_h	S_v	BCD
F_y, N	2	0.244	1.50	1936	−0.132	−0.280	−118	780.6
	4	0.239	1.19	3650	−0.678	−0.049	−156	1038
	6	0.164	1.27	5237	−1.61	−0.126	−181	1091
	8	0.112	1.36	6677	−2.16	0.125	−240	1017
M_z, N · m	2	0.247	2.56	−15.53	−3.92	−0.464	−12.5	−9.820
	4	0.234	2.68	−48.56	−0.46	−0.082	−11.7	−30.45
	6	0.164	2.46	−112.5	−2.04	−0.125	−6.00	−45.39
	8	0.127	2.41	−191.3	−3.21	−0.009	−4.22	−58.55
F_x, N	2	0.178	1.55	2193	0.432	0.000	25.0	605.0
	4	0.171	1.69	4236	0.619	0.000	70.6	1224
	6	0.210	1.67	6090	0.686	0.000	80.1	2136
	8	0.214	1.78	7711	0.783	0.000	104	2937

Source: Bakker et al. (1987).

For cornering stiffness (i.e., the initial slope of the cornering force–slip angle curve):

$$BCD = a_3 \sin \left[a_4 \arctan \left(a_5 F_z \right) \right] \tag{1.81}$$

where a_3, a_4, and a_5 are empirical coefficients.

For aligning stiffness (i.e., the initial slope of the self-aligning torque–slip angle curve) or longitudinal stiffness (i.e., the initial slope of the braking effort–skid curve):

$$BCD = \frac{a_3 F_z^2 + a_4 F_z}{e^{a_5 F_z}} \tag{1.82}$$

The shape factor C appears to be practically independent of F_z, and the average values for the car tire tested may be taken as follows (based on the data shown in Table 1.10):

For the cornering force–slip angle relationship, $C = 1.30$
For the self-aligning torque–slip angle relationship, $C = 2.40$
For the braking effort–skid relationship, $C = 1.65$

Table 1.11 Values of coefficients a_1 to a_8 for a car tire (F_z in kN).

	a_1	a_2	a_3	a_4	a_5	a_6	a_7	a_8
F_y, N	−22.1	1011	1078	1.82	0.208	0.000	−0.354	0.707
M_z, N · m	−2.72	−2.28	−1.86	−2.73	0.110	−0.070	0.643	−4.04
F_x, N	−21.3	1144	49.6	226	0.069	−0.006	0.056	0.486

Source: Bakker et al. (1987).

The stiffness factor B can be derived from

$$B = \frac{BCD}{CD} \tag{1.83}$$

The curvature factor E as a function of normal load F_z is given by

$$E = a_6 F_z^2 + a_7 F_z + a_8 \tag{1.84}$$

where a_6, a_7, and a_8 are empirical coefficients.

Table 1.11 gives the values of coefficients a_1 to a_8 for the same tire as in Table 1.10. It should be noted that in Eqs. (1.80)–(1.84), F_z is in kN.

Camber angle γ is found to have an influence on the relationships between cornering force and slip angle and self-aligning torque and slip angle, in the form of horizontal and vertical shifts, S_h and S_v (Bakker et al. 1987). The additional shifts due to camber angle γ may be expressed by

$$\begin{aligned} \Delta S_h &= a_9 \gamma \\ \Delta S_v &= \left(a_{10} F_z^2 + a_{11} F_z\right) \gamma \end{aligned} \tag{1.85}$$

where a_9, a_{10}, and a_{11} are empirical coefficients.

The change in stiffness factor ΔB is obtained by multiplying B by $(1 - a_{12}|\gamma|)$:

$$\Delta B = (1 - a_{12}|\gamma|)B \tag{1.86}$$

where a_{12} is an empirical coefficient.

The value of the self-aligning torque at high slip angles will change due to this change in stiffness factor B. To compensate for this effect, the curvature factor E for self-aligning torque M_z must be divided by $(1 - a_{13}|\gamma|)$.

The values of coefficients a_9 to a_{13} for the same tire as in Table 1.10 are given in Table 1.12.

Table 1.12 The values of coefficients a_9 to a_{13} for a car tire (camber angle in degrees).

	a_9	a_{10}	a_{11}	a_{12}	a_{13}
F_y, N	0.028	0.000	14.8	0.022	0.000
M_z, N · m	0.015	−0.066	0.945	0.030	0.070

Source: Bakker et al. (1987).

When brakes are applied during a turning maneuver, the tires on a vehicle develop both slip angles and skids. Under these circumstances, Eqs. (1.78) and (1.79) are inadequate for characterizing tire behavior. To characterize the combined effects of slip angle and skid on the cornering force, self-aligning torque, or braking effort, empirical weighting functions G are introduced, which when multiplied by the original functions given in Eqs. (1.78) and (1.79) produce the interactive effects of skid on cornering force and self-aligning torque, or of slip angle on braking effort (Pacejka 1996; Pacejka and Besselink 1997; Bayle et al. 1993). When the tire operates only with slip angle or skid, the weighting functions G take the value of one. However, when a tire operates under a given slip angle and at the same time its skid gradually increases, then the weighting function for cornering force F_y may first show a slight increase in magnitude, then reach its peak, followed by a continuous decrease. The weighting function G takes the following form:

$$G = D' \cos\left[C' \arctan\left(B'x\right)\right] \tag{1.87}$$

where B', C', and D' are empirical coefficients, and x is either slip angle or skid. For instance, if Eq. (1.87) represents the weighting function for determining the effect of skid on the cornering force F_y at a given slip angle, then x in Eq. (1.87) represents the skid of the tire. For details concerning the characterization of tire behavior under the combined effects of slip angle and skid, please refer to Pacejka (1996), Pacejka and Besselink (1997), and Bayle et al. (1993).

The discussions presented above are for characterizing the steady-state cornering behavior of tires. When a vehicle is in transient motion, such as when the steering wheel angle and/or braking effort vary with time during a turning maneuver, the slip angle and/or skid of the tire is in a transient state. The equations given previously may be inadequate for characterizing the transient response of the tire. Studies on the transient cornering behavior of tires have been made (Pacejka and Besselink 1997; van Zanten et al. 1989; Zhou et al. 1999).

Example 1.2 Using the Magic Formula, estimate the braking effort developed by a tire with a normal load of 6 kN (1349 lb), at a skid of -25%, and having empirical coefficients B, C, D, E, S_h, and S_v shown in Table 1.10.

Solution

For this case, the variables Y and X in the Magic Formula, Eqs. (1.78) and (1.79), represent the braking effort F_x and skid i_s, respectively. Note that skid i_s in the Magic Formula is expressed in percentage and considered to be a negative value and that the value of the arctan function should be expressed in radians.

$$F_x = D \sin\left[C \arctan\left(B(i_s + S_h) - E\{B(i_s + S_h)\right.\right.$$
$$\left.\left. - \arctan\left[B(i_s + S_h)\right]\}\right)\right] + S_v$$

Using the appropriate values of the empirical coefficients for a normal load of 6 kN (1349 lb) given in Table 1.10, the braking effort at a skid of -25% is calculated as follows:

$$F_x = 6090 \sin\left[1.67 \arctan\left(0.210\left(-25 + 0\right) - 0.686\{0.210(-25 + 0)\right.\right.$$
$$\left.\left. - \arctan\left[0.210(-25 + 0)\right]\}\right)\right] + 80.1$$
$$= 6090 \sin\{1.67 \arctan\left[-5.25 - 0.686(-5.25 + 1.3826)\right]\}$$
$$+ 80.1 = -5433\,\text{N}\,(-1221\,\text{lb})$$

1.5   Performance of Tires on Wet Surfaces

The behavior of tires on wet surfaces is of considerable interest from a vehicle safety point of view, as many accidents occur on slippery roads. The performance of tires on wet surfaces depends on the surface texture, water depth, tread pattern, tread depth, tread material, and operating mode of the tire (i.e., free rolling, braking, accelerating, or cornering). To achieve acceptable performance on wet surfaces, maintaining effective contact between the tire tread and the road is of importance, and there is no doubt about the necessity of removing water from the contact area as much as possible.

To maintain effective contact between the tire and the road, the tire tread should have a suitable pattern and depth to facilitate the flow of fluid from the contact area, and the surface of the pavement should have an appropriate texture to facilitate drainage as well. To provide good skid resistance, road surfaces must fulfill two requirements: an open macrotexture to facilitate gross draining, and micro-harshness to produce sharp points that can penetrate the remaining water film (Schallamach 1971).

The effects of tread pattern and speed on the braking performance of tires on various wet surfaces have been studied experimentally by a number of investigators. Figures 1.47 and 1.48 show the

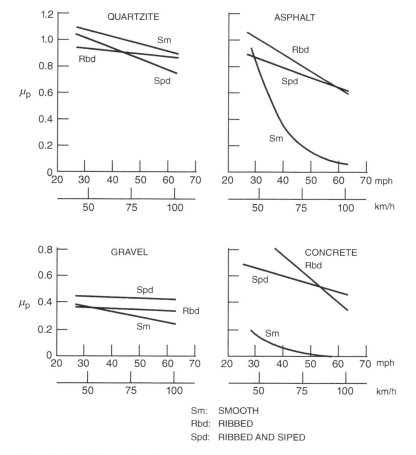

Sm: SMOOTH
Rbd: RIBBED
Spd: RIBBED AND SIPED

Figure 1.47   Effects of tread design on the peak value of road adhesion coefficient μ_p over wet surfaces. Reproduced with permission from *Mechanics of Pneumatic Tires*, edited by S.K. Clark, Monograph 122, National Bureau of Standards, 1971.

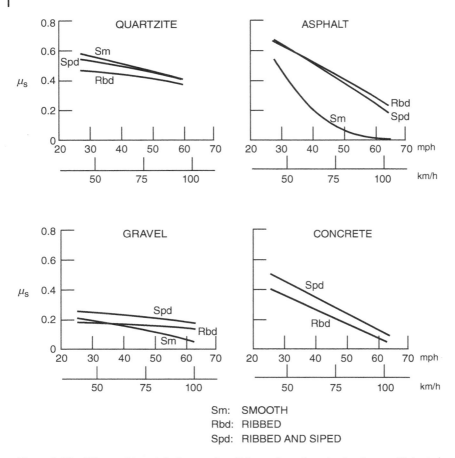

Figure 1.48 Effects of tread design on the sliding value of road adhesion coefficient μ_s over wet surfaces. Reproduced with permission from *Mechanics of Pneumatic Tires*, edited by S.K. Clark, Monograph 122, National Bureau of Standards, 1971.

variations of the peak values μ_p and the sliding values μ_s of the coefficient of road adhesion with speed for a smooth tire, a tire with ribs, and a tire with ribs and sipes on wet quartzite, asphalt, gravel, and concrete (Schallamach 1971). There is a marked difference in the coefficient of road adhesion between patterned tires, including the ribbed and siped tires, and smooth tires on wet asphalt and concrete surfaces. The tread pattern increases the value of the coefficient of road adhesion and reduces its speed dependency. In contrast, there is little pattern effect on wet quartzite surfaces, and a high level of road adhesion is maintained over the entire speed range. Thus, it can be concluded that the advantages of a patterned tire over a smooth tire are pronounced only on badly drained surfaces.

Tread pattern can function satisfactorily on a wet road only when the grooves and sipes constitute a reservoir of sufficient capacity, and its effectiveness decreases with the wear of the tread or the tread depth. The decline in value of the coefficient of road adhesion with the decrease of tread depth is more pronounced on smooth than on rough roads, as rough roads can provide better drainage.

When a pneumatic tire is braked over a flooded surface, the motion of the tire creates hydrodynamic pressure in the fluid. The hydrodynamic pressure acting on the tire builds up as the square of

Figure 1.49 Hydroplaning of a tire on flooded surfaces: (a) partial hydroplaning; (b) complete hydroplaning. Reproduced with permission from *Mechanics of Pneumatic Tires*, edited by S.K. Clark, Monograph 122, National Bureau of Standards, 1971.

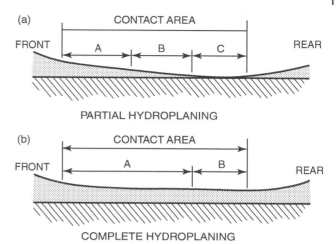

A - UNBROKEN WATER FILM
B - REGION OF PARTIAL BREAKDOWN OF WATER FILM
C - CONTACT ZONE

speed of the tire and tends to separate the tire from the ground. At low speeds, the front part of the tire rides on a wedge or a film of fluid. This fluid film extends backward into the contact area as the speed of the tire increases. At a particular speed, the hydrodynamic lift developed under the tire equals the vertical load, the tire rides completely on the fluid, and all contact with the ground is lost. This phenomenon is usually referred to as "hydroplaning" and is illustrated in Figure 1.49 (Clark 1971).

For smooth or close-patterned tires that do not provide escape paths for water and for patterned tires on flooded surfaces with a fluid depth exceeding the groove depth in the tread, the speed at which hydroplaning occurs may be determined based on the theory of hydrodynamics. It can be assumed that the lift component of the hydrodynamic force F_h is proportional to the tire–ground contact area A, fluid density ρ_f, and the square of the vehicle speed V (Horne and Dreher 1963; Horne and Joyner 1965):

$$F_h \propto \rho_f A V^2 \tag{1.88}$$

When hydroplaning occurs, the lift component of the hydrodynamic force is equal to the vertical load acting on the tire. The speed at which hydroplaning begins, therefore, is proportional to the square root of the nominal ground contact pressure W/A, which is proportional to the inflation pressure of the tire p_i. Based on this reasoning and on experimental data shown in Figure 1.50, the following formula was proposed by Horne and Joyner (1965) for predicting the hydroplaning speed V_p:

$$V_p = 10.35\sqrt{p_i} \text{ mph} \tag{1.89}$$

or

$$V_p = 6.34\sqrt{p_i} \text{ km/h} \tag{1.90}$$

where p_i is the inflation pressure of the tire in psi for Eq. (1.89) and in kPa for Eq. (1.90).

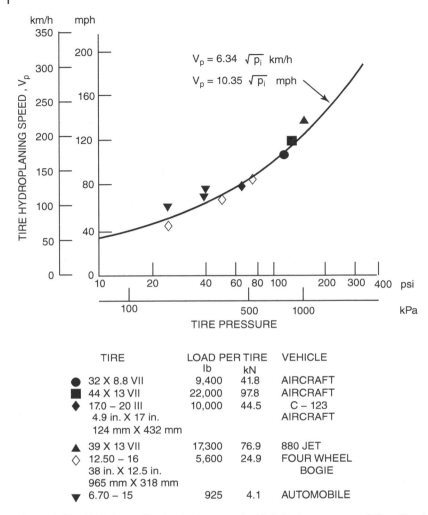

Figure 1.50 Variations of hydroplaning speed with inflation pressure of tires. Reprinted with permission from SAE paper 650145 © 1965 Society of Automotive Engineers, Inc.

For passenger car tires, the inflation pressure is usually in the range 193–248 kPa (28–36 psi). According to Eq. (1.90), the hydroplaning speed V_p for a tire at an inflation pressure of 193 kPa (28 psi) is approximately 88 km/h (54.7 mph), which is well within the normal operating range for passenger cars. For heavy trucks, the inflation pressure is usually in the range 620–827 kPa (90–120 psi). From Eq. (1.90), the hydroplaning speed V_p for a tire at an inflation pressure of 620 kPa (90 psi) is approximately 158 km/h (98 mph), which is beyond the normal range of operating speed for heavy trucks. This would suggest that hydroplaning may not be possible for heavy truck tires under normal circumstances. However, the tractive performance of truck tires is still significantly influenced by the presence of fluid on wet pavements.

For patterned tires on wet surfaces where the fluid depth is less than the groove depth of the tread, the prediction of the hydroplaning speed is more complex, and a generally accepted theory has yet

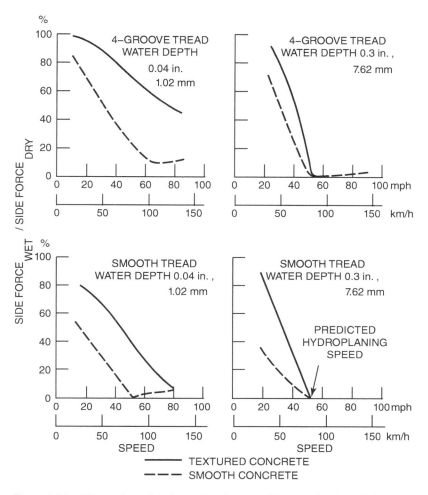

Figure 1.51 Effects of tread design and surface conditions on the degradation of cornering capability of tires on wet surfaces. Reprinted with permission from SAE paper 650145 © 1965 Society of Automotive Engineers, Inc.

to be evolved. The parameters found to be of significance to hydroplaning are pavement surface texture, pavement fluid depth, fluid viscosity, fluid density, tire inflation pressure, tire normal load, tire tread pattern, and tire tread depth.

The most important effect of hydroplaning is the reduction in the coefficient of road adhesion between the tire and the ground. This affects braking, steering control, and directional stability. Figure 1.51 shows the degradation of the cornering force of passenger car tires on two different wet surfaces at various speeds (Horne and Dreher 1963).

Because of the difference in design priorities, a noticeable difference in traction on wet pavements between truck and passenger car tires is observed. Figure 1.52 shows a comparison of the peak value μ_p and sliding value μ_s of the coefficient of road adhesion on wet pavements of a sample of three radial-ply truck tires and a corresponding sample of radial-ply passenger car tires with different tread depths (Segel 1984). It can be seen that the tractive performance of the truck tires tested is substantially poorer than that of the passenger car tires.

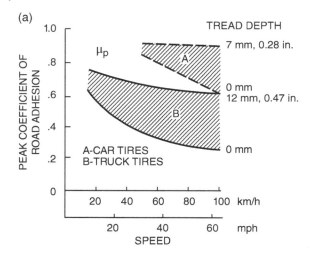

(a)

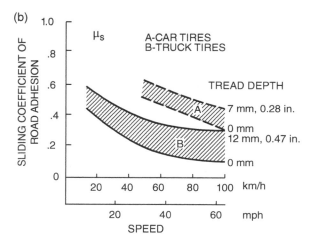

(b)

Figure 1.52 Comparisons of (a) the peak value of road adhesion coefficient μ_p and (b) the sliding value of road adhesion coefficient μ_s of car and truck tires on wet surfaces. *Source:* Segel (1984). Reproduced with permission of the University of Michigan Transportation Research Institute.

In the design of heavy truck tires, greater emphasis is placed on tread life. As a result, tread patterns and tread compounds for truck tires are different from those for passenger car tires. For instance, natural rubber as the base polymer for the tread is widely used for truck tires, whereas synthetic rubber-based compounds are universally adopted for passenger car tires. As mentioned previously, while natural rubber compounds offer higher abrasion resistance and lower hysteresis losses, synthetic rubber compounds provide a fundamentally higher value of coefficient of road adhesion, particularly on wet pavements. The substantial difference in tractive performance between car and truck tires results in a significant difference in stopping distance. For instance, it has been reported that on a wet, slippery road, the stopping distance for a heavy truck with tires ranging from the best available to the worst, but of a fairly typical type could be 1.65–2.65 times longer than that of a passenger car with normal high-grip tires (French 1989).

1.6 Ride Properties of Tires

Supporting the weight of the vehicle and cushioning it over surface irregularities are two of the basic functions of a pneumatic tire. When a normal load is applied to an inflated tire, the tire progressively deflects as the load increases. Figure 1.53 shows the static load–deflection relationship for a 5.60 × 13 bias-ply tire at various inflation pressures (Overton et al. 1969–1970). The type of diagram shown in Figure 1.53 is usually referred to as a lattice plot, in which the origin of each load–deflection curve is displaced along the deflection axis by an amount proportional to the inflation pressure. The relationship between the load and the inflation pressure for constant deflection can also be shown in the lattice plot. Figure 1.54 shows the interrelationship among the static load, inflation pressure, and deflections for a 165 × 13 radial-ply passenger car tire (Overton et al. 1969–1970). The lattice plots of the load–deflection data at various inflation pressures for tractor tires 11-36 and 7.50-16 are shown in Figures 1.55 and 1.56, respectively (Matthews and Talamo 1965). The load–deflection curves at various inflation pressures for a terra tire 26 × 12.00 - 12 are shown in Figure 1.57. The vertical load–deflection curves are useful in estimating the static vertical stiffness of tires.

In vehicle vibration analysis and ride simulation, the cushioning characteristics of a pneumatic tire may be represented by various mathematical models. The most widely used and simplest model representing the fundamental mode of vibration of the pneumatic tire consists of a mass element and a linear spring in parallel with a viscous damping element, as shown in Figure 1.58(a). Other models, such as the so-called "viscoelastic" model shown in Figure 1.58(b), have also been proposed.

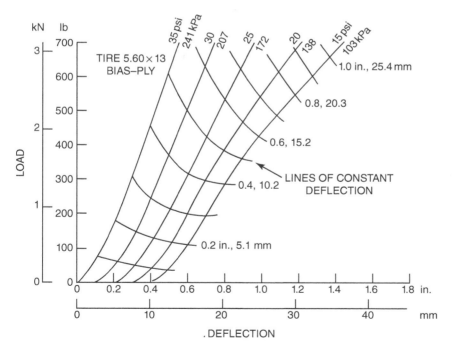

Figure 1.53 Static load–deflection relationships of a bias-ply car tire. *Source:* Overton et al. (1969–1970). Reproduced with permission of the Council of the Institution of Mechanical Engineers.

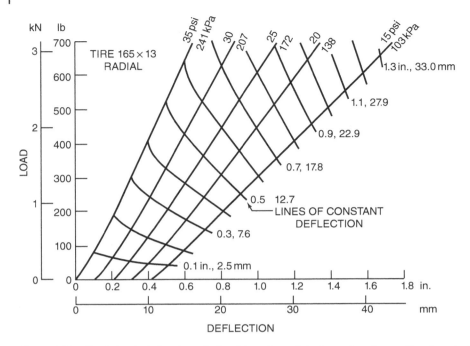

Figure 1.54 Static load–deflection relationships of a radial-ply car tire. *Source:* Overton et al. (1969–1970). Reproduced with permission of the Council of the Institution of Mechanical Engineers.

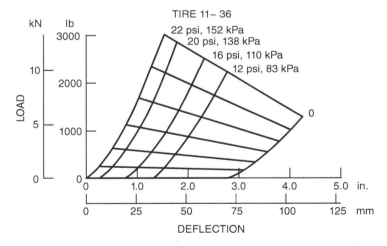

Figure 1.55 Static load–deflection relationships of a tractor tire 11-36. *Source:* Matthews and Talamo (1965). Reproduced with permission of the *Journal of Agricultural Engineering Research*.

Figure 1.56 Static load–deflection relationships of a tractor tire 7.50-16. *Source:* Matthews and Talamo (1965). Reproduced with permission of the *Journal of Agricultural Engineering Research.*

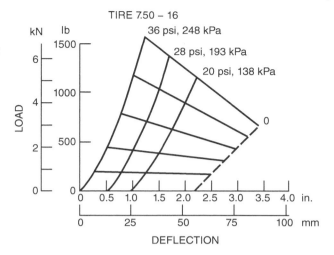

Figure 1.57 Static load–deflection relationships of a terra tire 26 × 12.00-12 for all-terrain vehicles.

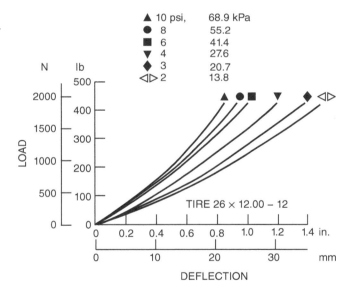

Depending upon the test conditions, three distinct types of tire vertical stiffness may be defined: static, non-rolling dynamic, and rolling dynamic stiffness.

1) Static stiffness

The static vertical stiffness of a tire is determined by the slope of the static load–deflection curves, such as those shown in Figures 1.53–1.57. For a given inflation pressure, the load–deflection characteristics for both radial- and bias-ply tires are approximately linear, except at relatively low values of load. Consequently, it can be assumed that the tire vertical stiffness is independent of load in the range of practical interest. Figure 1.59 shows the variation of the stiffness with inflation pressure for the 165 × 13 radial-ply tire. The values of stiffness shown are derived from the load–deflection curves shown in Figure 1.54 (Overton et al. 1969–1970). The values of the static vertical stiffness of the tractor tires 11-36 and 7.5-16, and those of the terra tire 26 × 12.00-12 at various inflation pressures are given in Table 1.13.

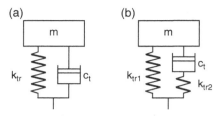

Figure 1.58 (a) A linear model and (b) a viscoelastic model for tire vibration analysis.

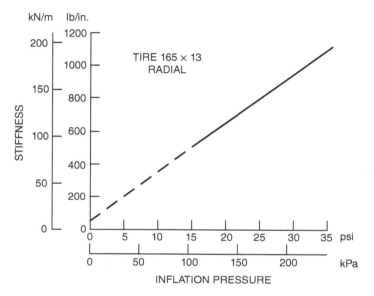

Figure 1.59 Variations of static stiffness with inflation pressure for a radial-ply car tire. *Source:* Overton et al. (1969–1970). Reproduced with permission of the Council of the Institution of Mechanical Engineers.

2) Non-rolling dynamic stiffness

The dynamic stiffness of a non-rolling tire may be obtaining using various methods. One of the simplest is the so-called drop test. In this test, the tire with a certain load is allowed to fall freely from a height at which the tire is just in contact with the ground. Consequently, the tire remains in contact with the ground throughout the test. The transient response of the tire is recorded. A typical amplitude decay trace is shown in Figure 1.60. The values of the equivalent viscous damping coefficient c_{eq} and the dynamic stiffness k_z of the tire can then be determined from the decay trace using the well-established theory of free vibration for a single-degree-of-freedom system:

$$c_{eq} = \sqrt{\frac{4m^2\omega_d^2\delta^2/\left(\delta^2 + 4\pi^2\right)}{1 - \left[\delta^2/\left(\delta^2 + 4\pi^2\right)\right]}} \qquad (1.91)$$

and

$$k_z = \frac{m\omega_d^2}{1 - \delta^2/\left(\delta^2 + 4\pi^2\right)} \qquad (1.92)$$

Table 1.13 Vertical stiffness of off-road tires.

Tire	Inflation pressure	Load	Static stiffness	Non-rolling dynamic stiffness (average)	Damping coefficient
11-36 (4-ply)	82.7 kPa	6.67 kN (1500 lb)	357.5 kN/m (24,500 lb/ft)	379.4 kN/m (26,000 lb/ft)	2.4 kN · s/m (165 lb · s/ft)
	(12 psi)	8.0 kN (1800 lb)	357.5 kN/m (24,500 lb/ft)	394.0 kN/m (27,000 lb/ft)	2.6 kN · s/m (180 lb · s/ft)
		9.34 kN (2100 lb)	–	423.2 kN/m (29,000 lb/ft)	3.4 kN · s/m (230 lb · s/ft)
	110.3 kPa	6.67 kN (1500 lb)	379.4 kN/m (26,000 lb/ft)	394.0 kN/m (27,000 lb/ft)	2.1 kN · s/m (145 lb · s/ft)
	(16 psi)	8.0 kN (1800 lb)	386.7 kN/m (26,500 lb/ft)	437.8 kN/m (30,000 lb/ft)	2.5 kN · s/m (175 lb · s/ft)
		9.34 kN (2100 lb)	394.0 kN/m (27,000 lb/ft)	423.2 kN/m (29,000 lb/ft)	2.5 kN · s/m (175 lb · s/ft)
7.5-16 (6-ply)	138 kPa	3.56 kN (800 lb)	175.1 kN/m (12,000 lb/ft)	218.9 kN/m (15,000 lb/ft)	0.58 kN · s/m (40 lb · s/ft)
	(20 psi)	4.45 kN (1000 lb)	175.1 kN/m (12,000 lb/ft)	233.5 kN/m (16,000 lb/ft)	0.66 kN · s/m (45 lb · s/ft)
		4.89 kN (1100 lb)	182.4 kN/m (12,500 lb/ft)	248.1 kN/m (17,000 lb/ft)	0.80 kN · s/m (55 lb · s/ft)
	193 kPa	3.56 kN (800 lb)	218.9 kN/m (15,000 lb/ft)	233.5 kN/m (16,000 lb/ft)	0.36 kN · s/m (25 lb · s/ft)
	(28 psi)	4.45 kN (1100 lb)	226.2 kN/m (15,500 lb/ft)	262.7 kN/m (18,000 lb/ft)	0.66 kN · s/m (45 lb · s/ft)
		4.89 kN (1300 lb)	255.4 kN/m (17,500 lb/ft)	277.3 kN/m (19,000 lb/ft)	0.73 kN · s/m (50 lb · s/ft)
26 × 12.00-12 (2-ply)	15.5 kPa (2.25 psi)	1.78 kN (400 lb)	51.1 kN/m (3500 lb/ft)	–	0.47 kN · s/m (32 lb · s/ft)
	27.6 kPa (4 psi)	1.78 kN (400 lb)	68.6 kN/m (4700 lb/ft)	–	0.49 kN · s/m (34 lb · s/ft)

Source: Matthews and Talamo (1965).

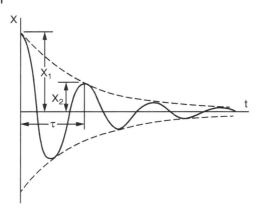

Figure 1.60 An amplitude decay record of a non-rolling tire obtained from a drop test.

ω_d is the damped natural frequency of the tire with mass m, and can be identified from the amplitude decay trace shown in Figure 1.60:

$$\omega_d = 2\pi/\tau \tag{1.93}$$

where τ is the period of damped oscillation shown in Figure 1.60.

δ is the logarithmic decrement, which is defined as the natural logarithm of the ratio of any two successive amplitudes, such as x_1 and x_2, shown in Figure 1.60:

$$\delta = \ln(x_1/x_2) \tag{1.94}$$

The drop test may also be performed utilizing a tire endurance testing machine consisting of a beam pivoted at one end and carrying the test tire with an appropriate load at the other end against a drum. To initiate the test, the beam is rotated slightly about the pivot and then is released from a position at which the tire is just in contact with the drum. The system is set in angular oscillation about the pivot of the beam. A decay trace for the amplitude of angular displacement is recorded. A set of equations for this torsional system, similar to that for a single-degree-of-freedom linear system described above, can be derived for determining the equivalent damping coefficient and non-rolling dynamic stiffness for the tire from the decay trace.

Table 1.13 shows the values of the non-rolling dynamic stiffness and the damping coefficient for the tractor tires 11-36 and 7.5-16 (Matthews and Talamo 1965), and the damping coefficient for the terra tire 26 × 12.00-12. The values of the damping coefficient for the 5.60 × 13 bias-ply and the 165 × 13 radial-ply car tires are given in Table 1.14 (Overton et al. 1969–1970).

3) Rolling dynamic stiffness

The rolling dynamic stiffness is usually determined by measuring the response of a rolling tire to a known harmonic excitation. The response is normally measured at the hub, and the excitation is given at the tread. By examining the ratio of output to input and the phase angle, it is possible to determine the dynamic stiffness and the damping coefficient of a rolling tire.

An alternative method for determining the dynamic stiffness of a tire is to measure its resonant frequency when rolling on a drum or belt. Figure 1.61 shows the values of the dynamic stiffness for various types of car tire obtained using this method (van Eldik Thieme and Pacejka 1971). It is shown that the dynamic stiffness of car tires decreases sharply as soon as the tire is rolling. However, beyond a speed of approximately 20 km/h (12 mph), the influence of speed becomes less noticeable.

Table 1.14 Damping coefficient of car tires.

Tire	Inflation pressure	Damping coefficient
Bias-ply	103.4 kPa (15 psi)	4.59 kN · s/m (315 lb · s/ft)
5.60 × 13	137.9 kPa (20 psi)	4.89 kN · s/m (335 lb · s/ft)
	172.4 kPa (25 psi)	4.52 kN · s/m (310 lb · s/ft)
	206.9 kPa (30 psi)	4.09 kN · s/m (280 lb · s/ft)
	241.3 kPa (35 psi)	4.09 kN · s/m (280 lb · s/ft)
Radial-ply	103.4 kPa (15 psi)	4.45 kN · s/m (305 lb · s/ft)
165 × 13	137.9 kPa (20 psi)	3.68 kN · s/m (252 lb · s/ft)
	172.4 kPa (25 psi)	3.44 kN · s/m (236 lb · s/ft)
	206.9 kPa (30 psi)	3.43 kN · s/m (235 lb · s/ft)
	241.3 kPa (35 psi)	2.86 kN · s/m (196 lb · s/ft)

Source: Overton et al. (1969–1970).

Table 1.15 shows the values of vertical stiffness of a sample of truck tires at rated loads and inflation pressures (Fancher et al. 1986). They were obtained when the tires were rolling at a relatively low speed. Values of the vertical stiffness for the truck tires shown in the table range from 764 to 1024 kN/m (4363 to 5850 lb/in.), and that the vertical stiffness of radial-ply truck tires is generally lower than that of bias-ply tires of similar size.

Figure 1.62 shows the variation of the dynamic stiffness of a 13.6 × 38 radial tractor tire with speed (Lines and Young 1989). The static load on the tire was 18.25 kN (4092 lb) and the inflation pressure was 138 kPa (20 psi). The dynamic stiffness of the tractor tire decreases sharply as soon as the tire begins rolling, similar to that for passenger car tires shown in Figure 1.61. The effects of inflation pressure on the dynamic stiffness of the same tire are shown in Figure 1.63. The variation of the damping coefficient with speed for the tractor tire is shown in Figure 1.64. It can be seen that beyond a speed of 1 km/h (0.6 mph), the damping coefficient drops rapidly until a speed of 5 km/h (3.1 mph) is reached, and then approaches an asymptote. The effects of inflation pressure on the damping coefficient are shown in Figure 1.65.

Attempts to determine the relationship between the static and dynamic stiffness of tires have been made, but no general conclusions have been reached. Some reports indicate that for passenger car tires, the rolling dynamic stiffness may be 10–15% less than the stiffness derived from static load–deflection curves, whereas for heavy truck tires, the dynamic stiffness is approximately 5% less than the static value. For tractor tires, it has been reported that the dynamic stiffness may be 26% lower than the static value. In simulation studies of vehicle ride, the use of the rolling dynamic stiffness is preferred.

Among various operation parameters, inflation pressure, speed, normal load, and wear have a noticeable influence on tire stiffness. Tire design parameters, such as the crown angle of the cords, tread width, tread depth, number of plies, and tire material, also affect the stiffness.

The damping of a pneumatic tire is mainly due to the hysteresis of tire materials. It is neither Coulomb-type nor viscous-type, and it appears to be a combination of both. However,

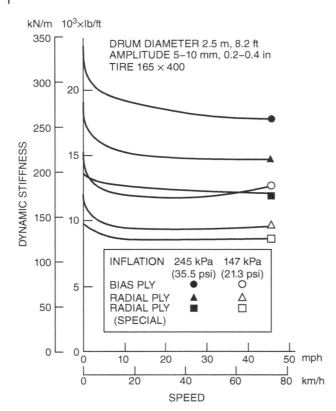

Figure 1.61 Effects of speed on rolling dynamic stiffness of car tires. Reproduced with permission from *Mechanics of Pneumatic Tires*, edited by S.K. Clark, Monograph 122, National Bureau of Standards, 1971.

an equivalent viscous damping coefficient can usually be derived from the dynamic tests mentioned previously. Its value is subject to variation, depending on the design and construction of the tire, as well as operating conditions. The damping of pneumatic tires made of synthetic rubber compounds is considerably less than that provided by a shock absorber.

To evaluate the overall vibrational characteristics of tires, tests may be carried out on a variable-speed rotating drum. The profile of the drum may be random, sinusoidal, square, or triangular. Experience has shown that the use of a periodic type of excitation enables rapid assessments to be made. Figure 1.66 shows the wheel hub acceleration as a function of frequency for a radial-ply and a bias-ply tire over a sinusoidal profile with 133 mm (5.25 in.) pitch and 6 mm (0.25 in.) peak-to-peak amplitude (Barson et al. 1967–1968). The transmissibility ratios in the vertical direction over a wide frequency range of a radial-ply and a bias-ply tire are shown in Figure 1.67 (Barson et al. 1967–1968). This set of results has been obtained using a vibration exciter. The vibration input is imparted to the tread of a non-rolling tire through a platform mounted on the vibration exciter.

It can be seen from Figures 1.66 and 1.67 that the transmissibility ratio for vertical excitation of the radial-ply tire is noticeably higher than that of the bias-ply tire in the frequency range of 60–100 Hz. Vibrations in this frequency range contribute to the passenger's sensation of "harshness". On the other hand, the bias-ply tire is significantly worse than the radial-ply tire in the frequency range of approximately 150–200 Hz. In this frequency range, vibrations contribute to induced tire noise, commonly known as "road roar" (French 1989).

Table 1.15 Vertical stiffness of truck tires at rated loads and inflation pressures.

Tire type	Tire construction	Vertical stiffness	
		kN/m	lb/in.
Unspecified 11.00-22/G	Bias-ply	1024	5850
Unspecified 11.00-22/F	Bias-ply	977	5578
Unspecified 15.00 × 22.5/H	Bias-ply	949	5420
Unspecified 11.00-20/F	Bias-ply	881	5032
Michelin Radial 11 R22.5 XZA (1/3 tread)	Radial-ply	874	4992
Michelin Radial 11 R22.5 XZA (1/2 tread)	Radial-ply	864	4935
Michelin Radial 11 R22.5 XZA	Radial-ply	831	4744
Unspecified 10.00-20/F	Bias-ply	823	4700
Michelin Radial 11 R22.5 XZA	Radial-ply	809	4622
Michelin Pilote 11/80R22.5 XZA	Radial-ply	808	4614
Unspecified 10.00-20/F	Bias-ply	788	4500
Michelin Pilote 11/80R22.5 XZA	Radial-ply	774	4418
Unspecified 10.00-20/G	Bias-ply	764	4363

Source: Fancher et al. (1986).

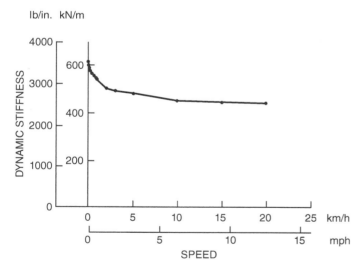

Figure 1.62 Effects of speed on rolling dynamic stiffness of a radial-ply tractor tire 13.6 × 38. *Source:* Lines and Young (1989). Reproduced with permission of the International Society for Terrain–Vehicle Systems.

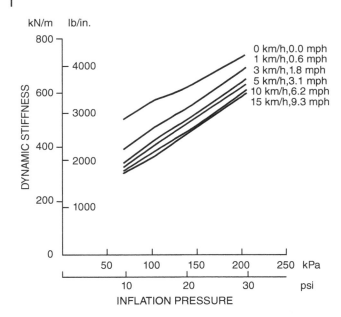

Figure 1.63 Effects of inflation pressure on rolling dynamic stiffness at various speeds of a radial-ply tractor tire 13.6 × 38. *Source:* Lines and Young (1989). Reproduced with permission of the International Society for Terrain–Vehicle Systems.

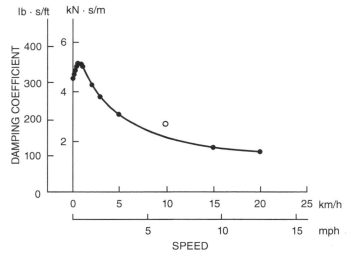

Figure 1.64 Effects of speed on damping coefficient of a radial-ply tractor tire 13.6 × 38. *Source:* Lines and Young (1989). Reproduced with permission of the International Society for Terrain–Vehicle Systems.

Figure 1.65 Effects of inflation pressure on damping coefficient at various speeds of a radial-ply tractor tire 13.6 × 38. *Source:* Lines and Young (1989). Reproduced with permission of the International Society for Terrain–Vehicle Systems.

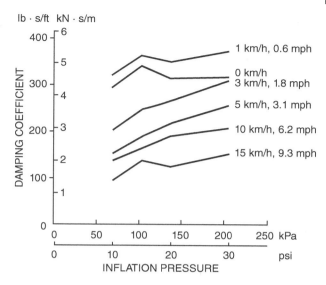

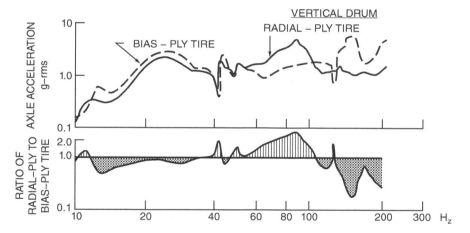

Figure 1.66 Vibration characteristics of a bias-ply and a radial-ply car tire subject to sinusoidal excitation. *Source:* Barson et al. (1967–1968). Reproduced with permission of the Council of the Institution of Mechanical Engineers.

1.7 Tire/Road Noise

With growing concerns over the impact of traffic noise on urban environments, tire/road noise has attracted increasing attention. Traffic noise is the collective sound generated from road vehicles. It primarily includes the sound emanating from tire–road interaction, engine–transmission and vehicle–air interaction (aerodynamic factors), braking, etc. At steady-speed driving, noise generated from tire–road interaction usually dominates (Sandberg 2001).

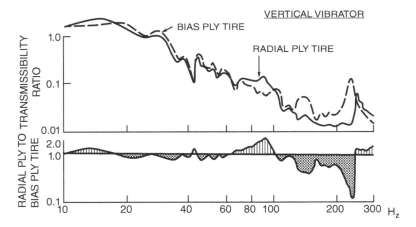

Figure 1.67 Transmissibility ratios of a bias-ply and a radial-ply car tire. *Source:* Barson et al. (1967–1968). Reproduced with permission of the Council of the Institution of Mechanical Engineers.

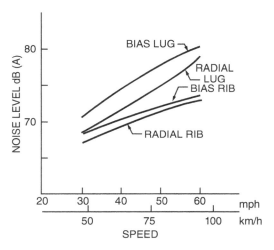

Figure 1.68 Effects of speed on noise generated by bias-ply and radial-ply truck tires. Reprinted with permission from SAE SP-729 © 1988 Society of Automotive Engineers, Inc.

Major mechanisms that generate tire/road noise are (Ford and Charles 1988):

1) Air pumping effect

As the tire rolls, air is trapped and compressed in the voids between the tread and the pavement. Noise is generated when the compressed air is released at high speed to the atmosphere at the exit of the contact patch.

2) Tread element vibrations

Tread elements impact the pavement as the tire rolls. When the elements leave the contact patch, they are released from a highly stressed state. These induce vibrations of the tread, which form a major source of tire noise. Carcass vibrations and the grooves and lug voids in the tread act like resonating pipes, also contributing to noise radiation from the tire.

Since the air pumping effect, vibrations of tread elements and carcass, etc. are related to speed, the noise level generated by a tire is a function of operating speed. Figure 1.68 shows the variations

Table 1.16 Effects of pavement texture on noise level generated by a bias-ply truck Tire.

Road surface	Noise level dB (A)
Moderately smooth concrete	70
Smooth asphalt	72
Worn concrete (exposed aggregate)	72
Brushed concrete	78

Source: Ford and Charles (1988).

Table 1.17 Exterior tire rolling noise limit values.

C1 (Passenger car tires)		C2 (Light utility vehicle tires)		C3 (Heavy commercial vehicle tires)	
Nominal tire section width, mm	Limit values in dB(A)	Category of use	Limit values in dB(A)	Category of use	Limit values in dB(A)
≤185	70	Normal tires	72	Normal tires	73
>185 ≤ 215	71	Traction tires	73	Traction tires	75
>215 ≤ 245	71				
>245 ≤ 275	72				
>275	74				

Source: EC Regulation 661/2009.

of noise level with speed for various types of truck tire, including both bias-ply and radial-ply tires with ribs or lugs, on a smooth pavement (Ford and Charles 1988). The results were obtained following the SAE J57 test procedure. The effect of pavement texture on the noise level generated by a bias-ply, ribbed truck tire at 80 km/h (50 mph) is shown in Table 1.16 (Ford and Charles 1988). Other factors that affect tire/road noise include tire types (tires for cars, light utility vehicles, and heavy commercial vehicles), tire section widths, and tire tread patterns (for four-season tires, snow tires, traction tires, etc.).

In the European Union, tire rolling noise level is required to be identified on the tire label (Regulation EC 661/2009). The noise levels are determined in accordance with the procedure specified in the implementing measures to the Regulation. It should be noted that the noise level in the Regulation refers to the noise of the tire measured outside the vehicle and hence not that experienced by the driver or passengers. This exterior rolling noise is measured in dB(A).

The exterior rolling noise limit values for various types of tires, according to EU Regulation 661/2009, are given in Table 1.17.

For type C1 snow tires, extra load tires, or reinforced tire (carcass designed to carry extra load than that of the standard tire), the limits are increased by 1 dB(A). For types C2 and C3 special use tires, the limits are increased by 2 dB(A). For type C2 traction snow tires, an additional 2 dB(A) are allowed. For all other C2 and C3 types of snow tires, an additional 1 dB(A) is allowed.

References

Bakker, E., Nyborg, L., and Pacejka, H.B. (1987). Tyre modelling for use in vehicle dynamic studies. Society of Automotive Engineers paper 870421.

Bakker, E., Pacejka, H.B., and Lidner, L. (1989). A new tire model with an application in vehicle dynamic studies. Society of Automotive Engineers paper 890087.

Barson, C.W., James, D.H., and Morcombe, A.W. (1967–1968). Some aspects of tire and vehicle vibration testing. *Proceedings of the Institution of Mechanical Engineers*, **182** (3B).

Bayle, P., Forissier, J.F., and Lafon, S. (1993). A new tyre model for vehicle dynamics simulations, *Proc. Automotive Technology International'93*. U.K. & International Press.

Clark, S.K. (1971). The contact between tire and roadway. In: *Mechanics of Pneumatic Tires* (ed. S.K. Clark). Monograph 122. Washington, DC: National Bureau of Standards.

Collier, B.L. and Warchol, J.T. (1980). The effect of inflation pressure on bias, bias-belted and radial tire performance. Society of Automotive Engineers paper 800087.

DeRaad, L.W. (1977). The influence of road surface texture on tire rolling resistance. Society of Automotive Engineers Special Publication **P-74**, Tire Rolling Losses and Fuel Economy: An R&D Planning Workshop.

Ellis, J.R. (1969). *Vehicle Dynamics*. London: Business Books.

Ellis, J.R. (1994). *Vehicle Handling Dynamics*. London: Mechanical Engineering Publications.

Ervin, R.D. (1975). Mobile measurement of truck tire traction. *Proc. of a Symposium on Commercial Vehicle Braking and Handling*, University of Michigan, Ann Arbor, MI. Highway Safety Research Institute.

Fancher, P.S. and Grote, P. (1971). Development of a hybrid simulation for extreme automobile maneuvers. *Proc. 1971 Summer Computer Simulation Conference*. Boston, MA.

Fancher, P.S., Ervin, R.D., Winkler, C.B., and Gillespie, T.D. (1986). A fact book of the mechanical properties of the components for single-unit and articulated heavy trucks. Report No. DOT HS 807 125, National Highway Traffic Safety Administration, U.S. Department of Transportation.

Ford, T.L. and Charles, F.S. (1988). Heavy duty truck tire engineering. The Thirty-Fourth L. Ray Buckendale Lecture, Society of Automotive Engineers **SP-729**.

French, T. (1959). Construction and behaviour characteristics of tyres. *Proceedings of the Institution of Mechanical Engineers, Automobile Division*, AD **14/59**.

French, T. (1989). *Tire Technology*. Bristol, UK: Adam Hilger.

Gough, V.E. (1956). Practical tire research. *SAE Transactions* **64**.

Gough, V.E. (1971). Structure of the tire. In: *Mechanics of Pneumatic Tires* (ed. S.K. Clark). Monograph 122. Washington, DC: National Bureau of Standards.

Hadekel, R. (1952). The mechanical characteristics of pneumatic tyres. S&T Memo No. 10/52, Ministry of Supply, London.

Harned, J.L., Johnston, L.E., and Sharpf, G. (1969). Measurement of tire brake force characteristics as related to wheel slip (antilock) control system design. *SAE Transactions* **78**, paper 690214.

Hartley, J.D.C. and Turner, D.M. (1956). Tires for high-performance cars. *SAE Transactions* **64**.

He, R., Sandu, C., Mousavi, H. et al. (2020). Updated standards of the International Society for Terrain–Vehicle Systems. *Journal of Terramechanics* **91**.

Horne, W.B. and Dreher, R.C. (1963). Phenomena of pneumatic tire hydroplaning. NASA TND-2056.

Horne, W.B. and Joyner, U.T. (1965). Pneumatic tire hydroplaning and some effects on vehicle performance. Society of Automotive Engineers paper 650145.

Hunt, J.D., Walter, J.D., and Hall, G.L. (1977). The effect of tread polymer variations on radial tire rolling resistance. Society of Automotive Engineers Special Publications **P-74**, Tire Rolling Losses and Fuel Economy: An R&D Planning Workshop.

Janssen, M.L. and Hall, G.L. (1980). Effect of ambient temperature on radial tire rolling resistance. Society of Automotive Engineers paper 800090.

Lines, J.A. and Young, N.A. (1989). A machine for measuring the suspension characteristics of agricultural tyres. *Journal of Terramechanics* **26** (3/4).

Matthews, J. and Talamo, J.D.C. (1965). Ride comfort for tractor operators, III: Investigation of tractor dynamics by analogue computer simulation. *Journal of Agricultural Engineering Research* **10** (2).

Moore, D.F. (1975). *The Friction of Pneumatic Tyres*. Amsterdam: Elsevier.

Nordeen, D.L. and Cortese, A.D. (1963). Force and moment characteristics of rolling tires. Society of Automotive Engineers paper 713A.

Overton, J.A., Mills, B., and Ashley, C. (1969–1970). The vertical response characteristics of the non-rolling tire. *Proceedings of the Institution of Mechanical Engineers*, **184** (2A) 2.

Pacejka, H.B. (1996). The tyre as a vehicle component. In: *Proc. of XXVI FISITA Congress* (ed. M. Apetaur). Prague, Czech Republic. FISITA (CD-ROM).

Pacejka, H.B. (2012). *Tire and Vehicle Dynamics*, 3e. Oxford, UK: Elsevier.

Pacejka, H.B. and Bakker, E. (1991). The Magic Formula tyre model. In: *Proc. 1st Colloquium on Tyre Models for Vehicle Analysis* (ed. H.B. Pacejka). Delft, Netherlands. Swets & Zeitlinger.

Pacejka, H.B. and Besselink, I.J.M. (1997). Magic Formula tyre model with transient properties. In: *Proc. 2nd International Colloquium on Tyre Models for Vehicle Dynamic Analysis* (eds. F. Bohm and H.-P. Willumeit). Berlin, Germany: Swets & Zeitlinger.

Robert Bosch GmbH (2018). *Bosch Automotive Handbook*, 10e. Hoboken, NJ: Wiley.

Sandberg, U. (2001). Tyre/road noise-myths and realities. *Proc. 2001 International Congress and Exhibition on Noise Control Engineering*. The Hague, Netherlands. Nederslands Akoestisch Genootschap.

Schallamach, A. (1971). Skid resistance and directional control. In: *Mechanics of Pneumatic Tires* (ed. S.K. Clark). Monograph 122. Washington, DC: National Bureau of Standards.

Segel, L. (1984). The mechanics of heavy-duty trucks and truck combinations. Engineering Summer Conferences, University of Michigan, Ann Arbor, MI.

Taborek, J.J. (1957). Mechanics of vehicles. *Machine Design*, May 30–Dec. 26.

van Eldik Thieme, H.C.A. and Pacejka, H.B. (1971). The tire as a vehicle component. In: *Mechanics of Pneumatic Tires* (ed. S.K. Clark). Monograph, 122. Washington, DC: National Bureau of Standards.

van Zanten, A., Ruf, W.D., and Lutz, A. (1989). Measurement and simulation of transient tire forces. Society of Automotive Engineers paper 890640.

Zhou, J., Wong, J.Y., and Sharp, R.S. (1999). A multi-spoke, three plane tyre model for simulation of transient behavior. *Vehicle System Dynamics* **31** (1).

Problems

1.1   A truck tire with vertical load of 24.78 kN (5570 lb) travels on a dry concrete pavement with a peak value of coefficient of road adhesion $\mu_p = 0.80$. The longitudinal stiffness of the tire during braking C_s is 224.64 kN/unit skid (55,000 lb/unit skid). Using the simplified theory described in Section 1.3, plot the relationship between the braking force and the skid of the tire up to skid $i_s = 20\%$.

1.2 Using the simplified theory described in Section 1.4.4, determine the relationship between the cornering force and the slip angle in the range 0–16° of the truck tire described in Problem 1.1. The cornering stiffness of the tire C_α is 132.53 kN/rad (520 lb/deg). There is no braking torque applied to the tire.

1.3 Determine the available cornering force of the truck tire described in Problems 1.1 and 1.2 as a function of longitudinal skid at a slip angle of 4° using the simplified theory described in Section 1.4.4. Plot the cornering force of the tire at a slip angle of 4° versus skid in the range 0–40%. The coefficient of road adhesion is 0.8.

1.4 A passenger car travels over a flooded pavement. The inflation pressure of the tires is 179.27 kPa (26 psi). If the initial speed of the car is 100 km/h (62 mph) and brakes are then applied, determine whether the vehicle will be hydroplaning.

1.5 An all-terrain vehicle weighs 3.56 kN (800 lb) and has four terra tires, each of which has a vertical stiffness of 52.54 kN/m (300 lb/in.) at an inflation pressure of 27.6 kPa (4 psi), and a stiffness of 96.32 kN/m (550 lb/in.) at a pressure of 68.9 kPa (10 psi). Estimate the fundamental natural frequencies of the vehicle in the vertical direction at the two inflation pressures. The vehicle has no spring suspension.

1.6 Using the Magic Formula described in Section 1.4.5, estimate the cornering force of a car tire at a normal load of 6 kN (1349 lb) with a slip angle of 5°. The values of the empirical coefficients in the Magic Formula for the tire are given in Table 1.10.

2

Mechanics of Vehicle–Terrain Interaction: Terramechanics

While transporting passengers and goods by vehicles on paved roads constitutes a significant part of surface transportation in a modern society, a wide range of human endeavors, in such fields as agriculture, construction, logging, mining, recreation, and defense operations, involves locomotion over unprepared terrain using off-road vehicles. With mankind's increasing interest in exploration of the universe, manned or unmanned rovers have been deployed to the Moon and Mars for surface exploration.

Systematic studies of the principles underlying the rational development of off-road vehicles have attracted growing interest, particularly since World War II. Research into the fundamentals for guiding the development of extraterrestrial rovers was initiated in earnest in the 1960s, leading to the successful deployment to the moon of the manned lunar roving vehicle by the U.S. National Aeronautics and Space Administration (NASA) and of the unmanned Lunokhod by the then Soviet Union in the 1970s. In recent years, unmanned robotic rovers have been successfully deployed to Mars by several countries.

The development of guiding principles for the optimal design of off-road vehicles or extraterrestrial rovers requires an in-depth understanding of the physical nature of interaction between the vehicle (or rover) and the terrain (or regolith). The mechanics of interaction of an off-road vehicle or an extraterrestrial rover with the terrain or regolith has now become known as "terramechanics" (Bekker 1956, 1960, 1969; Wong 2010). The role of terramechanics in the development of off-road vehicles or extraterrestrial rovers may be analogous to that of aerodynamics in the design of aircraft or spacecraft and to that of hydrodynamics in the development of marine craft. In this chapter, discussions will primarily focus on the fundamentals of the mechanics of interaction between an off-road vehicle and its operating environment – the terrain. Applications of terramechanics to evaluation of certain aspects of rover mobility on extraterrestrial surfaces will also be discussed.

In off-road operations, various types of terrain with differing behavior, ranging from desert sand through soft mud to fresh snow, may be encountered. The properties of the terrain quite often impose severe limitations to the mobility of off-road vehicles. An adequate knowledge of the engineering properties of the terrain and its response to vehicular loading is, therefore, essential to the proper development and design of an off-road vehicle for a given mission and environment. In this chapter, the measurement and characterization of terrain behavior will be discussed. On a given terrain, the performance of an off-road vehicle is, to a great extent, dependent on the way the vehicle interacts with the terrain. Consequently, an understanding of the mechanics of vehicle–terrain interaction is of importance to the proper selection of vehicle configuration and design parameters to meet specific operational requirements. A central issue in terramechanics is to establish a quantitative relationship between the performance and the design of an off-road vehicle for a given operating environment. Over the years, a variety of methods, ranging from empirically based to

Theory of Ground Vehicles, Fifth Edition. J.Y. Wong.
© 2022 John Wiley & Sons, Inc. Published 2022 by John Wiley & Sons, Inc.
Companion website: www.wiley.com/go/wong/TGV5e

physics-based, for predicting the performance of tracked and wheeled vehicles over unprepared terrain have been developed or proposed. Some of the representative ones will be presented in this chapter.

2.1 Applications of the Theory of Elasticity to Predicting Stress Distributions in the Terrain under Vehicular Loads

Certain types of terrain, such as mineral soils which cover part of the trafficable earth surface, may be compared to an ideal elastoplastic material with the stress–strain relationship shown in Figure 2.1. When the stress level in the terrain under vehicular load does not exceed a certain limit, such as that denoted by "*a*" in Figure 2.1, the terrain may exhibit elastic behavior. The idealization of the terrain as an elastic medium has found applications in the prediction of stress distribution in the soil, in connection with the study of soil compaction due to vehicular loads (Söhne 1958).

The prediction of stress distribution in an elastic medium subject to any specific load may be based on the analysis of the distribution of stresses under a point load. The method for calculating the stress distribution in a semi-infinite, homogeneous, isotropic, elastic medium subject to a vertical point load applied on the surface was first developed by Boussinesq. His solutions give the following expressions for the vertical stress σ_z at a point in the elastic medium defined by the coordinates shown in Figure 2.2:

$$\sigma_z = \frac{3W}{2\pi} z^3 \left(x^2 + y^2 + z^2\right)^{-5/2}$$

$$= \frac{3}{2\pi} \frac{1}{\left[1 + (r/z)^2\right]^{5/2}} \frac{W}{z^2}$$

or

$$\sigma_z = \frac{3W}{2\pi R^2} \left(\frac{z}{R}\right)^3 = \frac{3W}{2\pi R^2} \cos^3\theta \tag{2.1}$$

where

$$r = \sqrt{x^2 + y^2} \text{ and } R = \sqrt{z^2 + r^2}$$

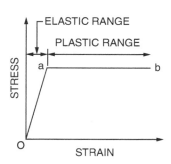

Figure 2.1 Behavior of an idealized elastoplastic material.

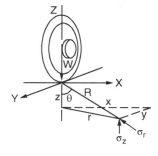

Figure 2.2 Stresses at a point in a semi-infinite elastic medium subject to a point load. *Source: Theory of Land Locomotion* by M.G. Bekker, copyright © 1956 by the University of Michigan. Reproduced with permission of the University of Michigan Press.

When polar coordinates are used, the radial stress σ_r (Figure 2.2) is given by

$$\sigma_r = \frac{3W}{2\pi R^2} \cos \theta \qquad (2.2)$$

Note that the stresses are independent of the modulus of elasticity of the material, and that they are only functions of the load applied and the distance from the point of application of the load. Equations (2.1) and (2.2) are valid only for calculating stresses at points not too close to the point of application of the load. The material in the immediate vicinity of the point load does not behave elastically.

Based on the analysis of stress distribution beneath a point load, the distribution of stresses in an elastic medium under a variety of loading conditions may be predicted using the principle of superposition. For instance, for a circular contact area having a radius r_0 and with a uniform pressure p_0 (Figure 2.3), the vertical stress at a depth z below the center of the contact area may be determined in the following way (Bekker 1956). The load acting upon the contact area may be represented by a combination of discrete point loads, $dW = p_0 \, dA = p_0 r dr d\theta$. Hence, in accordance with Eq. (2.1),

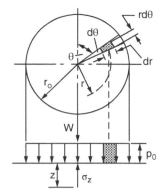

Figure 2.3 Vertical stresses in a semi-infinite elastic medium below the center of a circular loading area. *Source: Theory of Land Locomotion* by M.G. Bekker, copyright © 1956 by the University of Michigan. Reproduced with permission of the University of Michigan Press.

$$d\sigma_z = \frac{3}{2\pi} \frac{p_0 r dr d\theta}{\left[1 + (r/z)^2\right]^{5/2} z^2} \qquad (2.3)$$

The resultant vertical stress σ_z at a depth z below the center of the contact area is then equal to the sum of the stresses produced by point loads of $p_0 r dr d\theta$, and can be computed by a double integration (Bekker 1956):

$$\sigma_z = \frac{3}{2\pi} p_0 \int_0^{r_0} \int_0^{2\pi} \frac{r dr d\theta}{\left[1 + (r/z)^2\right]^{5/2} z^2} = 3p_0 \int_0^{r_0} \frac{r dr}{\left[1 + (r/z)^2\right]^{5/2} z^2}$$

By substituting $(r/z)^2 = u^2$, it is found that

$$\sigma_z = 3p_0 \int_0^{r_0/z} \frac{u du}{\left[1 + u^2\right]^{5/2}} = p_0 \left[1 - \frac{z^3}{(z^2 + r_0^2)^{3/2}}\right] \qquad (2.4)$$

The computation of the stresses at points other than those directly below the center of the contact area is rather involved and cannot be generalized by a simple set of equations. The stress distribution in an elastic medium under distributed loads over an elliptic area, like that applied by a tire, can be determined in a similar way.

Another case of interest from the vehicle viewpoint is the distribution of stresses in the elastic medium under the action of a strip load (Figure 2.4). Such a strip load may be considered as an idealization of that applied by a tracked vehicle. It can be shown that the stresses in an elastic medium due to a uniform pressure p_0 applied over a strip of infinite length and of constant width b (Figure 2.5) can be computed by the following equations (Bekker 1956):

$$\sigma_x = \frac{p_0}{\pi} (\theta_2 - \theta_1 + \sin \theta_1 \cos \theta_1 - \sin \theta_2 \cos \theta_2) \qquad (2.5)$$

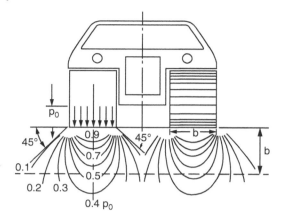

Figure 2.4 Distributions of vertical stresses in a semi-infinite elastic medium under a tracked vehicle. *Source: Theory of Land Locomotion* by M.G. Bekker, copyright © 1956 by the University of Michigan. Reproduced with permission of the University of Michigan Press.

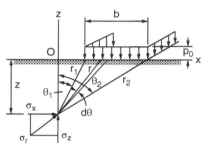

Figure 2.5 Stresses at a point in a semi-infinite elastic medium subject to a uniform strip load. *Source: Theory of Land Locomotion* by M.G. Bekker, copyright © 1956 by the University of Michigan. Reproduced with permission of the University of Michigan Press.

$$\sigma_z = \frac{p_0}{\pi}\left(\theta_2 - \theta_1 - \sin\theta_1\cos\theta_1 + \sin\theta_2\cos\theta_2\right) \quad (2.6)$$

$$\tau_{xz} = \frac{p_0}{\pi}\left(\sin^2\theta_2 - \sin^2\theta_1\right) \quad (2.7)$$

The points in the medium that experience the same level of stress form a family of isostress surfaces commonly known as pressure bulbs. The general characteristics of the bulbs of vertical pressure under a uniform strip load are illustrated in Figure 2.4. At a depth equal to the width of the strip, the vertical stress under the center of the loading area is approximately 50% of the applied pressure p_0 and it practically vanishes at a depth equal to twice the width of the strip. The boundaries of the bulbs of vertical pressure, for all practical purposes, may be assumed as being sloped at an angle of 45° with the horizontal shown in Figure 2.4 (Bekker 1956).

The use of the theory of elasticity for predicting stresses in a real soil produces approximate results only. Measurements have shown that the stress distribution in a real soil deviates from that computed using Boussinesq's equations (Söhne 1958). There is a tendency for the compressive stress in the soil to concentrate around the loading axis. This tendency becomes greater when the soil becomes more plastic due to increased moisture content or when the soil is less cohesive, such as sand. In view of this, various semi-empirical equations have been developed to account for the different behavior of various types of soil. Fröhlich introduced a concentration factor υ to Boussinesq's equations (Söhne 1958). The factor υ reflects the behavior of various types of soil in different conditions. Introducing the concentration factor, the expressions for the vertical and radial stress in the soil subject to a point load on the surface take the following forms:

$$\sigma_z = \frac{\upsilon W}{2\pi R^2}\cos^\upsilon\theta = \frac{\upsilon W}{2\pi z^2}\cos^{\upsilon+2}\theta \quad (2.8)$$

$$\sigma_r = \frac{\upsilon W}{2\pi R^2}\cos^{\upsilon-2}\theta = \frac{\upsilon W}{2\pi z^2}\cos^\upsilon\theta \quad (2.9)$$

Equations (2.8) and (2.9) are identical to Eqs. (2.1) and (2.2), respectively, if υ is equal to 3. The value of the concentration factor depends on the type of soil and its moisture content. Figure 2.6 shows the bulbs of radial stress σ_r under a point load in soils with different concentration factors (Söhne 1958).

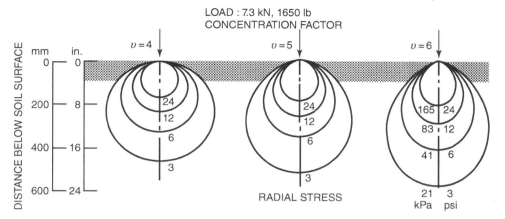

Figure 2.6 Distributions of radial stresses under a point load in soils with different concentration factors. *Source:* Söhne (1958). Reproduced with permission of *Agricultural Engineering*, ASAE.

A tire transfers its load to the soil surface usually not at one point, but through a finite area of contact. To determine the stress distribution in the soil due to tire loading, the actual size of the contact area and the pressure distribution over the contact patch must be known. Figure 2.7 shows the measured contact areas of a tire under different soil conditions (Söhne 1958). The rut becomes deeper with increasing porosity and moisture content of the soil. An approximately uniform pressure over the entire contact area may be assumed for tires without lugs in hard, dry soil. In soft soils, the pressure over the contact area varies with the depth of the rut. Usually, the contact pressure decreases toward the outside of the contact area and is more concentrated toward the center of the loading area. Representative pressure distributions over the contact area in hard, dry soil, in a moist, relatively dense soil, and in wet soil are shown in Figure 2.8(a)-(c), respectively.

Figure 2.7 Contact areas of a tire under different soil conditions. *Source:* Söhne (1958). Reproduced with permission of *Agricultural Engineering*, ASAE.

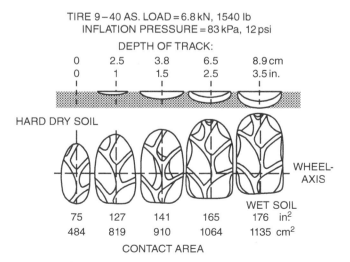

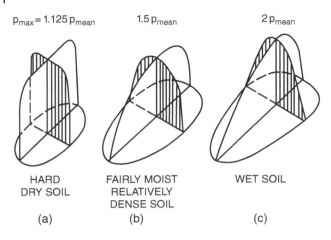

$p_{max} = 1.125 \, p_{mean}$ $1.5 \, p_{mean}$ $2 \, p_{mean}$

HARD
DRY SOIL

FAIRLY MOIST
RELATIVELY
DENSE SOIL

WET SOIL

(a)

(b)

(c)

Figure 2.8 Pressure distributions on the tire contact area under different soil conditions. *Source:* Söhne (1958). Reproduced with permission of *Agricultural Engineering*, ASAE.

Knowing the shape of the contact area and pressure distribution over the contact patch, it is possible to predict the distribution of stresses in a soil by employing Eq. (2.8) or (2.9), as proposed by Fröhlich. It has been reported by Söhne that the difference between the measured values and calculated ones obtained using Fröhlich's equations is approximately 25%, which may be regarded as reasonable for this type of study (Söhne 1969). Figure 2.9 shows the distributions of the major principal stress in a soil having normal field density and moisture under tires of different sizes and carrying different normal loads, but with the same inflation pressure (Söhne 1958). In the calculations, it is assumed that the concentration factor υ is 5, and that the pressure distribution over

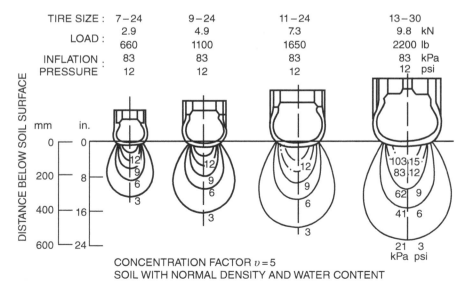

Figure 2.9 Distributions of major principal stresses in a soil under various tire loads. *Source:* Söhne (1958). Reproduced with permission of *Agricultural Engineering*, ASAE.

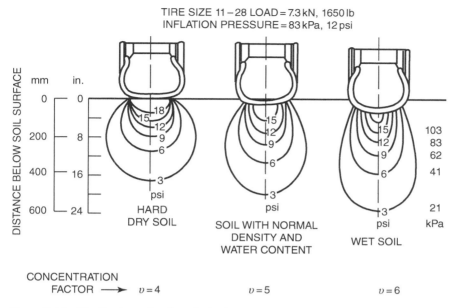

Figure 2.10 Distributions of major principal stresses under a tire for different soil conditions. *Source:* Söhne (1958). Reproduced with permission of *Agricultural Engineering*, ASAE.

the contact area is like that shown in Figure 2.8(b). It shows that for the same inflation pressure, the stress can penetrate much deeper with larger tires carrying higher loads. This is because the larger tire has a larger contact area. As a result, the stress at the same depth beneath the center of the contact patch increases as indicated by Eq. (2.4), although the pressure applied on the soil surface remains the same. This indicates that the stress distribution in a soil is a function of not only contact pressure but also contact area. It should also be mentioned that soil compaction is more closely related to the major principal stress than the vertical stress.

Figure 2.10 shows the effects of soil conditions on the shape of the pressure bulbs (Söhne 1958). In hard, dry, and dense soil, the lines of equal major principal stress are approximately circular. The softer the soil, the narrower the pressure bulbs become. This is because in soft soil, the soil can flow sideways so that the stress is more concentrated toward the center of the loading area.

For tires with lugs, such as tractor tires, the pressure distribution over the contact area differs from that shown in Figure 2.8. In hard, dry soil, the lugs of the tire carry the entire load. The pressure over the contact area of the lugs is three to four times higher than that of an equivalent tire without lugs. In a wet soil, because of the sinkage of the tire, there may be hardly any difference between the contact pressure under the lugs and that under the carcass. In this case, the distribution of contact pressure may be like that shown in Figure 2.8(c). In principle, the stress distribution in the soil under tires with lugs may be estimated following an approach like that described above. However, the computation will be more involved.

Example 2.1   The contact patch of a tire without lugs on a hard and dry soil may be approximated by a circular area of radius of 20 cm (7.9 in.). The contact pressure is assumed to be a uniform 68.95 kPa (10 psi). For this type of soil, the concentration factor v is assumed to be 4. Calculate the vertical stress σ_z in the soil at a depth of 20 cm directly below the center of the contact area.

Solution

When the concentration factor v is 4, the vertical stress σ_z at a point in the soil due to a point load W applied on the soil surface is expressed by

$$\sigma_z = \frac{4W}{2\pi R^2} \cos^4 \theta = \frac{4W}{2\pi} \frac{z^4}{(z^2 + r^2)^3}$$

$$= \frac{4W}{2\pi} \frac{1}{z^2 \left[1 + (r/z)^2\right]^3}$$

The vertical stress σ_z at a depth z directly below the center of a circular contact area of radius r_0 and with a uniform contact pressure p_0 is given by

$$\sigma_z = \frac{4p_0}{2\pi} \int_0^{r_0} \int_0^{2\pi} \frac{r \, dr \, d\theta}{z^2 \left[1 + (r/z)^2\right]^3}$$

$$= 4p_0 \int_0^{r_0/z} \frac{u \, du}{(1 + u^2)^3}$$

$$= p_0 \left\{ 1 - \frac{1}{\left[1 + (r_0/z)^2\right]^2} \right\}$$

where $u^2 = r^2/z^2$. For $p_0 = 68.95$ kPa, $r_0 = 20$ cm, and $z = 20$ cm,

$$\sigma_z = 68.95 \left\{ 1 - \frac{1}{\left[1 + (20/20)^2\right]^2} \right\}$$

$$= 51.7 \text{ kPa } (7.5 \text{ psi})$$

2.2 Applications of the Theory of Plastic Equilibrium to the Mechanics of Vehicle–Terrain Interaction

When the vehicular load applied to the terrain surface exceeds a certain limit, the stress level within a certain boundary of the terrain may reach that denoted by "*a*" on the idealized stress–strain curve shown in Figure 2.1. An infinitely small increase of stress beyond point "*a*" produces a rapid increase in strain, which constitutes plastic flow. The state that precedes plastic flow is usually referred to as plastic equilibrium. The transition from the state of plastic equilibrium to that of plastic flow represents the failure of the mass.

There are several criteria proposed for the failure of soils and other similar materials. A widely used one is commonly known as the Mohr–Coulomb failure criterion. It postulates that the material at a point will fail if the shear stress at that point in the medium satisfies the following condition:

$$\tau = c + \sigma \tan \phi \tag{2.10}$$

where τ is the shear strength of the material, c is the apparent cohesion of the material, σ is the normal stress on the sheared surface, and ϕ is the angle of internal shearing resistance (or angle of internal friction) of the material.

Cohesion of the material is the bond that cements particles together irrespective of the normal pressure exerted by one particle upon the other. On the other hand, particles of frictional masses can be held together only when a normal pressure exists between them. Thus, theoretically, the

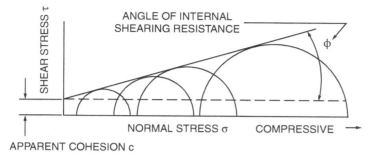

Figure 2.11 Mohr–Coulomb failure criterion.

shear strength of saturated clay and the like does not depend on the normal pressure, whereas the shear strength of dry sand increases with an increase of the normal pressure. For dry sand, therefore, the shear strength may be expressed by

$$\tau = \sigma \tan \phi \tag{2.11}$$

and for saturated clay and the like, it may take the form

$$\tau = c \tag{2.12}$$

Mineral soils (granular masses) that cover most of the trafficable earth surface, however, usually have both cohesive and frictional properties.

The meaning of the Mohr–Coulomb criterion may be illustrated with the aid of the Mohr circle of stress. If specimens of a soil are subject to different states of stress, for each mode of failure a Mohr circle can be constructed (Figure 2.11). If a straight line enveloping the set of Mohr circles is drawn, it can be expressed in the form of Eq. (2.10), with the cohesion of the soil being determined by the intercept of the enveloping line with the shear stress axis, and the angle of internal shearing resistance being represented by its slope. The Mohr–Coulomb criterion is simply that if a Mohr circle representing the state of stress at a point in the soil touches the enveloping line, failure will take place at that point.

The shear strength parameters c and ϕ in Eq. (2.10) may be measured by a variety of devices (Osman 1964; Wong 2010). The triaxial apparatus and the translational shear box are the most used in civil engineering. For vehicle mobility study, however, rectangular or annular shear plates, shown in Figure 2.12, are usually employed to simulate the shearing action of the vehicle running gear and to obtain the shear strength parameters of the terrain. This will be discussed further later in this chapter.

To illustrate the application of the Mohr–Coulomb criterion, let us consider the problem of plastic equilibrium of a prism in a semi-infinite mass (Figure 2.13). The prism of soil with unit weight γ_s, having depth z and width equal to unity, is in a state of incipient plastic failure due to lateral pressure, as shown in Figure 2.13. There are no shear stresses on the vertical sides of the prism; the normal stress on the base of the prism and that on the vertical sides are therefore the principal stresses. The prism may be set into a state of plastic equilibrium by two different operations: one is to stretch it, and the other is to compress it in the horizontal direction. If the prism is stretched, the normal stress on the vertical sides decreases until the conditions for plastic equilibrium are satisfied, while the normal stress on the bottom remains unchanged. Any further expansion merely causes a plastic flow without changing the state of stress. In this case, the weight of the soil assists in producing an expansion, and this type of failure is called the active failure. On the other hand, if

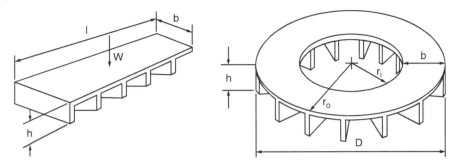

Figure 2.12 Rectangular and annular shear plates for measuring terrain shear strength parameters. *Source: Introduction to Terrain–Vehicle Systems* by M.G. Bekker, copyright © 1969 by the University of Michigan. Reproduced with permission of the University of Michigan Press.

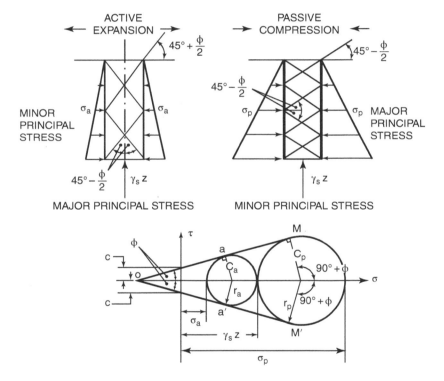

Figure 2.13 Active and passive failure of soil.

the prism of soil is compressed by the normal stress on the two vertical sides, while the normal stress at the bottom remains unchanged, then, in this case, lateral compression of the soil is resisted by its own weight. The resulting failure is called the passive failure. The two states of stress prior to plastic flow caused by expansion and compression of the soil prism are generally referred to as the Rankine active and passive state, respectively (Terzaghi 1966).

Both types of soil failure may be analyzed quantitatively by means of the Mohr circle, as shown in Figure 2.13. In the case of active failure, the normal stress σ on the base of the element at depth z ($\sigma = \gamma_s z$) is the major principal stress. Circle C_a therefore can be traced to represent the state of stress at that point. This circle is tangent to the lines OM and OM', which represent the Mohr–Coulomb failure criterion. The point of intersection between the circle and the horizontal axis of the Mohr diagram determines the minor principal stress, which is the normal stress on the vertical sides required to bring the mass at that point into active failure. This normal stress is called the active earth pressure σ_a. From the geometry of the Mohr diagram shown in Figure 2.13, the expression for the active earth pressure σ_a is given by

$$\sigma_a = \gamma_s z - 2 r_a$$

where r_a is the radius of circle C_a shown in Figure 2.13 and is expressed by

$$r_a = \frac{1}{1 - \sin \theta} (c \cos \phi + \sigma_a \sin \phi)$$

Therefore,

$$\sigma_a = \gamma_s z - \frac{2}{1 - \sin \phi} (c \cos \phi + \sigma_a \sin \phi)$$

and

$$
\begin{aligned}
\sigma_a &= \frac{\gamma_s z}{(1 + \sin \phi)/(1 - \sin \phi)} - \frac{2c \cos \phi/(1 - \sin \phi)}{(1 + \sin \phi)/(1 - \sin \phi)} \\
&= \frac{\gamma_s z}{\tan^2(45° + \phi/2)} - \frac{2c \tan(45° + \phi/2)}{\tan^2(45° + \phi/2)} \\
&= \gamma_s z \frac{1}{N_\phi} - 2c \frac{1}{\sqrt{N_\phi}}
\end{aligned}
\tag{2.13}
$$

where N_ϕ is equal to $\tan^2(45° + \phi/2)$ and is called the flow value.

Circle C_a touches the boundaries of failure, OM and OM', at a and a', respectively, as shown in Figure 2.13. This indicates that there are two planes sloped to the major principal stress plane on either side at an angle of $45° + \phi/2$, on which the shear stress satisfies the Mohr–Coulomb criterion. These planes are called surfaces of sliding, and the intersection between a surface of sliding and the plane of drawing is usually referred to as a shear line or slip line. It follows that there are two sets of slip lines sloped to the major principal stress on either side at an angle of $45° - \phi/2$. In the case of active failure, since the major principal stress is vertical, the slip line field comprises parallel lines sloped to the horizontal at $45° + \phi/2$, as shown in Figure 2.13.

As passive failure is caused by lateral compression, the normal stress σ acting on the bottom of the element ($\sigma = \gamma_s z$) is the minor principal stress. Circle C_p can, therefore, be drawn to represent the stress conditions of a point in the state of incipient passive failure, as shown in Figure 2.13. The point of intersection between the circle and the horizontal axis of the Mohr diagram determines the major principal stress, which is also the lateral, compressive stress on the vertical sides required to set the mass at that point into passive failure. This normal stress is referred to as the passive earth pressure σ_p. From the geometric relationships shown in Figure 2.13, the expression for the passive earth pressure σ_p is given by

$$\sigma_p = \gamma_s z + 2 r_p$$

where r_p is the radius of the circle C_p shown in Figure 2.13 and is expressed by

$$r_p = \frac{1}{1 - \sin \phi} (c \cos \phi + \gamma_s z \sin \phi)$$

Therefore,

$$
\begin{aligned}
\sigma_p &= \gamma_s z \frac{1 + \sin\phi}{1 - \sin\phi} + 2c \frac{\cos\phi}{1 - \sin\phi} \\
&= \gamma_s z \, \tan^2(45° + \phi/2) + 2c \, \tan(45° + \phi/2) \\
&= \gamma_s z N_\phi + 2c\sqrt{N_\phi}
\end{aligned}
\tag{2.14}
$$

For passive failure, since the major principal stress is horizontal, the slip line field is composed of parallel lines sloped to the horizontal at $45° - \phi/2$, as shown in Figure 2.13.

If a pressure q is applied to the soil surface, usually referred to as the surcharge, then the normal stress at the base of an element at depth z is

$$
\sigma = \gamma_s z + q
\tag{2.15}
$$

Accordingly, the active and passive pressures are given by

$$
\sigma_a = \gamma_s z \frac{1}{N_\phi} + q \frac{1}{N_\phi} - 2c \frac{1}{\sqrt{N_\phi}}
\tag{2.16}
$$

$$
\sigma_p = \gamma_s z N_\phi + q N_\phi + 2c\sqrt{N_\phi}
\tag{2.17}
$$

The action of a soil cutting device, such as a bulldozer blade, or a lug (grouser) of a wheel generally causes passive failure of terrain. Consequently, the theory of passive earth pressure has found applications in the prediction of the forces acting on a bulldozer blade and in the estimation of the tractive effort developed by a lug of a wheel, as shown in Figure 2.14.

Consider a vertical bulldozer blade being pushed against the soil. The soil in front of the blade will be brought into a state of passive failure. If the ratio of the width of the blade to the cutting depth is large, the problem may be considered as two-dimensional. Furthermore, if the blade is vertical and its surface is relatively smooth, then the normal pressure exerted by the blade on the soil will be the major principal stress and will be equal to the passive earth pressure σ_p. If there is no surcharge, the resultant force acting on the blade per unit width F_p may be calculated by integrating the passive earth pressure σ_p over the cutting depth h_b. From Eq. (2.14),

$$
\begin{aligned}
F_p &= \int_0^{h_b} \sigma_p \, dz = \int_0^{h_b} \left(\gamma_s z N_\phi + 2c\sqrt{N_\phi} \right) dz \\
&= \frac{1}{2} \gamma_s h_b^2 N_\phi + 2c h_b \sqrt{N_\phi}
\end{aligned}
\tag{2.18}
$$

If there is a surcharge q acting on the soil surface in front of the blade, the resultant force acting on the blade per unit width F_p may be expressed by

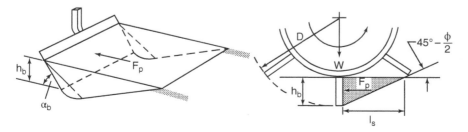

Figure 2.14   Interaction of a bulldozer blade and of a grouser of a wheel with soil.

$$F_p = \int_0^{h_b} \sigma_p \, dz = \int_0^{h_b} \left(\gamma_s z N_\phi + q N_\phi + 2c\sqrt{N_\phi} \right) dz$$

$$= \frac{1}{2} \gamma_s h_b^2 N_\phi + q h_b N_\phi + 2c h_b \sqrt{N_\phi} \tag{2.19}$$

It should be mentioned that for a blade of finite width, end effects would increase the total force acting on the blade.

A similar approach may be followed for estimating the tractive effort developed by the lugs of a cage wheel, such as that used in paddy fields or that attached to a tire as a traction-aid device on wet soils, as shown in Figure 2.14. In general, the lug may behave in one of two ways. If the spacing between the lugs is too small, the space between them may be filled up with soil, and shearing will occur across the lug tips. Under these conditions, the major effect of the lugs would be the increase of the effective diameter of the wheel. On the other hand, if the spacing between the lugs is large so that the soil fails in a manner shown in Figure 2.14, then the behavior of the lug will be like that of the bulldozer blade. When the ratio of the lug width to the penetrating depth is large and the lug surface is relatively smooth, the tractive effort per unit width developed by the lug in the vertical position can be estimated using Eq. (2.18). If the wheel rim and the lugs are of the same width, there will be a surcharge acting on the soil surface behind the lug due to the vertical load applied through the wheel rim. In this case, Eq. (2.19) is applicable. For cage wheels with lugs attached to tires as traction-aid devices, the wheel rim is relatively narrow, and the vertical load is mainly supported by the tire. Under these circumstances, the benefit of the surcharge would not be obtained. It should be pointed out that the shearing forces developed on the vertical surfaces on both sides of the lug would increase the total tractive effort, and that they should be taken into account when the penetration depth of the lug is relatively large.

Example 2.2 A traction-aid device with 20 lugs on a narrow rim is to be attached to the driven wheel of a vehicle to increase its traction. The outside diameter of the device measured from the lug tips is 1.72 m (5.6 ft). The lugs are 25 cm (10 in.) wide and penetrate 15 cm (6 in.) into the ground at the vertical position. Estimate the tractive effort that a lug can develop in the vertical position in a clayey soil with $c = 20$ kPa (2.9 psi), $\phi = 6°$, and $\gamma_s = 15.7$ kN/m^3 (100 lb/ft^3). The surface of the lug is relatively smooth, and the friction and adhesion between the lug surface and the soil may be neglected.

Solution

The spacing between two lugs at the tip is 27 cm (10.6 in.). The rupture distance l_s shown in Figure 2.14 with a penetration $h_b = 15$ cm (6 in.) is

$$l_s = \frac{h_b}{\tan(45° - \phi/2)} = 16.7 \text{ cm (6.6 in.)}$$

It indicates that the spacing between two adjacent lugs is large enough to allow the soil to fail in accordance with the Rankine passive failure. Since the rim of the device is narrow, the effect of surcharge may be ignored. The horizontal force acting on a lug in the vertical position is given by

$$F_p = b \left(\frac{1}{2} \gamma_s h_b^2 N_\phi + 2c h_b \sqrt{N_\phi} \right)$$

where b is the width of the lug.

Substituting the given data of the soil and of the traction-aided device into the above expression, the value of the tractive effort that the lug in the vertical position can develop is

$$F_p = 1.72 \text{ kN } (387 \text{ lb})$$

As the wheel rotates, the inclination as well as the penetration of the lug changes. Thus, the tractive effort developed by the lug varies with its angular position. Since more than one lug may be in contact with the terrain, the total tractive effort that the traction-aid device can develop is the sum of the horizontal forces acting on all lugs in contact with the ground.

There are limitations on the application of the simple earth pressure theory described above to the solution of practical problems. For instance, the surface of bulldozer blades is usually not smooth, as assumed in the simple theory. It has been found that the angle of soil–metal friction δ may vary from $11°$ for a highly polished, chromium-plated steel with dry sand to almost equal to the angle of internal shearing resistance of the soil for very rough steel surfaces (Osman 1964). Because of the existence of friction and/or adhesion between the soil and the blade surface, there will be shear stresses on the blade-soil interface when the soil adjoining the blade is brought into a state of plastic equilibrium. Consequently, the normal pressure on the contact surface will no longer be the principal stress, and the failure pattern of the soil mass will be as that shown in Figure 2.15(a). The soil mass in zone ABC is in the Rankine passive state, which is characterized by straight slip lines inclined to the horizontal at an angle of $45° - \phi/2$. Zone ABD adjacent to the blade is characterized by curved and radial slip lines and is usually called the radial shear zone. The shape of the curved slip lines, such as DB in Figure 2.15(a), can be considered, with sufficient accuracy, as being either a logarithmic spiral (for materials with

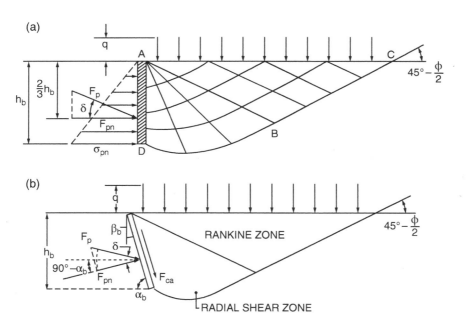

Figure 2.15 Failure patterns of soil in front of (a) a vertical and (b) an inclined cutting blade with rough surface. *Source: Theory of Land Locomotion* by M.G. Bekker, copyright © 1956 by the University of Michigan. Reproduced with permission of the University of Michigan Press.

frictional property) or an arc of a circle (for cohesive materials). In the presence of friction and/or adhesion between the blade and the soil, Eq. (2.14) can no longer be used to predict the passive earth pressure.

Referring to Figure 2.15(a), the normal component σ_{pn} of the passive earth pressure acting on a vertical rough blade at a depth z below A can be approximately expressed by a linear equation:

$$\sigma_{pn} = \gamma_s z K_{p\gamma} + q K_{pq} + c K_{pc} \tag{2.20}$$

where q is the surcharge, and $K_{p\gamma}$, K_{pq}, and K_{pc} are constants and are functions of the angle of internal shearing resistance of the soil and of the friction between the soil and the blade surface, but do not depend on z and γ_s. They may be computed by various methods, including the logarithmic spiral method and the friction circle method (Terzaghi 1966; Hettiaratchi and Reece 1974; Karafiath and Nowatzki 1978; McKyes 1985). The resultant force F_p will be at an angle δ to the normal of the blade, which is equal to the soil–metal friction angle, as shown in Figure 2.15(a).

In practice, bulldozer blades are usually not vertical, and the inclination of the lug of a wheel varies as the wheel rotates. If the blade or the lug is sloped to the horizontal at an angle α_b (or to the vertical at an angle $\beta_b = 90° - \alpha_b$), as shown in Figure 2.15(b), then the force F_{pn} normal to the blade will be

$$F_{pn} = \frac{1}{\sin \alpha_b} \int_0^{h_b} \left(\gamma_s z K_{p\gamma} + q K_{pq} + c K_{pc} \right) dz$$

$$= \frac{1}{2} \gamma_s h_b^2 \frac{K_{p\gamma}}{\sin \alpha_b} + \frac{h_b}{\sin \alpha_b} \left(q K_{pq} + c K_{pc} \right)$$

or

$$F_{pn} = \frac{1}{2} \gamma_s h_b^2 \frac{K_{p\gamma}}{\cos \beta_b} + \frac{h_b}{\cos \beta_b} \left(q K_{pq} + c K_{pc} \right) \tag{2.21}$$

Combining the normal component F_{pn} with the frictional component $F_{pn} \tan \delta$, the resultant force F_p, which acts at an angle δ to the normal on the contact surface, is given by

$$F_p = \frac{F_{pn}}{\cos \delta} = \frac{1}{2} \gamma_s h_b^2 \frac{K_{p\gamma}}{\sin \alpha_b \cos \delta} + \frac{h_b}{\sin \alpha_b \cos \delta} \left(q K_{pq} + c K_{pc} \right)$$

or

$$F_p = \frac{1}{2} \gamma_s h_b^2 \frac{K_{py}}{\cos \beta_b \cos \delta} + \frac{h_b}{\cos \beta_b \cos \delta} \left(q K_{pq} + c K_{pc} \right) \tag{2.22}$$

In addition to the soil–metal friction, there may be adhesion c_a between the soil and the surface of the blade. The adhesion force F_{ca} is expressed by

$$F_{ca} = \frac{h_b}{\sin \alpha_b} c_a = \frac{h_b}{\cos \beta_b} c_a \tag{2.23}$$

In addition to the methods described above, several other methods for predicting the passive earth pressure based on more rigorous applications of the theory of plastic equilibrium have been developed (Hettiaratchi and Reece 1974; Karafiath and Nowatzki 1978; McKyes 1985).

The theory of passive earth pressure also finds application in the prediction of the maximum load of a tracked vehicle that can be supported by soil or terrain without causing failure. The vertical load applied by a rigid track to the soil surface may be idealized as a strip load. When the load

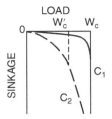

Figure 2.16 Load–sinkage relationships of a footing under different soil conditions.

is light, the soil beneath it may be in a state of elastic equilibrium, as mentioned previously. However, when the load is increased to a certain level, the soil beneath the track will pass into a state of plastic flow, and the sinkage of the track will increase abruptly. This may be illustrated by the load–sinkage curve C_1 shown in Figure 2.16. The initial part of the curve represents the elastic deformation and compression of the soil. The failure of the soil beneath the track may be identified by the transition of the curve into a steep tangent, such as at W_c of curve C_1 in Figure 2.16. The load per unit contact area that causes failure is usually called the bearing capacity of the soil.

At the point of failure, the soil beneath the track can be divided into three different zones, as shown in Figure 2.17(a). When the base of the track is relatively rough, which is usually the case, the existence of friction and adhesion limits the lateral movement of the soil immediately beneath the track. The soil in zone AA_1D is in a state of elastic equilibrium and behaves as if it were rigidly attached to the track. Both boundaries of the wedge-shaped soil body, AD and A_1D, may therefore be identified with the inclined blades discussed previously. However, in this case, the friction angle between the blade and the soil will be equal to the angle of internal shearing resistance of the soil, and the adhesion between the blade and the soil will be the same as the cohesion of the soil. ABD in Figure 2.17(a) is a radial shear zone, whereas ABC is the Rankine passive zone. As the track sinks, the wedge-shaped soil body AA_1D moves vertically downward. This requires that the slip line DB at point D has a vertical tangent. As mentioned previously, potential slip lines in the soil mass intersect each other at an angle

(a)

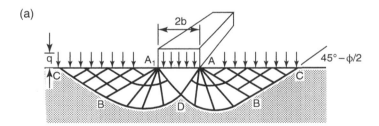

(b)

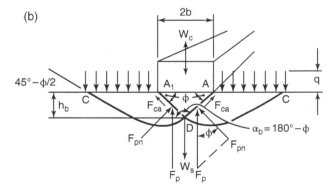

Figure 2.17 (a) Failure patterns under a strip load and (b) corresponding forces acting on the footing.
Source: Theory of Land Locomotion by M.G. Bekker, copyright © 1956 by the University of Michigan. Reproduced with permission of the University of Michigan Press.

of $90° - \phi$. AD and A_1D therefore must be sloped to the horizontal at an angle of ϕ, as shown in Figure 2.17(b). In other words, both boundaries AD and A_1D of the wedge-shaped soil body can be considered as inclined blades with an angle $\alpha_b = 180° - \phi$. The problem of determining the bearing capacity of the soil for supporting strip loads can then be solved using the passive earth pressure theory discussed previously (Terzaghi 1966).

The reaction F_p per unit length of the footing (representing an idealized track) shown in Figure 2.17(b), which acts at an angle ϕ to the normal on AD and A_1D, will be vertical, as the base angle of the wedge-shaped body AA_1D is equal to ϕ. From Eq. (2.22) and with $\alpha_b = 180 - \phi$, $\delta = \phi$, and $h_b = b \tan \phi$, F_p per unit length is expressed by

$$F_p = \frac{1}{2}\gamma_s b^2 K_{p\gamma} \frac{\tan \phi}{\cos^2 \phi} + \frac{b}{\cos^2 \phi}\left(qK_{pq} + cK_{pc}\right) \tag{2.24}$$

The adhesion force F_{ca} per unit length acting along AD and A_1D is

$$F_{ca} = \frac{b}{\cos \phi}c \tag{2.25}$$

The weight per unit length of the soil in zone AA_1D is

$$w_s = \gamma_s b^2 \tan \phi \tag{2.26}$$

The equilibrium of the soil mass in zone AA_1D requires that the sum of the vertical forces be equal to zero.

$$W_c + w_s - 2F_p - 2F_{ca} \sin \phi = 0 \tag{2.27}$$

where W_c is the critical load per unit length, which causes failure of the soil beneath it.

Substituting Eqs. (2.24), (2.25), and (2.26) into Eq. (2.27), the expression for W_c becomes (Terzaghi 1966)

$$\begin{aligned}
W_c &= \gamma_s b^2 K_{p\gamma} \frac{\tan \phi}{\cos^2 \phi} + \frac{2b}{\cos^2 \phi}\left(qK_{pq} + cK_{pc}\right) \\
&\quad + 2bc \tan \phi - \gamma_s b^2 \tan \phi \\
&= \gamma_s b^2 \tan \phi\left(\frac{K_{p\gamma}}{\cos^2 \phi} - 1\right) + 2bq\frac{K_{pq}}{\cos^2 \phi} \\
&\quad + 2bc\left(\frac{K_{pc}}{\cos^2 \phi} + \tan \phi\right)
\end{aligned} \tag{2.28}$$

If it is denoted that

$$\frac{1}{2}\tan \phi\left(\frac{K_{p\gamma}}{\cos^2 \phi} - 1\right) = N_\gamma$$

$$\frac{K_{pq}}{\cos^2 \phi} = N_q$$

and

$$\frac{K_{pc}}{\cos^2 \phi} + \tan \phi = N_c$$

then

$$W_c = 2\gamma_s b^2 N_\gamma + 2bqN_q + 2bcN_c \tag{2.29}$$

The parameters N_γ, N_q, and N_c, which are usually referred to as Terzaghi's bearing capacity factors, can be determined from $K_{p\gamma}$, K_{pq}, and K_{pc}, which are functions of ϕ. The variations of the

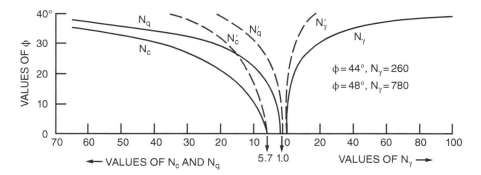

Figure 2.18 Variations of the Terzaghi bearing capacity factors with the angle of internal shearing resistance of soil. *Source:* Terzaghi (1966). Reproduced with permission of Wiley.

bearing capacity factors with ϕ are shown in Figure 2.18 (Terzaghi 1966). Equation (2.29) and the values of N_γ, N_q, and N_c given in Figure 2.18 are applicable only to dense soils of which the deformation preceding failure is small. There is no noticeable sinkage of the track until a state of plastic equilibrium is reached. This kind of failure is called general shear failure (Terzaghi 1966). For loose soils, failure is preceded by considerable deformation, and the relationship between the sinkage and the load is shown by curve C_2 in Figure 2.16. In this case, the critical load that causes failure of the soil is identified, somewhat arbitrarily, by the point where the curve passes into a steep and straight tangent, as point W'_c in Figure 2.16. This type of failure is usually referred to as local shear failure (Terzaghi 1966). Because of the compressibility of the loose soil, the critical load W'_c per unit length for local shear failure is different from that for general shear failure. In the calculation of the critical load for local shear failure, the shear strength parameters c' and ϕ' of the soil are assumed to be smaller than those for general shear failure (Terzaghi 1966):

$$c' = \frac{2}{3}c$$

and

$$\tan \phi' = \frac{2}{3} \tan \phi$$

Accordingly, the critical load W'_c per unit length for local shear failure is given by

$$W'_c = 2\gamma_s b^2 N'_\gamma + 2bqN'_q + \frac{4}{3}bcN'_c \tag{2.30}$$

The values of N'_γ, N'_q, and N'_c are lower than those of N_γ, N_q, and N_c, as can be seen from Figure 2.18.

Based on the theory of bearing capacity, the critical load W_{ct} of a tracked vehicle that may be supported by two tracks without causing failure of the soil can then be estimated by the following equations.

For general shear failure

$$
\begin{aligned}
W_{ct} &= 2lW_c \\
&= 4bl\left(\gamma_s bN_\gamma + qN_q + cN_c\right)
\end{aligned}
\tag{2.31}
$$

and for local shear failure

$$W'_{ct} = 2lW'_c$$
$$= 4bl\left(\gamma_s bN'_\gamma + qN'_q + \frac{2}{3}cN'_c\right) \tag{2.32}$$

where b and l are the width and length of each of the two tracks in contact with the terrain, respectively.

Equations (2.31) and (2.32) may shed light on the selection of track configurations from a bearing capacity point of view. Consider the case of general shear failure and assume that there is no surcharge q. Then in a dry sand (cohesion $c = 0$), the critical load W_{ct} of the vehicle is given by

$$W_{ct} = 4b^2\, l\gamma_s N_\gamma \tag{2.33}$$

This indicates that the load carrying capacity of a track in a frictional soil increases with the square of the track width. To increase the maximum load that the vehicle can carry without causing soil failure, it is, therefore, preferable to increase the track width than to increase the track length. This concept may be illustrated by the following example. Consider two tracked vehicles having the same ground contact area, but the track width of one vehicle is twice that of the other; that is, $b_1 = 2b_2$. Consequently, the contact length of the vehicle with the wider track will be half that of the other; that is, $l_1 = 0.5l_2$. According to Eq. (2.33), the ratio of the critical loads that the two vehicles can carry is given by

$$\frac{W_{ct_1}}{W_{ct_2}} = \frac{4b_1^2 l_1 \gamma_s N_\gamma}{4(0.5b_1)^2 2l_1 \gamma_s N_\gamma} = 2$$

This indicates that the critical load that the vehicle with the wider track can support is higher than that of the other, although both vehicles have the same ground contact area.

In cohesive soils, such as saturated clay ($\phi = 0$), the critical load W_{ct} is given by

$$W_{ct} = 4blcN_c \tag{2.34}$$

This indicates that under these circumstances, the critical load merely depends on the contact area of the track.

It should be emphasized that the use of the bearing capacity theory to predict the critical load that a tracked vehicle can carry without excessive sinkage provides, at best, only approximate results. This is because several simplifying assumptions have been made. For instance, the track is simplified as a rigid footing with uniform pressure distribution. In practice, the interaction between the running gear of an off-road vehicle and the terrain is much more complex than the earth pressure theory or the bearing capacity theory assumes.

Figures 2.19–2.22 show the flow patterns of sand beneath a wide rigid wheel under various operating conditions captured by photographic techniques (Wong and Reece 1966; Wong 1967, 2010). The flow patterns beneath a rigid wheel in the longitudinal plane depend on several factors, including wheel slip or skid, as defined by Eq. (1.5) or Eq. (1.30), respectively. There are normally two zones of soil flow beneath a rolling wheel, whether slipping or skidding. In one zone, the soil flows forward, and in the other, it flows backward. These two zones degenerate into a single backward zone at 100% slip (Figure 2.21), or a single forward zone for a locked wheel at 100% skid (Figure 2.22). It is interesting to note that a wedge-shaped soil body is formed in front of a locked wheel, and that it behaves like a vertical bulldozer blade AC. Figure 2.23

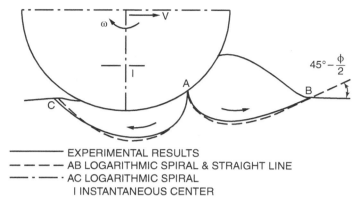

EXPERIMENTAL RESULTS
— — — — AB LOGARITHMIC SPIRAL & STRAIGHT LINE
— · — · AC LOGARITHMIC SPIRAL
I INSTANTANEOUS CENTER

Figure 2.19 Flow patterns and bow wave under the action of a driven roller in sand.

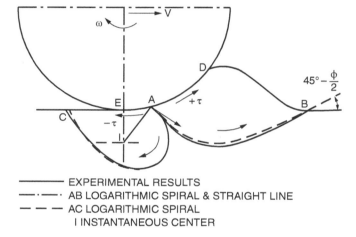

EXPERIMENTAL RESULTS
— · — · AB LOGARITHMIC SPIRAL & STRAIGHT LINE
— — — AC LOGARITHMIC SPIRAL
I INSTANTANEOUS CENTER

Figure 2.20 Flow patterns and bow wave under the action of a towed roller in sand.

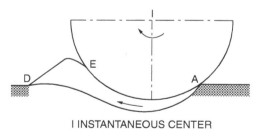

I INSTANTANEOUS CENTER

Figure 2.21 Flow patterns beneath a driven roller at 100% slip in sand.

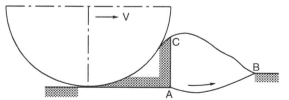

Figure 2.22 Flow patterns and soil wedge formed in front of a locked wheel with 100% skid in sand.

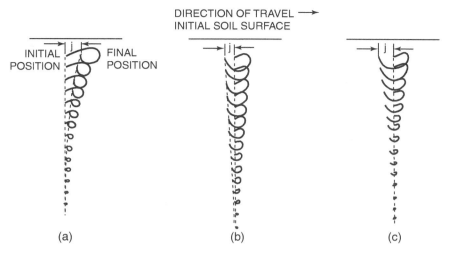

Figure 2.23 Trajectories of clay particles beneath a roller under various operating conditions: (a) towed, (b) driven at 37% slip, and (c) driven at 63% slip.

shows the trajectories of clay particles beneath a wide rigid wheel under various operating conditions, obtained using photographic techniques (Wong and Reece 1966; Wong 1967, 2010). The characteristics of the trajectories indicate that the soil is at first in the Rankine passive state when it is in front of an oncoming wheel. As the wheel is advancing, the soil beneath it is driven backward. Under a free-rolling, towed wheel, the final position of a soil particle is in front of its initial position (Figure 2.23(a)), whereas under a driven wheel with slip, its final position is behind its initial position (Figure 2.23(b) and (c)). The characteristics of the trajectories of soil particles further confirm the existence of two flow zones beneath a rolling wheel. The problem of wheel–soil interaction is complex in that the wheel rim represents a curved boundary, and that the interaction is influenced by a variety of design and operational parameters, including wheel slip or skid.

Attempts have been made to apply the theory of plastic equilibrium to the examination of the complex processes of vehicle–terrain interaction, such as wheel–soil interaction described above (Karafiath and Nowatzki 1978). In the analysis, a set of equations that combine the differential equations of equilibrium for the soil mass with the Mohr–Coulomb failure criterion is first established. The boundary conditions, such as the friction angle or, more generally, the direction of the major principal stress on the wheel–soil interface, as well as the contact angles and the separation angle of the front and rear flow zones shown in Figures 2.19 and 2.20, are then assumed or specified as input. The solution to the set of differential equations with the specified boundary conditions yields the geometry of the slip line field and associated stresses on the wheel–soil interface. As an example, Figure 2.24(a) and (b) show the slip line fields in the soil beneath a driven and a towed rigid wheel, respectively (Karafiath 1971). Based on the predicted normal and shear stresses on the wheel–soil interface, the motion resistance and the tractive effort developed by the wheel can be predicted.

In practice, the boundary conditions on the wheel–soil interface are complex and vary with the design and operational parameters of the wheel, as well as terrain conditions. This makes it very difficult, if not impossible, to assume or specify realistic boundary conditions at the outset. Because of the complexity of the problem, the approach developed so far for specifying the required boundary

(a)

(b)

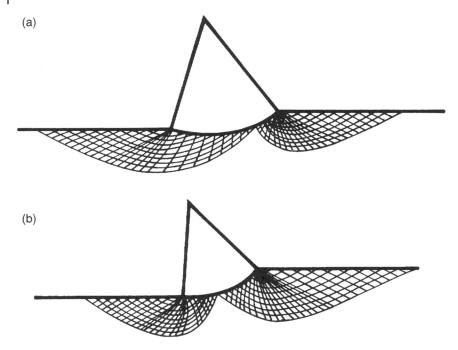

Figure 2.24 Slip line fields in soil beneath (a) a driven rigid wheel and (b) a towed rigid wheel predicted using the theory of plastic equilibrium. *Source:* Karafiath (1971). Reproduced with permission of International Society for Terrain–Vehicle Systems.

conditions is primarily empirical in nature (Karafiath and Nowatzki 1978). This indicates that the elaborate solution procedures based on the theory of plastic equilibrium for predicting the performance of vehicle running gear heavily rely on either empirical inputs or assumed boundary conditions (Wong 1979, 2010). Furthermore, the theory of plastic equilibrium assumes that the terrain behaves like an ideal elastoplastic medium (or a rigid, perfectly plastic material). This means that the terrain does not deform significantly until the stress within a certain boundary reaches the level at which failure occurs. Beyond this point, the strain increases rapidly, while the stress remains essentially constant. Although dense sand and the like may exhibit behavior similar to that of an ideal elastoplastic medium, a wide range of natural terrains encountered in off-road operations, such as farm soil, snow, and organic terrain, have a high degree of compressibility, and their behavior does not conform to that of an idealized material. Failure zones in these natural terrains under vehicular loads, therefore, do not develop in a manner similar to that assumed in the theory of plastic equilibrium, and the sinkage of the vehicle running gear is primarily due to compression and not plastic flow of the terrain material. It should also be mentioned that the theory of plastic equilibrium is mainly concerned with the prediction of the maximum load that causes failure of the soil mass but does not deal with the deformation of the terrain under vehicular load. In practice, the prediction of terrain deformation, expressed in terms of vehicle sinkage under vehicular load, is required.

Because of the issues described above, there are severe limitations to the applications of the theory of plastic equilibrium to the evaluation and prediction of off-road vehicle performance in the field (Wong 1979, 2010).

2.3   Empirically Based Models for Predicting Off-Road Vehicle Mobility

When developments of vehicle mobility models were initiated decades ago, the understanding of the physical nature and the mechanics of vehicle–terrain interaction was such that it was considered more practical to follow an empirical approach than any other types. Representatives of empirically based vehicle mobility models include the North Atlantic Treaty Organization (NATO) Reference Mobility Model (NRMM), first released in 1979 (Haley et al. 1979), and the model based on the mean maximum pressure (*MMP*) developed in the United Kingdom in 1972 (Rowland 1972). The *MMP* represents the mean value of the maximum normal pressures occurring under all wheel stations.

The general approach to the development of empirical models for predicting off-road vehicle mobility is to conduct tests of a select group of vehicles considered to be representative over a range of terrains of interest. The terrain is identified by simple measurements and field observations. The results of vehicle performance testing and the terrain characteristics identified are then empirically correlated. This can lead to the development of empirical relationships for evaluating terrain trafficability on the one hand, and vehicle mobility on the other. Representative empirically based models for predicting cross-country performance of off-road vehicles are outlined below.

2.3.1   NATO Reference Mobility Model (NRMM)

In the United States and some other NATO countries, military ground vehicle mobility has been evaluated using NRMM. Its first edition was released in 1979 (Haley et al. 1979) and its second edition NRMM II was released in 1992 (Ahlvin and Haley 1992). NRMM II has been revised and updated over the years since its release.

NRMM II is designed for predicting the steady-state operating capability of a given vehicle on a prescribed terrain. It consists of three primary modules:

- Performance prediction module, including cross-country performance submodule and on-road performance submodule
- Vehicle dynamics module for predicting vehicle ride quality on the pitch plane
- Obstacle-crossing performance module

In this section, discussions focus on the vehicle cross-country performance submodule. NRMM II uses the cone index (*CI*) or its derivatives measured with a cone penetrometer, together with the identification of the soil type using the Unified Soil Classification System (USCS) to characterize terrain properties. Table 2.1 shows soil classifications using the USCS.

The early design of the cone penetrometer consisted of a 30° circular cone with a 0.5 in.^{2} (3.23 cm^2) base area, a proving ring, and a dial gauge for indicating the force required to push the cone into the terrain (Figure 2.25). The recommended rate of penetration is approximately 1.2 in./s (3 cm/s). The force per unit cone base area is the *CI*. While the *CI* is assigned no dimensions in its applications to identifying terrain conditions for military ground vehicle operations, it is, in fact, the penetration force in pounds divided by the cone base area in square inches. Thus, it has the unit of pressure in psi (Society of Automotive Engineers 1967). With advances in electronics and computer technology, a variety of cone penetrometers using electronic (or electrical) sensors for monitoring the force and penetration depth, as well as computer technology for storing and processing measured data, has been developed (Wong 2010).

Table 2.1 Unified Soil Classification System (USCS).

Soil class	Symbol	Soil description
Coarse-grained soils	GW	Well-graded gravel, fine to coarse gravel
	GP	Poorly graded gravel
	GM	Silty gravel
	GC	Clayey gravel
	SW	Well-graded sand, fine to coarse sand
	SP	Poorly graded sand
	SM	Silty sand
	SC	Clayey sand
Fine-grained soils	ML	Silt
	CL	Clay of low plasticity, lean clay
	OL	Organic silt, organic clay
	MH	Silt of high plasticity, elastic silt
	CH	Clay of high plasticity, fat clay
	OH	Organic clay, organic silt
Highly organic soils	Pt	Peat

Source: ASTM D 2487.

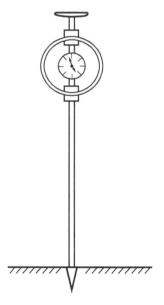

Figure 2.25 The basic component of a cone penetrometer.

The *CI* was originally developed during World War II by the U.S. Army Corps of Engineers Waterways Experiment Station (WES) to provide military intelligence and reconnaissance personnel with a simple means to assess vehicle mobility on a "go/no go" basis in fine- and coarse-grained soils. Fine-grained soils are silt or clayey soils in which 50% or more of grains by weight pass through a No. 200 (0.075 mm) sieve. Coarse-grained soils are those that more than 50% of grains by weight are retained on a No. 200 (0.075 mm) sieve.

In addition to the *CI*, other indices can be obtained using the cone penetrometer. For instance, for fine-grained soils or poorly drained wet sands, a remolding index (*RI*) can be obtained to evaluate changes in terrain strength that may occur under repeated vehicular traffic. The remolding test is conducted on a soil sample in a remolding cylinder. The *CI* of the soil sample before remolding is measured by the cone penetrometer. For fine-grained soils, 100 blows are then applied to the sample with a hammer of 2.5 lb dropping from a height of 12 in. and the *CI* is remeasured (Society of Automotive Engineers 1967). The *RI* is the ratio of the *CI* of the soil measured after remolding to that before remolding. Remolding may cause an increase or decrease in the strength of the terrain, depending on its type and conditions. The rating cone index (*RCI*), which is the product of the *RI* and the *CI* measured before remolding, is used to represent terrain strength under repeated vehicular traffic.

The *CI* or *RCI* is used to quantitatively identify terrain strength in NRMM for predicting cross-country performance of off-road vehicles, dependent on the type of terrain, such as coarse-grained soils or fine-grained soils. It should be noted that the *CI* or the *RCI* is a composite parameter reflecting the combined effects of compressive and shear strengths of the terrain and the friction and adhesion on the cone-terrain interface under a given set of conditions. It is difficult, however, to differentiate the contributions of individual factors to the *CI* or the *RCI*.

The procedure of NRMM II for predicting the cross-country performance of tracked and wheeled vehicles on fine-grained soils is briefly outlined in this section. It is used as the prime example to illustrate the empirical approach of NRMM II. It should be noted that in NRMM II the performance of each tracked or wheeled assembly of a vehicle is determined individually. Performance parameters, such as drawbar pull and motion resistance, of all assemblies of a vehicle may be summed to yield the performance of the entire vehicle.

1) Prediction of cross-country performance of tracked vehicles on fine-grained soils

The procedure begins with using the following empirical equation to calculate the mobility index (*MI*) of a tracked assembly which is considered to include a pair of identical tracks. The *MI* is expressed by

$$\text{Mobility index} = \left(\frac{\begin{array}{c}\text{contact} \\ \text{pressure} \times \begin{array}{c}\text{weight} \\ \text{factor}\end{array} \\ \text{factor}\end{array}}{\begin{array}{c}\text{track} \\ \text{factor}\end{array} \times \begin{array}{c}\text{grouser} \\ \text{factor}\end{array}} + \begin{array}{c}\text{bogie} \\ \text{factor}\end{array} - \begin{array}{c}\text{clearance} \\ \text{factor}\end{array} \right)$$
$$\times \begin{array}{c}\text{engine} \\ \text{factor}\end{array} \times \begin{array}{c}\text{transmission} \\ \text{factor}\end{array} \quad (2.35)$$

where

$$\text{Contact pressure factor} = \frac{\text{weight on tracked assembly (track pair), lb}}{\text{area of tracks in contact with ground, in.}^2}$$

(In calculating the area of tracks in contact with ground, it is assumed that a tracked vehicle has two identical powered tracks.)

Weight factor = 1.0, if the weight of tracked assembly is less than 50,000 lb (222.4 kN)

= 1.2, if the weight is in the range of 50,000 − 69,999 lb (222.4 − 311.4 kN)

= 1.4, if the weight is in the range of 70,000 − 99,999 lb (311.4 − 444.8 kN)

= 1.8, if the weight is greater than 100,000 lb (444.8 kN)

$$\text{Track factor} = \frac{\text{track width in.}}{100}$$

Grouser factor = 1.0, if grouser height < 1.5 in. (3.8 cm)

= 1.1, if grousers height ≥ 1.5 in. (3.8 cm)

$$\text{Wheel load factor} = \frac{\text{weight on tracked assembly, lb, divided by 10}}{\left(\begin{array}{c} \text{total number of road wheels} \\ \text{on tracks in contact with} \\ \text{ground} \end{array}\right) \times \left(\begin{array}{c} \text{area of one track} \\ \text{shoe, in.}^2 \end{array}\right)}$$

$$\text{Clearance factor} = \frac{\text{clearance, in.}}{10}$$

$$\text{Engine factor} = 1.0, \text{if} \frac{\text{hp}}{\text{ton}} \geq 10 \text{ for tracked assembly}$$

$$= 1.05, \text{if hp/ton} < 10 \text{ for tracked assembly}$$

$$\text{Transmission factor} = 1.0 \text{ for a transmission without a manual clutch}$$

$$= 1.05 \text{ for a transmission with a manual clutch}$$

Based on the *MI*, a vehicle mobility metric called the vehicle cone index (*VCI*) is calculated. The *VCI* represents the minimum strength of a soil in the critical layer that permits a given vehicle to successfully make a specific number of passes without immobilization. The critical layer varies with the type and weight of the vehicle and soil strength profile.

The value of one-pass vehicle cone index (*VCI$_1$*) for the tracked assembly on the fine-grained soil is calculated from the *MI* using the following empirical equation:

$$VCI_1 = 7.0 + 0.2MI - \left(\frac{39.2}{MI + 5.6}\right) \tag{2.36}$$

After the value of *VCI* is determined and that of *RCI* of the fine-grained soil is obtained, the drawbar pull coefficient (the ratio of drawbar pull to tracked assembly weight), as a quantitative measure of mobility, is predicted using an empirical equation, based on the excess soil strength $RCI_x = RCI - VCI$. The drawbar pull coefficient is a physically well defined, readily measurable, and widely accepted parameter for evaluating vehicle mobility.

The empirical equation for predicting the drawbar pull coefficient of the tracked assembly on fine-grained soils with $RCI_x \geq 0$ is given by

$$D_{snom}/W = A + \frac{B}{RCI_x + C} + D \tag{2.37}$$

where D_{snom}/W is drawbar pull coefficient at the nominal slip, which is the maximum slip of 100% on fine-grained soils (Ahlvin and Haley 1992); RCI_x is the excess soil strength (i.e., $RCI - VCI$), as noted previously; *A*, *B*, *C*, and *D* are empirical coefficients, dependent upon the type of soil defined by the USCS and the contact pressure factor (P_{FG}) of the tracked assembly, which is one of the factors for determining the *MI*, as shown in Eq. (2.35). The values of empirical coefficients are given in Table 2.2.

In Table 2.2, Soil group 1 – USCS types: SC and GC; 2 – USCS types: CH, MH, and OH; 3 – USCS types: ML, CL, and OL; 4 – USCS types: SM, SC, GM, and GC; 5 – Rock. For information on soil symbols, please refer to Table 2.1.

If the value of *RCI* of the terrain is lower than that of the vehicle cone index for one pass (*VCI$_1$*) or the excess soil strength (RCI_x) is negative, it would indicate that the assembly would be unable to propel itself on its first pass.

Table 2.2 Values of empirical coefficients A, B, C, and D for tracked assemblies.

Assembly type	P_{GF} (psi)	Soil group	Coefficients			
			A	B	C	D
Tracked	≥4	3, 4, 5	0.6512633	−4.906830	7.285463	0.02224646
Tracked	≥4	1, 2	0.6989994	−5.131209	6.992280	0.03483978
Tracked	<4	1, 2, 3, 4, 5	0.7241378	−4.838035	6.301396	0.04359810

Source: Ahlvin and Haley (1992).

2) Prediction of cross-country performance of wheeled vehicles on fine-grained soils

Like the empirical method for predicting tracked assembly performance described above, an empirical equation is used to calculate the *MI* of a wheeled assembly and is given below:

$$\text{Mobility index} = \left(\frac{\text{contact pressure} \times \text{weight factor}}{\text{tire factor} \times \text{grouser factor}} + \text{wheel load factor} - \text{clearance factor} \right) \qquad (2.38)$$
$$\times \text{engine factor} \times \text{transmission factor}$$

where

$$\text{Contact pressure factor} = \frac{\text{weight on wheeled assembly (axle), lb}}{\text{nominal tire width, in.} \times \text{outside radius of tire, in.} \times \text{no. of tires}}$$

Weight factor:	Weight range, lb	Weight factor equation
	<2000 (8.9 kN)	$\bar{Y} = 0.553\bar{X}/1000$
	2000–13 500 (8.9–60 kN)	$\bar{Y} = 0.033\bar{X}/1000 + 1.050$
	13 501–20 000 (60–88.9 kN)	$\bar{Y} = 0.142\bar{X}/1000 - 0.420$
	>20 000 (88.9 kN)	$\bar{Y} = 0.278\bar{X}/1000 - 3.115$

where $\bar{Y}$ = weight factor; $\bar{X}$ = weight on wheeled assembly, lb

$$\text{Tire factor} = \frac{10 + \text{tire width, in.}}{100}$$

$$\text{Grouser factor} = 1.0 \text{ without traction assist chains}$$
$$= 1.05 \text{ with traction assist chains}$$

$$\text{Wheel load factor} = \frac{\text{weight on wheeled assembly (axle) lb/1000}}{2}$$

$$\text{Clearance factor} = \frac{\text{clearance, in.}}{10}$$

$$\text{Engine factor} = 1.0, \text{if hp/ton} \geq 10 \text{ for wheeled assembly}$$

$$= 1.05, \text{if hp/ton} < 10 \text{ for wheeled assembly}$$

$$\text{Transmission factor} = 1.0 \text{ for a transmission without a manual clutch}$$

$$= 1.05 \text{ for a transmission with a manual clutch}$$

Like the empirical method for predicting tracked assembly performance described previously, the *MI* for wheeled assembly is used to determine the *VCI*. For instance, the one-pass vehicle cone index (VCI_1) on fine-grained soils is related to the *MI* by the following empirical equations.

For all unpowered wheeled assemblies and powered wheeled assemblies for which $MI \leq 115$

$$VCI_1 = 11.48 + 0.2MI - \left(\frac{39.2}{MI + 3.74}\right) \tag{2.39}$$

and for powered wheeled assemblies for which $MI > 115$

$$VCI_1 = 4.1MI^{0.446} \tag{2.40}$$

Lately, a correction factor that takes into account the effect of tire deflection has been introduced into the calculation of *VCI* and is expressed by $\sqrt[4]{0.15/(\delta/h)}$, where δ is tire deflection and h is the unloaded section height of the tire. The corrected value of VCI_1 is obtained by multiplying Eqs. (2.39) or (2.40) by the correction factor.

After the *VCI* of the wheeled assembly and the excess soil strength RCI_x for the fine-grained soil to be traversed are determined, the drawbar pull coefficient of the wheeled assembly at the nominal slip, which is the maximum slip of 100% mentioned previously, is predicted using the same empirical equation as that for the tracked assembly (i.e., Eq. (2.37)). However, the values of empirical coefficients *A, B, C*, and *D* for wheeled assemblies are different from those for tracked assemblies and are given in Table 2.3. They take into account whether chains or pads are installed on the tires.

In Table 2.3, Soil group 1 – USCS types: SC and GC; 2 – USCS types: CH, MH, and OH; 3- USCS types: ML, CL, and OL; 4 – USCS types: SM, SC, GM, and GC; 5 – Rock. For information on soil symbols, please refer to Table 2.1.

3) Prediction of cross-country performance of vehicles on muskeg (organic soils)

On muskeg, the one-pass vehicle cone index $VCI_{1(MK)}$ for tracked assemblies is given by

$$VCI_{1(MK)} = 13 + 0.0625 \left(\frac{W}{b + l}\right)$$

Table 2.3 Values of empirical coefficients *A, B, C,* and *D* for wheeled assemblies.

				Coefficients			
Assembly type	Chains or pads	P_{GF} (psi)	Soil group	A	B	C	D
Wheeled	None	≥ 4	4, 5	0.5500142	−4.558500	8.177059	0.03746007
Wheeled	None	≥ 4	1, 2, 3	0.6152356	−6.183363	9.258565	0.05261765
Wheeled	None	<4	1, 2, 3, 4, 5	0.6452267	−5.036329	7.350470	0.03994437
Wheeled	Chains	All	1, 2, 3, 4, 5	0.6452267	−5.036329	7.350470	0.03994437

Source: Ahlvin and Haley (1992).

and for wheeled assemblies is given by

$$VCI_{1(MK)} = 13 + 0.0535\left(\frac{W}{(b+d)n}\right)$$

where W is the weight on the assembly in lb; b is track width or tire section width in in.; l is the track length in contact with ground in in.; d is the tire undeflected diameter in in. at highway inflation pressure; and n is the number of tires on the assembly (axle).

Upon obtaining the vehicle cone index $VCI_{1(MK)}$, different empirical equations are used for predicting the drawbar pull coefficient at the nominal slip, which is 20% on muskeg, of the tracked assembly and wheeled assembly (Ahlvin and Haley 1992). The empirical equation for the tracked or wheeled assembly is based on the difference between the CI for muskeg and the corresponding VCI, known as the excess of soil strength for muskeg (organic soils). Details for predicting the drawbar pull coefficient of the tracked or wheeled assembly on muskeg are contained in NRMM II Users Guide (Ahlvin and Haley 1992).

4) Prediction of cross-country performance of vehicles on coarse-grained soils

While in the first edition of NRMM (NRMM I), VCI is used as a metric for evaluating vehicle mobility on coarse-grained soils, it is no longer the case in NRMM II. The U.S. Army WES coarse-gained soil numeric (sand numeric) method with modifications is used in NRMM II for predicting drawbar pull coefficient, as a quantitative measure of mobility (Ahlvin and Haley 1992).

For instance, for the wheeled assembly, the drawbar pull coefficient at the nominal slip, which is 15% on coarse-grained soils (Ahlvin and Haley 1992), is predicted using an empirical equation based on the sand numeric and takes into account the tire motion resistance coefficient on a hard surface. The sand numeric is a function of

- The CI of the coarse-grained soil
- The partial numeric, which is based on tire nominal section width, nominal section height, nominal diameter, deflection, and normal load, as well as the number of tires on the wheeled assembly (axle)
- The multiple-pass soil strength correction factor that takes into account the effect on terrain properties of repetitive loadings by passages of successive tires

On coarse-grained soils, the prediction of the drawbar pull coefficient bypasses the use of VCI as an intermediary parameter. Details for predicting the drawbar pull coefficient of the wheeled or tracked assembly on coarse-grained soils are contained in NRMM II Users Guide (Ahlvin and Haley 1992).

5) Prediction of cross-country performance of vehicles on shallow snow on frozen ground

On shallow snow on frozen ground, VCI is again not used for evaluating vehicle mobility. Instead, the maximum drawbar pull coefficient of a tracked or wheeled assembly is used as a quantitative measure for evaluating mobility (Ahlvin and Haley 1992). The maximum drawbar pull coefficient is the difference between the maximum thrust (tractive effort) coefficient (i.e. the ratio of the maximum thrust to the weight on the tracked or wheeled assembly) and the total motion resistance coefficient (i.e. the ratio of the total motion resistance to the weight on the tracked or wheeled assembly). The prediction of the maximum thrust coefficient is based on the Mohr–Coulomb equation for shear strength, Eq. (2.10).

For the tracked assembly, the maximum thrust coefficient is calculated from the snow shear strength (cohesion and angle of internal shearing resistance), track ground contact area, and weight on the tracked assembly. The total track motion resistance coefficient is the sum of two components:

- One component is calculated using an empirical equation based on density and depth of snow and track length.
- The other is the track motion resistance coefficient on a hard surface.

For the wheeled assembly, the maximum thrust coefficient is calculated from the snow shear strength, tire ground contact area, number of tires on the axle, and weight on the wheeled assembly. The total motion resistance coefficient is the sum of two components:

- One component is calculated using an empirical equation based on density and depth of snow and tire parameters (nominal section width, nominal diameter, and deflection).
- The other is the tire motion resistance coefficient on a hard surface.

It shows that predictions of the maximum drawbar pull coefficient of the tracked and wheeled assembly on shallow snow on frozen ground also bypass the use of *VCI* as an intermediary parameter, like that on coarse-grained soils. Details for predicting the maximum drawbar pull coefficient of the tracked or wheeled assembly on shallow snow on frozen ground are contained in NRMM II Users Guide (Ahlvin and Haley 1992).

Based on the information presented above, it appears that in NRMM II a collection of disparate empirical methodologies is utilized for predicting vehicle cross-country mobility on different types of terrain.

2.3.2 Empirical Models for Predicting Single Wheel Performance

For a single tire, empirical models for predicting its performance on cohesive and frictional soils, based on dimensionless prediction terms, referred to as soil–tire numerics, were developed by the U.S. Army WES (Freitag 1965; Turnage 1978). The clay–tire numeric N_c is for tires operating in purely cohesive soils (near-saturated clay), while the sand–tire numeric N_s is for tires operating in purely frictional soils (air-dry sand). These two numerics are defined as follows:

$$N_c = \frac{Cbd}{W} \times \left(\frac{\delta}{h}\right)^{1/2} \times \frac{1}{1 + (b/2d)} \tag{2.41}$$

and

$$N_s = \frac{G_s(bd)^{3/2}}{W} \times \frac{\delta}{h} \tag{2.42}$$

where b is the tire section width, C is the CI, d is the tire diameter, G_s is the sand penetration resistance gradient, h is the unloaded tire section height, W is the tire load, and δ is the tire deflection.

For tires operating in soils with both cohesive and frictional properties, a soil–tire numeric N_{cs} is used and it is defined as (Wismer and Luth 1972)

$$N_{cs} = \frac{Cbd}{W} \tag{2.43}$$

Based on test results obtained primarily in laboratory soil bins, these soil–tire numerics have been empirically correlated with two tire performance parameters: the drawbar coefficient μ, and the drawbar efficiency η at 20% slip. The drawbar coefficient (or drawbar pull coefficient) is defined as the ratio of drawbar pull to the normal load on the tire, while the drawbar efficiency is defined as the ratio of the drawbar power (i.e. the product of drawbar pull and forward speed of the tire) to the power input to the tire. Figure 2.26 shows the empirical relations between μ and η at 20% slip and the clay–tire numeric N_c. These relations were obtained on cohesive clays with tires ranging from 4.00-7 to 31 × 15.50 - 13, with loads from 0.23 to 20 kN (52–4500 lb), and with ratios of tire deflection to section height from 0.08 to 0.35. The *CI* values of these clays in the top 15 cm (6 in.) ranged from 55 to 390 kPa (8–56 psi). Figure 2.27 shows the relations between μ and η at 20% slip and the sand–tire numeric N_s. These empirical relations were based on test results obtained on a particular type of sand known as desert Yuma sand, with tires similar to those for Figure 2.26, with loads from 0.19 to 20 kN (42–4500 lb), and with ratios of tire deflection to section height from 0.15 to 0.35. The values of the penetration resistance gradient G for the desert Yuma sand ranged from 0.9 to 5.4 MPa/m (3.3–19.8 psi/in.). Figure 2.28 shows the empirical relations between μ and η at 20% slip and the soil–tire numeric N_{cs}. These relations were obtained on cohesive–frictional soils, with tires ranging from 36 to 84 cm (14–33 in.) in width and from 84 to 165 cm (33–65 in.) in diameter, and loads from 2.2 to 28.9 kN (495–6500 lb). These soils ranged from a tilled

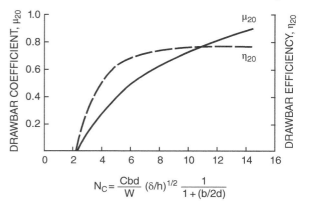

$$N_C = \frac{Cbd}{W}\,(\delta/h)^{1/2}\,\frac{1}{1+(b/2d)}$$

Figure 2.26 Empirical relations between drawbar coefficient and drawbar efficiency at 20% slip and clay–tire numeric N_c. *Source:* Turnage (1978). Reproduced with permission of the International Society for Terrain–Vehicle Systems.

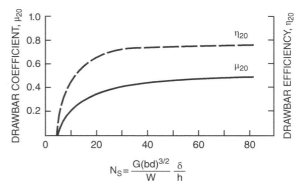

$$N_S = \frac{G(bd)^{3/2}}{W}\,\frac{\delta}{h}$$

Figure 2.27 Empirical relations between drawbar coefficient and drawbar efficiency at 20% slip and the sand–tire numeric N_s. *Source:* Turnage (1978). Reproduced with permission of the International Society for Terrain–Vehicle Systems.

soil with an average before-traffic cone index value of 130 kPa (19 psi) in a layer of 15 cm (6 in.) deep to an untilled soil with an average cone index value of 3450 kPa (500 psi) (Turnage 1978).

The empirical relations for predicting tire performance based on soil–tire numerics, particularly the sand–tire numeric, have undergone several revisions and updates since they were first proposed, as new experimental data emerged (Turnage 1984).

It should be noted that the empirical equations presented above are for predicting the performance of a single tire. If for a wheeled vehicle all its axles have the same tread (i.e. transverse distance between left- and right-side tires on the same axle), the succeeding tires (such as rear tires) will run in the ruts formed by the preceding tires (such as front tires) in straight-line motion. The effect of repetitive loadings on terrain behavior by successive passages of the vehicle running gear is commonly referred to as the multiple-pass effect. To predict the overall performance of the vehicle, the multiple-pass effect must be considered.

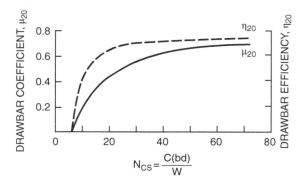

Figure 2.28 Empirical relations between drawbar coefficient and drawbar efficiency at 20% slip and the numeric N_{cs} for cohesive–frictional soils. *Source:* Turnage (1978). Reproduced with permission of the International Society for Terrain–Vehicle Systems.

2.3.3 Empirical Models Based on the Mean Maximum Pressure

Another empirical approach for evaluating off-road vehicle mobility is based on the concept of mean maximum pressure (*MMP*). It is defined as the mean value of the maxima occurring under all the wheel stations (Rowland 1972, 1975).

1) Tracked vehicles

Empirical equations for predicting the values of *MMP* for track systems with different design features are given below (Rowland 1975). For link and belt tracks on rigid roadwheels,

$$MMP = \frac{1.26W}{2n_r A_l b \sqrt{t_t D}} \text{ kPa} \tag{2.44}$$

and for belt tracks on pneumatic tired roadwheels,

$$MMP = \frac{0.5W}{2n_r A_l b \sqrt{D f_t}} \text{ kPa} \tag{2.45}$$

where W is the vehicle weight in kN, n_r is the number of wheel stations in one track, A_l is the rigid area of link (or belt track cleat) as a proportion of $b \times t_t$, b is the track or pneumatic tire width in m, t_t is the track pitch in m, D is the outer diameter of the roadwheel or pneumatic tire in m, and f_t is the radial deflection of the pneumatic tire under load in m.

Table 2.4 shows the values of *MMP* for various types of tracked vehicles (Rowland 1975; Larminie 1992). To evaluate whether a particular vehicle with a specific value of *MMP* will have adequate mobility over a specific terrain, a set of desirable values of the *MMP* for different types of terrain is proposed, as shown in Table 2.5 (Rowland 1975).

In the empirical Eqs. (2.44) and (2.45), terrain characteristics are not considered in the calculation of *MMP*. Thus, the values of *MMP* calculated are independent of terrain conditions. The pressure distribution under a track, and hence the actual value of *MMP*, however, is strongly influenced by terrain characteristics. It is shown that the values of *MMP* calculated using Rowland's empirical formulae are significantly different from those measured in many types of terrain (Wong 1994d, 2010; Wong et al. 2019). In addition, only a handful of vehicle design parameters are included in the empirical equations for determining the values of *MMP*, and a number of design features found to have noticeable influence on mobility are not taken into consideration (Wong 2010). Furthermore, Eqs. (2.44) and (2.45), together with Table 2.5, can be employed only to evaluate the soft ground mobility of tracked vehicles on a "go/no go" basis, and cannot be used to quantitatively predict the performance parameters of a vehicle, such as the motion resistance, thrust, drawbar pull, and tractive efficiency, under a given operating condition.

2) Wheeled vehicles

An empirical equation for predicting the value of *MMP* for the wheeled vehicle is given by (Rowland 1975; Rowland and Peel 1976):

Table 2.4 Values of the mean maximum pressure of a sample of tracked vehicles.

Vehicle	Track configuration	Weight, kN	Mean maximum pressure, kPa
Amphibious Carrier M29C Weasel	Link track	26.5	27
Armored Personnel Carrier M113	Link track	114	121
Articulated All-Terrain Carrier BV 202	Belt track, Pneumatic tire	42	33
Articulated All-Terrain Carrier BV 206	Belt track, Pneumatic tire	62	34
Infantry Fighting Vehicle Bradley M2/M3	Link track	219	145
Main Battle Tank Abrams M1A1	Link track	548	236
Main Battle Tank AMX 30	Link track	370	249
Main Battle Tank Challenger	Link track	608	282
Main Battle Tank Leopard I	Link track	436	235
Main Battle Tank Leopard II	Link track	542	222
Main Battle Tank Merkava	Link track	589	279
Main Battle Tank S-Tank	Link track	363	290
Main Battle Tank T62	Link track	373	256
Tractor Caterpillar D4	Link track	59	82
Tractor Caterpillar D7	Link track	131	80

Source: Rowland (1975), Larminie (1992).

Table 2.5 Desired values of the mean maximum pressure on various types of terrain.

Terrain	Mean maximum pressure (kPa)		
	Ideal (multiple-pass operation or good gradeability)	Satisfactory	Maximum acceptable (mostly trafficable at single-pass level)
Wet, fine-grained			
Temperate	150	200	300
Tropical	90	140	240
Muskeg	30	50	60
Muskeg floating mat and European bogs	5	10	15
Snow	10	25–30	40

Source: Rowland (1975).

$$MMP = \frac{KW}{2mb^{0.85}d^{1.15}(\delta/h)^{0.5}} \text{ kPa} \tag{2.46}$$

where W is vehicle weight in kN; m is the number of axles on the vehicle with a single tire on both the left- and right-hand sides; b is unloaded tire width in m; d is unloaded tire diameter in m; δ is tire deflection on a hard surface in m; h is tire section height in m; and K is a factor dependent on the proportion of axles driven given in Table 2.6. For instance, for a two-axle vehicle with $m = 2$ and proportion of axle driven of 1/2, then it will indicate that one of the

Table 2.6 Factor *K* for various configurations of drive axles.

	Proportion of axle driven				
	1	2/3	1/2	1/3	1/4
Number of axles on the vehicle *m*			*K*		
2	3.65	–	4.4	–	–
3	3.9	4.35	–	5.25	–
4	4.1	–	4.95	–	6.05
6	4.6	5.15	5.55	6.2	–

Source: Rowland (1975).

two axles of the vehicle is driven or the vehicle configuration is 4×2, as commonly known. From Table 2.6, for m = 2 and proportion of axle driven of 1/2, the value of *K* is 4.4.

Table 2.7 shows the values of *MMP* for various types of wheeled vehicle (Larminie 1992).

The *MMP* is employed as a metric to classify the mobility of military logistics wheeled vehicles by the British Ministry of Defence. In the British Defence Standard 23-6, Issue 4, with publication date of 1 November 2005, mobility classifications for military light trucks (payload less than 4 t or 39.24 kN), medium trucks (payload 4 to 8 t or 39.24 to 78.48 kN), and heavy trucks (payload heavier than 8 t or 78.48 kN) are given in Table 2.8. Mobility classes of military logistics vehicle are designated as follows:

- HMLC – High Mobility Load Carrier
- IMMLC – Improved Medium Mobility Load Carrier
- MMLC – Medium Mobility Load Carrier
- ILMLC – Improved Low Mobility Load Carrier
- LMLC – Low Mobility Load Carrier

With further analysis and evaluation of the *MMP* methodology, modifications to Eq. (2.46) for predicting the values of *MMP* of wheeled vehicles have been proposed (Larminie 1992; Hetherington 2001; Hetherington and White 2002; Priddy and Willoughby 2006).

Table 2.7 Values of the mean maximum pressure of a sample of wheeled vehicles.

Vehicle	Weight (kN)	Tire	Mean maximum pressure (kPa)
Armored Personnel Carrier Panhard VAB: VTT (4×4)	128	14×20	513
Armored Personnel Carrier Panhard VAB: VTT (6×6)	139	14×20	342
Light Armored Vehicle Canada Grizzly (6×6)	116	11.0×16	493
Light Armored Vehicle Canada LAV (USMC) (8×8)	117	11.0×16	394
Light Armored Vehicle Panhard M11 VBL (4×4)	35	9.00×16	252
Multipurpose Vehicle M998 Hummer (4×4)	38	$36 \times 12.5R16.5$	228
Truck 4t Bedford MK (4×4)	93	12×20	454
Truck 8t Bedford TM (4×4)	160	15.5/80R20	564
Truck 14t Bedford (6×6)	236	15.5/80R20	591
Truck DROPS Leyland-DAF MMLC (8×6)	295	18R22.5	578

Source: Larminie (1992).

Table 2.8 British Defence Standard 23-6 classifications of mobility of military logistics wheeled vehicles by the mean maximum pressure.

Criterion	Vehicle types	Mobility classes				
		HMLC	IMMLC	MMLC	ILMLC	LMLC
MMP (kPa)	Light truck	Less than 280	280–350	350–550	550–700	Greater than 700
	Medium truck	Less than 350	350–450	450–600	600–700	Greater than 700
	Heavy truck	Less than 350	350–450	450–600	600–700	Greater than 700

Source: British Defence Standard 23-6, Issue 4, publication date 1 November 2005 – Technology Guidance for Military Logistics Vehicles.

2.3.4 Limitations and Prospects for Empirically Based Models

Empirical models, such as NRMM has been used in the U.S. and some other NATO countries for evaluating cross-country performance of military ground vehicles. The model based on *MMP* has been used in the U.K. and other countries for evaluating or classifying military vehicle mobility. It should be noted that while empirical relations would be useful in evaluating the performance of vehicles with design features like those that have been tested under similar operating conditions, they cannot normally be extrapolated beyond the conditions upon which they were derived. Consequently, empirical models have inherent limitations. For instance, it is uncertain that they could play a useful role in evaluating vehicles with design features different from those that were used in tests or in predicting vehicle performance in operating environments beyond those in which tests were conducted. Furthermore, an entirely empirical approach is feasible only where the number of variables involved in the model is relatively small. If the model involves a large number of vehicle or terrain parameters, then an empirical approach may not necessarily be feasible or cost-effective. Furthermore, there are specific issues that would require attention in the applications of the empirically based model NRMM or the model based on the *MMP* presented above.

1) NATO reference mobility model (NRMM)

- Drawbar pull coefficient is a universally accepted metric for characterizing cross-country performance of off-road vehicles. In NRMM II, using empirically based *VCI* to predict vehicle drawbar pull coefficient on fine-grained soils, for instance, three empirical steps must be followed. These include firstly, empirically determining *MI* from vehicle design parameters, as outlined previously; secondly, calculating *VCI* from *MI* based on empirical data; and finally, determining the drawbar pull coefficient based on the difference between *RCI* of fine-grained soils and *VCI* (known as the excess soil strength) using empirical relations. The uncertainty of each of these three consecutive empirical steps would be propagated to the prediction of drawbar pull coefficient, likely leading to a higher order of uncertainty of the results (Wong et al. 2020).
- *VCI* is defined as the minimum strength of a soil in the critical layer that permits a given vehicle to successfully make a specific number of passes, usually 1 pass or 50 passes. In practice, it is difficult to experimentally validate the value of *VCI* for a given vehicle, as the determination of the critical layer is complex and usually requires subjective judgments. In addition, it is a challenging task to locate an area of natural terrain of which the properties and conditions are such that would allow a given vehicle to traverse exactly once or 50 times.
- In view of the limitations of the current version of the empirically based NRMM II, it is to be superseded by the physics-based Next Generation NATO Reference Mobility Model

(NG-NRMM). A NATO standardization recommendation (STANREC 4813) that provides guidance applicable to the development of NG-NRMM has been promulgated in July 2021. A NG-NRMM is defined to be any mobility modeling and simulation (M&S) capability that produces map-based probabilistic mobility predictions of ground and amphibious vehicles through interoperation of M&S tools that include: geographic information systems (GIS) software; 3-dimensional physics-based vehicle dynamics; terramechanics models for off-road operations; autonomous control M&S software, as well as uncertainty quantification (UQ) software for probabilistic M&S (McCullough et al. 2017a; NATO Standard AMSP-06 2021).

- In contrast with NRMM II using empirically based mobility metrics (such as *VCI* for operations on fine-grained soils and muskeg), physics-based mobility metrics that are physically well defined and readily measurable have been proposed for next generation vehicle mobility models. They include drawbar pull coefficient at 20% slip as a mobility metric for characterizing cross-country performance in the range of optimal tractive efficiency; drawbar pull coefficient at 80% slip as a mobility metric for characterizing traction limits; and the maximum possible speed in straight-line motion under steady-state operating conditions from one location to another (also known as the speed-made-good) as a mobility metric for characterizing vehicle traversability (Wong et al. 2020).

2) Models based on the mean maximum pressure (MMP)

- Terrain characteristics are not considered in determining the values of *MMP* for tracked or wheeled vehicles. Thus, they are independent of terrain conditions. Experimental evidence has shown that the normal pressure distribution under a track or a wheel, and hence the actual value of *MMP*, is strongly influenced by terrain characteristics (Wong 1994d; Wong et al. 2019). The measured values of *MMP* for a military tracked vehicle at various slips on a sandy terrain, a muskeg, and two shallow snow-covered terrains, together with the calculated values using Rowland's empirical formula Eq. (2.44), and the predicted values using the vehicle mobility model NTVPM, which will be described later in Section 2.6, are shown in Table 2.9 (Wong et al. 2019). The value of *MMP* calculated by Rowland's empirical formula for a given vehicle is a constant for all types of terrain and is significantly different from those measured on the sand terrain and the two shallow snow-covered terrains. This is because in Rowland's empirical formula, terrain characteristics are not taken into consideration. It is interesting to note that the values of *MMP* predicted by NTVPM are in reasonably close agreement with those measured. Table 2.10 shows a comparison of the average of all measured values of *MMP* at various slips on the four types of terrain, the calculated value using Rowland's empirical formula Eq. (2.44), and the average of all predicted values at various slips using NTVPM. It is shown that the ratios of average measured values of *MMP* to the calculated value of *MMP* by Rowland's empirical formula on the sandy terrain, muskeg, and the two shallow snow-covered terrains are 3.91, 0.94, 2.60, and 2.77, respectively. This indicates that the calculated value of *MMP* using Rowland's empirical formula is significantly different from the average measured values on the sandy terrain and the two shallow snow-covered terrains, and that only on muskeg is the calculated value using Rowland's empirical formula close to the measured average value. The ratios of average measured values of *MMP* to the average predicted values of *MMP* using NTVPM on the sandy terrain, muskeg, and the two shallow snow-covered terrains are 1.26, 1.20, 1.05, and 0.97, respectively. This indicates that the average values of *MMP* predicted by NTVPM are in closer agreement with the average measured values of *MMP* than the calculated value using Rowland's empirical formula on the sandy terrain and the two shallow snow-covered terrains.

- *MMP* can only be used for predicting vehicle mobility on a "go/no go" basis and is not capable of predicting vehicle performance metrics, such as drawbar pull coefficient and motion resistance coefficient.

Table 2.9 Measured *MMP*, predicted *MMP* by NTVPM, and calculated *MMP* by Rowland's empirical formula for a military tracked vehicle at different slips on various types of terrain.

Terrain type	Slip, %	Measured *MMP*, kPa	Predicted *MMP* by NTVPM, kPa	Calculated *MMP* by Rowland's formula, (Eq. (2.44)), kPa
LETE sand	1.8	317.6	312.2	100.3
	2.8	448.8	312.1	100.3
	5.9	522	311.1	100.3
	6.1	377.8	311.0	100.3
	7.1	432.6	310.6	100.3
	8.2	406	310.1	100.3
	10.8	311	309.1	100.3
	15.5	317.4	307.6	100.3
	Average	**391.7**	**310.5**	**100.3**
Petawawa muskeg A	0.5	100.6	83.5	100.3
	2.7	91.1	80.3	100.3
	3.0	101.3	79.9	100.3
	6.6	111.0	76.0	100.3
	8.1	90.3	74.7	100.3
	8.5	70.4	74.5	100.3
	Average	**94.1**	**78.2**	**100.3**
Petawawa snow A	0.0	225.6	250.6	100.3
	2.2	257.5	250.2	100.3
	2.4	246.5	250.1	100.3
	3.3	237.4	249.8	100.3
	4.8	287.5	249.3	100.3
	5.6	296.0	249.1	100.3
	8.3	246.4	245.7	100.3
	8.6	289.1	245.6	100.3
	Average	**260.8**	**248.8**	**100.3**
Petawawa snow B	4.5	285.2	287.1	100.3
	5.8	280.4	286.9	100.3
	5.9	280.5	286.9	100.3
	6.5	309.7	286.8	100.3
	6.6	261.5	286.8	100.3
	6.9	307.7	286.7	100.3
	7.2	242.7	286.6	100.3
	7.9	269.2	286.5	100.3
	9.6	300.7	286.1	100.3
	10.2	244.1	286.0	100.3
	Average	**278.2**	**286.6**	**100.3**

Source: Wong et al. (2019).

Table 2.10 Comparison of the average measured *MMP*, the calculated *MMP* by Rowland's empirical formula, and the average predicted *MMP* by NTVPM on various types of terrain.

Terrain type	Average measured *MMP*, kPa	Average predicted *MMP* by NTVPM, kPa	Calculated *MMP* by Rowland's formula (Eq. 2.44), kPa	Average measured *MMP*/average predicted *MMP* by NTVPM	Average measured *MMP*/calculated *MMP* by Rowland's formula (Eq.2.44)
LETE sand	391.7	310.5	100.3	1.26	3.91
Petawawa muskeg A	94.1	78.2	100.3	1.20	0.94
Petawawa snow A	260.8	248.8	100.3	1.05	2.60
Petawawa snow B	278.2	286.6	100.3	0.97	2.77

Source: Wong et al. (2019).

- Using *MPP* for classifying vehicle mobility is not necessarily conducive for vehicle designers to explore innovative ways of achieving high mobility in the field. Instead, it may encourage designers just to manipulate a handful of vehicle parameters in the empirical equations to meet the value of *MMP* specified (Hetherington 2001; Hetherington and White 2002; Wong et al. 2020).

2.4 Measurement and Characterization of Terrain Response

Owing to the limitations of theoretical models based on the theories of elasticity and plastic equilibrium, as well as empirically based models, such as the NATO Reference Mobility Model (NRMM) and the model based on the *MMP* described above, vehicle mobility models based on the study of the physical nature and the mechanics of vehicle running gear–terrain interaction have been developed for parametric analysis of off-road vehicle cross-country performance. These models are hereinafter referred to as the physics-based models. Several of these models are based on the measurement of terrain responses under loading conditions like those exerted by off-road vehicles (Bekker 1969; Wong 2010). The measurement and characterization of terrain responses to loadings similar to those applied by off-road vehicles are discussed in this section, while the mechanics of vehicle–terrain interaction and some of the physics-based models for off-road vehicle cross-country performance are described in subsequent sections.

A ground vehicle, through its running gear, applies normal load to the terrain surface, which results in sinkage giving rise to motion resistance, as shown in Figure 2.29. Also, the torque applied to the sprocket of a track or to the tire initiates shearing action between the running gear and the terrain surface, which results in the development of thrust and associated slip. The measurement of both the normal pressure–sinkage and shear stress–shear displacement relationships, therefore, is of importance to predicting the cross-country performance of off-road vehicles. Furthermore, when an off-road wheeled vehicle with all axles having the same tread is in straight-line motion, an element of the terrain under the wheel path is subject to the repetitive loading of successive wheels.

Similarly, for a tracked vehicle, an element of the terrain under the track path is subject to the repetitive loading of its roadwheels. To realistically predict the performance of an off-road vehicle, responses of the terrain to repetitive normal and shear loadings should be measured (Bekker 1969; Wong et al. 1982, 1984; Wong 2010). For vehicles with rubber tires or rubber tracks, part of the vehicle thrust is developed through rubber–terrain interaction. As a result, the rubber–terrain shearing characteristics should be measured. Over exceedingly soft terrain, the sinkage of the running gear may be greater than the vehicle ground clearance and the vehicle belly (hull) may be in contact with the terrain surface. The vehicle belly–terrain contact induces an additional drag, which is commonly known as the belly drag. To predict the belly drag, the vehicle belly–terrain shearing characteristics should be measured as well.

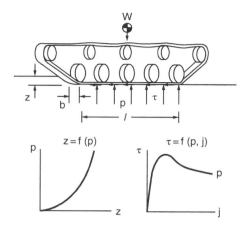

Figure 2.29 A simplified model for predicting tracked vehicle performance.

One of the well-known techniques for measuring the responses of terrain to loadings pertinent to vehicle mobility is that proposed by Bekker (Bekker 1960, 1969; Wong and Bekker 1985). This technique has become known as the bevameter technique. The original bevameter technique includes two basic sets of tests: a set of plate penetration tests and a set of shear tests. In the penetration test, a circular plate of suitable size is usually used to simulate the contact area of a tire or an equivalent track link, and the pressure–sinkage relationship of the terrain is measured. It is used to predict the normal pressure distribution on the vehicle–terrain interface, as well as vehicle sinkage and motion resistance. To minimize the uncertainty in applying the measured data to the prediction of the performance of full-size vehicles, it is preferable that the radius of the circular plate used in the test be of the same order of magnitude as the equivalent radius of the contact area of a tire or a track link (Wong et al. 1984; Wong 2010). In the shear test, a shear annulus (ring) shown in Figure 2.12 is usually used, as it is free from the "bulldozing effect" at the front of a rectangular shear plate. The shear stress–shear displacement relationships of the terrain under various normal pressures are measured. They provide the necessary input for predicting the shear stress distribution on the vehicle–terrain interface and the thrust–slip relationship of the vehicle. The bevameter technique has later been extended to include the measurements of rubber–terrain and belly-terrain shearing characteristics. The measurement of the responses of terrain to repetitive normal and shear loadings has also become part of the extended bevameter technique.

The basic features of a bevameter designed to carry out the tests described above are illustrated in Figure 2.30. A hydraulic ram is usually used to apply normal load to the sinkage plate in the pressure–sinkage test. The applied pressure and the resulting sinkage of the plate are recorded, as shown in Figure 2.30. In the shear test, under a range of normal pressures, the torque applied to the shear annulus, with suitable grouser arrangements to ensure internal shearing of terrain material, and its angular displacement are recorded, as shown in Figure 2.30, from which the shear stress–shear displacement relationship and the shear strength parameters of the terrain can be derived. The rubber–terrain or vehicle belly–terrain shearing characteristics can be measured using a shear annulus with its surface covered with the same material as that of the rubber tire, rubber track, or vehicle belly. Figure 2.31 shows a bevameter mounted in front of a vehicle for use in the field.

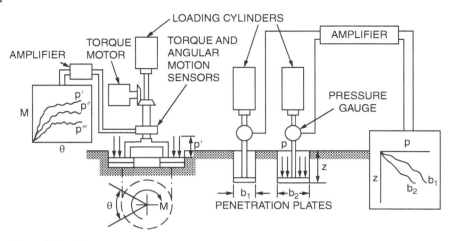

Figure 2.30 Schematic view of a bevameter for measuring terrain properties. *Source: Introduction to Terrain–Vehicle Systems* by M.G. Bekker, copyright © 1969 by the University of Michigan. Reproduced with permission of the University of Michigan Press.

Figure 2.31 A vehicle-mounted bevameter for used in the field.

2.4.1 Characterization of Pressure–Sinkage Relationships

After the pressure–sinkage data have been collected, they should be properly processed and characterized so that they may be integrated into an analytical framework for predicting the performance of off-road vehicles. Depending on the type and conditions of the terrain, different mathematical functions are used to characterize the pressure–sinkage relationship.

Figure 2.32 Pressure–sinkage relationships for various homogeneous soils. *Source: Introduction to Terrain–Vehicle Systems* by M.G. Bekker, copyright © 1969 by the University of Michigan. Reproduced with permission of the University of Michigan Press.

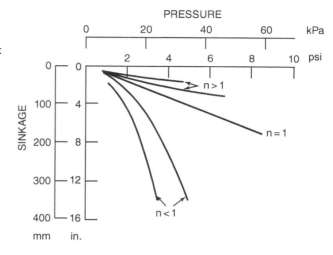

1) Homogeneous terrain

For a homogeneous terrain, if its pressure–sinkage relationship takes one of the forms shown in Figure 2.32, it may be characterized by the following equation proposed by Bekker (1969):

$$p = \left(\frac{k_c}{b} + k_\phi\right) z^n \tag{2.47}$$

where p is pressure; b is the radius of a circular plate or the width of a rectangular plate (with an aspect ratio of at least 4.5) used in the test; z is sinkage; and n, k_c, and k_ϕ are pressure–sinkage parameters.

Equation (2.47) is commonly referred to as the Bekker pressure–sinkage equation. It is inspired by an equation for characterizing the relationship of ground deformation and pressure applied by a structure or foundation in civil engineering soil mechanics (Bekker 1960). In the early stage of development of terramechanics, k_c is assumed to be related to the cohesive property of the soil, while k_ϕ is related to its frictional property. Later, it is recognized that parameters k_c and k_ϕ do not necessarily have any specific physical meaning. Their values are derived through fitting Eq. (2.47) to measured data.

The values of k_c, k_ϕ, and n can be derived from the results of a minimum of two tests with two sizes of plates having different radii (or widths of rectangular plates). The tests produce two curves:

$$p_1 = \left(\frac{k_c}{b_1} + k_\phi\right) z^n$$
$$p_2 = \left(\frac{k_c}{b_2} + k_\phi\right) z^n \tag{2.48}$$

On the logarithmic scale, the above equations can be rewritten as follows:

$$\log p_1 = \log\left(\frac{k_c}{b_1} + k_\phi\right) + n \log z$$
$$\log p_2 = \log\left(\frac{k_c}{b_2} + k_\phi\right) + n \log z \tag{2.49}$$

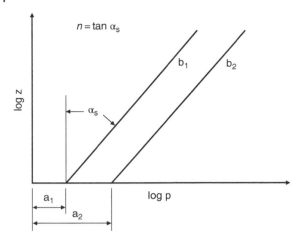

Figure 2.33 A graphical method for determining sinkage moduli and exponent.

They represent two parallel straight lines of the same slope on the log–log scale as shown in Figure 2.33. It is evident that $\tan\alpha_s = n$. Thus, the exponent of deformation n can be determined from the slope of the straight lines. At sinkage $z = 1$, the values of the normal pressure for the two sizes of plates are

$$(p_1)_{z=1} = \frac{k_c}{b_1} + k_\phi = a_1$$
$$(p_2)_{z=1} = \frac{k_c}{b_2} + k_\phi = a_2 \tag{2.50}$$

In the above equations, $(p_1)_{z=1}$ and $(p_2)_{z=1}$ are measured values, and the unknowns are k_c and k_ϕ. Thus, k_c and k_ϕ can be determined by the following equations:

$$k_\phi = \frac{a_2 b_2 - a_1 b_1}{b_2 - b_1}$$
$$k_c = \frac{(a_1 - a_2)b_1 b_2}{b_2 - b_1} \tag{2.51}$$

Owing to the non-homogeneity of the terrain in the field and possible experimental errors, the pressure–sinkage lines may not be quite parallel on the log–log scale in some cases. Thus, two values of the exponent of deformation may be produced. Under these circumstances, the value of n is usually taken as the mean of the two values obtained. The above-noted procedure relies on the skill of the investigator to plot the appropriate straight lines to represent the pressure–sinkage data on the log–log scale. Consequently, it is liable to error for inexperienced personnel. Furthermore, the values of k_c and k_ϕ are determined using only the pressures measured at sinkage $z = 1$.

To provide a more rational approach to deriving the values of n, k_c, and k_ϕ from measured pressure–sinkage data, a computerized procedure based on the weighted least-squares method has been developed and successfully used to process test data (Wong 1980, 2010). This method has been widely accepted for processing pressure–sinkage test data that would fit Eq. (2.47) and has been proven to be very reliable in comparison with other data processing methods (Jayakumar et al. 2014).

The values of k_c, k_ϕ, and n for a sample of terrains are given in Table 2.11 (Bekker 1969; Harrison 1975; Wong 1983; Wong and Huang 2006a; Wong et al. 2019).

Table 2.11 A sample of terrain values.

Terrain	Moisture content (%)	n	k_c lb/in.$^{n+1}$	k_c kN/m^{n+1}	k_ϕ lb/in.$^{n+2}$	k_ϕ kN/m^{n+2}	c lb/in.^{2}	c kPa	φ deg
Dry sand (Land Locomotion Lab., LLL)	0	1.1	0.1	0.99	3.9	1528.43	0.15	1.04	28°
Sandy loam (LLL)	15	0.7	2.3	5.27	16.8	1515.04	0.25	1.72	29°
Sandy loam	22	0.2	7	2.56	3	43.12	0.2	1.38	38°
Michigan (Strong, Buchele)	11	0.9	11	52.53	6	1127.97	0.7	4.83	20°
Sandy loam (Hanamoto)	23	0.4	15	11.42	27	808.96	1.4	9.65	35°
	26	0.3	5.3	2.79	6.8	141.11	2.0	13.79	22°
	32	0.5	0.7	0.77	1.2	51.91	0.75	5.17	11°
Clayey soil (Thailand)	38	0.5	12	13.19	16	692.15	0.6	4.14	13°
	55	0.7	7	16.03	14	1262.53	0.3	2.07	10°
Heavy clay (WES)	25	0.13	45	12.70	140	1555.95	10	68.95	34°
	40	0.11	7	1.84	10	103.27	3	20.69	6°
Lean clay (WES)	22	0.2	45	16.43	120	1724.69	10	68.95	20°
	32	0.15	5	1.52	10	119.61	2	13.79	11°
LETE sand (Wong)		0.79	32	102	42.2	5301	0.20	1.4	31.6°
Upland sandy loam (Wong)	44	0.74	10.1	26.8	14.6	1522	0.48	3.3	33.7°
Rubicon sandy loam (Wong)	43	0.66	3.5	6.9	9.7	752	0.54	3.7	29.8°
North Gower clayey loam (Wong)	46	0.73	16.3	41.6	24.5	2471	0.88	6.1	26.6°
Grenville loam (Wong)	24	1.01	0.008	0.06	20.9	5880	0.45	3.1	29.8°
Snow (Harrison)		1.6	0.07	4.37	0.08	196.72	0.15	1.03	19.7°
		1.6	0.04	2.49	0.10	245.90	0.09	0.62	23.2°
Snow (Sweden)		1.44	0.3	10.55	0.05	66.08	0.87	6	20.7°

Source: Bekker (1969), Harrison (1975), Wong (1983), Wong and Huang (2006a), Wong et al. (2019).

It should be noted that the Bekker pressure–sinkage equation, Eq. (2.47), indicates that the pressure p is dependent on k_c/b. In other words, the pressure p on a contact area required to achieve a specific sinkage z is dependent on the size of the contact area represented by b. This shows that for a larger contact area having higher value of b, the pressure p required to penetrate to the same depth z is lower than that with a smaller contact area having lower value of b. For some test data, particularly those obtained in the field, the value of k_c or k_ϕ derived from measured data through curve fitting may turn out to be negative. A negative value of k_c, for instance, implies that with a larger contact area having higher value of b, the pressure p required to penetrate to the same depth z is higher than that with a smaller contact area having lower value of b. This shows that the measured data demonstrate an opposite trend of pressure–sinkage relationship to that indicated by the Bekker equation. In other words, Eq. (2.47) does not really fit the trend of the measured pressure–sinkage relationship. For such a case, other pressure–sinkage relationships may be considered as an alternative, such as $p = k_{eq} z^n$, proposed by Russian researchers, where k_{eq} is a terrain parameter independent of the size of contact area (Bekker 1956).

It should also be pointed out that in Eq. (2.47) the parameters k_c and k_ϕ have variable dimensions, depending on the value of the exponent n. Influenced by the work of civil engineering soil mechanics and by experimental evidence, Reece proposed a new equation for the pressure–sinkage relationship (Reece 1965–1966):

$$p = \left(c k'_c + \gamma_s b k'_\phi \right) \left(\frac{z}{b} \right)^n \tag{2.52}$$

where n, k'_c, and k'_ϕ are the Reece pressure–sinkage parameters and non-dimensional; γ_s is the weight density of the terrain; and c is cohesion of the terrain. Other parameters, b, p, and z in Eq. (2.52) are the same as those in Eq. (2.47). A series of penetration tests were carried out to validate the principal features of the above equation. Plates with various widths and with an aspect ratio of at least 4.5 were used. The test results are shown in Figure 2.34. For a frictionless clay, the term k'_ϕ should be negligible. The results shown in Figure 2.34 indicate that this is the case, with the curves for clay collapsing to almost a single line for all plates regardless of their width when plotted against z/b. For dry, cohesionless sand, the term k'_c should be negligible. The equation thus suggests that pressure increases linearly with the width of the plate for a given value of z/b. Wetting the sand would not alter its value of ϕ, but would add a cohesive component. This would add a pressure term independent of width represented by the first term on the right-hand side of Eq. (2.52). This again seems to be borne out by test results.

Reece pointed out that while Eq. (2.52) differs from Eq. (2.47) only in the effect of the width b, this is sufficient to mark a radical improvement. The soil values k'_c and k'_ϕ in Eq. (2.52) are dimensionless, whereas k_c and k_ϕ in Eq. (2.47) have dimensions dependent on n. Furthermore, Eq. (2.52) seems to allow itself to fit in with the classical bearing capacity theory of soil mechanics. For instance, if the critical load W_c per unit length of the footing in Terzaghi's bearing capacity equation, Eq. (2.29), without surcharge is expressed in terms of the critical pressure W_c/b given below

$$W_c/b = 2\gamma_s b N_\gamma + 2c N_c \tag{2.53}$$

then both Eq. (2.53) and the Reece equation, Eq. (2.52), will have similar features. In dry, cohesionless sand, both equations show that increasing b causes a linear increase in the critical pressure W_c/b. On the other hand, in frictionless clay, both equations show that the increase of b has no effect on the critical pressure.

It should be noted, however, that Eq. (2.52) only applies to homogeneous terrain with respect to depth. For non-homogeneous terrains in depth or for layered terrains, the pressure–sinkage

Figure 2.34 Pressure–sinkage curves obtained using various sizes of rectangular plates in different soils. *Source:* Reece (1965–1966). Reproduced with permission of the Council of the Institution of Mechanical Engineers.

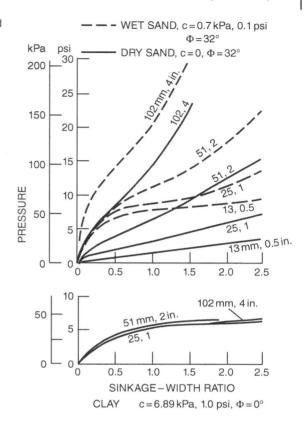

relationship obtained using a smaller-size plate may not be extrapolated to that for a larger-size contact area. This is because for the same ratio of z/b, a plate with a larger value of b must penetrate to a deeper layer of the terrain, which may have different engineering properties from those of a shallower layer.

2) Organic terrain (muskeg)

For a commonly encountered organic terrain (muskeg) in North America, there is a mat of living vegetation on the surface with a layer of saturated peat beneath it. A representative pressure–sinkage curve for the organic terrain obtained in the field is shown in Figure 2.35 (Wong et al. 1979, 1982; Wong 2010). Initially, the pressure increases with an increase in sinkage, but when the applied pressure (or load) reaches a certain level, the surface mat is broken. Since the saturated peat beneath the mat is often weaker than the mat and offers lower resistance, the pressure decreases with an increase of sinkage after the surface mat is broken, as shown in Figure 2.35. Based on experimental observations, a mathematical model for the failure of the surface mat has been developed (Wong et al. 1979; Wong 2010). In the model, it is assumed that the organic terrain consists of two layers: the surface mat and the peat beneath. The surface mat is idealized as a membrane-like structure, which means that it can only sustain a force of tension directed along the tangent to the surface and cannot offer any resistance to bending. The underlying peat is assumed to be a medium that offers a resistance proportional to its deformation in the vertical direction.

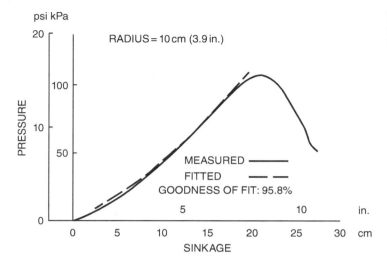

Based on this model, the pressure–sinkage relationship for an organic terrain up to the critical sinkage z_{cr}, where the breaking of the surface mat is initiated, is given by

$$p = k_p z + 4m_m z^2 / D_h \tag{2.54}$$

where p is the pressure, z is the sinkage, k_p is a stiffness parameter for the peat, m_m is a strength parameter for the surface mat, and D_h is the hydraulic diameter of the contact area (or sinkage plate), which is equal to $4A/L$, where A and L are the area and the perimeter of the contact patch, respectively.

A computerized procedure based on the least-squares method has been developed to derive the values of k_p and m_m from measured data (Wong et al. 1979, 1982; Wong 2010). Table 2.12 shows the values of k_p, m_m, and z_{cr} for two types of organic terrain found in the Petawawa area in Ontario, Canada. The value of k_p varies with the penetration rate. For Petawawa Muskeg A and B shown in Table 2.12, by increasing the penetration rate from 2.5 to 10 cm/s (1 to 4 in./s), the value of k_p increases from 407 to 471 kN/m^3 (1.5–1.7 lb/in.^{3}) and from 954 to 1243 kN/m^3 (3.5–4.6 lb/in.^{3}), respectively. The increase in the apparent stiffness is probably due to the movement of water within the saturated peat, which creates an additional hydrodynamic resistance related to the penetration rate.

Table 2.12 Values of k_p, m_m, and z_{cr} for a sample of organic terrains (muskeg).

Terrain type	Penetration rate		k_p		m_m		z_{cr}	
	cm/s	in./s	kN/m^3	lb/in.^{3}	kN/m^3	lb/in.^{3}	cm	in.
Petawawa	2.5	1	407	1.5	97	0.36	20	7.9
Muskeg A	10	4	471	1.7	112	0.41	17	6.7
Petawawa	2.5	1	954	3.5	99	0.36	21	8.3
Muskeg B	10	4	1243	4.6	99	0.36	22	8.7

Note: Data obtained using a circular plate of 10 cm (4 in.) in diameter.
Source: Wong et al. (1982), Wong (2010).

3) Snow-covered terrain with ice layers

In the northern temperate zone, the snow on the ground is often subject to the "melt–freeze" cycle during the winter season. Consequently, crusts (ice layers) of significant strength form at the surface of snow covers in open areas. With subsequent snowfall on top of the crusts, snow covers containing ice layers are formed. Figure 2.36 shows the pressure–sinkage data obtained in a snow cover in the Petawawa area, Ontario, Canada. It contains a significant ice layer at a depth of approximately 10 cm (4 in.) from the surface and with a frozen ground at the base (Wong and Preston-Thomas 1983b; Wong and Irwin 1992; Wong 2010).

It can be seen from Figure 2.36 that the pressure first increases gradually with sinkage as the snow within a certain boundary under the plate is deformed. When the penetration of the plate is close to the ice layer, the pressure increases rapidly with an increase of sinkage. When the applied pressure exceeds a certain level, the ice layer is broken, resulting in a sudden drop in pressure. After the ice layer is fractured, further penetration of the plate produces increasing deformation

Figure 2.36 Pressure–sinkage relationships for a snow-covered ground measured using two different sizes of circular plates.

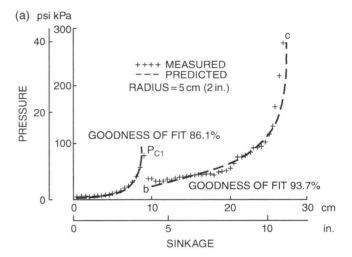

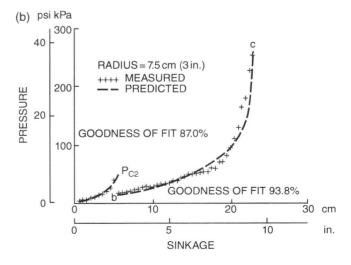

of the snow beneath the ice layer. As the plate approaches the frozen ground at the base of the snow cover, the pressure again increases rapidly, and the pressure–sinkage curve approaches an asymptote.

Based on the results shown in Figure 2.36, the pressure–sinkage relationship, before as well as after the failure of the ice layer, may be described by an exponential function of the following form (Wong and Preston-Thomas 1983b; Wong and Irwin 1992; Wong 2010):

$$z = z_\omega [1 - \exp(-p/p_\omega)]$$

or

$$p = p_\omega [-\ln(1 - z/z_\omega)] \tag{2.55}$$

where p is the pressure, z is the sinkage, z_ω defines the asymptote of the pressure–sinkage curve, and as a first approximation may be taken as the depth of the ice layer or that of the frozen ground, and p_ω is an empirical parameter which may be taken as 1/3 of the pressure where the sinkage z is 95% of the value of z_ω. For instance, the value of z_ω for defining the pressure–sinkage relationship in the range between 0 and 10 cm (0–4 in.) shown in Figure 2.36(a) is approximately 9 cm (3.5 in.). The pressure p at 95% of z_ω (8.6 cm or 3.4 in.) is approximately 58.8 kPa (8.5 psi). Therefore, p_ω is equal to 19.6 kPa (2.8 psi).

A computerized procedure based on the least-squares principle has been developed for deriving the values of z_ω and p_ω from measured data (Wong and Preston-Thomas 1983b; Wong 2010).

The pressures p_{c1} and p_{c2} that cause the failure of the ice layer shown in Figure 2.36(a) and (b) may be predicted using a method based on the theory of plasticity (Wong and Preston-Thomas 1983b; Wong 2010).

2.4.2 Characterization of the Response to Repetitive Normal Loading

As mentioned previously, when an off-road wheeled vehicle with all axles having the same tread (i.e. transverse distance between left- and right-side wheels on the same axle) is in straight-line motion, an element of the terrain is first subject to the load applied by the front wheel. When the front wheel has passed, the load on the terrain element is reduced to zero. Load is reapplied as the succeeding wheel rolls over it. Similarly, an element of the terrain under the track is subject to the repetitive loadings of the roadwheels of a tracked vehicle. The loading–unloading–reloading cycle for the terrain element continues until the rear wheel (or roadwheel) of the vehicle has passed over it. To predict the normal pressure distribution under a moving off-road vehicle and hence its sinkage and motion resistance, the response of the terrain to repetitive normal loadings must be measured, in addition to the pressure–sinkage relationship described previously.

Figures 2.37–2.39 show the response to repetitive normal load of a sandy terrain, an organic terrain, and a snow-covered terrain, respectively (Wong et al. 1982; Wong et al. 1984; Wong 2010). The pressure initially increases with sinkage along curve *OA*. However, when the load applies to the terrain is reduced at *A*, the pressure–sinkage relationship during unloading follows line *AB*. When the load is reapplied at *B*, the pressure–sinkage relationship follows almost the same path as that during unloading for the sandy terrain and the snow-covered terrain, as shown in Figures 2.37 and 2.39, respectively. For the organic terrain, however, when the load is reapplied at *B*, the pressure–sinkage relationship follows a different path from that during unloading, as shown in Figure 2.38. This indicates that a significant amount of hysteresis exists during the unloading–reloading cycle. When the reapplied load exceeds that at which the preceding unloading begins (point *A* in the figures), additional sinkage results. With the further increase of pressure, the

Figure 2.37 Response to repetitive normal load of a mineral terrain.

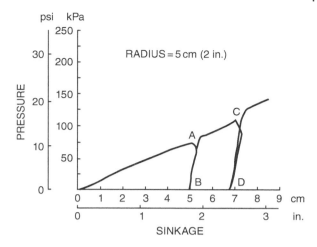

Figure 2.38 Response to repetitive normal load of an organic terrain (muskeg).

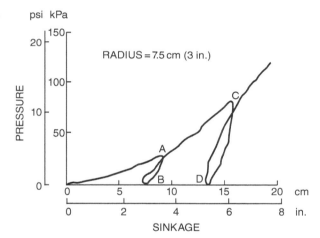

Figure 2.39 Response to repetitive normal load of snow-covered ground.

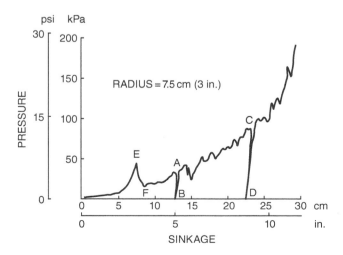

pressure–sinkage relation appears to follow the continuous loading path as *AC* shown in the figures. The characteristics of the second unloading–reloading cycle which begins at point *C* are like those of the first one.

Based on experimental observations, for the three types of terrain described above, the pressure–sinkage relationship during both unloading and reloading, such as *AB* and *BA* shown in the figures, may be approximated by a linear function that represents the average response of the terrain (Wong et al. 1984; Wong 2010):

$$p = p_u - k_u(z_u - z) \tag{2.56}$$

where p and z are the pressure and sinkage, respectively, during unloading or reloading; p_u and z_u are the pressure and sinkage, respectively, when unloading begins; and k_u is the pressure–sinkage parameter representing the average slope of the unloading–reloading line *AB*.

The slope of the unloading–reloading line *AB* (i.e. k_u) represents the degree of elastic rebound during unloading. If line *AB* is vertical, then during unloading, there is no elastic rebound, and the terrain deformation is entirely plastic.

From measured data, it is found that the parameter k_u is a function of the sinkage z_u where unloading begins. As a first approximation, their relationship may be expressed by (Wong et al. 1984; Wong 2010)

$$k_u = k_0 + A_u z_u \tag{2.57}$$

where k_0 and A_u are parameters characterizing the response of the terrain to repetitive loading and z_u is the sinkage where unloading begins. The values of k_0 and A_u for different types of terrain are given in Table 2.13.

The method described above for characterizing terrain response to repetitive normal loading is based on test data available. With more test data for a wider range of terrains becoming available, the characterization of terrain response to repetitive normal loading may evolve.

2.4.3 Characterization of Shear Stress–Shear Displacement Relationships

When a torque is applied to the tire or to the sprocket of a track, shearing action is initiated on the vehicle running gear–terrain interface, as shown in Figure 2.40. To predict vehicle thrust and associated slip, the shear stress–shear displacement relationship of the terrain is required, and this can be measured using the bevameter technique described previously. Figure 2.41 shows the shear

Table 2.13 Values of k_0 and A_u for various types of terrain.

Terrain type	k_0 kN/m^3	k_0 lb/in.^{3}	A_u kN/m^4	A_u lb/in.^{4}
LETE sand	0	0	503,000	47.07
Petawawa muskeg A	123	0.46	23,540	2.20
Petawawa muskeg B	147	0.54	29,700	2.78
Petawawa snow A	0	0	109,600	10.26
Snow (Sweden)	0	0	87,985	8.23

Source: Wong et al. (1984), Wong (2010).

Figure 2.40 Shearing action of a track and a wheel.

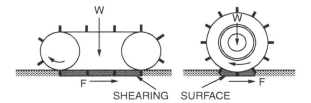

Figure 2.41 Shear stress–shear displacement relationships obtained using (a) a shear ring with outside diameter of 22.2 cm (8.75 in.), (b) a shear ring with outside diameter of 29.8 cm (11.75 in.), and (c) a rigid track of 13.2 × 71.1 cm (5.18 × 28 in.) in sand. *Source:* Wills (1963). Reproduced with permission of the *Journal of Agricultural Engineering Research.*

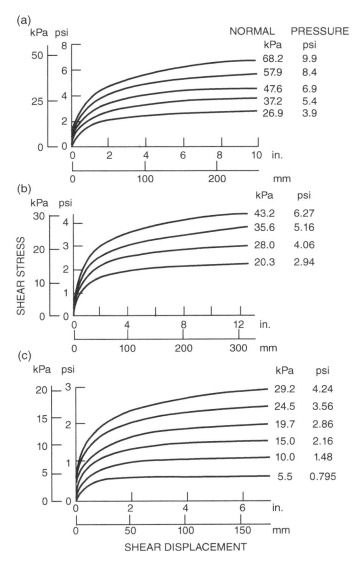

stress–shear displacement relationships for a sand under various normal pressures obtained using different shear devices (Wills 1963). If the maximum shear stress of the terrain is plotted against the corresponding normal pressure, a straight line may be obtained, as shown in Figure 2.42 (Wills 1963). The slope of the straight line determines the angle of internal shearing resistance ϕ, and

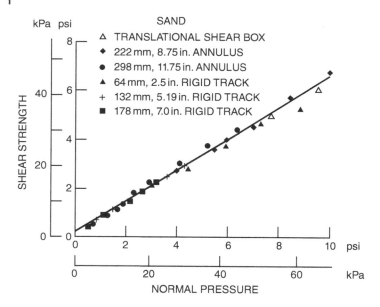

Figure 2.42 Shear strength of sand determined by various methods. *Source:* Wills (1963). Reproduced with permission of the *Journal of Agricultural Engineering Research.*

the intercept of the straight line with the shear stress axis determines the apparent cohesion *c* of the terrain, as discussed previously. The results shown in Figure 2.42 indicate that the shear strength determined by various shearing devices, including translational shear box, shear ring, rectangular shear plate, and rigid track, is comparable.

Based on a considerable amount of field data, it is found that there are three types of shear stress–shear displacement relationship commonly observed for the internal shearing of terrain, dependent upon the type of terrain and its conditions (Wong and Preston-Thomas 1983a; Wong 2010).

1) Loose sand, saturated clay, dry fresh snow, and most disturbed soils

The shear stress–shear displacement relationship for these types of soil exhibits characteristics shown in Figure 2.43. The shear stress initially increases rapidly with an increase in shear

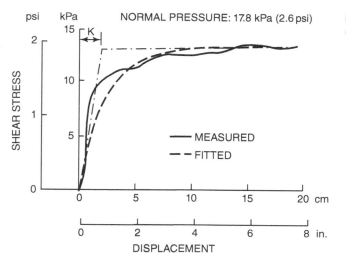

Figure 2.43 A shear curve of a simple exponential form.

displacement, and then approaches a constant value with a further increase in shear displacement. This type of shear stress–shear displacement relationship may be described by an exponential function of the following form proposed by Janosi and Hanamoto (1961).

$$\tau = \tau_{max}\left(1 - e^{-j/K}\right)$$
$$= (c + \sigma \tan \phi)\left(1 - e^{-j/K}\right) \tag{2.58}$$

where τ is the shear stress, j is the shear displacement, c and ϕ are the cohesion and the angle of internal shearing resistance of the terrain, respectively, and K is referred to as the shear deformation parameter.

K may be considered a measure of the magnitude of the shear displacement required to develop the maximum shear stress. The value of K determines the shape of the shear curve. Its value may be represented by the distance between the vertical axis and the point of intersection of the straight-line tangent to the shear curve at the origin and the horizontal line representing the maximum shear stress τ_{max}. The slope of the shear curve at the origin can be obtained by differentiating τ with respect to j in Eq. (2.58):

$$\left.\frac{d\tau}{dj}\right|_{j=0} = \left.\frac{\tau_{max}}{K}e^{-j/K}\right|_{j=0} = \frac{\tau_{max}}{K} \tag{2.59}$$

Thus, the value of K can be determined from the slope of the shear curve at the origin and τ_{max}. The value of K may also be taken as 1/3 of the shear displacement where the shear stress τ is 95% of the maximum shear stress τ_{max}.

In practice, shear curves, particularly those for natural terrains obtained in the field, are not smooth, as shown in Figure 2.43. The optimum value of K that minimizes the overall error in fitting Eq. (2.58) to the measured curve may be obtained from the following equation, based on the weighted least-squares principle (Wong 1980, 2010):

$$K = -\frac{\Sigma(1 - \tau/\tau_{max})^2 j^2}{\Sigma(1 - \tau/\tau_{max})^2 j[\ln(1 - \tau/\tau_{max})]} \tag{2.60}$$

where τ_{max} is the measured maximum shear stress, and τ and j are the measured shear stress and the corresponding shear displacement, respectively.

The standard method of least squares has also been used to process shear stress–shear displacement test data to fit Eq. (2.58) (Senatore and Iagnemma 2011). To obtain the best fitted values for K and τ_{max}, the following function for the sum of the differences between measured values and fitted values should be minimized,

$$f(K, \tau_{max}) = \sum\left(\tau - \tau_{max}\left(1 - e^{-j/K}\right)\right) \tag{2.61}$$

where τ and j are measured shear stress and the corresponding shear displacement, respectively.

The best fitted values of K and τ_{max} are obtained by the standard minimization procedure:

$$\frac{\partial f(K, \tau_{max})}{\partial K} = 0$$

$$\frac{\partial f(K, \tau_{max})}{\partial \tau_{max}} = 0$$

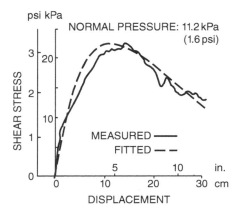

Figure 2.44 A shear curve exhibiting a peak and decreasing residual shear stress.

Based on the experimental data collected, the value of K varies from 1 cm (0.4 in.) for firm sandy terrain to 2.5 cm (1 in.) for loose sand and is approximately 0.6 cm (1/4 in.) for clay at maximum compaction (Reece 1965–1966; Wong 2010). For undisturbed, fresh snow, the value of K varies in the range from 2.5 to 5 cm (1–2 in.) (Harrison 1975). Available experimental results also suggest that the value of K may be a function of normal pressure. However, their precise relationship is yet to be determined.

2) Organic terrain

For organic terrain (muskeg or tundra) with a mat of living vegetation on the surface and saturated peat beneath, the shear stress–shear displacement relationship exhibits characteristics shown in Figure 2.44. The shear stress initially increases rapidly with the increase of shear displacement and reaches a "hump" of maximum shear stress where the "shear-off" of the surface mat is initiated. With a further increase of shear displacement, the shear stress continually decreases, as the peat beneath the mat offers a lower shearing resistance than the surface mat. This type of shearing behavior may be characterized by the following equation (Wong et al. 1979, 1982; Wong and Preston-Thomas 1983a; Wong 2010):

$$\tau = \tau_{max}(j/K_\omega) \exp (1 - j/K_\omega) \tag{2.62}$$

where K_ω is the shear displacement where the maximum shear stress τ_{max} occurs.

In many cases, the values of τ_{max} and K_ω may be directly identified from the measured shear curve. However, in some cases, the value of K_ω may not be distinct or easy to identify. Under these circumstances, the optimum value of K_ω that minimizes the overall error in fitting Eq. (2.62) to the measured curve may be obtained by solving the following equation, which is derived from the weighted least-squares principle (Wong et al. 1979, 1982; Wong and Preston-Thomas 1983a; Wong 2010):

$$\sum (\tau/\tau_{max})^2 \{ \ln (\tau/\tau_{max}) - [1 + \ln (j/K_\omega) - j/K_\omega]\}(K_\omega - j) = 0 \tag{2.63}$$

A computerized procedure has been developed to derive the optimum value of K_ω from measured data (Wong et al. 1979, 1982; Wong and Preston-Thomas 1983a; Wong 2010).

Based on the experimental data collected, the value of K_ω varies from 14.4 to 16.4 cm (5.7–6.5 in.) for various types of organic terrain tested in the Petawawa area, Ontario, Canada (Wong et al. 1979; Wong et al. 1982; Wong and Preston-Thomas 1983a; Wong 2010).

3) Compact mineral soils, seabed terrain, and sintered snow

For compact mineral soils, seabed terrain, and sintered snow, they may exhibit shearing characteristics shown in Figure 2.45 (Wong and Preston-Thomas 1983a; Asaf et al. 2006; Dai et al. 2010; Dai et al. 2015; Cho et al. 2019). The shear stress initially increases rapidly with shear displacement and reaches a "hump" of maximum shear stress at a specific shear displacement. With a further increase in shear displacement, the shear stress decreases and approaches a constant residual value. This type of shearing behavior may be characterized by the following function (Wong and Preston-Thomas 1983a; Wong 2010):

$$\tau = \tau_{max}K_r(1 + \{1/(K_r[1 - 1/e]) - 1\}\exp(1 - j/K_\omega)) \\ \cdot [1 - \exp(-j/K_\omega)]$$ (2.64)

where K_r is the ratio of the residual shear stress τ_r to the maximum shear stress τ_{max}, and K_ω is the shear displacement where the maximum shear stress τ_{max} occurs.

In many cases, the values of K_r, K_ω, and τ_{max} may be directly identified from the measured shear curve. However, in some cases, their values are not distinct or easy to identify. A computerized procedure for determining the optimum values of K_r, K_ω, and τ_{max} for a given measured shear curve has been developed, which is based on the least-squares principle (Wong and Preston-Thomas 1983a; Wong 2010).

Based on the field data collected for various types of firm, mineral terrain, the values of K_ω and K_r vary from 2.7 to 7.1 cm (1.1–2.8 in.) and from 0.38 and 0.72, respectively. For seabed terrain, the values of K_ω and K_r, in the range of 3.5 to 4.2 cm (1.4 to 1.7 in.) and in the range of 0.28 to 0.37, respectively, have been cited (Dai et al. 2010; Dai et al. 2015). For a sintered snow, the values of K_ω and K_r are approximately 2.2 cm (0.9 in.) and 0.66, respectively.

It is interesting to note that the tractive (braking) effort–longitudinal slip (skid) relationships for pneumatic tires on road surfaces described in Chapter 1 exhibit characteristics similar to those of the shear stress–shear displacement relationship shown in Figure 2.45. The peak value of tractive effort $\mu_p W$ and the sliding value $\mu_s W$ shown in Figure 1.16 are analogous to τ_{max} and τ_r shown in Figure 2.45, respectively.

As noted previously, in addition to the internal shearing behavior of terrain, rubber–terrain shearing characteristics will be a significant factor to the prediction of the cross-country performance of vehicles with rubber tires, rubber tracks, or tracks with rubber pads. Furthermore, if vehicle sinkage is large and the vehicle belly contacts the terrain surface, additional belly drag will result. To predict the belly drag, belly–terrain shearing characteristics must be known.

Figure 2.45 A shear curve exhibiting a peak and constant residual shear stress.

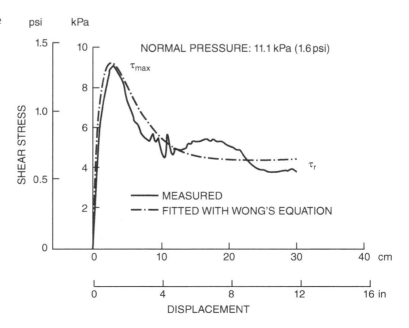

These indicate that to enable a comprehensive evaluation of vehicle mobility, rubber–terrain and vehicle belly–terrain shearing behavior should be measured and characterized accordingly. It is found that the shear stress–shear displacement relationship for rubber–terrain or vehicle belly material–terrain shearing can usually be described using Eq. (2.58). For rubber–snow shearing, dependent upon snow conditions, the shear stress–shear displacement relationship described by Eq. (2.64) is also observed.

During shear tests, as the shear ring is being rotated or the rectangular shear plate is being moved horizontally, additional sinkage of the shear ring or shear plate is observed. This additional sinkage due to shear loading (or shear displacement) is usually referred to as slip sinkage. The topic of slip sinkage will be further elaborated in Section 2.10.

2.4.4 Characterization of the Response to Repetitive Shear Loading

When an all-wheel drive vehicle with all axles having the same tread (i.e. transverse distance between left- and right-side wheels on the same axle) is in straight-line motion, an element of the terrain under the path of the vehicle running gear is subject to the repetitive shearing of successive tires. To realistically predict the shear stress distribution on the vehicle running gear–terrain interface and the thrust developed by the vehicle, the response of the terrain to repetitive shear loading should be measured. Figure 2.46 shows the response of a frictional terrain (a dry sand) to repetitive shearing under a constant normal load (Wong et al. 1984; Wong 2010). When the shear load is reduced from B to zero and is then reapplied at C, the shear stress–shear displacement relationship during re-shearing, such as CDE, is similar to that when the terrain is being sheared in its virgin state, such as OAB. This means that when re-shearing takes place after the completion of a loading–unloading cycle, the shear stress does not instantaneously reach its maximum value for a given normal pressure. Rather, a certain amount of additional shear displacement must take place before the maximum shear stress can be developed, like that when the terrain is being sheared in its virgin state.

Results of an investigation by Keira (1979) on the shearing force developed beneath a rectangular shear plate under a cyclic normal load lead to a similar conclusion. Figure 2.47 shows the variation of the shearing force beneath a rectangular shear plate on a dry sand subject to a vertical harmonic load at a frequency of 10.3 Hz. It indicates that during the loading portion of each cycle, the shear force does not reach its maximum value instantaneously ($S_{max} = P \tan \phi$, where P is the instantaneous value of the normal load and ϕ is the angle of internal shearing resistance). This is demonstrated by the fact that the slope of the normal load curve is steeper than that of the shearing force curve. During the unloading portion of the cycle, however, the shearing force decreases in proportion to the instantaneous value of the normal load.

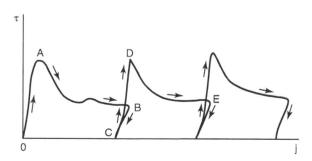

Figure 2.46 Response to repetitive shear loading of a dry sand.

Figure 2.47 Development of shear force under a rectangular shear plate subject to cyclic normal load on a dry sand. (*Source:* Keira 1979.)

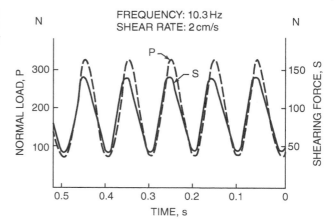

The response of the terrain under repetitive shear loading and its shearing behavior under cyclic normal loads described above have a significant effect on the development of the shear stress on the vehicle running gear–terrain interface. Figure 2.48(b) illustrates the development of the shear stress under a track when the response of the terrain to repetitive shear loading is considered for an idealized case, as compared with that when it is not taken into account, shown in Figure 2.48(a). Since the tractive effort developed by a track is the summation of the shear stress over the entire contact area, when the repetitive shearing characteristics of the terrain are taken into consideration, the predicted total tractive effort of the vehicle at a given slip may be considerably lower than that when they are not taken into account, particularly at low slips (Wong et al. 1984).

2.4.5 Bekker–Wong Terrain Parameters

The group of parameters for characterizing the pressure–sinkage relationship, terrain internal shearing behavior, rubber–terrain shearing characteristics, belly–terrain shearing behavior, and responses to repetitive normal and shear

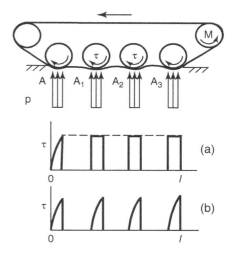

Figure 2.48 Development of shear stresses under a track on a frictional terrain predicted by (a) the conventional method and (b) the improved method taking into account the response of terrain to repetitive shear loading.

loadings of various types of terrain, measured using the bevameter and described above, has become known as the Bekker–Wong terrain parameters (Andrade et al. 2012; McCullough et al. 2017a, b; Edwards 2018; Huang et al. 2020). This group of the Bekker–Wong terrain parameters has been included in the Standards of the International Society for Terrain–Vehicle Systems (ISTVS) updated in 2020 (He et al. 2020) and generally consists of:

- Pressure–sinkage parameters
- Terrain internal shear parameters
- Rubber–terrain shear parameters
- Vehicle belly material–terrain shear parameters
- Parameters for characterizing responses of terrain to repetitive normal and shear loadings

In comparison with the *CI*, which is a composite parameter reflecting the combined effects of compressive and shear strengths of the terrain and the friction and adhesion on the cone–terrain interface, the Bekker–Wong terrain parameters provide more pertinent information on engineering properties of terrain for detailed studies of vehicle mobility. This facilitates the development of physics-based models for predicting various aspects of vehicle cross-country performance. As noted previously, pressure–sinkage parameters of the terrain provide the required information for predicting sinkage of the tire or track and the associated motion resistance of the vehicle. The internal shear parameters of the terrain provide the needed input for evaluating vehicle tractive capability and for predicting tractive effort–slip relationship of an off-road vehicle. The rubber–terrain shear parameters enable the evaluation of the contribution of rubber–terrain interaction to the overall tractive capability of a vehicle with rubber tires, rubber tracks, or rubber pads attached to metal tracks. The vehicle belly–terrain shear parameters will enable the evaluation of additional belly drag if vehicle sinkage is greater than ground clearance. Parameters for characterizing the response of terrain to repetitive normal and shear loadings will enable the evaluation of effects of terrain behavior under repetitive loadings on vehicle tractive performance. Furthermore, if the Bekker–Wong terrain parameters are collected in-situ and used as input to a physics-based vehicle mobility model, it will provide predictions that realistically reflect the operating conditions in the field.

It is recognized that the interaction between a vehicle and deformable terrain is complex and that developments of physics-based vehicle mobility models may follow a variety of approaches with different types of terrain input. The Bekker–Wong terrain parameters, nevertheless, provide more detailed terrain information for the development of physics-based vehicle mobility models, in comparison with the *CI* and the like.

The Bekker–Wong terrain parameters are used as input to physics-based vehicle mobility models, such as NTVPM for tracked vehicles with relatively short track pitch or rubber-belt tracks, RTVPM for tracked vehicles with relatively long track pitch, and NWVPM for off-road wheeled vehicles. These models will be discussed later in this chapter. The Bekker–Wong terrain parameters are also employed in many other types of vehicle mobility model, including those based on the multi-body dynamics (MBD) formulations (Edwards 2018).

To make use of the vast amount of *CI* data that have been collected from various parts of the globe over many decades, various methodologies are being explored for deriving the Bekker–Wong terrain parameters from the *CI* data (Huang et al. 2020).

2.5 A Simplified Physics-Based Model for the Performance of Tracked Vehicles

The track and the wheel constitute the two basic forms of running gear for off-road vehicles. The study of the mechanics of track–terrain and wheel–terrain interaction over unprepared terrain is, therefore, of fundamental interest. The objective is to establish physics-based models for predicting their performance in relation to their design parameters and terrain characteristics.

One of the earlier physics-based models for parametric analysis of track system performance was developed by Bekker (1956, 1960, 1969). In his model, it is assumed that the track in contact with the terrain is like a rigid footing. Making use of the pressure–sinkage relationship of the terrain measured by the bevameter described in the previous section, track sinkage and motion resistance due to compacting the terrain are predicted. Based on the shear stress–shear displacement relationship and the shear strength of the terrain, the thrust–slip relationship and the maximum traction of a track system are predicted.

2.5.1 Motion Resistance of a Track

As mentioned above, in the model developed by Bekker, the track is assumed to be like a rigid footing. The normal reaction exerted on the track by the terrain can then be equated to that beneath a sinkage plate at the same depth in a pressure–sinkage test. If the center of gravity of the vehicle is located at the midpoint of the track contact area, the normal pressure distribution may be assumed to be uniform, as shown in Figure 2.29. On the other hand, if the center of gravity of the vehicle is located ahead of or behind the midpoint of the track contact area, a sinkage distribution of trapezoidal form may be assumed.

Using the pressure–sinkage equation proposed by Bekker (Eq. (2.47)), for a track with uniform contact pressure, the sinkage z_0 is given by

$$z_0 = \left(\frac{p}{k_c/b + k_\phi} \right)^{1/n} = \left(\frac{W/bl}{k_c/b + k_\phi} \right)^{1/n} \tag{2.65}$$

where p is the normal pressure, W is the normal load on the track, and b and l are the width and length of the track in contact with the terrain, respectively.

The sinkage z_0 of the track predicted using Eq. (2.65) is caused by its normal load. In general, the sinkage of a vehicle running gear (track or wheel) due to its normal load alone is referred to as the static sinkage.

The work done in compacting the terrain and making a rut of width b, length l, and depth z_0 is given by

$$\begin{aligned} \text{work} &= bl \int_0^{z_0} p \, dz \\ &= bl \int_0^{z_0} \left(k_c/b + k_\phi \right) z^n dz \\ &= bl \left(k_c/b + k_\phi \right) \left(\frac{z_0^{n+1}}{n+1} \right) \end{aligned} \tag{2.66}$$

Substituting for z_0 from Eq. (2.65) yields

$$\text{work} = \frac{bl}{(n+1)\left(k_c/b + k_\phi \right)^{1/n}} \left(\frac{W}{bl} \right)^{(n+1)/n} \tag{2.67}$$

If the track is pulled a distance l in the horizontal direction, the work done by the towing force, which is equal to the magnitude of the motion resistance due to terrain compaction R_c, can be equated to the vertical work done in making a rut of length l, as expressed by Eq. (2.67):

$$R_c l = \frac{bl}{(n+1)\left(k_c/b + k_\phi \right)^{1/n}} \left(\frac{W}{bl} \right)^{(n+1)/n}$$

and

$$\begin{aligned} R_c &= \frac{b}{(n+1)\left(k_c/b + k_\phi \right)^{(1/n)}} \left(\frac{W}{bl} \right)^{(n+1)/n} \\ &= \frac{1}{(n+1)b^{1/n}\left(k_c/b + k_\phi \right)^{(1/n)}} \left(\frac{W}{l} \right)^{(n+1)/n} \end{aligned} \tag{2.68}$$

This is the equation for calculating the motion resistance due to terrain compaction of a track with uniform pressure distribution, based on Bekker's pressure–sinkage relationship. Expressions for motion resistance based on other pressure–sinkage relationships described in Section 2.4 may be derived in a similar way.

On soft terrain where vehicle sinkage is significant, Bekker suggested that a bulldozing resistance acting in the front of the track should be considered, in addition to the compaction resistance R_c. The bulldozing resistance may be calculated using the earth pressure theory described in Section 2.2 (Bekker 1960, 1969).

As the method described above is based on several simplifying assumptions, it can provide only a preliminary assessment of the tractive performance of tracked vehicles. For instance, the idealization of a track as a rigid footing is not realistic for tracked vehicles designed for high-speed operations, such as military tracked vehicles. For these vehicles to achieve high operating speeds, it is necessary to have relatively short track pitch to minimize speed fluctuations and associated vibrations due to the polygon (or chordal) effect of sprocket tooth–track link engagement. For military tracked vehicles to have adequate ability to ride over large obstacles, large diameter roadwheels with sufficient suspension travel are generally used. As a result of using a relatively small number (typically five to seven) of large diameter roadwheels and short track pitch, the normal pressure distribution under the track is nonuniform and significant pressure peaks are observed under the roadwheels. Figure 2.49 shows the normal pressure distributions measured at a depth of 0.23 m (9 in.) below the terrain surface under military tracked vehicles with different design features (Rowland 1972). Under the tracks of Comet and Sherman V, the normal pressure varies greatly, and the peak pressure is much higher than the average (nominal) ground pressure. With overlapping roadwheel arrangements, such as that in the German tank Panther A used in World War II, the fluctuations of normal pressure under the track are reduced. A similar situation is observed for Churchill V, which has nine small-diameter roadwheels.

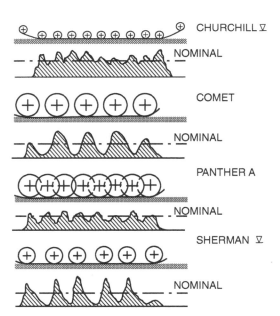

Figure 2.49 Measured normal pressure distributions at a depth of 23 cm (9 in.) below the soil surface under various military tracked vehicles. *Source:* Rowland (1972). Reproduced with permission of the International Society for Terrain–Vehicle Systems.

To improve the accuracy of predictions of the tractive performance of tracked vehicles, two advanced physics-based models have been developed, one for vehicles with rubber belt tracks or link tracks with relatively short track pitch, commonly used in high-speed military tracked vehicles, and the other for low-speed vehicles with relatively long track pitch, such as those used in agriculture and construction industry. These two models are discussed later in Sections 2.6 and 2.7, respectively.

2.5.2 Tractive Effort and Slip of a Track

The tractive effort of a track is produced by the shearing of the terrain, as shown in Figure 2.40. The maximum tractive effort F_{max} that can be developed by a track is determined by the shear strength of the terrain τ_{max}, as described by the Mohr–Coulomb equation, Eq. (2.10), and the contact area A:

$$
\begin{aligned}
F_{max} &= A\tau_{max} \\
&= A(c + p\tan\phi) \\
&= Ac + W\tan\phi
\end{aligned}
\tag{2.69}
$$

where A is the contact area of the track, W is the normal load, and c and ϕ are the apparent cohesion and the angle of internal shearing resistance of the terrain, respectively. In frictional soil, such as dry sand, the cohesion c is negligible; the maximum tractive effort, therefore, depends on the vehicle weight. The heavier the vehicle, the higher the tractive effort that it can develop. The dimensions of the track do not affect the maximum tractive effort. For dry sand, the angle of internal shearing resistance could be as high as 35°. The maximum tractive effort of a vehicle on dry sand can therefore be expected to be approximately 70% of the vehicle weight. In cohesive soil, such as saturated clay, the value of ϕ is low and cohesion c is the dominant shear strength component, the maximum tractive effort primarily depends on the contact area of the track, and the weight has little effect. Thus, the dimensions of the track are crucial in this case; the larger the contact area, the higher the thrust the track can develop.

It should be noted that Eq. (2.69) can only be used for predicting the maximum tractive effort of a tracked vehicle. In vehicle performance evaluation, it is, however, desirable to determine the variation of thrust with track slip over the full operating range. To predict the relationship between thrust and slip, it is necessary to examine the development of shear displacement beneath a track since shear stress is a function of shear displacement, as discussed in Section 2.4.3. The shear displacement at various points beneath a track is shown schematically in Figure 2.50 (Bekker 1956). At point 1, the grouser is just coming into contact with the terrain; it cannot develop the same shear displacement as the other grousers 2, 3, 4, and 5 since they have been shearing the terrain for varying periods of time. The amount of horizontal shear displacement j increases along the contact length and reaches

Figure 2.50 Development of shear displacement under a track. *Source: Theory of Land Locomotion* by M.G. Bekker, copyright © 1956 by the University of Michigan. Reproduced with permission of the University of Michigan Press.

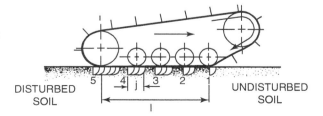

DISTURBED
SOIL

UNDISTURBED
SOIL

its maximum value at the rear of the contact area. To examine the development of shear displacement beneath a track quantitatively, the slip of a track ϕ must be defined first:

$$i = 1 - \frac{V}{rw} = 1 - \frac{V}{V_t} = \frac{V_t - V}{V_t} = \frac{V_j}{V_t} \tag{2.70}$$

where V is the actual forward speed of the track, V_t is the theoretical speed which can be determined from the angular speed ω and the radius r of the pitch circle of the sprocket, and V_j is the speed of slip of the track with reference to the ground. When the vehicle is slipping, V_j will be in the direction opposite that of vehicle motion. On the other hand, when the vehicle is skidding, V_j will be in the same direction as that of vehicle motion. The definition for the slip of a track given by Eq. (2.70) is the same as that of a tire given by Eq. (1.5). Since the track cannot be stretched, the speed of slip V_j is the same for every point of the track in contact with the terrain. The shear displacement j at a point located at a distance x from the front of the contact area shown in Figure 2.51 can be determined by

$$j = V_j t \tag{2.71}$$

where t is the contact time of the point in question with the terrain and is equal to x/V_t.

Rearranging Eq. (2.71), the expression for shear displacement j becomes

$$j = \frac{V_j x}{V_t} = ix \tag{2.72}$$

This indicates that the shear displacement beneath a flat track increases linearly from the front to the rear of the contact area, as shown in Figure 2.51. Since the development of shear stress is related to shear displacement as discussed previously, the shear stress distribution along the contact length can be found. For instance, to determine the shear stress developed at a point located at a distance A from the front of the contact area, the shear displacement at that point should first be calculated using Eq. (2.72). Making use of the shear stress–shear displacement relationships obtained from shear tests, such as those shown in Figures 2.43–2.45, or from the semiempirical equations, such as Eqs. (2.58), (2.62), and (2.64), the shear stress at that point can then be determined. As an example, the shear stress distribution beneath a track on a particular type of terrain at a given slip is shown in Figure 2.51. The total tractive effort developed by a track at a given slip is represented by the area beneath the shear stress curve in Figure 2.51. Alternatively, if Eq. (2.58) is used to

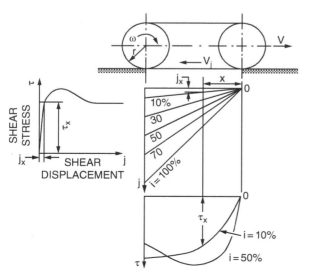

Figure 2.51 Distributions of shear displacement and shear stress under a track.

describe the shear stress–shear displacement relationship, the total tractive effort of a track can be calculated as follows:

$$F = b \int_0^l \tau \, dx$$
$$= b \int_0^l (c + p \tan \phi)\left(1 - e^{-j/K}\right) dx \tag{2.73}$$

The above equation indicates that the tractive effort of a track depends on the normal pressure distribution over the contact area, among other factors. For a uniform normal pressure distribution, p is independent of x and equal to W/bl. In this case, the total tractive effort of a track is determined by

$$F = b \int_0^l \left(c + \frac{W}{bl} \tan \phi\right)\left(1 - e^{-ix/K}\right) dx$$
$$= (Ac + W \tan \phi)\left[1 - \frac{K}{il}\left(1 - e^{-il/K}\right)\right] \tag{2.74}$$

Equation (2.74) expresses the functional relationship of tractive effort, vehicle design parameters, terrain parameters, and track slip. If the slip is 100%, then Eqs. (2.74) and (2.69) are practically identical. Among the vehicle design parameters, the contact length of the track deserves special attention. Consider two tracked vehicles with identical ground contact area and normal load (i.e., $A_1 = A_2$ and $W_1 = W_2$) operating over the same terrain. However, the track length of one vehicle is twice that of the other (i.e., $l_1 = 2l_2$). To keep the total contact area the same, the width b_1 of the track with length l_1 is half that of the other (i.e., $b_1 = 0.5b_2$). If these two tracked vehicles are to develop the same tractive effort, then from Eq. (2.74), the slip of the vehicle with contact length l_1 will be half that of the other with contact length l_2. It may be concluded, therefore, that, in general, a shorter track will slip more than a longer one if they are to develop the same tractive effort.

The above analysis is applicable to predicting the tractive effort of a track with uniform normal pressure distribution. In practice, the normal pressure distribution is seldom uniform, as mentioned previously. It is, therefore, of interest to assess the effect of normal pressure distribution on the tractive effort developed by a track. This problem has been investigated by Wills (1963), among others. Consider the case of the multipeak sinusoidal pressure distribution described by

$$p = \frac{W}{bl}\left(1 + \cos\frac{2n\pi x}{l}\right) \tag{2.75}$$

where n is the number of periods, as shown in Figure 2.52(b).

In a frictional soil, the shear stress developed along the contact length can be expressed by

$$\tau = \frac{W}{bl} \tan \phi\left(1 + \cos\frac{2n\pi x}{l}\right)\left(1 - e^{-ix/K}\right) \tag{2.76}$$

and hence the tractive effort is given by

$$F = b \int_0^l \frac{W}{bl} \tan \phi\left(1 + \cos\frac{2n\pi x}{l}\right)\left(1 - e^{-ix/K}\right) dx$$
$$= W \tan \phi\left[1 + \frac{K}{il}\left(e^{-il/K} - 1\right) + \frac{K\left(e^{-il/K} - 1\right)}{il\left(1 + 4n^2 K^2 \pi^2 / i^2 l^2\right)}\right] \tag{2.77}$$

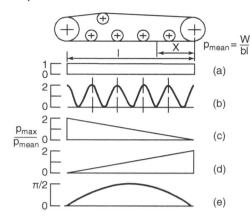

Figure 2.52 Various types of idealized normal pressure distribution under a track. *Source:* Wills 1963. Reproduced with permission of the *Journal of Agricultural Engineering Research.*

The tractive effort of a track with other types of normal pressure distribution can be evaluated in a similar way. In the case of normal pressure increasing linearly from front to rear (i.e., $p = 2(w/bl)(x/l)$) as shown in Figure 2.52(c), the tractive effort of a track in a frictional soil is given by

$$F = W \tan \phi \left[1 - 2 \left(\frac{K}{il} \right)^2 \left(1 - e^{-il/K} - \frac{il}{K} e^{-il/K} \right) \right]$$

(2.78)

In the case of normal pressure increasing linearly from rear to front (i.e., $p = 2(W/bl)(l - x)/l$), as shown in Figure 2.52(d), the tractive effort in a frictional soil is calculated by

$$F = 2W \tan \phi \left[1 - \frac{K}{il} \left(1 - e^{-il/K} \right) \right]$$
$$- W \tan \phi \left[1 - 2 \left(\frac{K}{il} \right)^2 \left(1 - e^{-il/K} - \frac{il}{K} e^{-il/K} \right) \right]$$

(2.79)

In the case of sinusoidal distribution with maximum pressure at the center and zero pressure at the front and rear end (i.e., $p = (W/bl)(\pi/2) \sin(\pi x/l)$), as shown in Figure 2.52(e), the tractive effort in a frictional soil is determined by

$$F = W \tan \phi \left[1 - \frac{e^{-il/K} + 1}{2(1 + i^2 l^2 / \pi^2 K^2)} \right]$$

(2.80)

Figure 2.53 shows the variation of the tractive effort with slip of a track with various types of normal pressure distribution discussed above. It is shown that the normal pressure distribution has a noticeable effect on the development of tractive effort, particularly at low slips.

Example 2.3 Two tracked vehicles with the same gross weight of 135 kN (30 350 lb) travel over a terrain, which is characterized by $n = 1.6$, $k_c = 4.37$ kN/m$^{2.6}$(0.07 lb/in.$^{2.6}$), $k_\phi = 196.72$ kN/m$^{3.6}$(0.08 lb/in.$^{3.6}$), $K = 5$ cm (2 in.), $c = 1.0$ kPa (0.15 psi), and $\phi = 19.7°$. Both vehicles have the same ground contact area of 7.2 m^2 (77.46 ft^2). However, the width b and contact length l of the tracks of the two vehicles are not the same. For vehicle A, $b = 1$ m (3.28 ft) and $l = 3.6$ m (11.8 ft), and for vehicle B, $b = 0.8$ m (2.62 ft) and $l = 4.5$ m (14.76 ft). Estimate the motion resistance due to terrain compaction and the thrust–slip characteristics of these two vehicles. In the calculations, uniform ground contact pressure may be assumed.

Solution

(a) Motion resistance of vehicle A

$$\text{Sinkage}: z_0 = \left(\frac{p}{k_c/b + k_\phi} \right)^{1/n} = \left(\frac{135.0/7.2}{4.37/1 + 196.72} \right)^{0.625}$$
$$= 0.227 \text{ m } (9 \text{ in.})$$

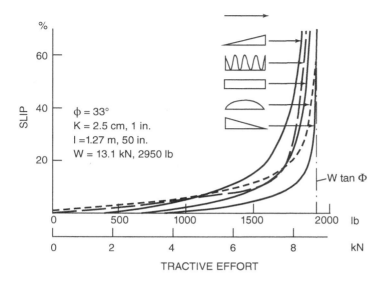

Figure 2.53 Effects of normal pressure distributions on the tractive effort–slip relationships of a track on sand. *Source:* Wills (1963). Reproduced with permission of the *Journal of Agricultural Engineering Research.*

$$\text{Compaction resistance} : R_c = 2b \left(k_c/b + k_\phi\right) \frac{z_0^{n+1}}{n+1}$$

$$= \frac{2 \times 1 \times 201.09 \times 0.227^{2.6}}{2.6}$$

$$= 3.28 \text{ kN (738 lb)}$$

(b) Motion resistance of vehicle B

$$\text{Sinkage} : \quad z_0 = 0.226 \text{ m (9 in.)}$$

$$\text{Compaction resistance} : \quad R_c = 2.60 \text{ kN (585 lb)}$$

(c) Thrust–slip characteristics of vehicle A

$$F = (2blc + W \tan \phi) \left[1 - \frac{K}{il} \left(1 - e^{-il/K}\right)\right]$$

$$= F_{max} \left[1 - \frac{K}{il} \left(1 - e^{-il/K}\right)\right]$$

$$F_{max} = 2 \times 1 \times 3.6 \times 1 + 135 \times 0.358$$

$$= 7.2 + 48.34 = 55.54 \text{ kN}(12,486 \text{ lb})$$

The thrust of vehicle A at various slips is given in Table 2.14.

(d) Thrust–slip characteristics of vehicle B

The maximum thrust of vehicle B will be the same as that of vehicle A since the contact area and the weight of the two vehicles are identical. The thrust–slip relationship will, however, be

Table 2.14 Thrust–slip relationships for vehicles A and B.

Vehicle type	Slip i (%)	5	10	20	40	60	80
A	Thrust						
	kN	40.54	47.82	51.68	53.62	54.25	54.57
	lb	9,114	10,750	11,618	12,054	12,196	12,268
B	Thrust						
	kN	43.32	49.37	52.46	54.0	54.51	54.77
	lb	9,739	11,099	11,794	12,140	12,254	12,313

different, as the contact lengths of the two vehicles are not the same. The thrust of vehicle B at various slips is given in Table 2.14.

The performance of vehicle B is somewhat better than that of vehicle A in the terrain specified. For instance, the compaction resistance of vehicle B is approximately 20.7% lower than that of vehicle A. At 10% slip, the thrust of vehicle B is approximately 3.2% higher than that of vehicle A.

2.6 An Advanced Physics-Based Model for the Performance of Vehicles with Flexible Tracks

To provide vehicle designers with comprehensive and realistic tools for performance and design evaluation of off-road vehicles, from a traction perspective, a series of advanced physics-based models have been developed. These include models for evaluating vehicles with flexible tracks, vehicles with long-pitch link tracks, and off-road wheeled vehicles. These models are based on the study of the physical nature and the detailed analysis of the mechanics of vehicle running gear–terrain interaction. The basic features and practical applications of the model for vehicles with flexible tracks are outlined in this section, while those of the others are discussed in subsequent sections.

For high-speed tracked vehicles, such as military fighting and logistics vehicles and off-road transport vehicles, link tracks (segmented metal tracks) with relatively short track pitch or rubber-belt (band) tracks are commonly used. The short-pitch link track system typically has a ratio of roadwheel diameter to track pitch in the range of 4–6, a ratio of roadwheel spacing (i.e., distance between the centers of two adjacent roadwheels) to track pitch in the range of 4–7, and a ratio of sprocket pitch diameter to track pitch of the order of 4. The short-pitch link track and the rubber-belt (band) track are hereinafter referred to as the "flexible track" and it can be idealized as a flexible and extensible belt in the analysis of track–terrain interaction. For a flexible track, the assumption that the track is equivalent to a rigid footing used in the simplified model described in Section 2.5 is not adopted.

To provide a method for realistically evaluating the effects of vehicle design features and of terrain conditions on the cross-country performance of vehicles with flexible tracks, an advanced physics-based model known as NTVPM (Nepean Tracked Vehicle Performance Model) has been developed (Wong et al. 1984, 2019; Wong 2010). It takes into account all major design parameters of the vehicle, including the track system configuration, number of roadwheels, roadwheel dimensions and spacing, track dimensions and geometry, initial track tension, track longitudinal elasticity (stiffness), suspension characteristics, location of the center of gravity, arrangements for the sprockets, idlers and supporting rollers, and vehicle belly (hull) shape (for the analysis of vehicle belly–terrain interaction, when track sinkage is greater than vehicle ground clearance and vehicle belly is in contact with the terrain surface). All pertinent Bekker–Wong terrain parameters, as discussed in Section 2.4.4, are considered. NTVPM can be employed in performance and design evaluation of both single-unit and two-unit articulated tracked vehicles (Wong 1986, 1992a, 1992c, 1994a, 1994b, 1994c, 1995, 1997, 1999; Wong and Preston-Thomas 1986, 1988).

NTVPM can be used to predict normal and shear stress distributions on the track–terrain interface, and the external motion resistance, tractive effort (thrust), drawbar pull, and tractive efficiency of the vehicle as functions of track slip. The basic features of the model have been validated by means of full-scale vehicle tests on various types of terrain, including sandy terrain, organic terrain, and snow-covered terrain, which will be presented in Section 2.6.4. The model is particularly suited to the evaluation of competing designs, optimization of design parameters, and selection of vehicle candidates for a given mission and environment. It has been employed in assisting vehicle manufacturers in the development of new products and governmental agencies in evaluating vehicle candidates in many countries (Wong 1992a, 1995; Wong and Huang 2006b; Wong et al. 2018, 2019). The basic approach to the development of the model NTVPM is outlined below.

2.6.1 Approach to the Prediction of Normal Pressure Distribution under a Track

As noted previously, in the development of the model, the track is modeled as a flexible belt. This idealization is reasonable for rubber-belt tracks and for link tracks with relatively short track pitch. A schematic of the track–roadwheel system on a deformable terrain under steady-state operating conditions is shown in Figure 2.54.

When a tracked vehicle rests on a hard surface, the track lies flat on the ground. In contrast, when the vehicle travels over a deformable terrain, the normal load applied through the track system causes the terrain to deform. The track segments between the roadwheels take up load and as a result they deflect and have the form of a curve. The actual length of the track in contact with the terrain between the front and rear roadwheels increases in comparison with that when the track rests on firm ground. This causes a reduction in the sag of the top run of the track and a change in track tension. The elongation of the track under tension is taken into consideration in the analysis.

The deflected track in contact with the terrain may be divided into two sections (Figure 2.54(b)): one in contact with both the roadwheel and the terrain (such as segments *AC* and *FH*) and the other in contact with the terrain only (such as segment *CF*). The shape of the track segment in contact with the roadwheel, such as *AC*, is defined by the shape of the roadwheel, whereas the shape of the track segment in contact with the terrain only, such as *CF*, is determined by the track tension and roadwheel spacing and the pressure–sinkage relationship and response to repetitive loading of the terrain.

Along segment AB, the pressure exerted on the terrain increases from *A* to *B*. From *B* to *D* the pressure decreases corresponding to the unloading portion of the repetitive loading cycle in Figures 2.37, 2.38, or 2.39. Along segment *DE*, the pressure increases again, corresponding to the reloading portion of the repetitive loading cycle shown in Figures 2.37, 2.38, or 2.39.

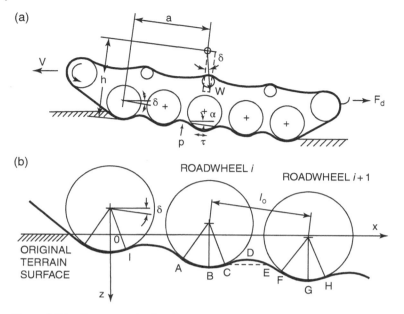

Figure 2.54 Geometry of a flexible track system in contact with deformable terrain.

Beyond point E, which is at the same level as point B, the sinkage is higher than that at B. As a result, the pressure increases and the sinkage of the succeeding roadwheel will be greater than that of the preceding roadwheel. This causes the vehicle to assume a nose-up attitude. Beyond point G, the pressure exerted on the terrain decreases again and another unloading–reloading cycle begins.

Over highly compressible terrain, such as deep snow, track sinkage may be greater than the ground clearance of the vehicle. If this occurs, the belly (hull) of the vehicle will be in contact with the terrain surface and will support part of the vehicle weight. This will reduce the load carried by the tracks and will adversely affect the traction of the vehicle on a terrain with a significant frictional component in its shear strength. Furthermore, the contact of the belly with the terrain will give rise to an additional drag component (the belly drag). The characteristics of vehicle belly–terrain interaction have been taken into consideration. In the model, the characteristics of the independent suspension of the roadwheels are fully considered as well. Torsion bar, hydropneumatic, or other types of independent suspension, with either linear or nonlinear load–deflection relationship, can be realistically simulated.

Based on the understanding of the physical nature of track–terrain interaction described above, a set of equations for the equilibrium of forces and moments acting on the track–roadwheel system, including the independently suspended roadwheels, and for the evaluation of the overall track length is derived. They establish the relationship between the shape of the deflected track in contact with the terrain and vehicle design parameters and terrain characteristics. The solution of this set of equations defines the sinkage of the roadwheels, the inclination of the vehicle body, the track tension, and the track shape in contact with the terrain. From these and considering the pressure–sinkage relationship and the response to repetitive loading of the terrain, the normal pressure distribution under a moving tracked vehicle is predicted.

2.6.2 Approach to the Prediction of Shear Stress Distribution under a Track

The tractive performance of a tracked vehicle is closely related to both its normal pressure and shear stress distributions on the track–terrain interface. To predict the shear stress distribution under a track, the shear stress–shear displacement relationship, the shear strength, and the response to repetitive shear loading of the terrain, as discussed in Section 2.4, are used as input. The shear stress at a given point on the track–terrain interface is a function of the shear displacement, measured from the point where shearing (or re-shearing) begins, as well as the normal pressure at that point. The shear displacement developed under a flexible track, shown in Figure 2.55, may be determined from the analysis of the slip velocity V_j like that described in Section 2.5. The slip velocity V_j of a point P on a flexible track relative to the terrain surface is the tangential component of the absolute velocity V_a shown in Figure 2.55. The magnitude of the slip velocity V_j is expressed by

$$
\begin{aligned}
V_j &= V_t - V\cos\alpha \\
&= r\omega - r\omega(1-i)\cos\alpha \\
&= r\omega[1 - (1-i)\cos\alpha]
\end{aligned}
\tag{2.81}
$$

where r and ω are the pitch radius and angular speed of the sprocket, respectively, i is the slip of the track, α is the angle between the tangent to the track at point P and the horizontal, V_t is the theoretical speed of the vehicle (i.e., $V_t = r\omega$), and V is the actual forward speed of the vehicle.

The shear displacement j along the track–terrain interface is given by

$$
\begin{aligned}
j &= \int_0^t r\omega[1 - (1-i)\cos\alpha]\, dt \\
&= \int_0^l r\omega[1 - (1-i)\cos\alpha]\frac{dl}{r\omega} \\
&= l - (1-i)x
\end{aligned}
\tag{2.82}
$$

where l is the distance along the track between point P and the point where shearing (or re-shearing) begins, and x is the corresponding horizontal distance between point P and the initial shearing (or re-shearing) point.

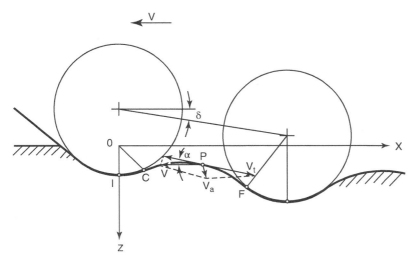

Figure 2.55 Slip velocity of a point on a flexible track in contact with deformable terrain.

If the shear stress–shear displacement relationship of the terrain is described by Eq. (2.58), then the shear stress distribution may be expressed by

$$\tau(x) = [c + p(x)\tan\phi]\left\{1 - \exp\left[-\left(\frac{l - (1-i)x}{K}\right)\right]\right\} \tag{2.83}$$

where $p(x)$ is the normal pressure on the track, which is a function of x.

In using Eq. (2.83) to predict the shear stress distribution under a flexible track, the response to repetitive shear loading of the terrain discussed in Section 2.4 and the shearing characteristics of the terrain under varying normal pressure should be taken into consideration (Wong et al. 1984; Wong 2010).

As noted above, the shear stress on the track–terrain interface varies along the contact length. Consequently, the track tension changes along the track length. As described in Section 2.6.1, the shape of the track segment in contact with the terrain only, such as CF shown in Figure 2.54(b), is a function of the track tension in that segment. Thus, the normal pressure distribution is affected by the presence of shear stress. An iterative procedure is therefore introduced to predict the normal pressure and shear stress distributions under a flexible track at a given slip under steady-state operating conditions. The basic steps of this procedure are outlined below:

- Using the method described in Section 2.6.1, calculate the initial normal pressure distribution under the track when no shear stress is present on the track–terrain interface.
- Following the approach described in this section, calculate the shear stress distribution under the track at a given slip using the initial normal pressure distribution predicted at the preceding step.
- Based on the shear stress distribution determined in the preceding step, calculate the tractive effort distribution along the track–terrain interface by integrating the shear stress over the corresponding contact areas. Determine the difference between the track tensions at either end of each track segment between adjacent roadwheels, such as that between C and F shown in Figure 2.54b. Calculate the average track tension in each track segment between adjacent roadwheels in contact with the terrain.
- Using the average track tension in each track segment between adjacent roadwheels to initiate the iterative process by recalculating the normal pressure distribution at the presence of shear stress. Check whether the conditions that define the equilibrium of forces and moments acting on the track system and the overall track length are satisfied.
- Repeat the steps described above until the errors in the equilibrium of forces and moments and in the overall track length are less than the prescribed values. When the conditions are satisfied, the normal pressure and shear stress distributions beneath the track at a given slip under steady-state operating conditions are defined.

2.6.3 Prediction of Motion Resistance and Drawbar Pull as Functions of Track Slip

When the normal pressure and shear stress distributions under a tracked vehicle at a given slip have been determined, the tractive performance of the vehicle can be predicted. The tractive performance of an off-road vehicle is usually characterized by its motion resistance, tractive effort, and drawbar pull (the difference between the tractive effort and motion resistance) as functions of slip.

On a level terrain, the external motion resistance R_t of the track can be determined from the horizontal component of the normal pressure acting on the track in contact with the terrain. For a vehicle with two tracks, R_t is given by

$$R_t = 2b \int_0^{l_t} p \sin \alpha \, dl \tag{2.84}$$

where b is the contact width of the track, l_t is the contact length of the track, p is the normal pressure, and α is the angle between a track element and the horizontal.

If the track sinkage is greater than the ground clearance of the vehicle, the belly (hull) will be in contact with the terrain, giving rise to an additional drag, known as the belly drag R_{be}. It can be determined from the horizontal components of the normal and shear stresses acting on the belly–terrain interface, and is expressed by

$$R_{be} = b_b \left[\int_0^{l_b} p_b \sin \alpha_b \, dl + \int_0^{l_b} \tau_b \cos \alpha_b \, dl \right] \tag{2.85}$$

where b_b is the contact width of the belly, l_b is the contact length of the belly, α_b is the angle between the belly and the horizontal, and p_b and τ_b are the normal pressure and shear stress on the belly–terrain interface, respectively.

On a level terrain, the tractive effort of the vehicle can be calculated from the horizontal component of the shear stress acting on the track in contact with the terrain. For a vehicle with two tracks, F is given by

$$F = 2b \int_0^{l_t} \tau \cos \alpha \, dl \tag{2.86}$$

where τ is shear stress on the track–terrain interface.

Since both the normal pressure p and shear stress τ are functions of track slip, the track motion resistance R_t, belly drag R_{be} (if any), and tractive effort F vary with slip.

For a track with rubber pads, part of the tractive effort is generated by rubber–terrain shearing. To predict the tractive effort developed by the rubber pads, the portion of the vehicle weight supported by the rubber pads, the area of the rubber pads in contact with the terrain, and the characteristics of rubber–terrain shearing are taken into consideration.

The tractive effort F calculated by Eq. (2.86) is due to the shearing action of the track across the grouser tips. For a track with high grousers, additional thrust will be developed due to the shearing action on the vertical surfaces on either side of the track. This additional thrust may be estimated using a method proposed by Reece (1965–1966).

On a level terrain, the drawbar pull F_d is the difference between the total tractive effort (including the thrust developed by vertical shearing surfaces on both sides of the tracks) and the total external motion resistance of the vehicle (including the belly drag, if any), and is expressed by

$$F_d = F - R_t - R_{be} \tag{2.87}$$

From Eq. (2.87), the relationship between drawbar pull F_d and track slip i can be determined.

2.6.4 Experimental Substantiation

To validate the basic features of NTVPM, the tractive performances of a single-unit and a two-unit articulated tracked vehicle were measured over various types of terrain and compared with those predicted using NTVPM (Wong et al. 1984; Wong 2010). Figure 2.56(a)–(c) show comparisons between the measured and predicted normal pressure distributions under the track pad of an armored personnel carrier on a sandy terrain, an organic terrain (muskeg), and a snow-covered terrain, respectively. For engineering properties of these terrains, please refer to Wong et al. (1984) and Wong (2010).

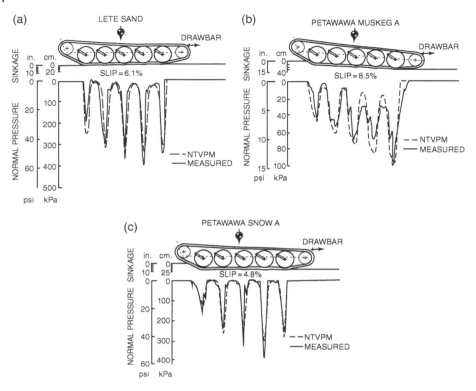

Figure 2.56 Comparisons of the measured normal pressure distributions under the track pad of a tracked armored personnel carrier and the predicted ones using NTVPM on (a) sandy terrain, (b) organic terrain (muskeg), and (c) snow-covered terrain.

Figure 2.57(a)–(c) show comparisons between the measured and the predicted drawbar performances of a tracked armored personnel carrier over a sandy terrain, an organic terrain (muskeg) and a snow-covered terrain, respectively. Figures 2.56 and 2.57 show that there is a reasonably close agreement between the predicted and measured normal pressure distributions and drawbar performances. Thus, the basic features of NTVPM have been substantiated.

2.6.5 Applications to Parametric Analysis and Design Optimization

With its basic features substantiated by field test data, NTVPM can then be applied to realistically evaluating the performance and design of vehicles with flexible tracks (Wong 1986, 1992a, 1992b, 1994a, 1994b, 1994c, 1995, 1997, 1999; Wong and Preston-Thomas 1986; Wong and Irwin 1992). Since NTVPM considers all major vehicle design features, as well as pertinent terrain characteristics, it is a useful tool for the vehicle designer in evaluating and optimizing vehicle designs from a traction perspective, as well as for the procurement manager in selecting vehicle candidates for a given mission and environment.

To demonstrate the capability of NTVPM, examples of employing it to evaluate the effects of vehicle configuration, initial track tension, suspension setting, and longitudinal location of the center of gravity of vehicle sprung weight on vehicle performance are presented (Wong and Preston-Thomas 1988; Wong and Huang 2005, 2006a, 2008; Wong 2007, 2010). As an example, to evaluate the effects

Figure 2.57 Comparisons of the measured drawbar performances of a tracked armored personnel carrier and the predicted ones using NTVPM on (a) sandy terrain, (b) organic terrain (muskeg), and (c) snow-covered terrain.

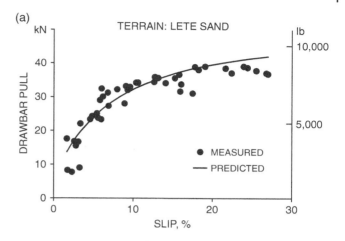

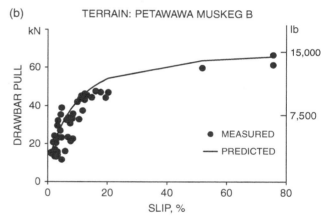

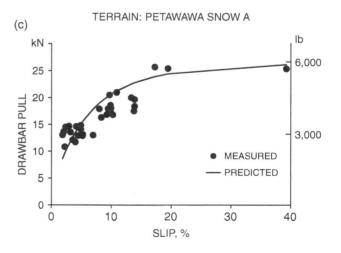

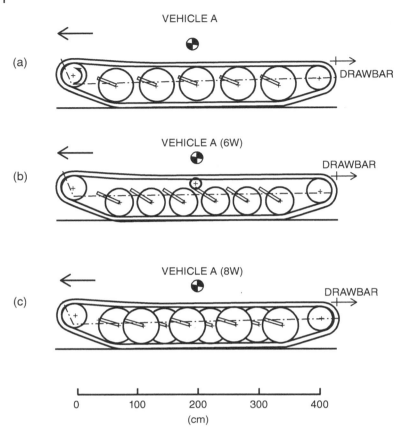

Figure 2.58 Schematic diagrams for the roadwheel-track systems of (a) vehicle A with five roadwheels, (b) vehicle A (6W) with six roadwheels, and (c) vehicle A (8W) with eight overlapping roadwheels.

of vehicle configuration and initial track tension on vehicle performance, three vehicle configurations are selected: vehicle A with five roadwheels, resembling a widely used armored personnel carrier, vehicle A (6W) with six roadwheels, and vehicle A (8W) with eight overlapping roadwheels, as shown in Figure 2.58(a), (b) and (c), respectively.

In the overlapping roadwheel system, the centers of the two rows of roadwheels on a track are shifted relative to each other in the longitudinal direction. It is intended to provide a more uniform normal pressure distribution under the track, while maintaining sufficient suspension travel for obstacle crossing. This overlapping roadwheel system was widely used in fighting vehicles produced in Germany during World War II. While these three vehicle configurations are different in design, their basic parameters, such as vehicle weight, track ground contact length and width, ground clearance, and fundamental natural frequency in heave, are the same, so as to provide a common basis for comparison.

The effects of initial track tension (i.e., the tension in the track system when the vehicle is stationary on a level, hard ground) on the mobility of these three vehicle configurations are evaluated and compared. Note that if the initial track tension is low, the track will be loose. Consequently, track segments between roadwheels cannot support much load and vehicle weight is primarily supported by the track segments (links) immediately under the roadwheels. In this case, a tracked

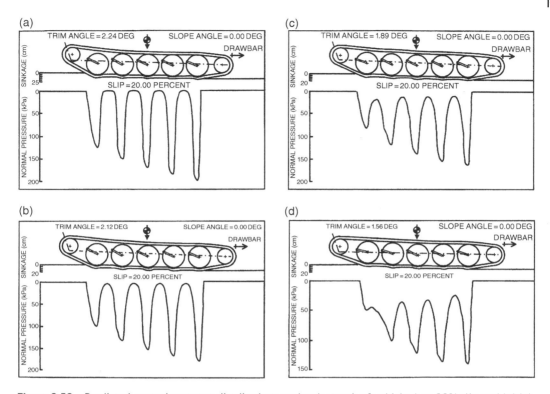

Figure 2.59 Predicted normal pressure distributions under the track of vehicle A at 20% slip and initial track tension coefficients of (a) 2.5%, (b) 10%, (c) 20%, and (d) 40% on a clayey soil with high moisture content.

vehicle in essence behaves like a multi-wheel vehicle. On the other hand, if the initial track tension is high, the track will be tight. Track segments between roadwheels can support substantial load. This reduces the peak normal pressures under the track.

Figures 2.59–2.61 show the normal pressure distributions under the tracks of vehicle A, vehicle A (6W), and vehicle A (8W), respectively, at initial track tension coefficient (i.e., the ratio of initial track tension to vehicle weight) of 2.5, 10, 20, and 40% and at 20% slip, on a clayey soil with high moisture content (Wong and Huang 2008). It can be seen from the figures that as the number of roadwheels or the initial track tension coefficient increases, the normal pressures on track segments between adjacent roadwheels increase, indicating that they take up more load. As a result, the peak normal pressure and the *MMP* decrease.

Figure 2.62 shows the variation of *MMP* for the three vehicle configurations with the initial track tension coefficient predicted using NTVPM. With the reduction in *MMP*, roadwheel sinkage and track motion resistance decrease, leading to improved performance. Figure 2.63 shows the variation of the drawbar pull coefficient (i.e., the ratio of the drawbar pull to vehicle weight) with the initial track tension coefficient for the three vehicle configurations. The results indicate that both the number of roadwheels and the initial track tension coefficient have significant effects on tractive performance. For instance, if the number of roadwheels is increased from five to eight, coupled with an increase of the initial track tension coefficient from 10 to 40%, the drawbar

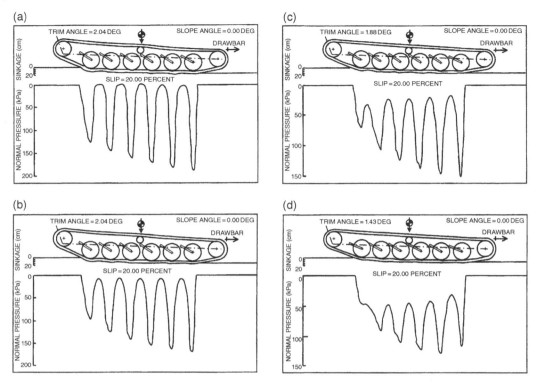

Figure 2.60 Predicted normal pressure distributions under the track of vehicle A (6W) at 20% slip and initial track tension coefficients of (a) 2.5%, (b) 10%, (c) 20%, and (d) 40% on a clayey soil with high moisture content.

pull coefficient and tractive efficiency at 20% slip will increase by 93.6 and 62.4%, respectively (Wong and Huang 2008). This shows that by a combination of increasing the number of road-wheels and the initial track tension coefficient, considerable improvements in vehicle mobility on the clayey soil may be achieved.

This research finding on the effects of initial track tension on tracked vehicle mobility has led to the development of a central initial track tension regulating system controlled by the driver. This remotely controlled device enables the driver to conveniently increase the initial track tension when traversing soft terrain is anticipated. It also enables the driver to conveniently reduce the initial track tension when the vehicle has passed over the soft patch, to minimize the wear and tear of the track system due to high initial track tension. The initial track tension regulating system has been installed in a new generation of military vehicles produced in several countries (Wong 1992a, 1995). The central initial track tension regulating system for improving tracked vehicle mobility is analogous to the central tire inflation system for improving wheeled vehicle mobility. Retrofitting existing tracked vehicles with the central initial track tension regulating system would be one of the cost-effective means to improve their soft ground mobility.

On exceedingly soft terrain, such as deep fresh snow, track sinkage may exceed vehicle ground clearance and the vehicle belly (hull) may be in contact with the terrain surface. The interaction between vehicle belly and terrain may significantly affect vehicle mobility. The effect

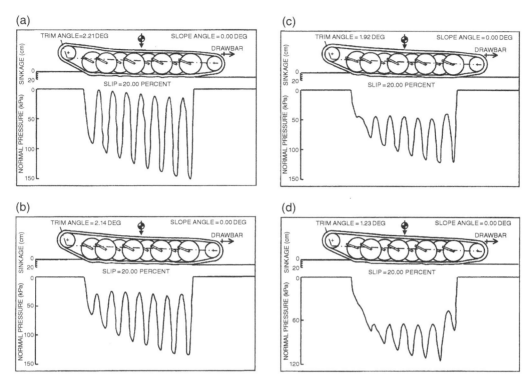

Figure 2.61 Predicted normal pressure distributions under the track of vehicle A (8W) at 20% slip and initial track tension coefficients of (a) 2.5%, (b) 10%, (c) 20%, and (d) 40% on a clayey soil with high moisture content.

Figure 2.62 Variations of the mean maximum pressure at 20% slip with the initial track tension coefficient of vehicle A, vehicle A (6W), and vehicle A (8W) on a clayey soil with high moisture content.

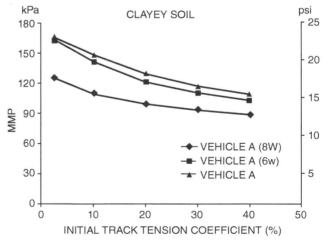

of belly–terrain interaction on vehicle performance depends on the attitude of the belly, as shown in Figure 2.64. If the belly takes a nose-down attitude, then most of the belly (except the front part) will not be in contact with the terrain surface. The angle between the belly and the horizontal is referred to as the belly trim angle and is designated as negative for a nose-down attitude. On the

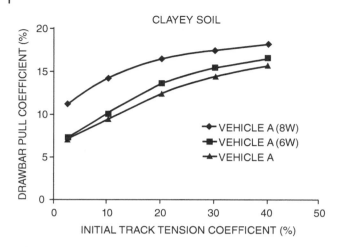

Figure 2.63 Variations of the drawbar pull coefficient at 20% slip with the initial track tension coefficient of vehicle A, vehicle A (6W), and vehicle A (8W) on a clayey soil with high moisture content.

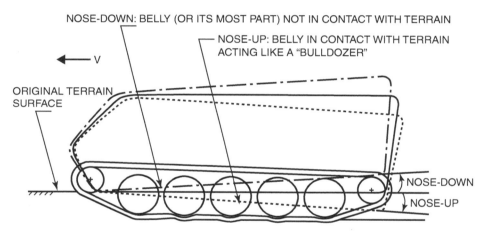

Figure 2.64 Effects of vehicle belly attitude on belly–terrain interaction, when track sinkage is greater than vehicle ground clearance.

other hand, if the belly takes a nose-up attitude (or the trim angle is positive), then the belly may be in full contact with the terrain surface and support part of the vehicle weight. This reduces the load applied to the tracks, hence the tractive effort on most terrains exhibiting frictional behavior. Furthermore, the shear force on the belly–terrain interface, caused by the sliding of the belly on the terrain, acts against the motion of the vehicle. In addition, the belly may act as a bulldozer pushing terrain material forward, like a snowplow, hence inducing additional bull-dozing drag. This indicates that if the belly is in contact with the terrain surface and takes a nose-up attitude, it will have adverse effects on vehicle mobility by reducing traction and inducing additional drags.

As noted previously, NTVPM takes into account the effect of belly–terrain interaction on vehicle performance. The following examples illustrate the application of NTVPM to evaluating the effect of suspension setting and of longitudinal location of the center of gravity of vehicle sprung weight on the mobility of vehicle A with five roadwheels (shown in Figure 2.58(a)) in deep snow with an average depth of 120 cm (Wong and Preston-Thomas 1988; Wong 2007, 2010).

Figure 2.65 Effects of suspension settings on drawbar pull coefficients of vehicle A in deep snow. *Source:* Wong and Preston-Thomas (1988). Reproduced with permission of the Council of the Institution of Mechanical Engineers.

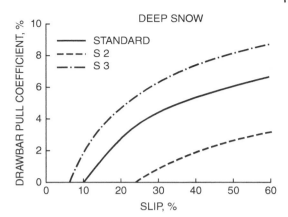

Figure 2.65 shows the effects of suspension settings on drawbar pull coefficients of vehicle A on the deep snow (Wong and Preston-Thomas 1988). The settings of the three suspension configurations examined are given in Table 2.15. The basic difference between them is in the setting of the torsion arm angles under no-load conditions. For the standard suspension configuration of vehicle A, the initial torsion arm angle is set at 43° below the horizontal for all five roadwheel stations, as shown in Table 2.15. For suspension configuration S2, the torsion arm angle is set in a decreasing order from 51.6° at the front (first) roadwheel station to 34.4° at the rear (fifth) roadwheel station, while maintaining an angle of 43° for the torsion arm at the middle (third) roadwheel station. This setting results in a nose-up attitude for the vehicle belly when the vehicle is stationary on level, hard ground. As noted previously, this has adverse effects on vehicle mobility in deep snow, as indicated in Figure 2.65. For suspension configuration S3, the torsion arm angle is set in an increasing order from 34.4° at the front (first) roadwheel station to 51.6° at the rear (fifth) roadwheel station, while maintaining an angle of 43° for the torsion arm at the middle (third) roadwheel station. This setting results in a nose-down attitude of the vehicle belly. As mentioned previously, this improves vehicle mobility in deep snow, in comparison to that with the standard suspension configuration and configuration S2, as shown in Figure 2.65.

Table 2.15 Torsion arm settings for the standard suspension and suspensions S2 and S3.

	Initial torsion arm angle under no load (below the horizontal), degrees		
	Suspension configuration		
Roadwheel station	Standard	S2	S3
1	43	51.6	34.4
2	43	47.3	38.7
3	43	43.0	43.0
4	43	38.7	47.3
5	43	34.4	51.6

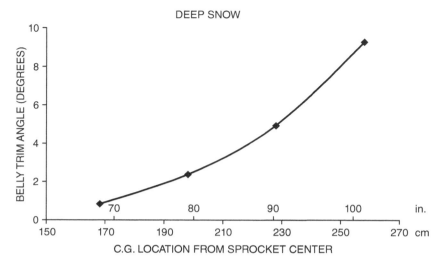

Figure 2.66 Variations of the belly trim angle with the longitudinal location of sprung weight center of gravity of vehicle A at 20% slip on deep snow.

The longitudinal location of the center of gravity (CG) of vehicle sprung weight affects the attitude of vehicle belly, and hence vehicle performance in deep snow. Figure 2.66 shows the variation of the belly trim angle with the longitudinal location of the C.G. of the sprung weight of vehicle A in deep snow (Wong 2007, 2010). The longitudinal location of the center of gravity is expressed in terms of the distance from the center of the sprocket, which is located at the front, as shown in Figure 2.58(a). The positive belly trim angle increases with the rearward shift of the center of gravity. This indicates that the vehicle belly takes an increasingly nose-up attitude as the center of gravity shifts toward the rear. Figure 2.67 shows the variation of the drawbar pull coefficient

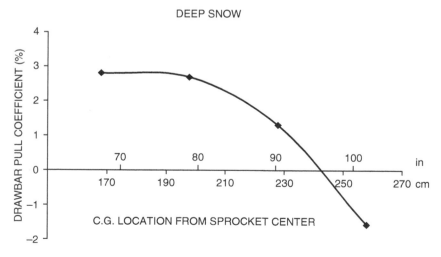

Figure 2.67 Variations of the drawbar pull coefficient with the longitudinal location of sprung weight center of gravity of vehicle A at 20% slip on deep snow.

at 20% slip with the longitudinal location of the sprung weight center of gravity at the initial track tension coefficient of 10% in deep snow. The drawbar pull coefficient decreases noticeably with the shifting of the center of gravity rearward. If the center of gravity is located at a distance more than approximately 240 cm from the sprocket center in the longitudinal direction, the drawbar pull coefficient will be negative, indicating that the vehicle will be unable to propel itself and will become immobile.

The examples presented above demonstrate that NTVPM can be a useful tool for vehicle designers in quantitatively evaluating competing designs and in optimizing vehicle design parameters for a given mission and operating environment, from a traction perspective. It is also shown that the application of NTVPM may lead to design innovations, as well as new avenues of approach to the development of tracked vehicles with flexible tracks.

2.7 An Advanced Physics-Based Model for the Performance of Vehicles with Long-Pitch Link Tracks

For low-speed tracked vehicles, such as those used in agriculture and construction industry, rigid-link tracks with relatively long track pitch are commonly used. The use of the long-pitch track is intended to achieve more uniform pressure distribution under the track. This type of track system has a ratio of roadwheel diameter to track pitch as low as 1.2 and a ratio of roadwheel spacing to track pitch typically 1.5. Consequently, the model NTVPM described in the previous section is not suitable for this type of vehicle.

An advanced physics-based model known as RTVPM (Rigid-Link Tracked Vehicle Performance Model) has, therefore, been developed for performance and design evaluation of tracked vehicles with long-pitch link tracks (Gao and Wong 1994; Wong and Gao 1994; Wong 1998). This model takes into account all major design parameters of the vehicle, including vehicle weight, location of the center of gravity, number of roadwheels, location of roadwheels, roadwheel dimensions and spacing, locations of sprocket and idlers, supporting roller arrangements, track dimensions and geometry, initial track tension, and drawbar hitch location. As the track links are rigid, the track is assumed to be inextensible. For most low-speed tracked vehicles, the roadwheels are not sprung and hence are rigidly connected to the track frame. RTVPM, like NTVPM, uses the Bekker–Wong terrain parameters for characterizing terrain properties.

2.7.1 Basic Approach

The model RTVPM treats the track as a system of rigid links connected with frictionless pins, as shown in Figure 2.68. As noted previously, the roadwheels, supporting rollers, and sprocket are assumed to be rigidly attached to the vehicle frame. The center of the front idler, however, is assumed to be mounted on a pre-compressed spring.

In the analysis, the track system is divided into four sections: the upper run of the track supported by rollers; the lower run of the track in contact with the roadwheels and the terrain; the section in contact with the idler; and the section in contact with the sprocket. By considering the force equilibrium of various sections of the track system, the interaction between the lower run of the track and the terrain, and the boundary conditions for various track sections, a set of equations can be formulated. The solutions to this set of equations determine the sinkage and inclination of the track system, the normal and shear stress distributions on the track–terrain interface, and

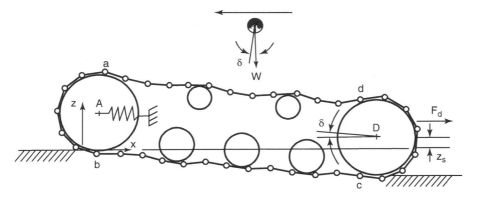

Figure 2.68 Schematic of a track system with long-pitch rigid links. *Source:* Gao and Wong (1994). Reproduced with permission of the Council of the Institution of Mechanical Engineers.

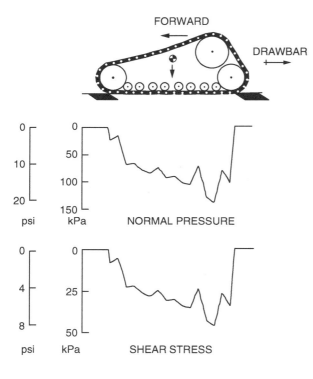

Figure 2.69 Normal and shear stress distributions under a vehicle with long-pitch rigid link tracks predicted using the model RTVPM.

the track motion resistance, thrust, drawbar pull, and tractive efficiency of the vehicle as functions of track slip. Figure 2.69 shows the predicted normal pressure and shear stress distributions under a tracked vehicle with eight roadwheels on a clayey soil.

2.7.2 Experimental Substantiation

The basic features of RTVPM have been substantiated with available field test data. Figures 2.70 and 2.71 show a comparison of the measured and predicted drawbar pull coefficient (ratio of drawbar

Figure 2.70 Comparisons of the measured and predicted drawbar pull coefficient of a vehicle with long-pitch link tracks on a dry, disked sandy loam using the model RTVPM. *Source:* Gao and Wong (1994). Reproduced with permission of the Council of the Institution of Mechanical Engineers.

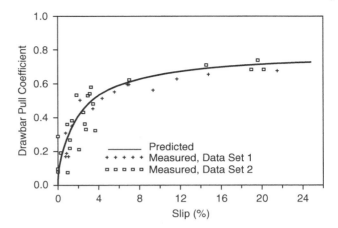

Figure 2.71 Comparisons of the measured and predicted tractive efficiency of a vehicle with long-pitch link tracks on a dry, disked sandy loam using the model RTVPM. *Source:* Gao and Wong (1994). Reproduced with permission of the Council of the Institution of Mechanical Engineers.

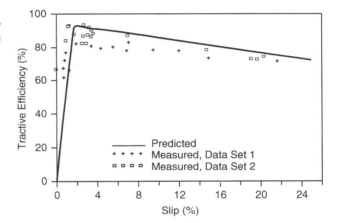

pull to vehicle weight) and tractive efficiency (the ratio of the product of drawbar pull and vehicle speed to the power input to the drive sprockets), respectively, for a heavy tracked vehicle used in the construction industry (Gao and Wong 1994). The vehicle has a total weight of 329 kN (73 966 lb). It has eight roadwheels of diameter 26 cm (10.2 in.) on each of the two tracks, and the average spacing between roadwheels is 34 cm (13.4 in.). The track pitch is 21.6 cm (8.5 in.), and the track width is 50.8 cm (20 in.). The terrain is a dry, disked sandy loam, with an angle of internal shearing resistance of 40.1° and a cohesion of 0.55 kPa (0.08 psi). The measured data shown in the figures are provided courtesy of Caterpillar, Peoria, IL, USA.

Figure 2.72 shows a comparison of the predicted (a) and measured (b) normal and shear stress distributions under a rigid-link track system with four roadwheels on a loosely cultivated sand (Wong 1998). The track contact length is 1.27 m (50 in.), track pitch is 0.149 m (5-7/8 in.), track width is 0.254 m (10 in.), and normal load is 13.12 kN (2450 lb).

It shows that the drawbar performance and stress distributions on the track–terrain interface of the vehicles predicted using RTVPM are reasonably close to the measured data. This suggests that the model can provide realistic predictions of the performance of vehicles with long-pitch link tracks in the field.

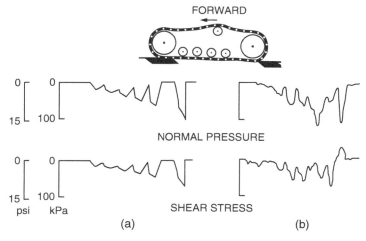

Figure 2.72 Comparisons of (a) the predicted and (b) the measured normal and shear stress distributions under a track system with long-pitch rigid links on a loosely cultivated sand, using the model RTVMP. *Source:* Wong (1998). Reproduced with permission of the Council of the Institution of Mechanical Engineers.

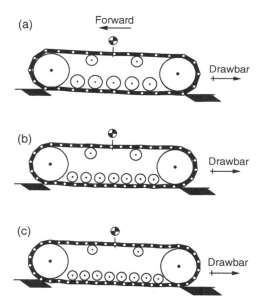

Figure 2.73 Three track systems with different numbers of roadwheels used in the study of the effects of the ratio of roadwheel spacing to track pitch on tractive performance. *Source:* Wong (1998). Reproduced with permission of the Council of the Institution of Mechanical Engineers.

2.7.3 Applications to Parametric Analysis and Design Optimization

The practical applications of the model RTVPM to design evaluation may be demonstrated through an example of the study of the optimum roadwheel spacing to track pitch ratio for vehicles with long-pitch link tracks.

To evaluate the effects of the ratio of roadwheel spacing to track pitch, the tractive performances of three track system configurations with five, seven, and eight roadwheels shown in Figure 2.73 and with tracks of various pitches were predicted using the model RTVPM. It was found that for given overall dimensions of a track system, the ratio of roadwheel spacing to track pitch is one of the design parameters that have significant effects on its tractive performance. Figures 2.74 and 2.75 show the variations of the drawbar pull coefficient and tractive efficiency at 20% slip with the ratio of roadwheel spacing to track pitch on a clayey soil, respectively (Wong 1998). If the ratio of roadwheel spacing to track pitch is similar, the tractive performances of the track systems with different numbers of roadwheels ranging from five to eight will be similar. This conclusion is further supported by the observation that the normal pressure distributions under the track systems with different numbers of roadwheels are similar for similar ratios of roadwheel spacing to track pitch (S/P). Figures 2.76 and 2.77 show the normal pressure distributions under

Figure 2.74 Effects of the ratio of roadwheel spacing to track pitch on the drawbar pull coefficient of track systems with different numbers of roadwheels on a clayey soil. *Source:* Wong (1998). Reproduced with permission of the Council of the Institution of Mechanical Engineers.

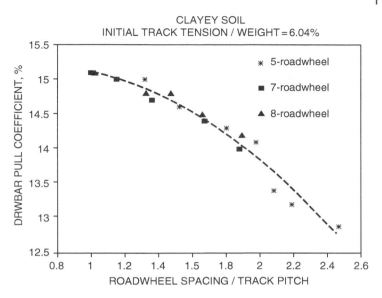

Figure 2.75 Effects of the ratio of roadwheel spacing to track pitch on the tractive efficiency of track systems with different numbers of roadwheels on a clayey soil. *Source:* Wong (1998). Reproduced with permission of the Council of the Institution of Mechanical Engineers.

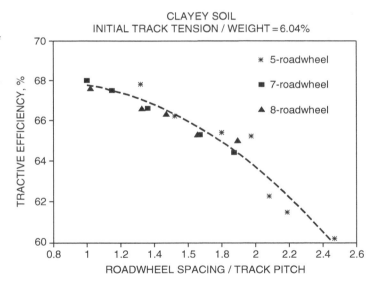

the track systems with five and eight roadwheels, respectively, at various ratios of roadwheel spacing to track pitch on the clayey soil. For similar ratios of roadwheel spacing to track pitch, the normal pressure distributions under the track system with five roadwheels have similar characteristics to those under the track system with eight roadwheels. Note that as the ratio decreases, the fluctuations of normal pressure under the track decrease. This indicates that if the ratio of roadwheel spacing to track pitch is lowered, the normal pressure under the track system is more uniformly distributed, which leads to improvements in the tractive performance of the vehicle.

It should also be noted that within a certain range of the ratio of roadwheel spacing to track pitch, the drawbar pull coefficient and tractive efficiency vary only slightly. For instance, the drawbar pull coefficient and tractive efficiency at 20% slip change marginally if the ratio of roadwheel spacing to track

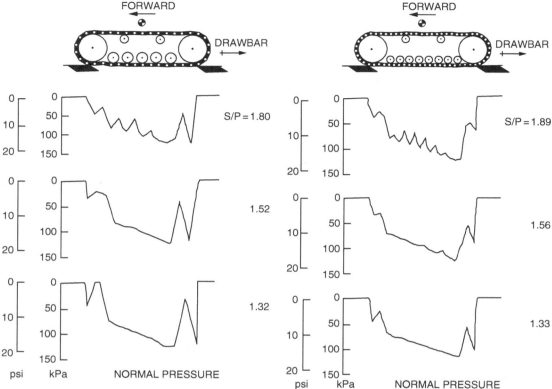

Figure 2.76 Normal pressure distributions under a track system with five roadwheels at various ratios of roadwheel spacing to track pitch on a soft soil. *Source:* Wong (1998). Reproduced with permission of the Council of the Institution of Mechanical Engineers.

Figure 2.77 Normal pressure distributions under a track system with eight roadwheels at various ratios of roadwheel spacing to track pitch in a soft soil. *Source:* Wong (1998). Reproduced with permission of the Council of the Institution of Mechanical Engineers.

pitch varies from 1.3 to 1 on the clayey soil, as shown in Figures 2.74 and 2.75, respectively. This implies that the designer would have a certain flexibility in selecting the appropriate track pitch or roadwheel spacing that, on the one hand, can ensure good tractive performance and, on the other hand, can minimize the fluctuations of vehicle speed due to the polygon (or chordal) effect of the sprocket tooth–track link engagement. It can be shown that the speed variation due to the polygon effect is given by

$$\delta_s = 1 - \sqrt{1 - \left(\frac{P}{D}\right)^2} \tag{2.88}$$

where δ_s is vehicle speed variation, P is the track pitch, and D is the pitch diameter of the sprocket.

For most agricultural and industrial tracked vehicles currently in use, the ratio of the sprocket pitch diameter to track pitch varies approximately from 3.7 to 4.3. Therefore, vehicle forward speed fluctuations will be in the range of 3.72–2.75%. If the speed variation is limited to 2.75% and the sprocket pitch diameter is 0.928 m (36.5 in.), then the track pitch should be 0.216 m (8.5 in.). On the clayey soil, to ensure good drawbar performance, the ratio of roadwheel spacing to track pitch should be in the range of 1.3 to 1. Consequently, the roadwheel spacing should be in the range of 0.281–0.216 m (11–8.5 in.).

The above-noted example illustrates one of the practical applications of RTVPM to the selection of optimum design parameters of vehicles with long-pitch link tracks. The model RTVPM has been successfully employed in assisting vehicle manufacturers in the development of new products.

2.8 Physics-Based Models for the Cross-Country Performance of Wheels (Tires)

2.8.1 Motion Resistance of a Rigid Wheel

While pneumatic tires have long replaced rigid wheels in most off-road wheeled vehicles, the mechanics of a rigid wheel over unprepared terrain is still of interest, as a pneumatic tire may behave like a rigid rim in soft terrain under certain conditions. Furthermore, rigid wheels are used on unmanned rovers for surface exploration of the moon and Mars, on tractors for operation on paddy field, and as traction-aid devices attached to pneumatic tires for operation on wet terrains.

One of the earlier models for predicting the motion resistance of a rigid wheel is that proposed by Bekker (1956, 1960, 1969). In developing the model, it is assumed that the terrain reaction at all points on the contact patch is purely radial and is equal to the normal pressure beneath a horizontal plate at the same depth in a pressure–sinkage test. As shown in Figure 2.78, the equilibrium equations of a towed rigid wheel can be written as follow:

$$R_c = b \int_0^{\theta_0} \sigma r \sin \theta \, d\theta \tag{2.89}$$

$$W = b \int_0^{\theta_0} \sigma r \cos \theta \, d\theta \tag{2.90}$$

where R_c is the motion resistance, W is the vertical load, σ is the normal pressure, and b and r are the width and radius of the wheel, respectively.

Since it is assumed that the normal pressure σ acting on the wheel rim is equal to the normal pressure p beneath a plate at the same depth z, then $\sigma r \sin\theta \, d\theta = p \, dz$ and $\sigma r \cos\theta \, d\theta = p \, dx$. Using the pressure–sinkage relationship defined by Eq. (2.47), Eq. (2.89) becomes

$$R_c = b \int_0^{z_0} \left(\frac{k_c}{b} + k_\phi \right) z^n dz$$
$$= b \left[\left(\frac{k_c}{b} + k_\phi \right) \frac{z_0^{n+1}}{n+1} \right] \tag{2.91}$$

Figure 2.78 Simplified wheel–soil interaction model. *Source: Theory of Land Locomotion* by M.G. Bekker, copyright © 1956 by the University of Michigan. Reproduced with permission of the University of Michigan Press.

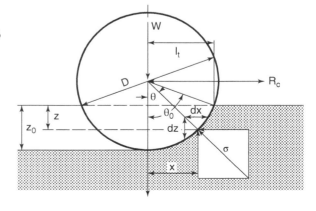

The value of R_c calculated by Eq. (2.91) is equivalent to the vertical work done per unit length in pressing a plate of width b into the ground to a depth of z_0. The assumption for the stress distribution made by Bekker implies that the motion resistance of a rigid wheel is due to the vertical work done in making a rut of depth z_0. The motion resistance R_c is referred to as the compaction resistance.

Using Eq. (2.91) to calculate the compaction resistance, the static sinkage z_0 expressed in terms of wheel parameters and terrain properties must be determined first. From Eq. (2.90),

$$W = -b \int_0^{z_0} p \, dx = -b \int_0^{z_0} \left(\frac{k_c}{b} + k_\phi\right) z^n dx \tag{2.92}$$

From the geometry shown in Figure 2.78,

$$x^2 = [D - (z_0 - z)](z_0 - z) \tag{2.93}$$

where D is the wheel diameter.
For small sinkages,

$$x^2 = D(z_0 - z) \tag{2.94}$$

and

$$2x \, dx = -D \, dz \tag{2.95}$$

Substituting Eq. (2.95) into Eq. (2.92), one obtains

$$W = b \left(k_c/b + k_\phi\right) \int_0^{z_0} \frac{z^n \sqrt{D}}{2\sqrt{z_0 - z}} \, dz \tag{2.96}$$

Let $z_0 - z = t^2$, then $dz = -2t \, dt$ and

$$W = b \left(\frac{k_c}{b} + k_\phi\right) \sqrt{D} \int_0^{\sqrt{z_0}} (z_0 - t^2)^n dt \tag{2.97}$$

Expanding $(z_0 - t^2)^n$ and only taking the first two terms of the series $(z_0^n - n z_0^{n-1} t^2 + n(n-1) z_0^{n-2} t^4/2 - n(n-1)(n-2) z_0^{n-3} t^6/6 + n(n-1)(n-2)(n-3) z_0^{n-4} t^8/24 + \cdots)$, one obtains

$$W = \frac{b\left(k_c/b + k_\phi\right)\sqrt{z_0 D}}{3} z_0^n (3 - n) \tag{2.98}$$

Rearranging Eq. (2.98), it becomes

$$z_0^{(2n+1)/2} = \frac{3W}{b\left(k_c/b + k_\phi\right)(3 - n)\sqrt{D}}$$

or

$$z_0 = \left[\frac{3W}{b(3 - n)\left(k_c/b + k_\phi\right)\sqrt{D}}\right]^{[2/(2n+1)]} \tag{2.99}$$

Substituting Eq. (2.99) into Eq. (2.91), the compaction resistance R_c becomes

$$R_c = \frac{1}{(3-n)^{(2n+2)/(2n+1)}(n+1)\,b^{1/(2n+1)}\left(k_c/b + k_\phi\right)^{1/(2n+1)}}$$
$$\cdot \left(\frac{3W}{\sqrt{D}}\right)^{(2n+2)/(2n+1)} \tag{2.100}$$

It can be seen from Eq. (2.100) that to reduce the compaction resistance, it seems more effective to increase the wheel diameter D than the wheel width b, as D enters the equation in higher power than b. Note that Eq. (2.100) is derived from Eq. (2.98), which is obtained using only the first two terms of a series to represent $(z_0 - t^2)^n$ in Eq. (2.97). As a result, Eq. (2.100) works well only for values of n up to about 1.3. Beyond that, the error in predicting the compaction resistance R_c increases, and when the value of n approaches 3, R_c approaches infinity – an obvious anomaly. For values of n greater than 1.3, the first five terms in the series should be taken to represent the function $(z_0 - t)^n$ in Eq. (2.97) for the integration. This will greatly improve the accuracy in the prediction of the compaction resistance R_c.

Bekker pointed out that acceptable predictions may be obtained using Eq. (2.100) for moderate sinkages (i.e., $z_0 \leq D/6$), and that the larger the wheel diameter and the smaller the sinkage, the more accurate the predictions are (Bekker 1969). He also mentioned that predictions based on Eq. (2.100) for wheels smaller than 50 cm (20 in.) in diameter becomes less accurate, and that predictions of sinkage based on Eq. (2.99) in dry, sandy soil are not accurate if there is significant slip sinkage (Bekker 1969). The issue of slip sinkage will be discussed in Section 2.10.

Experimental evidence shows that the actual normal pressure distribution beneath a rigid wheel is different from that assumed in the theory described above, as shown in Figure 2.79 (Onafeko and Reece 1967; Wong 1967; Wong and Reece 1967). According to the Bekker model described above, the maximum normal pressure should occur at the lowest point of contact (bottom-dead-center) where the sinkage is a maximum. Experimental results, however, show that the maximum normal pressure occurs in front of the bottom-dead-center, and that its location varies with slip i, as shown in Figure 2.79. The maximum normal pressure occurs at the junction of the two flow zones, as shown in Figures 2.19 and 2.20 (Wong 1967; Wong and Reece 1967). Based on test data on various types of sandy terrain, the variation of the angular position of the maximum normal pressure θ_m measured from the vertical (or the bottom-dead-center) with slip i, in relation to the contact angle of the wheel θ_o shown in Figure 2.78 may be expressed by the following equation (Wong and Reece 1967):

$$\frac{\theta_m}{\theta_o} = c_1 + c_2 i \tag{2.101}$$

where c_1 and c_2 are coefficients and for sandy terrains their values are given in Table 2.16.

The variation of normal pressure distribution with slip implies that the motion resistance should be expected as a function of slip. This indicates that the actual interaction between the wheel and the terrain is much more complicated than that assumed in the simplified model described above.

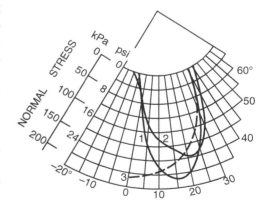

Figure 2.79 Comparisons of the measured normal pressure distributions on a rigid wheel at different slips with the predicted one using the simplified wheel–soil interaction model: curve 1 – measured at 3.1% slip; curve 2 – measured at 35.1%; and curve 3 – predicted using the simplified model.

Table 2.16 Values of coefficients c_1 and c_2 for various types of sandy terrain.

Soil type	Angle of internal shearing resistance $\phi°$	Cohesion c		Weight density γ_s		Coefficient		Source
		kPa	lb/in.^{2}	g/cm^3	lb/in.^{3}	c_1	c_2	
Compact sand	33.3	0.69	0.10	1.59	0.0575	0.43	0.32	Onafeko and Reece (1967)
Loose sand	31.1	0.83	0.12	1.33	0.048	0.18	0.32	
Sand	36.0	0.69	0.10	1.71	0.0617	0.285	0.32	Hegedus (1965)
Dry sand	24.0					0.38	0.41	Sela (1964)

In soft terrain where wheel sinkage is significant, Bekker suggested that a bulldozing resistance acting in front of the wheel should be taken into consideration, in addition to the compaction resistance R_c given by Eq. (2.100). The bulldozing resistance may be estimated using the earth pressure theory described in Section 2.2.

2.8.2 Motion Resistance of a Pneumatic Tire

The motion resistance of a pneumatic tire depends on its mode of operation. If the ground is sufficiently soft and the sum of the inflation pressure p_i and the pressure produced by the stiffness of the carcass p_c is greater than the maximum pressure that the terrain can support at the lowest point of the tire circumference, the tire will remain round like a rigid rim, as shown in Figure 2.80. This is usually referred to as the rigid mode of operation. On the other hand, if the terrain is firm and the tire inflation pressure is low, a portion of the circumference of the tire will be flattened. This is referred to as the elastic mode of operation. When predicting the motion resistance of a tire, it is necessary first to determine whether the pneumatic tire behaves like a rigid rim or an elastic wheel under a given operating condition. If the tire behaves like a rigid rim, using Bekker's pressure–sinkage relationship, the normal pressure at the lowest point of contact (bottom-dead-center) p_g is

$$p_g = \left(k_c/b + k_\phi\right)z_0^n \tag{2.102}$$

Substituting Eq. (2.99) into the above equation, the expression for p_g becomes

$$p_g = \left(k_c/b + k_\phi\right)^{1/(2n+1)} \left[\frac{3W}{(3-n)b\sqrt{D}}\right]^{2n/(2n+1)} \tag{2.103}$$

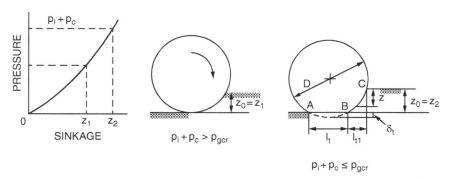

Figure 2.80 Behavior of a pneumatic tire in different operating modes.

If the sum of the inflation pressure p_i and the pressure due to carcass stiffness p_c is greater than the pressure defined by Eq. (2.103), which may be called the critical pressure p_{gcr}, the tire will remain round like a rigid wheel (Wong 1972). Under this condition, the motion resistance due to compacting the terrain can be predicted using Eq. (2.100). On the other hand, if the sum of p_i and p_c is less than p_{gcr} calculated from Eq. (2.103), a portion of the circumference of the tire will be flattened, and the contact pressure on the flat portion will be equal to $p_i + p_c$. In this case, the static sinkage of the tire z_0 can be determined by the following equation if Bekker's pressure–sinkage equation is used:

$$z_0 = \left(\frac{p_i + p_c}{k_c/b + k_\phi}\right)^{1/n} \tag{2.104}$$

Substituting Eq. (2.104) into Eq. (2.91), the expression for the motion resistance of an elastic wheel due to compacting the terrain becomes

$$R_c = b\left(k_c/b + k_\phi\right)\left(\frac{z_0^{n+1}}{n+1}\right)$$
$$= \frac{b(p_i + p_c)^{(n+1)/n}}{(n+1)\left(k_c/b + k_\phi\right)^{1/n}} \tag{2.105}$$

In practice, the pressure p_c exerted on the terrain due to carcass stiffness is difficult to determine, as it varies with the inflation pressure and normal load of the tire. As an alternative, Bekker proposed to use the average ground pressure p_{gr} of a tire on hard ground to represent the sum of p_i and p_c. The average ground pressure p_{gr} for a specific tire at a given normal load and inflation pressure can be derived from the generalized deflection chart, as shown in Figure 2.81. Using the chart, for a given normal load W and a specific tire inflation pressure p_i, one can determine tire deflection. From tire deflection one can obtain the ground contact area A, as indicated in the figure. The average ground pressure of the tire p_{gr} is equal to the load carried by the tire divided by the corresponding ground contact area A. As an example, Figure 2.82 shows the relationship between the average ground pressure p_{gr} and the inflation pressure p_i for a 11.00 R16XL tire at various normal loads. It appears that for the tire shown, the pressure p_c exerted on the ground due to carcass stiffness is not a constant, and that its value varies with inflation pressure and load. It is interesting to note from Figure 2.82 that when the tire load and inflation pressure are within certain ranges, the average ground pressure p_{gr} is lower than the inflation pressure p_i. Using the average ground pressure p_{gr} to represent the sum of p_i and p_c, Eqs. (2.104) and (2.105) can be rewritten as

$$z_0 = \left(\frac{p_{gr}}{k_c/b + k_\phi}\right)^{1/n} \tag{2.106}$$

$$R_c = \frac{bp_{gr}^{(n+1)/n}}{(n+1)\left(k_c/b + k_\phi\right)^{1/n}} \tag{2.107}$$

For tires that are wide in comparison with the diameter, such as terra tires and rolligons, care must be taken in using Eqs. (2.106) and (2.107)

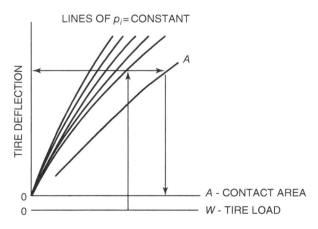

Figure 2.81 Generalized deflection chart for a tire.

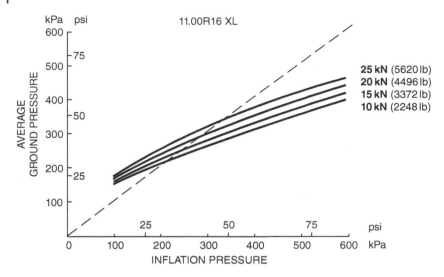

Figure 2.82 Variations of average ground pressure with inflation pressure and normal load for an off-road tire.

to predict the static sinkage and compaction resistance. For these types of tire, the smaller dimension of the loading area (i.e. the denominator of k_c in the pressure sinkage equation, Eq. (2.47)) is not necessarily the width of the tire, and the contact length l_t shown in Figure 2.80 may well be the smaller dimension of the contact patch. This indicates that to predict the performance of this type of tire, the contact length l_t must be determined by considering the vertical equilibrium of the tire. An approximate method for analyzing the performance of this type of tire is given below.

As a first approximation, it may be assumed that the contact length l_t is a function of tire deflection δ_t shown in Figure 2.80:

$$l_t = 2\sqrt{D\delta_t - \delta_t^2} \tag{2.108}$$

When l_t is less than the width of the tire, its value should be used as the denominator of k_c in calculating the static sinkage z_0:

$$z_0 = \left(\frac{p_{gr}}{k_c/l_t + k_\phi}\right)^{1/n} \tag{2.109}$$

The normal load W on the tire is supported by the ground pressure p_{gr} on the flat portion AB, as well as by the reaction on the curved portion BC shown in Figure 2.80. As a first approximation, BC may be assumed to be a circular arc with radius $r = D/2$. The vertical reaction W_{cu} along BC may be determined following an approach like that for analyzing a rigid wheel described in the previous section:

$$W_{cu} = -b \int_0^{z_0} p \, dx = -b\left(k_c/l_t + k_\phi\right) \int_0^{z_0} \frac{z^n \sqrt{D} \, dz}{2\sqrt{z_0 + \delta_t - z}} \tag{2.110}$$

Denote $z_0 + \delta_t - z = t^2$; then $dz = -2t \, dt$ and

$$W_{cu} = b\left(k_c/l_t + k_\phi\right)\sqrt{D} \int_{\sqrt{\delta_t}}^{\sqrt{z_0 + \delta_t}} \left(z_0 + \delta_t - t^2\right)^n dt \tag{2.111}$$

By expanding $(z_0 + \delta_t - t^2)^n$ into a series and taking only the first two terms of the series, one obtains

$$
\begin{aligned}
W_{cu} &= b(k_c/l_t + k_\phi)\sqrt{D} \int_{\sqrt{\delta_t}}^{\sqrt{z_0 + \delta_t}} \left[(z_0 + \delta_t)^n - n(z_0 + \delta_t)^{n-1}t^2 \right] dt \\
&= \left[b(k_c/l_t + k_\phi)\sqrt{D}(z_0 + \delta_t)^{n-1} \right] \\
&\quad \times \frac{\left[(3-n)(z_0 + \delta_t)^{3/2} - (3-n)\delta_t^{3/2} - 3z_0\sqrt{\delta_t} \right]}{3}
\end{aligned}
\tag{2.112}
$$

The equilibrium equation for the vertical forces acting on the tire is

$$
W = bp_{gr}l_t + W_{cu} \tag{2.113}
$$

It can be seen that the normal reaction of a given tire is a function of δ_t, l_t, and z_0, and that the relationships among l_t, z_0, and δ_t are governed by Eqs. (2.108) and (2.109). This indicates that for a given tire with known normal load, there is a particular value of tire deflection δ_t that satisfies Eq. (2.113) over a specific terrain. In principle, the tire deflection δ_t, therefore, can be determined by solving Eqs. (2.108), (2.109), (2.112), and (2.113) simultaneously. In practice, however, it is more convenient to follow an iterative procedure to determine the value of tire deflection. In the iteration process, a value of δ_t is first assumed and is substituted into Eq. (2.108) to calculate the contact length l_t. Then use is made of Eq. (2.109) to calculate the static sinkage z_0. With the values of δ_t, l_t, and z_0 known, the normal reaction of the tire for the assumed value of δ_t can be determined. If the assumed value of δ_t is a correct one, the calculated normal reaction should be equal to the given normal load. If not, a new value of δ_t should be assumed, and the whole process should be repeated until convergence is achieved. After the correct value of δ_t is obtained, the appropriate contact length l_t and static sinkage z_0 can be calculated using Eqs. (2.108) and (2.109). The compaction resistance can then be determined by

$$
R_c = b(k_c/l_t + k_\phi)\left(\frac{z_0^{n+1}}{n+1} \right) \tag{2.114}
$$

A pneumatic tire in the elastic mode of operation deforms. As a result, in addition to the compaction resistance, energy is dissipated in the hysteresis of tire material and in other internal losses, which appears as a resisting force acting on the tire. The resistance due to tire deformation depends on tire design, construction, and material, and on operating conditions. The value of this resistance is usually determined experimentally. Bekker and Semonin proposed the following equation for predicting the motion resistance due to tire deformation (Bekker and Semonin 1975):

$$
R_h = \left[3.581bD^2 p_{gr}\epsilon(0.0349\alpha - \sin 2\alpha) \right]/\alpha(D - 2\delta_t) \tag{2.115}
$$

where p_{gr} is the average ground pressure, and b, D, and δ_t are the tire width, diameter, and deflection, respectively. The parameters α and ϵ are calculated as follows:

$$
\alpha = \cos^{-1}[(D - 2\delta_t)/D] \tag{2.116}
$$

and

$$
\epsilon = 1 - \exp(-k_e\delta_t/h) \tag{2.117}
$$

where α is the contact angle in degrees, h is the tire section height, and k_e is a parameter related to tire construction (bias-ply or radial-ply). The value of k_e is 15 for bias-ply tires and 7 for radial-ply tires.

When the tire sinkage is significant, Bekker suggested that a bulldozing resistance also be considered in the calculation of the total motion resistance of a tire.

The method for predicting the contact geometry on the tire–soil interface presented above is based on simplifying assumptions. To improve predictions, a three-dimensional tire model based on the finite element method (FEM) has been developed for the study of tire–terrain interaction (Nakashima and Wong 1993).

The methods described above are for the prediction of a single tire (wheel). For a vehicle, if both the front and rear axles have the same tread (i.e. transverse distance between left- and right-side tires on the same axle), the rear tires will travel in the ruts formed by the front tires in straight-line motion. To predict the overall tractive performance of a multi-axle wheeled vehicle, the response of the terrain to repetitive normal and shear loading should be considered.

Example 2.4 A pneumatic tire 11.00 R16XL is to be installed on an off-road wheeled vehicle. The tire has a diameter of 97.5 cm (38.4 in.), a section height of 28.4 cm (11.2 in.), and a width of 28 cm (11 in.). It is to carry a load of 20 kN (4496 lb). The vehicle is to operate on a soil with pressure–sinkage parameters $n = 1$ and $k_\phi = 680$ kN/m^3(2.5 lb/in.^{3}). Two inflation pressures, 100 and 200 kPa (14.5 and 29 psi), are proposed. The relationships between the inflation pressure p_i and the average ground pressure p_{gr} for the tire under various normal loads are shown in Figure 2.82. Compare the static sinkage and compaction resistance of the tire at the two inflation pressures proposed.

Solution

On the soil specified, the critical pressure p_{gcr} for the tire can be determined using Eq. (2.103):

$$p_{gcr} = [k_\phi]^{1/(2n+1)} \left[\frac{3W}{(3-n)b\sqrt{D}} \right]^{2n/(2n+1)}$$
$$= 200 \text{ kPa (29 psi)}$$

(a) From Figure 2.82 for a normal load of 20 kN (4496 lb) at an inflation pressure $p_i = 100$ kPa (14.5 psi), the average ground pressure p_{gr} is 170 kPa (24.7 psi). Since $p_{gcr} > p_{gr}$, the tire is operating in the elastic mode, and the lower part of the tire in contact with the terrain is flattened. Using Eq. (2.106), the static sinkage z_0 is given by

$$z_0 = \left(\frac{P_{gr}}{k_\phi} \right)^{1/n} = 0.25 \text{ m (10 in.)}$$

and using Eq. (2.91), the compaction resistance R_c is given by

$$R_c = b(k_\phi) \left(\frac{z_0^{n+1}}{n+1} \right)$$
$$= 5.95 \text{ kN (1338 lb)}$$

(b) From Figure 2.82, for a normal load of 20 kN (4496 lb) at an inflation pressure $p_i = 200$ kPa (29 psi), the average ground pressure p_{gr} is 230 kPa (33.4 psi). Since $p_{gcr} < p_{gr}$, the tire behaves like a rigid wheel. Using Eq. (2.99), the static sinkage z_0 is given by

$$z_0 = \left[\frac{3W}{b(3-n)k_\phi\sqrt{D}} \right]^{2/(2n+1)}$$
$$= 0.294 \text{ m (11.6 in.)}$$

and using Eq. (2.91), the compaction resistance R_c is given by

$$R_c = b(k_\phi)\left(\frac{z_0^{n+1}}{n+1}\right)$$
$$= 8.23\,\text{kN}\,(1850\,\text{lb})$$

It shows that the compaction resistance of the tire at an inflation pressure of 200 kPa (29 psi) is approximately 38.3% higher than that at an inflation pressure of 100 kPa (14.5 psi).

2.8.3 Tractive Effort and Slip of a Wheel (Tire)

To evaluate the relationship between the tractive effort and slip of a rigid wheel, the development of shear displacement along the wheel–soil interface must be determined first. The slip i of a rigid wheel is defined in a similar way as that by Eq. (1.5) or by Eq. (2.70), with r as the radius of the rigid wheel. The shear displacement developed along the contact area of a rigid wheel may be determined based on the analysis of the slip velocity V_j. For a rigid wheel, the slip velocity V_j of a point on the rim relative to the terrain is the tangential component of the absolute velocity at the same point, as illustrated in Figure 2.83 (Wong and Reece 1967). The slip velocity V_j of a point on the rim defined by angle θ is shown in Figure 2.83 and its magnitude can be expressed by

$$V_j = r\omega[1 - (1 - i)\cos\theta] \tag{2.118}$$

It shows that the slip velocity for a rigid wheel varies with angle θ and slip i.

The shear displacement j along the wheel–soil interface is given by

$$j = \int_0^t V_j\,dt = \int_\theta^{\theta_0} r\omega[1 - (1 - i)\cos\theta]\frac{d\theta}{\omega}$$
$$= r[(\theta_0 - \theta) - (1 - i)(\sin\theta_0 - \sin\theta)] \tag{2.119}$$

where θ_0 is the entry angle that defines the angle where a point on the rim comes into contact with the terrain, as shown in Figure 2.83.

Based on the relationship between the shear stress and shear displacement discussed previously, the shear stress distribution along the contact area of a rigid wheel can be determined. For instance, using Eq. (2.58), the shear stress distribution may be described by

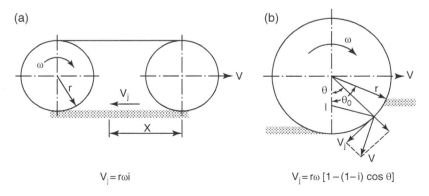

Figure 2.83 Development of shear displacement under (a) a track with rigid links and (b) a rigid wheel.

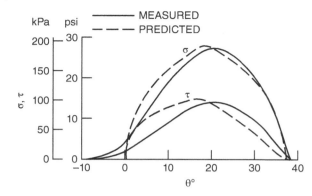

Figure 2.84 Comparisons of the measured and predicted normal and shear stress distributions on the contact area of a rigid wheel at 22.1% slip on compact sand.

$$\tau(\theta) = [c + p(\theta)\tan\phi]\left(1 - e^{-j/K}\right)$$

$$= [c + p(\theta)\tan\phi]\left[1 - \exp^{-(r/K)[\theta_0 - \theta - (1-i)(\sin\theta_0 - \sin\theta)]}\right] \qquad (2.120)$$

The normal pressure distribution along a rigid wheel $p(\theta)$ may be estimated by a variety of methods. Figure 2.84 shows a comparison of the measured shear stress distribution on the contact area of a rigid wheel at 22.1% slip on a compact sand and the predicted one following the procedures described in the reference Wong and Reece (1967). The location of the maximum normal pressure is shifted forward from the bottom-dead-center and is determined using Eq. (2.101), with coefficients $c_1 = 0.43$ and $c_2 = 0.32$ from Table 2.16 for operating on compact sand.

By integrating the horizontal component of tangential stress over the entire contact area, the total tractive effort F can be determined:

$$F = rb\int_0^{\theta_0}\tau(\theta)\cos\theta\,d\theta \qquad (2.121)$$

The vertical component of shear stress on the contact area supports part of the vertical load on the wheel. This fact has been neglected in the simplified wheel–soil interaction model shown in Figure 2.78. In a more complete analysis of wheel–soil interactions, the effect of shear stress should be taken into consideration, and the equations for predicting the tractive performance of a rigid wheel are given by the following (Wong and Reece 1967; Wong 2010):

$$\text{For vertical load } W = rb\left[\int_0^{\theta_0} p(\theta)\cos\theta\,d\theta + \int_0^{\theta_0}\tau(\theta)\sin\theta\,d\theta\right] \qquad (2.122)$$

$$\text{For drawbar pull } F_d = rb\left[\int_0^{\theta_0}\tau(\theta)\cos\theta\,d\theta - \int_0^{\theta_0} p(\theta)\sin\theta\,d\theta\right] \qquad (2.123)$$

$$\text{For wheel torque } M_\omega = r^2b\int_0^{\theta_0}\tau(\theta)\,d\theta \qquad (2.124)$$

Equation (2.120) is for the prediction of shear stress distribution along the contact area of a driven rigid wheel. For a free-rolling, towed rigid wheel, the shear stress distribution has different characteristics, and is shown in Figure 2.85. The shear stress changes direction at a particular

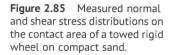

Figure 2.85 Measured normal and shear stress distributions on the contact area of a towed rigid wheel on compact sand.

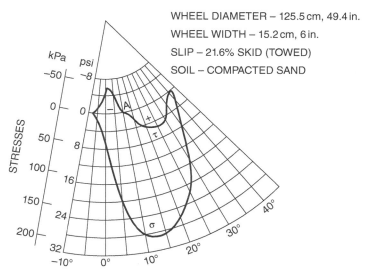

point on the wheel–soil interface, which may be called the transition point (Onafeko and Reece 1967; Wong and Reece 1967). This transition point corresponds to that where the two flow zones in the soil beneath a towed wheel meet each other, as shown in Figure 2.20. Under the action of section *AD* of the rim, the soil in the region *ABD* moves upward and forward, while the rim rotates around the instantaneous center *I*. The soil, therefore, slides along *AD* in such a way as to produce shear stress in the direction opposite to that of wheel rotation, which is denoted as positive. Between *A* and *E*, the soil moves forward slowly, while the wheel rim moves forward relatively fast. In this region, the shear stress acts in the direction of wheel rotation, which is denoted as negative. As a result, the resultant moment about the wheel center due to the shear stress acting on the rim of a free-rolling, towed rigid wheel is zero.

The method for predicting the tractive effort of a pneumatic tire depends on its mode of operation. If the average ground pressure p_{gr} is greater than the critical pressure p_{gcr} defined by Eq. (2.103), the tire will behave like a rigid wheel, and the shear displacement, shear stress, tractive effort, drawbar pull, and wheel torque can be predicted using Eqs. (2.119), (2.120), (2.121), (1.23), and (1.24), respectively.

If p_{gr} is less than p_{gcr}, then a pneumatic tire will deflect, as noted previously. For a pneumatic tire with deflection and operating on deformable terrain with sinkage, the contact geometry of the tire with the terrain is complex. As a result, the determination of its rolling radius for defining longitudinal slip would be different from that for a track, a rigid wheel, or a pneumatic tire on paved road. For a track, slip *i* is defined by Eq. (2.70), where *r* is the sprocket pitch radius which has a fixed value. For a rigid wheel, as noted previously, *r* is the radius of the rigid wheel. For a pneumatic tire operating on paved road, slip *i* is defined by Eq. (1.5), where *r* is the rolling radius of a loaded, free-rolling tire with zero input torque, moving along a straight path. According to the American National Standards Institute/American Society of Agricultural Engineers Standard ANSI/ASAE S296.5 DEC2003 (R2018), reaffirmed as an American National Standard in 2018, there are four possible conditions for specifying pneumatic tire rolling radius *r* for defining its longitudinal slip *i*:

- A self-propelled condition on a nondeforming surface (recommended for rolling circumference data)
- A self-propelled condition on the test surface (including deformable surface)

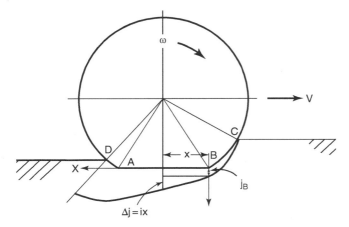

Figure 2.86 Development of shear displacement beneath a tire in the elastic operating mode.

- A towed condition on a nondeforming surface
- A towed condition on the test surface (including deformable surface)

The condition used to define tire rolling radius should always be stated in citing the slip.

Based on the results of a study on the rolling radius and slip of an agricultural tire over loose soil using an explicit finite element formulation, the use of the tire rolling radius r under towed condition (i.e. with zero gross traction force) for defining longitudinal slip i is proposed (Rubinstein et al. 2018).

When a pneumatic tire deflects, a portion of the tire circumference is assumed to be flattened, as shown in Figure 2.86. Under these circumstances, the shear displacement developed along BC in Figure 2.86 may be determined in the same way as that described earlier for a rigid wheel. For the flat portion AB, the slip velocity is a constant, like that beneath a rigid track described in Section 2.5.2. The increase in shear displacement Δj along section AB is proportional to the-slip of the tire i and the distance x between the point in question and point B, and is expressed by

$$\Delta j = ix \tag{2.125}$$

The cumulative shear displacement j_x at a distance x from point B is then given by

$$j_x = j_B + \Delta j = j_B + ix \tag{2.126}$$

where j_B is shear displacement at point B, which can be determined using Eq. (2.119).

The shear displacement along section AD, which is due to the elastic rebound of the terrain upon unloading, may again be determined in the same way as that for a rigid wheel. The development of shear displacement beneath a deflected tire is illustrated in Figure 2.86.

After the shear displacement on the tire–terrain interface has been determined, the corresponding shear stress distribution can be defined using an appropriate shear stress–shear displacement relationship for the terrain under consideration. The tractive effort can then be calculated by integrating the horizontal component of the shear stress over the entire contact area.

Figure 2.87 shows the measured normal and shear stress distributions on the contact patch of a tractor tire (11.5-15) at 10% slip on a sandy loam (Krick 1969). For a pneumatic tire operating in the elastic mode, the stress distribution on the tire–terrain interface is generally more uniform than that under a rigid wheel.

Figure 2.87 Measured normal and shear stress distributions on the contact patch of a tractor tire on sandy loam. *Source:* Krick (1969). Reproduced with permission of International Society for Terrain–Vehicle Systems.

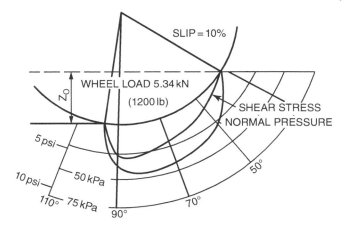

A physics-based model, known as NWVPM (Nepean Wheeled Vehicle Performance Model), that incorporates the procedures for predicting tire performance described above, has been developed for predicting the overall tractive performance of multi-axle wheeled vehicles (Wong 2010). The basic features of NWVPM are outlined in the following section.

2.9 A Physics-Based Model for the Performance of Off-Road Wheeled Vehicles

As noted previously, a model known as NWVPM has been developed for performance and design evaluation of multi-axle, off-road wheeled vehicles (Wong 2010). This method considers all major design parameters of the vehicle and of the tire. Vehicle parameters considered include vehicle weight, location of the center of gravity, number of axles, type of axle (driven or nondriven), axle load, axle spacing, axle suspension stiffness, axle clearance, number of pairs of tires on an axle, and tread of the axle (i.e. transverse distance between left- and right-side tires on the same axle). Tire parameters considered include diameter, tread width, section height, ratio of lug area to carcass area, height and width of the lug, inflation pressure and average ground pressure (i.e. p_{gr} defined in Section 2.8.2), and tire construction (bias or radial, for determining tire motion resistance due to internal hysteresis losses). Like NTVPM and RTVPM, NWVPM uses the Bekker–Wong terrain parameters for characterizing terrain behavior.

2.9.1 Basic Approach

In the development of NWVPM, the following are taken into consideration:

- The methods presented in Section 2.8 are employed to predict the operating mode, sinkage, motion resistance, and tractive effort of the tires on the vehicle.
- If the tread (i.e. transverse distance between left- and right-side tires on the same axle) for all axles is the same, the tires, except for those on the front axle, run in the ruts formed by the preceding tires in straight-line motion. The response of terrain to repetitive normal and shear loading is considered in predicting the performance of succeeding tires. Provision is made in NWVPM

for accommodating the case where succeeding tires partly travel in the ruts formed by preceding tires and partly travel on undisturbed terrain.

- The dynamic load transfer between axles due to drawbar pull is taken into consideration in predicting the overall performance of the vehicle. Effects of axle suspension stiffness on the dynamic load distribution among axles are also considered.

NWVPM can be used to evaluate effects of vehicle design and tire parameters on the tractive performance of wheeled vehicles over various types of terrain in straight-line motion under steady-state operating conditions.

2.9.2 Experimental Substantiation

The basic features of NWVPM have been substantiated with available field test data (Bekker 1985). Figures 2.88 and 2.89 show a comparison between the measured and predicted drawbar performances of a tractor on a plowed field and on a stubble field, respectively (Wong 2010). The two curves shown in each of the two figures represent those predicted using the upper and lower bounds of terrain values measured. The drawbar performance predicted using NWVPM is reasonably close

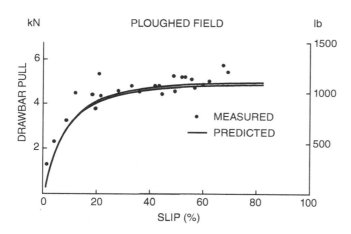

Figure 2.88 Comparisons of the predicted and measured drawbar performances of a tractor on a plowed field.

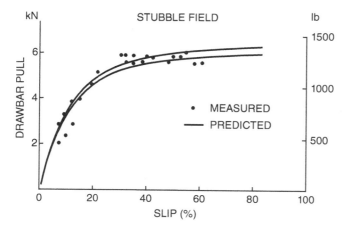

Figure 2.89 Comparisons of the predicted and measured drawbar performances of a tractor on a stubble field.

to that measured. This demonstrates that NWVPM can provide realistic predictions of the performance of off-road wheeled vehicles in the field.

2.9.3 Applications to Parametric Analysis

To demonstrate its applications, NWVPM was employed to evaluate the effects of two different types of tire with various inflation pressures on the performances of a two-axle, all-wheel drive vehicle on two types of terrain (Wong 2010). The static loads on the front and rear axles are 39.23 and 33.33 kN (8820 and 7493 lb), respectively. The axle suspension stiffness for the front axle and that for the rear axle are 2.42 and 1.57 kN/cm (1382 and 896 lb/in.), respectively. The wheelbase of the vehicle is 2.08 m (82 in.). The height of vehicle center of gravity and that of the drawbar hitch are 1.112 and 0.662 m (44 and 26 in.) above the ground, respectively.

Figures 2.90 and 2.91 show the variations of drawbar pull coefficient with slip for the vehicle with two types of tire, 12.5/75R20 and 11R16, with different combinations of inflation pressures for the front and rear tires on a medium soil and on a clayey soil, respectively. The first number indicated

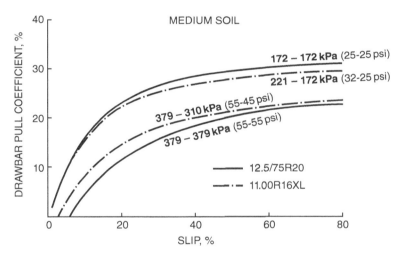

Figure 2.90 Variations of drawbar pull coefficient with slip of a four-wheel drive vehicle with two types of tire at different combinations of inflation pressures for front and rear tires on a medium soil.

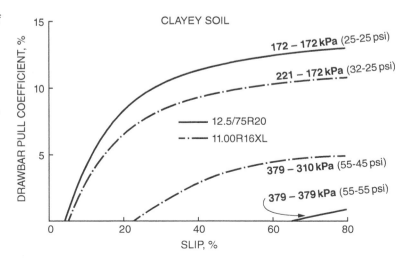

Figure 2.91 Variations of drawbar pull coefficient with slip of a four-wheel drive vehicle with two types of tire at different combinations of inflation pressures for front and rear tires on a clayey soil.

on the curves represents the inflation pressure on the front tire, whereas the second number represents that on the rear tire. The type of tire, inflation pressure combination of the front and rear tires, and terrain conditions have noticeable effects on vehicle performance. NWVPM has been employed to evaluate the performance of a variety of off-road vehicles for governmental agencies in North America and elsewhere. It has also been used to evaluate the mobility of lunar rover wheels (Wong and Asnani 2008).

2.10 Slip Sinkage

As noted previously, the phenomenon that the slip of a vehicle running gear (track or wheel) causes additional sinkage, which is referred to as slip sinkage, has long been recognized. For instance, it is observed that a tracked vehicle develops increasing sinkage with the increase of slip and that the sinkage at the rear is higher than that at the front, causing a tail-down attitude for the vehicle. This happens even when the load transfer due to drawbar pull is counteracted by a sliding weight to move the center of gravity forward (Reece 1965–1966). The slip sinkage effect is also observed during shear tests using either a rectangular shear plate or an annular shear ring shown in Figure 2.12. With a constant normal load on the shear plate or shear ring, as soon as the horizontal movement of the rectangular shear plate or the angular movement of the annular shear ring is initiated, additional sinkage of the rectangular shear plate or the annular shear ring occurs. In other words, whenever shearing action between the shear plate or the shear ring and the terrain takes place, additional sinkage will result. Figures 2.92(a)–(c) show the variations of sinkage with horizontal movement (shear displacement) of a rectangular shear plate of 30 in. (76.2 cm) long by 2 in. (5.1 cm) wide with lugs 1/4 in. (0.6 cm) high under various normal pressures on clay, dry sand, and wet sand, respectively (Reece 1965–1966). The sinkage at zero horizontal movement represents the static sinkage due to normal load (pressure) applied to the shear plate. In the figure, variations of horizontal (shear) force with horizontal movement under various normal pressures are also shown. It shows that even though the shear force approaches an asymptote, the sinkage continues increasing with the increase of horizontal movement. In other words, slip sinkage increases continually with the increase of shear displacement, even though the shear force (stress) approaches an asymptote. Thus, it indicates that slip sinkage is primarily related to shear displacement and not to shear stress. The test data shown in Figures 2.92(a)–(c) are consistent with the observation of the tail-down attitude of a tracked vehicle. As discussed in Section 2.5.2, at the front contact point of the track, the track link just meets the terrain, and its shear displacement is zero. As a result, the slip sinkage at the front is zero. As the vehicle moves forward, the track link eventually takes the position at the rear of the track, where the shear displacement is a maximum, as shown in Figure 2.50. Consequently, the slip sinkage is the highest at the rear and the vehicle takes the tail-down attitude.

2.10.1 Physical Nature of Slip Sinkage

The physical nature of slip sinkage may be illustrated by the soil flow beneath a wheel, as shown in Figures 2.19 and 2.21 (Wong 1967). Figure 2.19 shows that under a driven rigid wheel with slip, there are two zones of soil flow in the longitudinal plane. In the front flow zone, soil flows forward and forms a bow wave causing a bulldozing resistance. In the rear flow zone, soil flows backward. This causes the soil being transported from beneath to behind the wheel. With increasing slip, the backward flow zone expands, leading to the increase in wheel sinkage (Wong 1967). These two flow zones degenerate into a single backward zone at 100% slip, with the wheel no longer moving

Figure 2.92 Variations of slip sinkage and horizontal (shear) force with horizontal movement (shear displacement), for a 30 in. (76.2 cm) long and 2 in. (5.1 cm) wide shear plate with lugs 1/4 in. (0.6 cm) high, on (a) clay, $\phi = 0$, $c = 6.9$ kPa (1.0 psi); (b) dry sand, $\phi = 32°$, $c = 0$; (c) wet sand, $\phi = 32°$, $c = 0.69$ kPa (0.1 psi). *Source:* Reece (1965–1966). Reproduced with permission of the Council of the Institution of Mechanical Engineers.

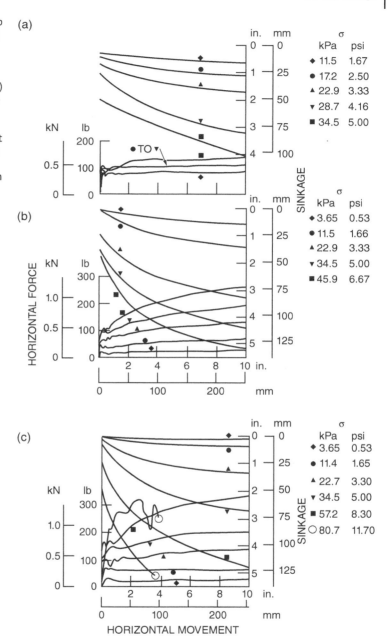

forward, as shown in Figure 2.21. Under these circumstances, the wheel behaves like an excavator to transfer the soil from beneath to behind the wheel. If the wheel slip is maintained at 100% over a period, the wheel sinkage will continuously increase, until, for instance, the vehicle belly contacts the terrain surface, which may prevent the sinkage of the wheel from increasing, or the torque applied to the wheel can no longer overcome the resisting torque and the wheel stops rotating. The shifting of soil from beneath to behind the wheel caused by slip is the physical process that causes slip sinkage. It should be mentioned that in wheel–soil interaction, soil also flows laterally. The soil flow in the longitudinal plane in comparison with that in the lateral direction is related to wheel width. With the increase of wheel width, soil flow in the longitudinal plane becomes more significant.

While the physical nature of slip sinkage is qualitatively illustrated by the process of soil flow beneath a slipping wheel, quantitative prediction of slip sinkage, as a function of terrain properties and wheel design and operational parameters, is a challenging and complex task. Stimulated by the incident of immobilization in 2009 on the Martian surface of a rover known as the Spirit, deployed by the U.S. National Aeronautics and Space Administration (NASA), efforts have been made to use the discrete element method to quantitatively evaluate slip sinkage of the rover wheel at high slips (Knuth et al. 2012; Johnson et al. 2015, 2017). The purpose is to better understand the immobilization process, which may lead to the development of measures that would prevent similar immobilization from occurring in the future. It should be mentioned that studies also show that slip sinkage increases with the contact time between the slipping wheel and soil. This indicates that slip sinkage, in general, is not only dependent on slip, but also on the contact time between a slipping wheel and terrain.

The applications of the discrete element method to the study of vehicle (or rover) mobility will be discussed in Section 2.12.2. It should be noted that the discrete element method requires lengthy computation, even using high-performance computing facilities. Studies also indicate that the ability to quantitatively predict vehicle mobility using the discrete element method depends, to a great extent, on whether the input micromechanical parameters of the elements (particles) realistically and adequately represent terrain behavior (Johnson et al. 2015).

2.10.2 Simplified Methods for Predicting Slip Sinkage

Since the early days of development of terramechanics, attempts have been made to develop methods for predicting slip sinkage that can readily be used in engineering practice (Reece 1965–1966). A number of simplified methods without requiring lengthy computation have been proposed over the years. In the following, three simplified methods for predicting slip sinkage are briefly outlined as examples (Bekker 1969; Lyasko 2010; Ding et al. 2014).

1) The method proposed by Bekker

Based on experimental observations, Bekker proposed that the relationship of shear stress τ, normal pressure p, shear displacement j, and slip sinkage z_j may be expressed as follows (Bekker 1969):

$$\frac{\tau_{max}}{p - p_{crit}} = \frac{j}{z_j} \tag{2.127}$$

where p_{crit} is the Terzaghi bearing capacity value given by

$$p_{crit} = cN_c + \gamma_s \left[N_q \left(z_0 + z_j \right) + 0.5bN_\gamma \right] \tag{2.128}$$

where b is the width of the contact area; N_c, N_q, and N_γ are the Terzaghi bearing capacity factors and γ_s is weight density of the soil, which are the same as those used in Eq. (2.29); z_0 is the static sinkage obtained using the Bekker pressure–sinkage equation, Eq. (2.47), and expressed by

$$z_0 = \left(p / \left[(k_c/b) + k_\phi \right] \right)^{1/n} \tag{2.129}$$

Combining Eqs. (2.127) and (2.128), the general expression for slip sinkage z_j is given by (with $\tau_{max} = c + p \tan\phi$)

$$z_j = \frac{j\left[p - cN_c - \gamma_s\left(N_q z_0 + 0.5bN_\gamma\right)\right]}{c + p\,tan\phi + \gamma_s N_q j} \qquad (2.130)$$

By adding Eqs. (2.129) and (2.130), the total sinkage z is obtained by

$$z = z_0 + z_j \qquad (2.131)$$

In applying the method outlined above to predicting slip sinkage, the following should be noted:

- The Terzaghi bearing capacity value p_{crit}, given by Eq. (2.128), is originally intended for use in the design of foundations or earthworks in civil engineering. Its basic assumption is that the soil is incompressible (Reece 1965–1966). A wide range of terrains that off-road vehicles encounter in operation are deformable.
- The Terzaghi bearing capacity factors, N_c, N_q, and N_γ in Eq. (2.128), are related to the value of the angle of friction ϕ of the soil, as shown in Figure 2.18. Consequently, an error in determining the value of ϕ will have a noticeable impact in the prediction of slip sinkage z_j (Bekker 1969).
- Equation (2.127) indicates that if $p < p_{crit}$, slip sinkage z_j would be negative. This inequality means that if the average normal pressure that the vehicle running gear exerts on the soil is less than p_{crit}, no slip sinkage z_j should be expected. Experiments have shown that this is not entirely correct for loose frictional soils (Bekker 1969).
- This method for predicting slip sinkage has not been adequately substantiated with test data.

2) The method proposed by Lyasko

This method is based on the following assumptions (Lyasko 2010):

- The magnitude of the work done for overcoming the external motion resistance of a vehicle running gear (track or wheel) is equal to the vertical work done in compacting the soil, taking into account the additional sinkage caused by slip.
- The vertical work done is predicted from the normal load on the running gear (wheel or track) and its total sinkage, including both static sinkage and slip sinkage. The additional sinkage due to slip can be determined from a consideration of soil deformation propagation in depth, soil particles trajectories, and soil volume changes under the vehicle running gear in the shearing process.

Based on these assumptions and from a detailed analysis, the relationship between the external motion resistance R_{ci} at slip $i > 0$ and the external motion resistance R_c at slip $i = 0$ is expressed by

$$R_{ci} = R_c \times \frac{1 + i}{1 - 0.5i} = R_c \times K_{ss} \qquad (2.132)$$

$$\text{and } K_{ss} = \frac{1 + i}{1 - 0.5i} \qquad (2.133)$$

Equation (2.133) indicates that the coefficient K_{ss} is only dependent on slip and is independent of the type of terrain or the type of vehicle running gear (track or wheel). It also indicates that at $i = 100\%$, the predicted external motion resistance R_{ci} is four times the external motion resistance R_c at slip $i = 0$.

As the external motion resistance is due to the vertical work done on soil compaction, which is proportional to the total sinkage of the vehicle running gear, the total sinkage z at slip $i > 0$ is expressed by

$$z = z_0 + z_j = K_{ss} \times z_0 \tag{2.134}$$

where z_0 and z_j are the static sinkage (or the sinkage at slip $i = 0$) and slip sinkage at slip $i > 0$, respectively.

Equation (2.134) indicates that at slip $i = 100\%$, the total sinkage z predicted by the proposed method is four times the static sinkage z_0.

Equations (2.132) and (2.134) have been evaluated with test data in the slip range from 0 to 30%. The test data include the measured average values of motion resistance coefficient (the ratio of motion resistance to normal load) in the slip range from 0 to 30% of a select group of agricultural tractor drive tires on firm soil, tilled soil, and soft or sandy soil. The measured values of rut depth in the slip range of 0 to 30% of a tractor with steel tracks on a firm loam and a tilled loam, and of a tractor with rubber tracks on a California wet untilled Yolo loam are also used in the evaluation (Lyasko 2010).

Figure 2.93 shows the variations of the measured (shown by points) and predicted (using Eq. (2.132) and represented by lines) motion resistance coefficient with slip in the range of 0 to 30% for a select group of agricultural tractor drive tires on concrete (designated by 1), firm soil (2), tilled soil (3), and soft or sandy soil (4) (Lyasko 2010). On concrete, the motion resistance of a tire is primarily due to hysteresis losses in tire materials caused by carcass deflection while rolling and is not sensitive to slip, as shown by 1 in the figure. On deformable terrain, tire deflection is generally lower than that on concrete and the motion resistance due to hysteresis losses of tire materials is accordingly lower. It is usually less significant in comparison with that due to tire–terrain interaction. The values of motion resistance coefficient of tire on deformable terrain shown in the figure are taken to be entirely due to terrain deformation and tire hysteresis losses are neglected. From the data shown in the figure, the ratio of motion resistance coefficient at 30% slip to that at 0% slip on firm soil, tilled soil, and soft or sandy soil are close to the value of K_{ss} of 1.53 determined by Eq. (2.133) with slip $i = 30\%$. This indicates that Eq. (2.132) provides reasonable

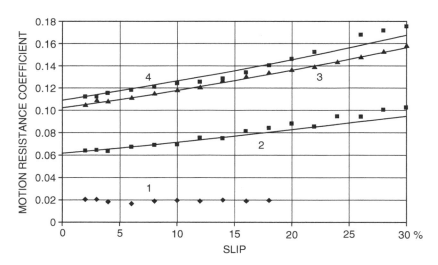

Figure 2.93 Comparisons of the measured (points) and predicted (lines) variations of the motion resistance coefficient with slip in the range of 0 to 30% for a select group of agricultural tractor drive tires on various types of surface: 1 – concrete; 2 – firm soil; 3 – tilled soil; and 4 – soft or sandy soil. *Source:* Lyasko (2010). Reproduced with permission of the International Society for Terrain–Vehicle Systems.

Figure 2.94 Variations of the measured rut depth with slip in the range from 0 to 30% for tractors with various types of track on different types of soil: 1 – a tractor with rubber belt tracks on a California wet untilled Yolo loam; 2 – a tractor with steel tracks on a firm loam; and 3 – a tractor with steel tracks on a tilled loam. *Source:* Adapted from Lyasko (2010). Reproduced with permission of the International Society for Terrain–Vehicle Systems.

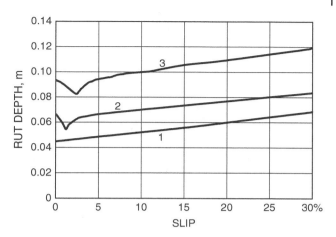

predictions of the ratio of motion resistance coefficient at 30% slip to that at 0% slip for the select group of agricultural tractor drive tires on the three types of farm soil.

Figure 2.94 shows the variations of the measured rut depth with slip in the range of 0 to 30% for a tractor with rubber belted tracks on a California wet untilled Yolo loam (Line 1), a tractor with steel tracks on a firm loam (2), and a tractor with steel tracks on a tilled loam (3) (Lyasko 2010). The ratio of rut depth at a given slip to that under static condition may be considered representative of the ratio of sinkage at a given slip to that under static condition. It should be mentioned that under normal circumstances rut depth may be slightly different from sinkage measured from the lowest point of running gear–terrain contact, due to elastic rebound of terrain material after passage of the running gear or terrain material being shifted from below to behind the running gear, as discussed previously. For Line 1 shown in Figure 2.94, the ratio of rut depth at 30% slip to that at 0% slip is close to the K_{ss} value of 1.53 calculated using Eq. (2.133) with slip $i = 30\%$. For Lines 2 and 3, variations of rut depth with slip follow a similar trend as that defined by Eq. (2.134). This indicates that for the three cases shown in Figure 2.94, the test data on the effects of slip on rut depth correlate reasonably well with that predicted based on Eq. (2.134) over the range of slip from 0 to 30%.

Based on the information presented above, the method for predicting the effects of slip on motion resistance coefficient or rut depth (sinkage) proposed by Lyasko may be summarized as follows:

- It appears that the method predicts reasonably well the effects of slip on the external motion resistance of a group of agricultural tractor drive tires on a range of farm soils, as well as on the rut depth of a group of tracked tractors on various types of farm soils, in the range of slip from 0 to 30%.

- It is not known that the method can be applied to a wider spectrum of soils, such as fine-grained soils and organic soils (as defined by the Unified Soil Classification System shown in Table 2.1) that may be encountered in cross-country operations.

- It is not shown that the method can be applied to predicting external motion resistance or rut depth at slips beyond 30%. For agricultural tractors and the like, to maintain operations with desirable drawbar (tractive) efficiency (see Section 4.1.2 Drawbar (Tractive) Efficiency), slip of tires or tracks is normally kept around 20%. However, for a variety of other off-road vehicles, including military vehicles and rovers for extraterrestrial surface exploration, they may operate at slips higher than 30%. Consequently, the capability for predicting the effects of slip on external motion resistance and rut depth beyond 30% slip is of practical interest.

- As there are no test data at slip $i = 100\%$ presented (Lyasko 2010), the validity of predictions by Eq. (2.133) that external motion resistance or rut depth at 100% slip is four times that at zero slip cannot be ascertained.

3) The approach proposed by Ding et al.

This approach is different from the others outlined above in that it incorporates the effects of slip into the pressure–sinkage relationship of terrain (regolith simulant) for predicting the performance of vehicle running gear (Ding et al. 2014).

Based on test data obtained using a range of rover wheels at various slips on a lunar regolith simulant, a new form of pressure–sinkage relationship is proposed as follows:

$$p = K_{st} z^{N_{st}} \tag{2.135}$$

where p is pressure; K_{st} is the stiffness modulus of terrain; z is the total sinkage, including both static sinkage and slip sinkage; and N_{st} is an exponent which is a variable to reflect the influence of a number of factors, including slip (or skid), normal load, and dimensions and design features of the wheel (such as the number of lugs and lug height).

To reflect the effects of slip on the pressure–sinkage relationship, the exponent N_{st} may be expressed by a linear, quadratic, or cubic function of slip i. If N_{st} is represented by a quadratic function of slip i, then it is expressed as follows:

$$N_{st} = n_0 + n_1 i + n_2 i^2 \tag{2.136}$$

Based on test data, for a rigid rover wheel with radius of 157.35 mm (6.2 in.), width of 165 mm (6.5 in.), and 30 lugs of height 5 mm (0.2 in.), the values of the coefficients are: $n_0 = 0.860$, $n_1 = 0.865$, and $n_2 = -0.250$. For a rover wheel with radius of 157.35 mm (6.2 in.), width of 165 mm (6.5 in.), and no lugs, the values of the coefficients are: $n_0 = 0.878$, $n_1 = 0.969$, and $n_2 = -0.629$. As noted previously, these values are for predicting the total wheel sinkage, including slip sinkage, up to slip $i = 60\%$. Details of this approach to predicting the effects of slip (or skid) on pressure–sinkage relationship of terrain are given in the paper by Ding et al. (2014).

It is noted that the values of parameters K_{st} and N_{st} in Eqs. (2.135) and (2.136) are derived from fitting the equations (Eqs. (2.135) and (2.136)) to experimental data. Thus, these values are only applicable to specific rover wheels and terrain (regolith simulant). It is uncertain that these values can be applied to other rover wheels with different dimensions and design features on different types of terrain (regolith simulant).

In summary, the following observations may be made on the proposed simplified methods for predicting slip sinkage:

- For the method proposed by Bekker, the prediction of slip sinkage is based on the shear displacement of vehicle running gear and using the Terzaghi bearing capacity factors as reference. As Terzaghi bearing capacity theory is primarily for the design of foundations or earthworks in civil engineering, it is uncertain that it is suitable for evaluating off-road vehicle mobility on unprepared natural terrain, which in many cases is highly deformable. Furthermore, the method has not yet been adequately substantiated with test data.
- It is shown that variations of external motion resistance with slip up to 30% predicted by the method proposed by Lyasko correlate reasonably well with test data obtained using a select group of agricultural tractor drive tires on a range of farm soils (coarse-grained soils). Variations of measured rut depth with slip up to 30% on farm soils for a group of tractors, with rubber belt tracks and steel tracks, demonstrate a similar trend as that predicted by Eq. (2.134). It is not known, however, whether the method can be applied to predicting the effects of slip on external

motion resistance or rut depth over a wider spectrum of terrains (such as fine-grained soils and organic soils) or for slips beyond 30%.
- For the approach proposed by Ding et al., the effects of slip are incorporated into the pressure–sinkage relationship of terrain (regolith simulant) for predicting rover wheel performance at various slips. The values of these pressure–sinkage parameters are obtained through curve fitting to test data for specific rover wheels and terrain (regolith simulant). Thus, it is uncertain that the values of the pressure–sinkage parameters of terrain (regolith simulant) so obtained can be applied to other rover wheels with different dimensions or design features.
- Further analytical and experimental studies would be required for developing an improved simplified method that can generally be applied to predicting slip sinkage of off-road vehicles over a wide spectrum of terrains and at slips up to 100%.

2.11 Applications of Terramechanics to the Study of Mobility of Extraterrestrial Rovers and their Running Gears

Terramechanics can be applied to evaluating the mobility of extraterrestrial rovers and their running gears. In this section, some examples are presented to demonstrate its applications. As most extraterrestrial rovers have used wheels as their running gears, discussions will focus on the mobility of wheeled rovers.

2.11.1 Predicting the Performance of Rigid Rover Wheels on Extraterrestrial Surfaces Based on Test Results Obtained on Earth

In the development of extraterrestrial rovers or their wheels, it is desirable to test their performances under the gravities of extraterrestrial surfaces where they are to be deployed. This can be achieved, for instance, by conducting tests of rovers or their wheels in a bin of regolith simulant in an aircraft, while it performs parabolic flight maneuvers to produce the required levels of gravity, as shown in Figure 2.95 (Kobayashi et al. 2010). This is, however, costly and the duration available for conducting tests under a specific gravity is rather short, usually in terms of seconds.

The current practice for testing the performances of rovers or their wheels is to conduct tests on earth on a simulant appropriate to the extraterrestrial surface of interest. In these tests, the normal force applied by the rover or by its wheel to the simulant corresponds to that expected on the extraterrestrial surface (Freitag et al. 1970). For instance, on the lunar surface, the gravity is 1/6 of that on earth. Consequently, during tests conducted on earth, the normal force of the rover or its wheel exerts on the simulant is reduced to 1/6 of that on earth. It should be noted, however, that the simulant during these tests is still subject to earth gravity. This raises the question as to whether the performance of the rover or its wheel measured in this type of test on earth represents that on the lunar surface, because the regolith on the lunar surface is subject to lunar gravity, while the simulant used in the tests on earth is subject to earth gravity (Wong and Asnani 2008).

If a method can be developed that will predict the performances of rovers or their running gears on extraterrestrial surfaces based on test results obtained on earth under earth gravity, it would make a significant contribution to the method of testing of extraterrestrial rovers or their running gears by reducing or even eliminating the needs for testing them under the gravity of the extraterrestrial body of interest.

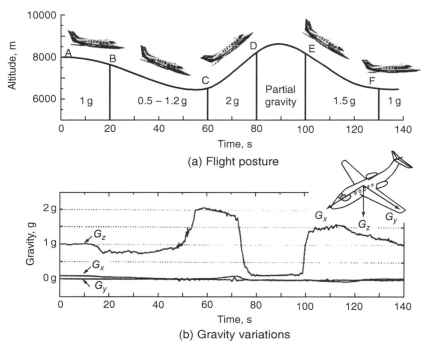

Figure 2.95 Parabolic flight maneuvers for testing the performance of a rigid rover wheel under various gravity conditions. *Source:* Kobayashi et al. (2010). Reproduced with permission of the International Society for Terrain–Vehicle Systems.

There are two issues involved in conducting such tests on earth:

- One is the selection of an appropriate simulant to represent the regolith on the extraterrestrial surface for conducting tests on earth. This issue is beyond the scope of this section.
- The other is to determine the effects of gravity on the performance of the rover or its running gear. This will form the basis for estimating the performance of a rover or its running gear on an extraterrestrial surface based on test results obtained on a simulant on earth. This is the topic to be addressed in this section.

Rigid wheels are widely used in unmanned rovers for extraterrestrial surface exploration. In the following, discussions will focus on the prediction of sinkage, motion resistance, and maximum thrust (tractive effort) of rigid rover wheels on extraterrestrial surfaces under various gravities, based on test data obtained on earth under earth gravity. However, predictions of the performances of flexible wheels for the manned lunar roving vehicle, used in NASA's Apollo missions 15, 16, and 17 in the 1970s, will also be discussed later in this section (Wong and Asnani 2008).

1) Characterization of mechanical properties of terrain (regolith, simulant)

As noted in Section 2.4.1, inspired by the classical bearing capacity theory of soil mechanics, Reece (1965–1966) proposed the following pressure–sinkage equation for terrain, Eq. (2.52):

$$p = \left(ck_c' + \gamma_s bk_\phi' \right) \left(\frac{z}{b} \right)^n$$

where γ_s is the weight density of the terrain. All other parameters in the Reece equation are defined previously in Section 2.4.1.

If the weight density γ_s of the terrain is expressed as the product of mass density γ_{mt} and acceleration due to gravity g, then the Reece equation may be rewritten as

$$p = \left(\frac{ck'_c}{b^n} + \frac{\gamma_{mt}gk'_\phi}{b^{n-1}} \right) z^n = \left(K_c + K_\phi g \right) z^n \tag{2.137}$$

where $K_c = \dfrac{ck'_c}{b^n}$ and $K_\phi = \dfrac{\gamma_{mt}k'_\phi}{b^{n-1}}$.

Equation (2.137) shows that the pressure-sinkage relationship is affected by gravity g. It implies, however, that for a given terrain when subject to different gravities, the change in the pressure–sinkage relationship is only due to the change in gravity g in the second term within the parenthesis on the right-hand side of Eq. (2.137), and that parameters n, K_c, and K_ϕ remain unchanged. The issue of possible effects of gravity on the values of these pressure–sinkage parameters will be reviewed later in Subsection 7). When pressure–sinkage tests are conducted on earth, g in Eq. (2.137) is the gravity g_e (9.81 m/s^2) on earth's surface.

Equation (2.137) provides a basis for evaluating the effects of gravity on rigid rover wheel sinkage and motion resistance due to compaction.

2) Predicting the effects of gravity on rigid rover wheel sinkage

A) Maintaining the same wheel normal force exerted on the surface under various gravities

As noted previously, this is the current practice in testing a rover or its wheel on earth. For instance, when testing the lunar roving vehicle wheel on earth, the normal force exerted by the wheel on the simulant was reduced to 1/6 of that on earth, as the gravity on the lunar surface is 1/6 of that on the earth surface (Freitag et al. 1970).

Using the simplified model for rigid wheel–soil interaction proposed by Bekker and described in Section 2.8.1, together with the Reece pressure–sinkage equation (i.e. Eq. (2.137)), the sinkage z_e of a rigid wheel on the simulant under earth gravity g_e is expressed by

$$z_e = \left[\frac{3W_e}{b(3-n)\left(K_c + K_\phi g_e\right)\sqrt{D}} \right]^{2/(2n+1)} \tag{2.138}$$

where W_e is the normal force of the wheel exerted on the simulant under earth gravity g_e. Other parameters in the equation have been defined previously.

It should be mentioned that the sinkage caused by normal load exerted by the wheel determined by Eq. (2.138) is referred to as the static sinkage, as noted previously.

On the extraterrestrial surface under gravity g_{ex}, the static sinkage z_{ex} of a rigid wheel is expressed by

$$z_{ex} = \left[\frac{3W_{ex}}{b(3-n)\left(K_c + K_\phi g_{ex}\right)\sqrt{D}} \right]^{2/(2n+1)} \tag{2.139}$$

where W_{ex} is the normal force of the wheel exerted on the extraterrestrial surface under gravity g_{ex}.

The ratio of the rigid wheel static sinkage z_{ex} under gravity g_{ex} to the static sinkage z_e under earth gravity g_e is given by

$$\frac{z_{ex}}{z_e} = \frac{\left[\dfrac{3W_{ex}}{b(3-n)\left(K_c + K_\phi g_{ex}\right)\sqrt{D}}\right]^{2/(2n+1)}}{\left[\dfrac{3W_e}{b(3-n)\left(K_c + K_\phi g_e\right)\sqrt{D}}\right]^{2/(2n+1)}} \tag{2.140}$$

For a given wheel, if its normal force W_e exerted on the simulant under earth gravity g_e is the same as the normal force W_{ex} exerted on the regolith on extraterrestrial surface under gravity g_{ex} (i.e., $W_e = W_{ex}$), and both the simulant and regolith are dry with negligible cohesion (i.e., the values of c and K_c are insignificant) and the values of n and K_ϕ do not change with gravity, then Eq. (2.140) can be simplified to

$$\frac{z_{ex}}{z_e} = \left[\frac{g_e}{g_{ex}}\right]^{2/(2n+1)} \tag{2.141}$$

It should be mentioned that on extraterrestrial surfaces that have so far been explored, the regolith is dry. As a result, neglecting its cohesion c and hence the value of K_c may be justified, at least as a first approximation. The issue of possible effects of gravity on the values of n and K_ϕ will be reviewed later in Subsection 7).

Equation (2.141) indicates that the ratio z_{ex}/z_e can be estimated by the ratio g_e/g_{ex} raised to the power of $2/(2n + 1)$.

For instance, if the regolith on the lunar surface has the same values of pressure–sinkage parameters as those of the simulant DLR-A with $n = 0.63$ (Krenn and Hirzinger 2008), then the ratio of static sinkage of a rigid wheel on the lunar surface to that on earth, z_{ex}/z_e, can be estimated using Eq. (2.141), with $g_e / g_{ex} = 6$,

$$z_{ex}/z_e = (g_e/g_{ex})^{2/(2n+1)} = (6)^{2/(1.26+1)} = 4.88.$$

Similarly, if the regolith on the Martian surface has the same properties as those of the simulant DLR-A with $n = 0.63$, then the ratio of static sinkage of a rigid wheel on the Martian surface to that on earth, z_{ex}/z_e, can be estimated using Eq. (2.141), with $g_e / g_{ex} = 2.63$,

$$z_{ex}/z_e = (g_e/g_{ex})^{2/(2n+1)} = (2.63)^{2/(1.26+1)} = 2.35.$$

The analysis presented above indicates that it is feasible to estimate the static sinkage of a rigid wheel on extraterrestrial surface based on that measured on earth, provided that the exponent n of the pressure–sinkage relation is known, and its value does not change with gravity. It should be noted that the value of n is usually derived from test data through curve fitting, as discussed in Section 2.4.1, and that it is subject to measurement and data processing errors. The measurement and data processing errors in determining the value of n would lead to uncertainty of using Eq. (2.141) for estimating the static sinkage ratio z_{ex}/z_e.

To minimize the uncertainty for estimating the static sinkage ratio z_{ex}/z_e, another approach is explored. It is based on the mass carried by the rover wheel on earth being identical to that on the extraterrestrial surface, as presented below.

B) Maintaining identical mass carried by the wheel under various gravities

In this case, based on Eq. (2.140), the ratio of the rigid wheel static sinkage z_{ex} under gravity g_{ex} to that on earth z_e under gravity g_e is given by

$$\frac{z_{ex}}{z_e} = \frac{\left[\dfrac{3mg_{ex}}{b(3-n)\left(K_c + K_\phi g_{ex}\right)\sqrt{D}}\right]^{2/(2n+1)}}{\left[\dfrac{3mg_e}{b(3-n)\left(K_c + K_\phi g_e\right)\sqrt{D}}\right]^{2/(2n+1)}} \tag{2.142}$$

With identical mass m carried by the wheel under various gravities and with the conditions of the simulant being dry with negligible cohesion (i.e., the values of c and K_c are insignificant and can be neglected), and the values of n and K_ϕ not changing with gravity, Eq. (2.142) can be simplified to

$$\frac{z_{ex}}{z_e} = 1 \tag{2.143}$$

Equation (2.143) indicates that with identical mass carried by the wheel, its static sinkage on any extraterrestrial surface under any gravity g_{ex} is the same as that on the earth surface under earth gravity g_e. It will be shown in the next section that based on test data available, Eq. (2.143) may also be applied to estimating the total sinkage of the wheel, which is the sum of the static sinkage and slip sinkage, on any extraterrestrial surface based on that observed on earth. This finding will provide a useful guidance for the method of testing of rover wheels, at least as a first approximation. Equation (2.143) indicates that the total wheel sinkage measured on earth will be the same as that on the moon, Mars, or any other extraterrestrial body, provided that the mass carried by the wheel on any extraterrestrial surface is identical to that on earth and that the values of pressure–sinkage parameters of the regolith on the extraterrestrial surface and those of the simulant used on earth are the same.

3) Evaluation of predicted rigid rover wheel sinkages under various gravities with test data

The finding that with identical mass carried by the wheel, its sinkage under various gravities will be the same, as expressed by Eq. (2.143), was evaluated with test data reported in the paper by Kobayashi et al. (2010). The tests were conducted with a rigid wheel in a soil bin on the ground and in an aircraft undergoing parabolic fight maneuvers to produce various levels of gravity, as shown in Figure 2.95. The rigid wheel had a diameter of 150 mm (approximately 6 in.) and a width of 80 mm (approximately 3 in.). The mass carried by the wheel was 10 kg (equivalent to approximately 22 lb of weight on earth surface). A lunar soil simulant (LSS) and a sand known as Toyoura sand were used in the tests. The basic properties of these two types of soil with relative densities of 50 and 70% measured on earth are given in Table 2.17. The shear strength parameters, cohesion c and angle of internal shearing resistance (angle of internal friction) ϕ, of the soils were obtained from drained triaxial compression tests. The soil was contained in a bin with length of 600 mm (approximately 24 in.), width 200 mm (approximately 8 in.) and depth of 100 mm (approximately 4 in.). The values of

Table 2.17 Basic properties of two types of soil used in the tests.

Soil type	Relative density D_r (%)	Bulk density ρ		Void ratio e	Cohesion c		Internal friction angle, $\phi°$
		g/cm^3	lb/ft^3		kPa	psi	
Lunar soil simulant (LSS)	50	1.71	106.75	0.72	1.07	0.16	40.1
	70	1.82	113.62	0.62	2.78	0.40	44.6
Toyoura sand	50	1.47	91.77	0.80	2.08	0.30	38.2
	70	1.54	96.14	0.73	2.66	0.39	40.7

Source: Kobayashi et al. (2010).

parameters n, K_c, and K_ϕ of the Reece pressure–sinkage equation were, however, not measured either on the ground or in the aircraft under various gravities.

Table 2.17 shows that the internal friction is the dominant component of the shear strength for the two types of soil with two relative densities and that the cohesion is relatively insignificant. This appears to satisfy the conditions that the cohesion c and hence K_c can be neglected in deriving Eq. (2.143) (i.e. $z_{ex}/z_e = 1$).

Two sets of experiments were performed (Kobayashi et al. 2010). One set of experiments was carried out on the ground (earth surface) on the two types of soil with two relative densities shown in Table 2.17. Wheel sinkage under self-propelled conditions (without drawbar load) on the ground were measured, with normal loads of $1/6\,W$, $1/2\,W$, $3/4\,W$, $1\,W$, and $2\,W$ (normal load of $1\,W$ is equivalent to $10\,kg \times 9.81\,m/s^2$ on earth's surface). The wheel sinkage at normal load of $1\,W$ (with mass 10 kg) measured on the ground on each of the two types of soil with two relative densities were used as the wheel sinkages z_e under earth gravity g_e for determining sinkage ratios z_{ex}/z_e under various gravity ratios g_{ex}/g_e. The other set of experiments with wheel mass of 10 kg was performed in the soil bin installed in the aircraft undergoing parabolic flight maneuvers to produce various levels of gravity. Wheel sinkages under self-propelled conditions were measured under gravities of $1/6\,g$, $1/2\,g$, $3/4\,g$, $1\,g$, and $2\,g$ ($g = 9.81\,m/s^2$). It should be noted that under self-propelled conditions, wheel slip occurred. Consequently, the wheel sinkage measured was the total wheel sinkage, including both the static sinkage and the slip sinkage. During tests on the ground and in the aircraft, the wheel was driven by a motor at a constant rotating speed of 0.314 rad/s (3 rpm).

The two sets of test data obtained under conditions described above were processed using the procedures described in Wong (2012) and Wong and Kobayashi (2012). Based on the time histories of measured total sinkage and slip of the wheel obtained in the tests, Figure 2.96 shows the variations of the total sinkage z_e of the wheel obtained on the ground under earth gravity and those obtained in the aircraft z_{ex} at various gravities ($1/2\,g$, $3/4\,g$, $1\,g$, and $2\,g$) on the Toyoura sand with relative density of 70%. It shows that when the wheel is stationary with slip = 0, wheel sinkage z_e on the ground and that in the aircraft at various gravities z_{ex} are close together, which is consistent with Eq. (2.143) (i.e. $z_{ex}/z_e = 1$). When the wheel is under self-propelled conditions, wheel slip develops. However, the variations with slip of the total sinkage z_e of the wheel under earth gravity and that in the aircraft under various gravities z_{ex} exhibit the same trend and that both essentially follow the same fitted line obtained using the least squares method, as shown in Figure 2.96. The goodness-of-

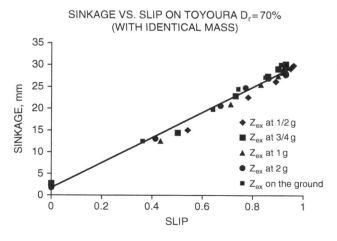

SINKAGE VS. SLIP ON TOYOURA D_r=70%
(WITH IDENTICAL MASS)

♦ Z_{ex} at 1/2 g
■ Z_{ex} at 3/4 g
▲ Z_{ex} at 1 g
● Z_{ex} at 2 g
■ Z_{ex} on the ground

Figure 2.96 Relationships between sinkage and slip under earth gravity on the ground and that under various gravities in the aircraft on Toyoura sand with relative density of 70% (the least squares fit having a coefficient of determination $R^2 = 0.988$).

Table 2.18 Summary of average values of measured wheel sinkage ratios z_{ex}/z_e under various gravity ratios.

		Gravity ratio g_{ex}/g_e			
		1/2	3/4	1	2
Soil type	Relative density D_r (%)	Average wheel sinkage ratio z_{ex}/z_e			
Lunar soil simulant (LSS)	50%	1.05	1.09	1.08	1.08
	70%	0.84	0.94	1.0	0.8
Toyoura sand	50%	1.10	0.96	1.01	1.03
	70%	1.07	1.09	1.01	1.0
Average measured value for all soils		1.02	1.02	1.03	0.98

Source: Wong and Kobayashi (2012).

fit of the straight line to test data is excellent with a value of coefficient of determination $R^2 = 0.988$ ($R^2 = 1$ indicating a perfect fit). This indicates that under self-propelled conditions, the average values of z_{ex} obtained in the aircraft under various gravities are close (or equal) to that of z_e under earth gravity, which again is consistent with Eq. (2.143). Variations of the total sinkage with slip of the rover wheel under self-propelled conditions at various gravities (1/2 g, 3/4 g, 1 g, and 2 g) on the Toyoura sand with relative density of 50% and that on the LSS with relative densities of 50 and 70% show similar characteristics to that presented in Figure 2.96 (Wong and Kobayashi 2012).

Based on the average values of the measured total sinkage for all slips at various gravities on the LSS and Toyoura sand with relative densities of 50 and 70%, a summary of the average values of the measured sinkage ratios z_{ex}/z_e for various gravity ratios 1/2, 3/4, 1, and 2 is presented in Table 2.18. It should be mentioned that the number of data points and the values of total wheel sinkage measured under gravity of 1/6 g were small. Thus, the average measured sinkage ratios z_{ex}/z_e for gravity ratio 1/6 would be uncertain and may not be representative. For these reasons, the measured sinkage ratios z_{ex}/z_e for gravity ratio $g_{ex}/g_e = 1/6$ on the two types of soil with two relative densities are not presented in Table 2.18.

The test data shown in Table 2.18 were plotted in Figure 2.97 (Wong 2012; Wong and Kobayashi 2012). They indicate that the average value of measured total wheel sinkage ratio z_{ex}/z_e varies in the range of 0.8–1.10 in the range of gravity ratio g_{ex}/g_e from 1/2 to 2, on all types of soil tested. The average values of measured total sinkage ratios on all types of soil tested under gravity ratios 1/2, 3/4, 1, and 2 are 1.02, 1.02, 1.03, and 0.98, respectively, as shown in the last row of Table 2.18. They are close to 1 (one), which is consistent with that predicted using Eq. (2.143), at least as a first approximation.

In summary, despite the probable errors in measurements of wheel sinkage during tests, the assumption of the values of pressure–sinkage parameters of the Reece equation not being affected by gravity, etc., the experimental evidence presented above substantiates the predictions using Eq. (2.143). It indicates that with identical mass, the total rigid wheel sinkage on the moon, Mars, or other planets under self-propelled conditions will be the same as that measured on earth's surface, at least as a first approximation. This provides a useful guidance for performance testing of extraterrestrial rovers and their running gear in the future and for predicting total rigid wheel sinkage on extraterrestrial surfaces based on test data obtained on earth under earth gravity.

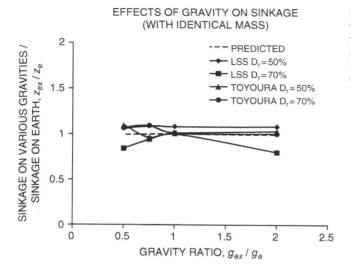

EFFECTS OF GRAVITY ON SINKAGE
(WITH IDENTICAL MASS)

Figure 2.97 Comparisons of the variations of the average measured total sinkage ratio z_{ex}/z_e with gravity ratio g_{ex}/g_e with that predicted having identical mass carried by the wheel under various gravities.

4) Predicting the effects of gravity on the motion resistance of rigid rover wheels with identical mass

Based on the analysis of motion resistance due to terrain compaction of a rigid wheel described in Section 2.8.1, and making use of the Reece equation for the pressure–sinkage relationship (i.e., Eq. (2.137)), one obtains the following equation for determining the motion resistance due to terrain compaction of a rigid wheel R_c under gravity g:

$$
\begin{aligned}
R_c &= b \left(K_c + K_\varphi g \right) \left(\frac{z^{n+1}}{n+1} \right) \\
&= \left(\frac{b \left(K_c + K_\varphi g \right)}{n+1} \right) \left[\frac{3W}{b \left(3-n \right) \left(K_c + K_\varphi g \right) \sqrt{D}} \right]^{(2n+2)/(2n+1)} \\
&= \left[\frac{1}{(3-n)^{(2n+2)/(2n+1)} \, (n+1) \, b^{1/(2n+1)} \left(K_c + K_\varphi g \right)^{1/(2n+1)}} \right] \times \left[\frac{3W}{\sqrt{D}} \right]^{(2n+2)/(2n+1)}
\end{aligned}
\tag{2.144}
$$

Following the approach to the analysis of the sinkage ratio of rigid wheels with identical mass m for a given gravity ratio discussed previously, the ratio of the rigid wheel motion resistance R_{cex} under gravity g_{ex} to that R_{ce} under gravity g_e is given by

$$
\frac{R_{cex}}{R_{ce}} = \frac{\left[\dfrac{1}{(3-n)^{(2n+2)/(2n+1)}(n+1)b^{1/(2n+1)} \left(K_c + K_\phi g_{ex} \right)^{1/(2n+1)}} \right] \times \left[\dfrac{3mg_{ex}}{\sqrt{D}} \right]^{(2n+2)/(2n+1)}}{\left[\dfrac{1}{(3-n)^{(2n+2)/(2n+1)}(n+1)b^{1/(2n+1)} \left(K_c + K_\phi g_e \right)^{1/(2n+1)}} \right] \times \left[\dfrac{3mg_e}{\sqrt{D}} \right]^{(2n+2)/(2n+1)}}
\tag{2.145}
$$

With identical mass m carried by the wheel under various gravities and with the conditions of the simulant being dry with negligible cohesion (i.e., the values of c and K_c are insignificant and can be neglected) and of the values of n and K_ϕ not changing with gravity, Eq. (2.145) can be simplified to

$$\frac{R_{cex}}{R_{ce}} = \frac{g_{ex}}{g_e} \tag{2.146}$$

Equation (2.146) indicates that with identical mass carried by the wheel, the motion resistance ratio R_{cex}/R_{ce} is simply equal to the gravity ratio g_{ex}/g_e.

It should be mentioned that the motion resistance ratio predicted using Eq. (2.146) is based on the motion resistance caused by terrain compaction and that it does not include the effects of wheel slip. However, it will be shown in the next section that based on test data available, Eq. (2.146) may also be applied to estimating the motion resistance ratio R_{cex}/R_{ce} under self-propelled conditions on which wheel slip occurred. This is similar to the case that Eq. (2.143) can be applied to estimating the ratio of total sinkage with slips under various gravity ratios. This finding will have a significant impact on guiding the testing of the motion resistance of rover wheels. It indicates that the motion resistance R_{cex} of a rigid wheel under self-propelled conditions on an extraterrestrial surface with gravity g_{ex} is equal to the motion resistance R_{ce} under self-propelled conditions measured on earth under earth gravity g_e multiplied by the gravity ratio g_{ex}/g_e, provided that the mass carried by the wheel on any extraterrestrial surface is identical to that on earth and that the values of pressure–sinkage parameters of the regolith on the extraterrestrial surface and those of the simulant used on earth are the same. As noted previously, the issue of possible effects of gravity on the values of pressure–sinkage parameters will be reviewed later in Subsection 7).

5) Evaluation of predicted rigid rover wheel motion resistances under various gravities with test data

The finding that with identical mass carried by the wheel, its motion resistance under gravity g_{ex} is equal to that under gravity g_e multiplied by the gravity ratio g_{ex}/g_e, as expressed by Eq. (2.146) was evaluated with test data reported in the paper by Kobayashi et al. (2010).

Two sets of experiments were performed. One set of experiments was carried out in the soil bin installed in the aircraft undergoing parabolic flight maneuvers to produce various levels of gravity. Under self-propelled conditions, the driving torques applied to the wheel under gravities of $1/6\,g$, $1/2\,g$, $3/4\,g$, $1\,g$, and $2\,g$ ($g = 9.81$ m/s^2) in the aircraft on two types of soil with relative densities of 50 and 70% shown in Table 2.17 were measured. The other set of experiments was performed on the ground (earth surface), for measuring wheel torques under self-propelled conditions with normal loads $1/6\,W$, $1/2\,W$, $3/4\,W$, $1\,W$, and $2\,W$. The measured driving torques obtained under wheel load of $1\,W$ (10 kg $\times 9.81$ m/s^2) on the ground on the two types of soil with two different relative densities were used as the wheel torques under earth gravity g_e for determining the torque ratios under various gravity ratios g_{ex}/g_e. It should be noted that the torque applied to the wheel under self-propelled conditions is proportional to its motion resistance and that wheel slip occurred during tests. In the following, the ratios of wheel toque measured under gravity g_{ex} to that under gravity g_e are represented by the ratio of motion resistance R_{cex}/R_{ce}. Thus, the measured motion resistance ratio includes the effects of wheel slip, similar to that for the measured total wheel sinkage ratio discussed previously.

The test data were processed using the procedures described in Wong (2012) and Wong and Kobayashi (2012). A summary of the average values of measured wheel motion resistance ratios R_{cex}/R_{ce}, representing the wheel torque ratios under self-propelled conditions, for gravity ratios of 1/6, 1/2, 3/4, 1, and 2, is presented in Table 2.19.

The test data shown in Table 2.19 were plotted in Figure 2.98 (Wong 2012; Wong and Kobayashi 2012). They show that the average values of measured wheel motion resistance ratios R_{cex}/R_{ce} on all soils under gravity ratios of 1/6, 1/2, 3/4, 1, and 2 are 0.15, 0.51, 0.79 1.01, and 2.13, respectively, as

Table 2.19 Summary of average measured motion resistance ratios R_{cex}/R_{ce} under various gravities.

Soil type	Relative density D_r (%)	Gravity ratio g_{ex}/g_e				
		1/6	1/2	3/4	1	2
		Average measured motion resistance ratio R_{cex}/R_{ce}				
Lunar soil simulant (LSS)	50%	0.15	0.47	0.84	1.04	1.86
	70%	0.19	0.54	0.84	1.04	2.27
Toyoura sand	50%	0.13	0.47	0.69	0.91	2.22
	70%	0.12	0.57	0.78	1.03	2.15
Average measured value for all soils		0.15	0.51	0.79	1.01	2.13

Source: Wong and Kobayashi (2012).

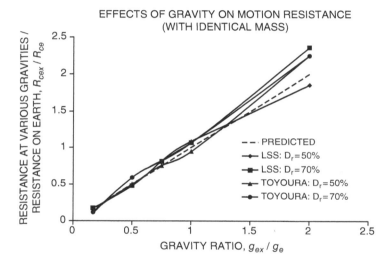

Figure 2.98 Comparisons of the variations of the average measured motion resistance ratio R_{cex}/R_{ce} with gravity ratio g_{ex}/g_e with that predicted having identical mass carried by the wheel under various gravities.

shown in the last row of Table 2.19. They are close to the gravity ratios of 0.166, 0.50, 0.75, 1, and 2, respectively, which is consistent with those predicted using Eq. (2.146).

In summary, despite the probable errors in measurements of wheel torque under self-propelled conditions with wheel slipping, the neglect of soil cohesion in the analysis, etc., the experimental evidence presented above appears to substantiate the predictions using Eq. (2.146), at least as a first approximation. It indicates that for a rigid wheel with identical mass on the extraterrestrial surface and on earth's surface, the motion resistance R_{cex} under self-propelled conditions on the extraterrestrial surface under gravity g_{ex} will be equal to the motion resistance R_{ce} under self-propelled conditions measured on earth under earth gravity g_e multiplied by the gravity ratio g_{ex}/g_e. This provides a useful guidance for performing tests of extraterrestrial rovers and their running gears in the future and for predicting rigid rover wheel motion resistance on extraterrestrial surfaces based on test data obtained on earth under earth gravity.

As the motion resistance of the running gear is a major factor for determining the propulsive power required for a given operating speed range, the ability of properly assessing the motion resistance on an extraterrestrial surface will have an impact on the proper selection of the capacity of the power plant for an extraterrestrial rover.

6) Predicting the effects of gravity on the maximum thrust of a rover or its running gear

Based on the Mohr–Coulomb equation, Eq. (2.10), the maximum thrust of a vehicle or its running gear may be determined using Eq. (2.69). This equation may be rewritten using the product of the mass carried by the rover or its running gear m and gravity g to represent the normal load W on the rover or its running gear as follows:

$$F_{max} = cA + W \tan \phi = cA + mg \tan \phi \qquad (2.147)$$

where F_{max} is the maximum thrust that a rover or its running gear can develop; c is the cohesion of the simulant or regolith; A is the ground contact area of the rover or its running gear; and ϕ is the angle of internal shearing resistance (or angle of internal friction) of the simulant or regolith.

When the rover or its running gear is subject to earth gravity g_e, the maximum thrust F_{emax} is expressed as

$$F_{emax} = cA + mg_e \tan \phi \qquad (2.148)$$

With identical mass carried by the rover or its running gear on an extraterrestrial surface with gravity g_{ex}, the maximum thrust F_{exmax} is given by:

$$F_{exmax} = cA + mg_{ex} \tan \phi \qquad (2.149)$$

On a dry simulant or on a dry extraterrestrial surface, the cohesion c is insignificant, so the term cA in Eq. (2.148) or Eq. (2.149) can be omitted. If the angle of internal shearing resistance ϕ is not affected by or is insensitive to gravity, the ratio of F_{exmax} to F_{emax} is given by

$$\frac{F_{exmax}}{F_{emax}} = \frac{g_{ex}}{g_e} \qquad (2.150)$$

Results of experimental study indicate that effects of gravity, ranging from $1/6\,g$ to $2\,g$, on the angle of repose of the Toyoura sand, with properties presented in Table 2.17, are negligible and that its angle of response is close to its angle of internal shearing resistance (Nakashima et al. 2011). This supports the assumption made in deriving Eq. (2.150) that the angle of internal shearing resistance ϕ is not affected by gravity.

Equation (2.150) indicates that with identical mass carried by the rover or its running gear, the maximum thrust ratio of F_{exmax} under gravity g_{ex} to F_{emax} under gravity g_e is simply equal the gravity ratio g_{ex}/g_e. This means that with identical mass carried by a rover or its running gear, the maximum thrust F_{exmax} on an extraterrestrial surface with gravity g_{ex} is equal to the maximum thrust F_{emax} on earth's surface under earth gravity g_e multiplied by the gravity ratio g_{ex}/g_e. It is noted that the effect of gravity on the maximum thrust of a rover or its running gear is the same as that on the motion resistance of a rigid wheel, as indicated by Eq. (2.146).

7) Effects of gravity on pressure–sinkage parameters

In the analysis of the effects of the gravity ratio on the sinkage ratio and on the motion resistance ratio of a rigid wheel outlined previously, it is assumed that the values of pressure–sinkage parameters n, K_c, and K_ϕ of the Reece equation (i.e., Eq. (2.137)) are not affected by gravity. This issue cannot be directly evaluated, as these pressure–sinkage parameters were not measured either on the ground under earth gravity or in the aircraft under various levels of gravity (Kobayashi et al. 2010).

Among the three pressure–sinkage parameters, K_c is related to cohesion c. As the cohesion c was low for the soils used in the tests described above and the regolith on extraterrestrial surfaces that have been explored so far has been found dry, it would be justified to ignore the effects of K_c in the analysis presented above. The reasonably close correlations between experimental data and predicted wheel sinkage ratios using Eq. (2.143) or predicted wheel motion resistance ratios using Eq. (2.146), based on the assumption that the values of pressure–sinkage parameters are not affected by gravity, do lend support to the notion that gravity would have insignificant effects on the values of pressure–sinkage parameters n and K_ϕ. In other words, based on the reasonably close correlations between test data and predictions, it may be inferred that gravity does not have significant effects on the values of the pressure–sinkage parameters, at least as a first approximation. It should be pointed out, however, that the overall pressure–sinkage relationship is still greatly influenced by gravity, because of the presence of gravity g in the second term within the parentheses on the right-hand side of the Reece equation (i.e., Eq. (2.137)).

To further evaluate the findings on the effects of gravity on rigid wheel sinkage and motion resistance described above, a study using the discrete element method is presented below.

8) Study of the effects of gravity on rigid wheel sinkage and motion resistance using the discrete element method

To further evaluate the findings on the effects of gravity on rigid wheel sinkage and motion resistance presented above, a study of using the two-dimensional discrete element method was performed (Nakashima and Kobayashi 2014). Simulations were conducted on the rigid wheel sinkage and motion resistance under various gravities, for determining the relations of wheel sinkage and motion resistance ratios with gravity ratios. In the study, the mass and dimensions of the rigid wheel and the specifications of the soil bin are identical to those used in the tests described in Subsections 3) and 5) above. The parameters of the discrete elements used in the simulations are selected using the Toyoura sand as a reference.

The focus of the study using the discrete element method is to determine the ratio of wheel sinkage under gravity g_{ex} to that under earth gravity g_e, as well as the ratio of motion resistance under gravity g_{ex} to that under g_e, for identical mass carried by the wheel, and to compare the results obtained using the discrete element method with those predicted using Eqs. (2.143) and (2.146), respectively, as well as the test data. It should be noted that the absolute value of wheel sinkage or motion resistance under any given gravity is not of interest in this study. This means that even if the values of the parameters of discrete elements for characterizing the properties of the sand used in simulations by the discrete element method are different from those of the LSS or the Toyoura sand with two relative densities used in the tests described above, predictions by the discrete element method of the relations between wheel sinkage ratio or motion resistance ratio and gravity ratio are appropriate or valid for comparison with findings given by Eq. (2.143) and by Eq. (2.146), respectively, as well as the test data described previously.

Table 2.20 shows a summary of the average wheel sinkage ratios z_{ex}/z_e for various gravity ratios, 1/6, 1/2, 3/4, 1, and 2, obtained using the discrete element method (Nakashima and Kobayashi 2014). The sinkage ratios under various gravity ratios shown in the table are the average of the results of three simulation runs. For comparison, predicted sinkage ratios z_{ex}/z_e at various gravity ratios g_{ex}/g_e based on Eq. (2.143), and the average wheel sinkage ratios z_{ex}/z_e at various gravity ratios g_{ex}/g_e obtained from tests presented in Subsection 3) are also shown in the table.

It is shown that the results of the relation between wheel sinkage ratios and gravity ratios obtained using the discrete element method correlate reasonably well with those predicted using Eq. (2.143) and test data.

Table 2.21 shows a summary of the average wheel motion resistance ratios R_{cex}/R_{ce} for various gravity ratios, 1/6, 1/2, 3/4, 1, and 2, obtained using the discrete element method (Nakashima and Kobayashi 2014). The wheel motion resistance ratios under various gravity ratios shown in the table

Table 2.20 Summary of average wheel sinkage ratios under various gravity ratios predicted using the discrete element method in comparison with that using Eq. (2.143) and with test data.

	Gravity ratio g_{ex}/g_e				
	1/6	1/2	3/4	1	2
Source	Wheel sinkage ratio z_{ex}/z_e				
Discrete element method	1.08	1.03	1.01	1.00	1.02
Eq. (2.143)	1	1	1	1	1
Average measured value for all soils[a]	–	1.02	1.02	1.03	0.98

[a] Presented in Table 2.18.

Table 2.21 Summary of average wheel motion resistance ratios under various gravity ratios predicted using the discrete element method in comparison with that using Eq. (2.146) and with test data.

	Gravity ratio g_{ex}/g_e				
	1/6	1/2	3/4	1	2
Source	Wheel motion resistance ratio R_{cex}/R_{ce}				
Discrete element method	0.12	0.49	0.81	1.00	1.97
Eq. (2.146)	0.166	0.50	0.75	1	2.00
Average measured value for all soils[a]	0.15	0.51	0.79	1.01	2.13

[a] Presented in Table 2.19.

are the average of the results of three simulation runs. For comparison, predicted wheel motion resistance ratios R_{cex}/R_{ce} at various gravity ratios g_{ex}/g_e based on Eq. (2.146), and the average wheel motion resistance ratios at various gravity ratios g_{ex}/g_e obtained from tests presented in Subsection 5) are also shown in the table.

It is shown that the results of the relations between wheel motion resistance ratios and gravity ratios obtained using the discrete element method correlate reasonably well with those predicted using Eq. (2.146) and test data, at least as a first approximation.

In summary, the results obtained using the discrete element method are reasonably close to those obtained using Eqs. (2.143) for wheel sinkage ratio z_{ex}/z_e and (2.146) for motion resistance ratios R_{cex}/R_{ce}, as well as the test data. It can be stated, therefore, that the results obtained by the discrete element method do support the findings expressed by Eqs. (2.143) and (2.146).

9) Summary

Based on the test data obtained in an aircraft undergoing parabolic flight maneuvers to produce various levels of gravity and on simulation results obtained using the discrete element method, it can be stated that the following findings based on terramechanics have been generally substantiated:

- With identical mass carried by a rigid wheel, its sinkage z_{ex} under gravity g_{ex} on an extraterrestrial surface, such as on the surface of the moon, Mars, or any other planet, is the same as the sinkage

z_e on earth's surface under earth gravity g_e, or in other words, $z_{ex}/z_e = 1$, at least as a first approximation.

- With identical mass carried by a rigid wheel, the ratio of its motion resistance R_{cex} under gravity g_{ex} on an extraterrestrial surface, such as that on the surface of the moon, Mars, or any other planet, to the motion resistance R_{ce} on earth's surface under earth gravity g_e is equal to the gravity ratio g_{ex}/g_e, or in other words, $R_{cex}/R_{ce} = g_{ex}/g_e$, at least as a first approximation.

In addition, it is found that with identical mass carried by a rover or its running gear, the maximum thrust ratio of F_{exmax} under gravity g_{ex} to F_{emax} under gravity g_e is simply equal the gravity ratio g_{ex}/g_e. This is similar to the finding on the relationship between the motion resistance ratio of a rigid wheel and the gravity ratio.

The findings described above provide a useful methodology for predicting the performances of rovers or their running gears on extraterrestrial surfaces, based on their performances on earth's surface under earth gravity. The ability of properly assessing the motion resistance and maximum thrust of a rover on an extraterrestrial surface will provide useful guidance for the selection of the appropriate capacity of the power plant for a rover operating on an extraterrestrial surface.

2.11.2 Performances of Lunar Roving Vehicle Flexible Wheels Predicted Using the Model NWVPM and Correlations with Test Data

The physics-based model NWVPM described in Section 2.9 was originally developed for design and performance evaluation of terrestrial off-road wheeled vehicles with either rigid or flexible wheels. As noted previously, for the manned lunar roving vehicle used in NASA's Apollo missions 15, 16, and 17, flexible wheels were employed. To assess the potential of NWVPM for evaluating the mobility of future generations of extraterrestrial rovers with flexible wheels, a study in collaboration with the Mechanical Components Branch, Glenn Research Center, NASA, was carried out. NWVPM was used to predict the performances of two flexible wheel candidates, a wire-mesh wheel and a hoop-spring wheel, for the manned lunar roving vehicle (Wong and Asnani 2008). The predicted performances were compared with test data. The test data were obtained in a laboratory soil bin under earth gravity and documented in the Technical Report M-70-2, prepared for George C. Marshall Space Flight Center, NASA, by the US Army WES (Freitag et al. 1970).

The purpose of this study was to assess the capability of NWVPM for future applications to evaluating performances of manned extraterrestrial rovers with flexible wheels. This investigation was not intended for evaluating the performances of flexible wheel candidates on the lunar surface under lunar gravity. The methodology for predicting rover wheel performance on lunar or any extraterrestrial surface based on test results obtained on earth is discussed in Section 2.11.1.

The wire-mesh wheel and the hoop-spring wheel were considered as the most promising candidates, among the eight types of wheel evaluated, for the manned lunar roving vehicle for NASA's Apollo missions (Asnani et al. 2009). The wire-mesh wheel was eventually chosen for the vehicle deployed to the moon on the Apollo 15, 16, and 17 missions in the 1970s.

1) The wire-mesh wheel and the hoop-spring wheel for the manned lunar roving vehicle

The baseline version of the wire-mesh wheel used in this study was similar in design but slightly larger in diameter than that installed on the manned lunar roving vehicle deployed to the moon shown in Figure 2.99. The wheel consists of a hub formed by an aluminum disc attached to the traction drive, an aluminum rim with a woven zinc-coated, spring-steel wire-mesh carcass, and an inner titanium bump-stop ring. The hub is riveted to the rim of the wire-meshed wheel. For flotation, the external surface of the carcass is covered with titanium tread strips arranged in a herringbone pattern. Since some of the data on the baseline version of the wire-mesh wheel required as input to NWVPM, such as the shearing characteristics between the zin-coated wire and terrain and that between the titanium tread strips and terrain, are not available in the test report (Freitag et al. 1970), two modified versions of the wheel, designated as the Boeing–General Motors (GM) IV and VI, were used in this correlation study. These two modified versions have basic structures as the baseline version described above. However, their surfaces are covered with fabric (Gray tape) coated with sand and their test data are available in the test report. As a result, the traction of these two modified versions can be predicted from the internal shearing characteristics of the simulant (sand) used in tests. This made it possible to use NWVPM to predict their performances and to compare the predictions with test data.

The major difference between the Boeing–GM IV and VI wheels is that, in the latter, 50% of the wire structure is removed, and hence it is more flexible and has greater contact area and lower contact (ground) pressure than the former under the same normal load. As shown in Table 2.22, the Boeing–GM IV wheel has a contact pressure on a hard surface of 13.31 kPa (1.93 psi) at a normal load of 311 N (70 lb), whereas for the Boeing–GM VI wheel, two sets of data are available, one with a contact pressure of 4.42 kPa (0.64 psi) and the other with 4.04 kPa (0.59 psi) under the same normal load. The difference in contact pressure between the Boeing–GM IV and VI leads to a significant difference in tractive performance,

Figure 2.99 The basic structure of the wire-mesh wheel for the manned lunar roving vehicle for NASA's Apollo 15, 16, and 17 missions to the moon. *Source:* NASA.

Table 2.22 Loads, contact areas, and contact pressures of Boeing–GM IV and VI wheels and Bendix I wheel used in the study.

Wheel type	Normal load		Carcass diameter		Contact area		Contact pressure	
	N	lb	cm	in.	cm^2	in.^{2}	kPa	psi
Boeing–GM IV	311	70	102.3	40.3	234.1	36.3	13.31	1.93
Boeing–GM VI	311	70	99.8	39.3	710.2	110.1	4.42	0.64
Boeing–GM VI	311	70	100.3	39.5	778.8	120.7	4.04	0.59
Bendix I	67	15	101.6	40	243.5	37.7	2.76	0.40
Bendix I	133	30	101.6	40	519.4	80.5	2.58	0.37
Bendix I	311	70	101.6	40	722.6	112.0	3.93	0.57

Source: Wong and Asnani (2008).

Figure 2.100 The basic structure of the hoop-spring wheel designated as Bendix I. *Source:* NASA.

which will be discussed later. The loaded section height and loaded section width of the two wire-mesh wheels used in this study are given in the paper by Wong and Asnani (2008).

The hoop-spring wheel, designated as Bendix I wheel, consists of a titanium outer band, hoop-spring elements, and a rigid aluminum hub with spokes, as shown in Figure 2.100. The surface of this wheel is coated with sand and no treads are installed. The traction is derived from the internal shearing of the sand, like that of the Boeing–GM IV and VI wheels described above. The basic parameters of the Bendix I wheel are also presented in Table 2.22. As can be seen from the table, under a normal load of 311 N, it has a contact pressure of 3.93 kPa (0.57 psi).

It should be noted that the normal load of 311 N (70 lb) on the Boeing–GM IV and VI wheels and on the Bendix I wheel for performance testing under earth gravity and shown in Table 2.22 is based on the load expected on the lunar surface, as on the lunar surface the gravity is 1/6 that on earth's surface.

2) Soil conditions for tests

The wire-mesh wheels (Boeing–GM IV and VI) and the hoop-spring wheel (Bendix I) were tested under earth gravity in a soil bin in the laboratory of the US Army WES (Freitag et al. 1970). The lunar simulant used in the tests was air-dry sand, designated as sand S_1, with an average moisture content of 0.5% and dry density of 1.484 g/cm^3 (92.64 lb/ft^3). The average values of the Bekker pressure–sinkage parameters (n, k_c, and k_ϕ) and shear strength parameters (cohesion c and angle of internal shearing resistance ϕ) are presented in Table 2.23.

The Bekker pressure–sinkage equation, Eq. (2.47), indicates that if the value of k_c is positive, the increase of b (the radius of a circular contact area or the width of a rectangular contact area) will cause a decrease in the pressure required for a tire or a track to sink to a given depth z. If the value of k_c is negative, as in the case of sand S_1 shown in Table 2.23, the increase of b will cause an increase of the pressure required for a tire or a track to sink to a given depth z, as mentioned previously in Section 2.4.1.

Table 2.23 Average values of the Bekker pressure–sinkage parameters and shear strength parameters of sand S_1.

	Pressure-sinkage parameters				Shear strength parameters	
		k_c		k_ϕ		
n	kN/m^{n+1}	lb/in.$^{n+1}$	kN/m^{n+2}	lb/in.$^{n+2}$	c	ϕ
0.91	−0.66	−0.13	754.13	3.73	0	27.4°

Source: Wong and Asnani (2008).

To predict the thrust-slip (or drawbar pull-slip) relationship of the wheels, the shear deformation parameters K in the shear stress–shear displacement relationship expressed by Eq. (2.58) is required. The parameter K for sand S_1 was, however, not given in the test report (Freitag et al. 1970). To obtain a reasonable value of K required as input to NWVPM, effort was made at the Glenn Research Center of NASA to replicate the conditions under which the original shear tests were performed at the US Army WES. Using a sand with similar particle size distribution, density, and moisture content to those of sand S_1, and employing a shear ring (annulus) with the same dimensions as the one originally used in the test, a series of shear tests was conducted at normal pressures corresponding to those under the wheels examined in this study. Based on test results, it was estimated that the value of K for sand S_1 is approximately 0.5 cm (0.2 in.). Accordingly, this estimated value of K was used as input to NWVPM in the simulations, the results of which are used in the study of the correlation between predictions by NWVPM and test results.

3) Correlations between the predicted and measured performances of the wire-mesh and hoop-spring wheels

The ratio of the drawbar pull to normal load of a wheel (commonly referred to as the drawbar pull coefficient, as noted previously) on level ground is widely used as an indicator for the tractive capability of a wheel. At 20% slip, the wheel is usually in the desirable range of operating efficiency. Accordingly, the drawbar pull coefficient at 20% slip, P_{20}/W, is used to represent the performance of the wheel in this study. The predicted values of the drawbar pull coefficient at 20% slip of the Boeing–GM IV and VI wheels and the Bendix I wheel, together with those measured, are shown in Table 2.24.

As shown in Table 2.24, the ratios of the predicted drawbar pull coefficient at 20% slip to that measured for the Boeing–GM IV and VI wheels at 311 N vary from 106.4 to 108.1%, whereas those for the Bendix I wheel at normal load from 67 N to 311 N vary between 87.7 and 102.2%. Considering the probable errors in measurements and other factors, the correlations between predictions by NWVPM and test data are considered reasonable.

4) Summary

- The tractive performances, as represented by the drawbar pull coefficient at 20% slip, P_{20}/W, of the Boeing–GM IV and VI wheels and the Bendix I wheel predicted by NWVPM on sand S_1 correlate reasonably well with that measured.

Table 2.24 Predicted and measured drawbar pull coefficient at 20% (P_{20}/W) slip of the Boeing–GM IV and VI wheels and of the Bendix I wheel.

	Normal load		Contact pressure		Predicted P_{20}/W	Measured P_{20}/W	Predicted/ Measured
Wheel type	N	lb	kPa	psi	%	%	%
Boeing–GM IV	311	70	13.31	1.93	29.8	28	106.4
Boeing–GM VI	311	70	4.42	0.64	40.9	38.4	106.5
Boeing–GM VI	311	70	4.04	0.59	41.5	38.4	108.1
Bendix I	67	15	2.76	0.40	37.2	42.4	87.7
Bendix I	133	30	2.58	0.37	45.2	45.8	98.7
Bendix I	311	70	3.93	0.57	47.3	46.3	102.2

Source: Wong and Asnani (2008).

- Results of this correlation study indicate that NWVPM has the potential as an engineering tool for the prediction of the tractive performance of flexible wheel candidates for future generations of extraterrestrial rovers. Further evaluation of NWVPM under a wider range of operating environments, however, would be useful.
- The sinkage, motion resistance, and the maximum thrust of a flexible rover wheel on extraterrestrial surfaces may be predicted using the methodology outlined in Section 2.11.1 based on the test results obtained on earth.
- There has been a hypothesis that methods developed for predicting the performances of heavily loaded off-road vehicles of large sizes may not be applicable to lightly loaded vehicles of small sizes. This may be so for empirically based models. As it is well known, empirical relations, in general, should not be extrapolated beyond the conditions upon which they were derived. Consequently, it is indeed uncertain that empirical models based on test data obtained with heavily loaded vehicles of large sizes could be applicable to predicting the performances of lightly load vehicles of small sizes. For NWVPM, it is a physics-based model. Consequently, it should be applicable to both heavily loaded off-road wheeled vehicles of large sizes and lightly loaded ones of small sizes, provided that, for instance, the values of pressure–sinkage parameters used as input to NWVPM are in the appropriate ranges of contact pressures and sinkages obtained using suitable sizes of sinkage plates. The results of this study provide evidence for supporting this view. Similar findings are found for the physics-based model NTVPM for tracked vehicles, which was originally developed for predicting the performances of heavily loaded tracked vehicles of large sizes. Results of a study on applying NTVPM to predicting the performance of a lightly loaded robotic track system of small size, with normal loads in the range of 125 N to 190 N and track contact length of 250 mm and contact width of 100 mm, lead to similar conclusions (Wong et al. 2015).

In addition to models NWVPM and NTVPM with capabilities for predicting the tractive performances of extraterrestrial rovers with wheels and tracks, respectively, other models for simulating the operations of rovers and their running gears have been developed. For instance, a model known as the Artemis (Adams-based Rover Terramechanics and Mobility Interaction Simulator) has been developed for simulating mobility and for traverse planning of NASA's Mars exploration rovers, such as the Spirit, Opportunity, and Mars Science Laboratory known as the Curiosity (Zhou et al. 2014). The basic approach of the Artemis to predicting rover performance follows that of classical terramechanics, as presented in previous sections of this chapter. It has been demonstrated that the Artemis can provide reasonable predictions of sinkage, torque, and drawbar pull of Mars exploration rovers at wheel slips up to approximately 70%. Beyond that, the predictive capability of the Artemis deteriorates, as the classical terramechanics approach has not yet been well suited for predicting vehicle (rover) performance at high slips (Johnson et al. 2017).

2.12 Finite Element and Discrete Element Methods for the Study of Vehicle–Terrain Interaction

The finite element method (FEM) and the discrete (distinct) element method (DEM) have been introduced to the study of vehicle–terrain interaction in recent years.

2.12.1 The Finite Element Method

The FEM was originally developed for analysis of the response of a continuum to external loading, such as in structural analysis. Its applications have later been expanded to cover many fields in engineering, including the study of vehicle–terrain interaction.

The basic concept of the finite element method is the idealization of a continuum (such as tire carcass or soil mass) as an assemblage of a finite number of elements. For a two-dimensional continuum, the finite elements may be in the form of triangles or quadrilaterals. For three-dimensional analysis, the finite elements may be tetrahedral, rectangular prisms, or hexahedra. Triangular or quadrilateral shell or membrane elements are used in modeling tires. These elements are interconnected at joints, which are called nodes or nodal points. Simple functions, such as polynomials, are chosen to approximate the distribution of displacements within each element. To describe the behavior of the element under load, the stiffness matrix is formulated for each element. The stiffness relates the displacement to the applied force at nodal points. The stiffness matrix consists of the coefficients of the equilibrium equations derived from the material and geometric properties of an element and usually obtained using the variational principle of mechanics (such as the principle of minimum potential energy). The assembly of the overall stiffness matrix for the entire body from individual element stiffness matrices and of the overall force or load vector from element nodal force vectors is then carried out. For a given load applied to the continuum, the equilibrium equations noted above are solved for the unknown displacements of the nodal points. Based on the predicted displacements of the nodal points, the deformation of the element can be computed. With the properties of the material known or given, the stresses and strains of the elements can be predicted. The finite element method offers the flexibility of assigning different mechanical properties to different elements, thus providing a means to examine engineering problems involving non-homogenous or anisotropic media.

Early work on applying the finite element method to the study of wheel (tire)–soil interaction assumes that the soil is either a linear or nonlinear elastic continuum. The normal and shear stress distributions on the wheel–soil interface are required as input for initiating the solution process (Perumpral et al. 1971; Yong and Fattah 1976). As described in Section 2.8, when the normal and shear stress distributions on the wheel–soil contact patch are specified, the performance of the wheel, expressed in terms of motion resistance, thrust, drawbar pull, and tractive efficiency, is completely defined. Thus, in early work the role of the finite element method is restricted to predicting the distributions of stress and deformation in the soil mass for given normal and shear stress distributions on the wheel (tire)–soil interface (Wong 1977).

Research on the application of the finite element method to the study of wheel (tire)–terrain interaction has advanced considerably since then. The normal and shear stress distributions on the wheel–terrain interface, hence the tractive performance of the wheel (tire), can be predicted for a given load and under specific operating conditions using the finite element method.

To accommodate the behavior of various types of terrain, several constitutive models have been introduced. Among them, the Drucker–Prager cap model and the Cam–Clay critical-state soil model are two of the widely used, although the Mohr–Coulomb yield model is also employed in some cases (Liu and Wong 1996; Liu et al. 1999; Seta et al. 2003; Fervers 2004; Zhang et al. 2005).

Several comprehensive, commercially available finite element codes considered to be applicable to the study of machine–terrain interactions have become widely available in recent years. All of these have facilitated advancements in the application of the finite element method to the investigation of the physical nature of wheel–terrain interaction and to the prediction of off-road vehicle cross-country performance.

VERTICAL STRESS (kPa)

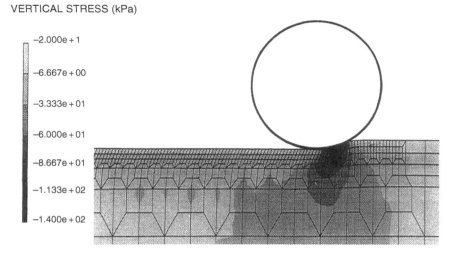

Figure 2.101 Finite element mesh and its deformed patterns and the distributions of the vertical stress on loose sand beneath a rigid wheel with diameter of 1.245 m (49 in.), width of 0.305 m (12 in.), and normal load of 9.28 kN (2086 lb) at 3.1% slip.

A modified Cam–Clay critical-state soil model in conjunction with a new nonlinear elastic law implemented in the finite element code MARC has been employed to predict the tractive performance of a rigid wheel and that of a pneumatic tire (Liu and Wong 1996; Liu et al. 1999). The tractive performance of the rigid wheel with diameter of 1245 mm (49 in.), width of 305 mm (12 in.), and normal load of 9.28 kN (2086 lb) on loose sand have been predicted and compared with test data. The performance of a 12.5/75R20 tire at an inflation pressure of 345 kPa (50 psi) and with normal load of 9.23 kN (2075 lb) on the Ottawa sand has also been predicted and compared with experimental data (Liu and Wong 1996; Liu et al. 1999). In both cases, a vertical load is applied and then the wheel (tire) moves horizontally at low speeds and at various slips, until a steady-state condition is reached.

Figure 2.101 shows the deformed patterns of the finite element mesh and the vertical stress on loose sand under the rigid wheel described above. A comparison of the predicted drawbar pull coefficient (i.e. the ratio of drawbar pull to wheel load) of the rigid wheel on loose sand as a function of slip obtained using the finite element code MARC with that measured is presented in Figure 2.102. Figure 2.103 shows a comparison of the predicted drawbar pull coefficient of the 12.5/75R20 tire on the Ottawa sand as a function of slip obtained using the finite element code MARC with that measured. It is shown that there is a reasonably close agreement between the predicted and measured relationships between drawbar pull coefficient and slip for both the rigid wheel and the pneumatic tire. However, there is only a qualitative agreement between the measured and predicted normal and shear stress distributions on the contact patch of the rigid wheel or of the pneumatic tire (Liu and Wong 1996; Liu et al. 1999).

In addition to the examples presented above, the finite element method has also been applied to the study of soil compaction under vehicular load (Xia 2011) and of tire–snow interaction (Shoop 2001; Lee 2011).

To provide guidance for improving tread pattern designs of winter tires, a three-dimensional analysis of the mechanics of tire–snow interaction has been performed (Seta et al. 2003). In the

Figure 2.102 Comparisons of the measured and predicted drawbar performance of the rigid wheel on loose sand.

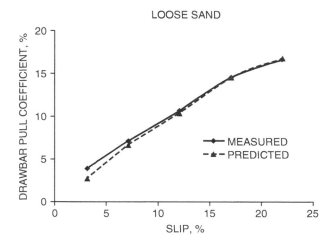

Figure 2.103 Comparisons of the measured and predicted drawbar performance of the 12.5/75R20 tire on the Ottawa sand.

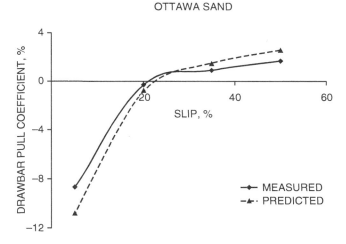

study, the tire is modeled using the finite element method, while the snow is modeled using the finite volume method (FVM).

Coupling between the tire and snow is by the coupling element. The snow is assumed to be a homogeneous elastoplastic material. The Mohr–Coulomb yield model, as discussed in Section 2.2, is adopted. The shear strength parameters of the snow are cohesion of 0.016 mPa (2.3 psi) and angle of internal shearing resistance of 31°. The tractive performance of a 195/65R15 tire on snow, with a normal load of 4.0 kN (900 lb), at an inflation pressure of 200 kPa (29 psi) and operating at a speed of 60 km (37 mph) with 30% slip, has been simulated using the computer program MSC.Dytran. The number of elements used to represent the tire in three dimensions is nearly 60,000. The number of three-dimensional elements used for the snow domain is approximately 70,000. A supercomputer was used to perform the simulations. Tables 2.25 and 2.26 show the measured and predicted traction forces of the 195/65R15 tire with different groove widths and groove angles, respectively. Despite a noticeable difference between the predicted and

Table 2.25 Comparison of the predicted and measured tire traction forces for different groove widths on snow.

Groove width	Predicted traction force			Measured traction force		
	kN	lb	Index	kN	lb	Index
Narrow	0.60	135	100	0.80	180	100
Wide	0.65	146	108	0.89	200	111

Source: Seta et al. 2003.

Table 2.26 Comparison of the predicted and measured tire traction forces for different groove angles on snow.

Groove angle, degrees	Predicted traction force			Measured traction force		
	kN	lb	Index	kN	lb	Index
0^a	0.69	155	100	0.94	211	100
30^b	0.65	146	94	0.92	207	97
-30^c	0.68	153	98	0.93	209	99

Source: Seta et al. (2003).
[a] Straight grooves parallel to the tire rotating axis.
[b] V-shape grooves in the direction of tire rotation, with both sides at 30° with the tire rotating axis.
[c] Reverse V-shape grooves in the direction of tire rotation, with both sides at 30° with the tire rotating axis.

measured traction forces in terms of absolute value, predictions are in good qualitative agreement with experimental data, when comparing the predicted and measured traction forces in terms of index, as shown in the tables (Seta et al. 2003).

In addition to applying the finite element method to the study of wheel (tire)–terrain interaction, efforts have also been made in applying the method to the study of track–terrain interaction. In an earlier study, the track is simplified to a rigid footing and the normal pressure distribution on the track–terrain interface is assumed to be either uniform or of trapezoidal form (Karafiath 1984). Ratios of shear stress to normal pressure are specified at the outset. As discussed previously in Sections 2.6 and 2.7, idealizing the track as a rigid footing is an oversimplification. A track system is a complex mechanical system. For instance, a segmented metal track, commonly used in military fighting or logistics vehicles, consists of many metal track links connected by pins (or with rubber bushings). It interacts with road wheels and through them with the suspension system, in addition to the terrain. Its operation is influenced by many design factors, such as road wheel arrangements, suspension characteristics, initial track tension, locations of the sprocket and idler, configuration of supporting rollers on the top run of the track, etc. Therefore, the application of the finite element method to the analysis of the track–terrain interaction is much more complicated than that by assuming that the track is equivalent to a rigid footing.

Based on the examples of the studies of wheel (tire)–terrain interaction using the finite element method presented above and research results reported in the literature, it appears that the capability of the finite element method in elucidating certain aspects of the physical nature of the wheel (tire)–terrain interaction has been demonstrated. It is shown that in general there is a reasonable qualitative agreement between predictions obtained using the finite element method and test data.

Hence, it could be a useful tool for the investigation of certain aspects of the mechanics of wheel (tire)–terrain interaction.

For further development of the application of the finite element method to the prediction of vehicle performance on deformable terrain, the following are suggested:

- Improving the modeling of mechanical properties of terrain elements to realistically reflect terrain behavior in the field under its natural state is of importance.
- Acquisition of reliable data on terrain element parameters as input to the finite element method is of significance. Without realistic values of input terrain element parameters, the validity of predictions by the finite element method could not be assured.

As noted previously, the application of the finite element method to the study of vehicle–terrain interaction is based on the premise that the terrain can be considered as a continuum and that the general principles of continuum mechanics are applicable. The application of the finite element method to granular soils, therefore, has inherent limitations, such as in simulating plastic flow of sandy soil under the action of a vehicle running gear. For instance, in sand significant soil flow beneath a wheel has been observed, as shown in Figures 2.19–2.22. To simulate the interaction between a wheel (tire) or a track and granular soil, the discrete (distinct) element method has been introduced.

2.12.2   The Discrete (Distinct) Element Method

The DEM was originally developed for the study of rock mechanics (Cundall and Strack 1979). Its applications have since been extended to other fields of engineering, including the study of vehicle–terrain interaction.

The concept of the discrete (distinct) element method is the representation of the soil (terrain) as an assemblage of discrete elements. At the basic level, the discrete elements are assumed to be of circular shape for a two-dimensional analysis and of spherical form for a three-dimensional simulation. Based on such an idealization, the characteristics of wheel (track)–soil interaction can be analyzed by examining the interacting forces between the wheel (track) and adjacent elements and that between the contacting elements. Elements in contact with the wheel (track) surface receive contact forces from the wheel (track). Elements not in contact with the wheel (track) surface receive contact forces from other contacting elements. The magnitude of the contact force is determined by the relative displacement and relative velocity of the contacting elements. It should be noted that with the use of elements of circular or spherical shape, for two-dimensional or three-dimensional analysis, respectively, the interlocking behavior of soil particles cannot be simulated. In more advanced applications of the discrete element method, elements of ellipsoidal, poly-ellipsoidal, polyhedral, or tri-sphere shape are used (Knuth et al. 2012; Johnson et al. 2015).

The discrete element method in its basic form assumes that each element has stiffness (in each of the normal and tangential directions) characterized by a spring constant k, and possesses damping (in each of the normal and tangential directions) characterized by a viscous damping coefficient η. However, in some versions of the method, Coulomb damping is assumed. Between the wheel (track) surface (or container wall) and adjacent elements or between contacting elements, it is assumed that friction exists in the tangential direction, which is characterized by a coefficient of friction μ. The maximum tangential force is limited by the product of the coefficient of friction and the normal force.

The mechanical interaction between two contacting elements or between an element and the wheel (track) surface (or container wall) is schematically shown in Figure 2.104 (Tanaka et al.

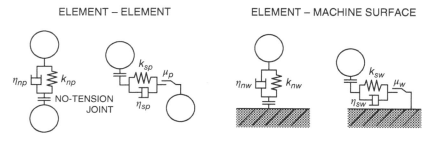

Figure 2.104 Mechanical interactions between two contacting elements and between an element and machine surface. *Source:* Tanaka et al. (2000). Reproduced with permission of the International Society for Terrain–Vehicle Systems.

2000). It should be noted that the spring elements in both normal and tangential directions shown in the figure cannot sustain tensile force. In other words, between contacting elements or between the wheel (track) surface (or container wall) and adjacent elements, there are no tension joints. This indicates that the cohesion between contacting elements and the adhesion between the wheel (track) surface (or container wall) and adjacent elements are not included in the basic model shown in Figure 2.104. However, in more advanced models, the cohesion between contacting elements and the adhesion between the wheel (track) surface (or container wall) and adjacent elements are taken into consideration.

To determine whether contact occurs between the wheel (track) surface (or container wall) and adjacent elements or between elements, their geometrical relationships are examined. If the distance between the wheel (track) surface (or container wall) and the center of an adjacent element is less than the radius of the element (assuming the element is of circular shape for a two-dimensional analysis or of spherical form in a three-dimensional simulation), contact between the wheel (track) surface (or container wall) and the element is considered to have established. On the other hand, if the distance between the centers of two elements is less than the sum of the radii of these two elements, contact between them is considered to have occurred. The position of each element is determined step by step at selected time intervals. The contact forces on each element determine the motion of the element. As an example, for a two-dimensional analysis, Figure 2.105 shows the positions of two contacting elements i and j at time t (Tanaka et al. 2000). The direction of motion of each element during time Δt is indicated by an arrow. Displacements of elements i and j in the X and Y directions during time Δt are expressed by Δu_i, Δv_i, Δu_j, and Δv_j, respectively.

As noted previously, contact forces acting on an element are assumed to be related to the relative displacement or relative velocity (if viscous damping is assumed) of two contacting elements or between the wheel (track) surface (or container wall) and an adjacent element. For instance, the normal component of the relative displacement between two contacting elements determines the normal force due to elastic deformation. The normal component of the relative velocity determines the normal component of the viscous damping force. The tangential component of the relative displacement induces a tangential force due to elastic deformation. This tangential force, however, cannot exceed the maximum frictional force between two contacting elements. The tangential component of the relative velocity determines the tangential component of the viscous damping force. For a two-dimensional analysis, based on the resultant force acting on the element from all other contacting elements, three equations of motion for each element can be established:

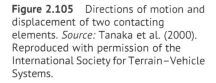

Figure 2.105 Directions of motion and displacement of two contacting elements. *Source:* Tanaka et al. (2000). Reproduced with permission of the International Society for Terrain–Vehicle Systems.

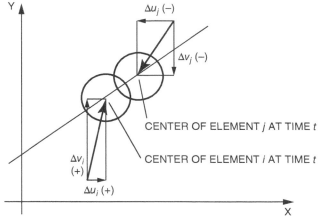

two for the linear motions of the mass center in the X and Y directions and one for the rotation about the mass center of the element. From these equations of motion, the linear accelerations of the mass center in the X and Y directions and the angular acceleration about the mass center of the element can be determined. By integrating the accelerations over the time interval and adding the values to the velocities at the previous time step, the velocity of the element in each direction at the current step can be obtained. To enhance the stability of solution, incremental displacements during the time interval are taken as the average of the incremental displacements at the previous step and the integrations of the velocities over the time interval.

By repeating these procedures for all elements involved and over the duration specified, the interacting forces on the wheel (track)–soil interface can be determined and the movements of soil particles (represented by the discrete elements) under the action of the wheel (track) can be identified. Usually, the solution process is initiated by specifying the vertical load, forward speed, and slip of the vehicle running gear (wheel or track).

The discrete element method has been applied to the study of the interaction between vehicle running gear and soil. For instance, a two-dimensional simulation of the interaction between a lugged, rigid wheel of a lunar micro-rover on lunar regolith simulant has been performed, as schematically shown in Figure 2.106 (Nakashima et al. 2007). In this simulation, the lunar regolith simulant is represented by circular elements. The parameters of the lugged, rigid wheel and of the

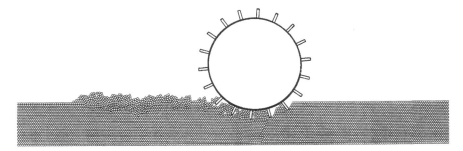

Figure 2.106 Simulations of the operation of a lugged, rigid wheel of a lunar micro-rover using the discrete element method. *Source:* Nakashima et al. (2007). Reproduced with permission of the International Society for Terrain–Vehicle Systems.

Table 2.27 Parameters of the lugged, rigid wheel of a lunar rover and of the discrete elements used in the simulation using the discrete element method.

Parameters	Values
Number of elements used in the simulation	6986
Diameter of element, mm	4.0
Mass density of element, g/cm^3	1.6
Diameter of wheel, mm	200
Mass of wheel, g	500
Diameter of lug element, mm	2.5
Vertical load on wheel, N	9.8, 19.6
Traction load on wheel, N	Varied
Angular velocity of wheel, rad/s	1.38
Duration of soil consolidation, s	1
Duration of vertical sinkage, s	1
Simulation time for wheel travel, s	45.0
Time increment, s	0.0001
Normal spring constants[a], N/m	10 000
Tangential spring constants[a], N/m	500
Normal damping coefficient[a], N s/m	8.97
Tangential damping coefficient[a], N s/m	2.01
Friction coefficient between soil elements	0.9
Friction coefficient for wheel, lug, and wall contact[b]	0.5

[a] Constants/coefficients are assumed to be the same for contact between elements, between elements and wheel or lug surface, and between elements and soil container wall.
[b] Friction coefficients are assumed to be the same for contact between elements, between elements and wheel or lug surface, and between elements and soil container wall.
Source: Nakashima et al. (2007). Reproduced with permission of the International Society for Terrain–Vehicle Systems.

elements used in the simulation are given in Table 2.27. It is found that there is a qualitative agreement between the test results and predictions obtained using the discrete element method, with respect to the effects of lug height, lug thickness, number of lugs, and wheel diameter on tractive performance of the wheel (Nakashima et al. 2007).

The immobilization of a NASA Mars exploration rover (MER), known as the Spirit, on the Martian surface in 2009 has stimulated a number of simulation studies of the digging (excavating) of regolith by the rover wheel at high slips (Knuth et al. 2012; Johnson et al. 2015; Johnson et al. 2017). In a three-dimensional simulation, the Hertzian contact force model is used to characterize the normal contact force, which is based on a nonlinear force–displacement relationship where the contact area between particles increases with the normal load (Knuth et al. 2012). The normal damping force is proportional to the normal component of the relative velocity between two contacting elements. The tangential contact force is determined by the tangential contact stiffness, which is dependent on the normal force, and the tangential component of the relative displacement of contacting elements. The tangential damping force is proportional to the tangential component of the relative velocity of contacting elements. To examine the significance of particle shape, spherical, ellipsoidal, and poly-ellipsoidal particles are used in the simulation. Spherical particles are often used in large-scale discrete element simulations for shortening computation time. However, the chief drawbars of using spherical particles are their inability to simulate the interlock of contacting particles and the dilating of the surrounding particles caused by particle rotation. It is found that the simulation results of rover wheel digging at high slips using the poly-ellipsoid particles correlate well, in driving torque and sinkage, with laboratory experimental data using an actual MER wheel on lunar regolith simulant known as JSC-1a (Johnson Space Center Simulant No. 1a).

As noted in Section 2.11.2, the capabilities of the simulation model Artemis are limited to simulating cross-country performance and traverses of rovers at slips up to 70%. To extend the applications of the Artemis to simulating performance and traverses of rovers in the range of slip higher than 70%, attempts have been made to use the three-dimensional discrete element method to simulate the operation of the rover wheel in the high slip range (Johnson et al. 2015, 2017). In the simulation, the particle packing density and contact friction coefficient are calibrated (tuned) using laboratory experimental data of the rover wheel at slips of 50 and 70%. The corresponding calibrated values of packing density and contact friction coefficient are 0.62 and 0.8, respectively. Monodispersed tri-sphere particles with radius for circumscribed sphere of 3.9 mm are used to allow particle interlocking in comparison with spherical particles. The use of polyhedral particles for further increasing particle interlocking is planned for future simulations. The intent of the study is to obtain data through three-dimensional discrete element simulations on drawbar pull, wheel torque, and sinkage in the high slip range beyond 70% for use in the Artemis. This would combine the ability of the discrete element method to simulate the performance and traverse of rovers at high slips with the computational efficiency of the Artemis (Johnson et al. 2017).

Based on the studies outlined above and on a review of the literature, the applications of the discrete element method to evaluating vehicle (rover) mobility may be summarized as follows (Tanaka et al. 2000; Horner et al. 2001; Nakashima and Oida 2004; Shigeta and Aruga 2005; Nakashima et al. 2007; Nakashima et al. 2011; Knuth et al. 2012; Nakashima and Kobayashi 2014; Johnson et al. 2015; Nakashima et al. 2015; Johnson et al. 2017; Du et al. 2017; Edwards 2018; Jiang et al. 2018):

- Even two-dimensional simulations with relatively simply particle geometries can provide meaningful qualitative information on certain aspects of the physical nature of vehicle (rover)–terrain interaction, such as the wheel digging phenomenon at high slips, and on comparing vehicle (rover) mobility on a relative basis. As noted previously, the physical nature of wheel digging phenomenon at high slips has been studied using photographic techniques to demonstrate the soil being shifted from beneath to behind the wheel, as shown in Fig. 2.21 (Wong 1967).
- The ability to develop quantitative representations or predictions of vehicle (rover) mobility improves in direct proportion to the degree that micromechanical properties considered to be significant are included in discrete element simulations.
- Micromechanical properties that are considered to have noticeable impacts on simulation results include interparticle contact friction and cohesion, particle interlocking, packing density, particle shape, size, and distribution, etc.
- It is estimated that to conduct a realistic three-dimensional simulation of full-scale vehicle (machine)–terrain interaction, the number of elements required may be in the order of 10^6–10^8. This would require lengthy computation and high-performance computing facilities. In many cases, due to limited computing resources available, a much smaller number of elements than that desired is used. This would lead to uncertainty and/or lower fidelity of the simulation results.
- To address the issue of computation time, it has been suggested that the use of a combination of the discrete elements and the finite elements to model terrain be explored. Particularly, it

is proposed that discrete elements be used to represent the upper layer of the terrain, while coarse finite elements be used to model the lower layer.

- Many current discrete element models include few of the micromechanical properties of soil particles that are known to affect soil mechanical behavior. Many of the models are based on simplifying assumptions. For instance, in many studies, spherical particles with relatively large diameters are used in simulations instead of using more realistic particle shapes and sizes; uniform particle-size distributions are used instead of using more realistic particle-size distributions; and two-dimensional models are used to simulate three-dimensional processes.
- A comprehensive examination of how closely a discrete element model needs to replicate the micromechanical features of a granular assembly has yet to be performed.

References

Ahlvin, R.B. and Haley, P.W. (1992). NATO Reference Mobility Model Edition II, NRMM II Users Guide. Technical Report GL-92-19. U.S. Army Corps of Engineers Waterways Experiment Station, Vicksburg, MS.

Andrade, J., Lindermann, R., Iagnemma, K., et al. (2012). XTerramechanics: Integrated simulation of planetary surface missions. Final Report. Keck Institute for Space Studies, California Institute of Technology, Pasadena, CA.

ANSI/ASAE S296.5 DEC2003 (R2018). (2018). *General terminology for traction of agricultural traction and transport devices and vehicles.* American National Standards Institute/American Society of Agricultural Engineers.

Asaf, Z., Rubinstein, D., and Shmulevich, J. (2006). Evaluation of link-track performances using DEM. *Journal of Terramechanics* **43** (2).

Asnani, V., Delap, D., and Creager, C. (2009). The development of wheels for the lunar roving vehicle. *Journal of Terramechanics* **46** (3).

Bekker, M.G. (1956). *Theory of Land Locomotion.* Ann Arbor, MI: University of Michigan Press.

Bekker, M.G. (1960). *Off-the-Road Locomotion.* Ann Arbor, MI: University of Michigan Press.

Bekker, M.G. (1969). *Introduction to Terrain–Vehicle Systems.* Ann Arbor, MI: University of Michigan Press.

Bekker, M.G. (1985). The effect of tire tread in parametric analysis of tire–soil systems. National Research Council of Canada Report No. 24146, Division of Energy, National Research Council of Canada.

Bekker, M.G. and Semonin, E.V. (1975). Motion resistance of pneumatic tires. *Journal of Automotive Engineering* **6** (2).

Cho, S.-G., Park, S., Oh, C. et al. (2019). Design optimization of deep-seabed pilot miner system with coupled relations between constraints. *Journal of Terramechanics* **83** (3).

Cundall, P.A. and Strack, O.D. (1979). Discrete numerical model for granular assemblies. *Geotechnique* **29** (1).

Dai, Y., Liu, S., Li, C. et al. (2010). Development of a fast simulation model for dynamic analysis of the 1000m deep ocean mining system. *World Journal of Modelling and Simulation* **6** (2).

Dai, Y., Zhu, X., and Chen, L.S. (2015). A new multi-body dynamic model for seafloor miner and its trafficability evaluation. *International Journal of Simulation Modelling* **14** (4).

Ding, L., Gao, H., Deng, Z. et al. (2014). New perspective on characterizing pressure-sinkage relationship of terrains for estimating interaction mechanics. *Journal of Terramechanics* **52**.

Du, Y., Gao, J., Jiang, L., and Zhang, Y. (2017). Numerical analysis on tractive performance of off-road wheel steering on sand using discrete element method. *Journal of Terramechanics* **71**.

Edwards, B. (2018). Co-simulation of MBD models with DEM code to predict mobility on soft soil. *Proc. 2018 NDIA Ground Vehicle Systems Engineering and Technology Symposium*, Novi, MI.

Fervers, C.W. (2004). Improved FEM simulation model for tire-soil interaction. *Journal of Terramechanics* **41** (2–3).

Freitag, D.R. (1965). A Dimensional Analysis of the Performance of Pneumatic Tires on Soft Soils. Technical Report 3-688. U.S. Army Corps of Engineers Waterways Experiment Station, Vicksburg, MS.

Freitag, D.R., Green, A.J., and Melzer K.J. (1970). Performance evaluation of wheels for lunar vehicles. Technical Report M-70-2. U.S. Army Corps of Engineer Waterways Experiment Station, Vicksburg, MS.

Gao, Y. and Wong, J.Y. (1994). The development and validation of a computer-aided method for design evaluation of tracked vehicles with rigid links. *Proceedings of the Institution of Mechanical Engineers, Part D: Journal of Automobile Engineering* **208** (D3).

Haley, P.W., Jurkat, M.P., and Brady, P.M., Jr. (1979). NATO Reference Mobility Model, Edition I, Users Guide, Vol. 1, Operational Modules. Technical Report 12503. U.S. Army Tank-Automotive Research and Development Command, Warren, MI.

Harrison, W.L. (1975). Vehicle Performance over Snow. Technical Report 268. U.S. Army Cold Regions Research and Engineering Laboratory, Hanover, NH.

He, R., Sandu, C., Mousavi, H. et al. (2020). Undated standards of the International Society for Terrain–Vehicle Systems. *Journal of Terramechanics* **91**.

Hegedus, E. (1965). Pressure distribution under rigid wheels. *Transactions of the American Society of Agricultural Engineers* **8** (3).

Hetherington, J.G. (2001). The applicability of the MMP concept in specifying off-road mobility for wheeled and tracked vehicles. *Journal of Terramechanics* **38** (2).

Hetherington, J.G. and White, J.N. (2002). An investigation of pressure under wheeled vehicles. *Journal of Terramechanics* **39** (2).

Hettiaratchi, D.R.P. and Reece, A.R. (1974). The calculation of passive soil resistance. *Geotechnique* **24** (3).

Horner, D.A., Peters, J.F., and Carrillo, A. (2001). Large scale discrete element modeling of vehicle-soil interaction. *Journal of Engineering Mechanics, Proceedings of American Society of Civil Engineers* **127** (10).

Huang, W., Wong, J.Y., Preston-Thomas, J., and Jayakumar, P. (2020). Predicting terrain parameters for physics-based vehicle mobility models from cone index data. *Journal of Terramechanics* **88**.

Janosi Z. and Hanamoto, B. (1961). The analytical determination of drawbar pull as a function of slip for tracked vehicles in deformable soils. *Proc. 1st Int. Conf. on the Mechanics of Soil–Vehicle Systems*, Torino, Italy.

Jayakumar, P., Melanz, D., MacLennan, J. et al. (2014). Scalability of classical terramechanics models for lightweight vehicle applications incorporating stochastic modeling and uncertainty propagation. *Journal of Terramechanics* **54**.

Jiang, M., Dai, Y., Cui, L., and Xi, B. (2018). Experimental and DEM analyses on wheel-soil interaction. *Journal of Terramechanics* **76**.

Johnson, J.B., Kulchitsky, A.V., Duvoy, P. et al. (2015). Discrete element method simulations of Mars exploration rover wheel performance. *Journal of Terramechanics* **62**.

Johnson, J.B., Duvoy, P., Kulchitsky, A.V. et al. (2017). Analysis of Mars exploration rover wheel mobility processes and the limitations of classical terramechanics models using discrete element method simulations. *Journal of Terramechanics* **73**.

Karafiath, L.L. (1971). Plasticity theory and stress distribution beneath wheels. *Journal of Terramechanics* **8** (2).

Karafiath, L.L. (1984). Finite element analysis of ground deformation beneath moving track loads. *Proc. 8th Int. Conf. of the International Society for Terrain–Vehicle Systems* 1, Cambridge, England.

Karafiath, L.L. and Nowatzki, E.A. (1978). *Soil Mechanics for Off-Road Vehicle Engineering.* Aedermannsdorf, Switzerland: Trans Tech Publications.

Keira, H.M.S. (1979). Effects of vibration on the shearing characteristics of soil engaging machinery. PhD thesis, Carleton University, Ottawa, Canada.

Knuth, M.A., Johnson, J.B., Hopkins, M.A. et al. (2012). Discrete element modeling of a Mars exploration rover wheel in granular material. *Journal of Terramechanics* **49** (1).

Kobayashi, T., Fujiwara, Y., Yamakawa, J. et al. (2010). Mobility performance of a rigid wheel in low gravity environments. *Journal of Terramechanics* **47** (4).

Krenn, R. and Hirzinger, G. (2008). Simulation of rover locomotion on sandy terrain – modeling, verification, and validation. *Proc. 10th ESA Workshop on Advanced Space Technologies for Robotics and Automation.* ASTRA.

Krick, G. (1969). Radial and shear stress distribution under rigid wheels and pneumatic tires operating on yielding soils with consideration of tire deformation. *Journal of Terramechanics* **6** (3).

Larminie, J.C. (1992). Modifications to the mean maximum pressure system. *Journal of Terramechanics* **29** (2).

Lee, J.H. (2011). Finite element modeling of interfacial forces and contact stresses of pneumatic tire on fresh snow for combined longitudinal and lateral slips. *Journal of Terramechanics* **48** (3).

Liu, C.H. and Wong, J.Y. (1996). Numerical simulations of tire–soil interaction based on critical state soil mechanics. *Journal of Terramechanics* **33** (5).

Liu, C.H., Wong, J.Y., and Mang, H.A. (1999). Large strain finite element analysis of sand: model, algorithm and application to numerical simulation of tire–sand interaction. *Computers and Structures* **74** (3).

Lyasko, M. (2010). Slip sinkage effect in soil–vehicle mechanics. *Journal of Terramechanics* **47** (1).

McCullough, M., Jayakumar, P., Dasch, J., and Gorsich, D. (2017a). The Next Generation NATO Reference Mobility Model development. *Journal of Terramechanics* **73**.

McCullough, M., Jayakumar, P., Preston-Thomas, J. et al. (2017b). Simple terramechanics models and their demonstration in the Next Generation NATO Reference Mobility Model. *Proc. 2017 NDIA Ground Vehicle Systems Engineering and Technology Symposium*, Novi, MI.

McKyes, E. (1985). *Soil Cutting and Tillage, Developments in Agricultural Engineering* **7**. Amsterdam, Netherlands: Elsevier Science.

Nakashima, H. and Kobayashi, T. (2014). Effects of gravity on rigid rover wheel sinkage and motion resistance assessed using two-dimensional discrete element method. *Journal of Terramechanics* **53**.

Nakashima, H. and Oida, A. (2004). Algorithm and implementation of soil–tire contact analysis code based on dynamic FE-DE method. *Journal of Terramechanics* **41** (2–3).

Nakashima, H. and Wong, J.Y. (1993). A three-dimensional tire model by the finite element method. *Journal of Terramechanics* **30** (1).

Nakashima, H., Fujii, H., Oida, A. et al. (2007). Parametric analysis of lugged wheel performance for a lunar microrover by means of DEM. *Journal of Terramechanics* **44** (2).

Nakashima, H., Shioji, Y., Kobayashi, T. et al. (2011). Determining the angle of repose of sand under low-gravity conditions using discrete element method. *Journal of Terramechanics* **48** (1).

Nakashima, H., Yoshida, T., Wang, X.L. et al. (2015). Comparison of gross tractive effort of a single grouser in two-dimensional DEM and experiment. *Journal of Terramechanics* **62**.

NATO Standard AMSP-06. (2021). Guidance for standards applicable to the development of Next Generation NATO Reference Mobility Models (NG-NRMM), Edition A, Version 1.

Onafeko, O. and Reece, A.R. (1967). Soil stresses and deformations beneath rigid wheels. *Journal of Terramechanics* **4** (1).

Osman, M.S. (1964). The mechanics of soil cutting blades. *Journal of Agricultural Engineering Research* **9** (4).

Perumpral, J.V., Liljedal, J.B., and Perloff, W.H. (1971). A numerical method for predicting the stress distribution and soil deformation under a tractor wheel. *Journal of Terramechanics* **8** (1).

Priddy, J.D. and Willoughby, W.E. (2006). Clarification of vehicle cone index with reference to mean maximum pressure. *Journal of Terramechanics* **43** (2).

Reece, A.R. (1965–1966). Principles of soil–vehicle mechanics. *Proceedings of the Institution of Mechanical Engineers* **180** (2A).

Rowland, D. (1972). Tracked vehicle ground pressure and its effect on soft ground performance. *Proc. 4th Int. Conf. of the International Society for Terrain–Vehicle Systems* 1, Stockholm, Sweden.

Rowland, D. (1975). A review of vehicle design for soft-ground operation. *Proc. 5th Int. Conf. of the International Society for Terrain–Vehicle Systems* 1, Detroit, MI.

Rowland, D. and Peel, J.W. (1976). Soft ground performance prediction and assessment for wheeled and tracked vehicles. *Proc. Conference on Off-Highway Vehicles, Tractors and Equipment*, 28–29 October 1975. Institution of Mechanical Engineers Conference Publication CP11/75.

Rubinstein, D., Shmulevich, I., and Frenckel, N. (2018). Use of explicit finite-element formulation to predict the rolling radius and slip of an agricultural tire during travel over loose soil. *Journal of Terramechanics* **80**.

Sela, A.D. (1964). The shear to normal stress relationship between a rigid wheel and dry sand. Paper presented at the *Annual Meeting of American Society of Agricultural Engineers*.

Senatore, C. and Iagnemma, K.D. (2011). Direct shear behaviour of dry, granular soil for low normal stress with application to lightweight robotic vehicle modelling. *Proc. 17th Int. Conf. of the International Society for Terrain–Vehicle Systems*, Blacksburg, VA.

Seta, E., Kamegawa, T., and Nakajima, Y. (2003). Prediction of snow/tire interaction using explicit FEM and FVM. *Tire Science and Technology TSTCA* **31** (3).

Shigeta, Y. and Aruga, K. (2005). Application of 3D distinct element method to track shoe model. *Proc. 15th Int. Conf. of the International Society for Terrain–Vehicle Systems*, Japan.

Shoop, S.A. (2001). Finite element modeling of tire-terrain interaction. Report TR-01-16. U.S. Army Corps of Engineers ERDC/CRREL, Hanover, NH.

Society of Automotive Engineers (1967). SAE J939 (Off-Road Vehicle Mobility Evaluation).

Söhne, W. (1958). Fundamentals of pressure distribution and soil compaction under tractor tires. *Agricultural Engineering* May issue.

Söhne, W. (1969). Agricultural engineering and terramechanics. *Journal of Terramechanics* **6** (4).

Tanaka, H., Momozu, M., Oida, A., and Yamazaki, M. (2000). Simulation of soil deformation and resistance at bar penetration by the distinct element method. *Journal of Terramechanics* **37** (1).

Terzaghi, K. (1966). *Theoretical Soil Mechanics*. New York: Wiley.

Turnage, G.W. (1978). A synopsis of tire design and operational considerations aimed at increasing in-soil tire drawbar performance. *Proc. 6th Int. Conf. of the International Society for Terrain–Vehicle Systems* 2, Vienna, Austria.

Turnage, G.W. (1984). Prediction of in-sand tire and wheeled vehicle drawbar performance. *Proc. 8th Int. Conf. of the International Society for Terrain–Vehicle Systems* 1, Cambridge, England.

Wills, B.M.D. (1963). The measurement of soil shear strength and deformation moduli and a comparison of the actual and theoretical performance of a family of rigid tracks. *Journal of Agricultural Engineering Research* **8** (2).

Wismer, R.D. and Luth, H.J. (1972). Off-road traction prediction for wheeled vehicles. American Society of Agricultural Engineers paper 72-619.

Wong, J.Y. (1967). Behaviour of soil beneath rigid wheels. *Journal of Agricultural Engineering Research* **12** (4).

Wong, J.Y. (1972). Performance of the air cushion–surface contacting hybrid vehicle for overland operation. *Proceedings of the Institution of Mechanical Engineers* **186** (50/72).

Wong, J.Y. (1977). Discussion on 'prediction of wheel–soil interaction and performance using the finite element method'. *Journal of Terramechanics* **14** (4).

Wong, J.Y. (1979). Review of "soil mechanics for off-road vehicle engineering". *Canadian Geotechnical Journal* **16** (3) and *Journal of Terramechanics* **16** (4).

Wong, J.Y. (1980). Data processing methodology in the characterization of the mechanical properties of terrain. *Journal of Terramechanics* **17** (1).

Wong, J.Y. (1983). Evaluation of soil strength measurements. Report No. 22881. Division of Energy, National Research Council of Canada.

Wong, J.Y. (1986). Computer-aided analysis of the effects of design parameters on the performance of tracked vehicles. *Journal of Terramechanics* **23** (2).

Wong, J.Y. (1992a). Optimization of the tractive performance of articulated tracked vehicles using an advanced computer simulation model. *Proceedings of the Institution of Mechanical Engineers, Part D: Journal of Automobile Engineering* **206** (D1).

Wong, J.Y. (1992b). Expansion of the terrain input base for the Nepean Tracked Vehicle Performance Model NTVPM to accept Swiss Rammsonde data from deep snow. *Journal of Terramechanics* **29** (3).

Wong, J.Y. (1992c). Computer-aided methods for the optimization of the mobility of single-unit and two-unit articulated tracked vehicles. *Journal of Terramechanics* **29** (4).

Wong, J.Y. (1994a). Computer simulation models for evaluating the performance and design of tracked and wheeled vehicles. *Procdings of the 1st North American Workshop on Modeling the Mechanics of Off-Road Mobility*, sponsored by the U.S. Army Research Office and held at the U.S. Army Corps of Engineers Waterways Experiment Station, Vicksburg, MS.

Wong, J.Y. (1994b). Terramechanics – its present and future. *Proc. 6th European ISTVS Conference and 4th OVK Symposium on Off-Road Vehicles in Theory and Practice*, Vienna, Austria.

Wong, J.Y. (1994c). Computer-aided methods for design evaluation of track systems. *SAE Transactions, Section 2, Journal of Commercial Vehicles*, Society of Automotive Engineers paper 941675.

Wong, J.Y. (1994d). On the role of mean maximum pressure as an indicator of cross-country mobility for tracked vehicles. *Journal of Terramechanics* **31** (3).

Wong, J.Y. (1995). Application of the computer simulation model NTVPM-86 to the development of a new version of the infantry fighting vehicle ASCOD. *Journal of Terramechanics* **32** (1).

Wong, J.Y. (1997). Dynamics of tracked vehicles. *Vehicle System Dynamics* **28** (2 and 3).

Wong, J.Y. (1998). Optimization of design parameters of rigid-link track systems using an advanced computer-aided method. *Proceedings of the Institution of Mechanical Engineers, Part D: Journal of Automobile Engineering* **212** (D3).

Wong, J.Y. (1999). Computer-aided methods for design evaluation of tracked vehicles and their applications to product development. *International Journal of Vehicle Design* **22** (1 and 2).

Wong, J.Y. (2007). Development of high-mobility tracked vehicles for over snow operations. Keynote paper, *Proc. of the International Society for Terrain–Vehicle Systems Joint North America, Asia–Pacific Conference and Annual Meeting of Japanese Society for Terramechanics*, Fairbanks, Alaska, June 23–26; also, *Journal of Terramechanics* **46** (4).

Wong, J.Y. (2010). *Terramechanics and Off-Road Vehicle Engineering*, 2e. Oxford, England: Elsevier.

Wong, J.Y. (2012). Predicting the performances of rigid rover wheels on extraterrestrial surfaces based on test results obtained on earth. *Journal of Terramechanics* **49** (1).

Wong, J.Y. and Asnani, V.M. (2008). Study of the correlation between the performances of lunar vehicle wheels predicted by the Nepean wheeled vehicle performance model and test data. *Proceedings of the Institution of Mechanical Engineers, Part D: Journal of Automobile Engineering* **222** (11).

Wong J.Y. and Bekker, M.G. (1985). Terrain–Vehicle Systems Analysis. Monograph, Department of Mechanical and Aerospace Engineering, Carleton University, Ottawa, Canada.

Wong, J.Y. and Gao, Y. (1994). Applications of a computer-aided method to parametric study of tracked vehicles with rigid links. *Proceedings of the Institution of Mechanical Engineers, Part D: Journal of Automobile Engineering* **208** (D4).

Wong, J.Y. and Huang, W. (2005). Evaluation of the effects of design features on tracked vehicle mobility using an advanced computer simulation model. *International Journal of Heavy Vehicle Systems* **12** (4).

Wong, J.Y. and Huang, W. (2006a). An investigation into the effects of initial tracked tension on soft ground mobility of tracked vehicles using an advanced computer simulation model. *Proceedings of the Institution of Mechanical Engineers, Part D: Journal of Automobile Engineering* **220** (6).

Wong, J.Y. and Huang, W. (2006b). Study of detracking risks of track systems. *Proceedings of the Institution of Mechanical Engineers, Part D: Journal of Automobile Engineering* **220** (9).

Wong, J.Y. and Huang, W. (2008). Approaches to improving the mobility of military tracked vehicles on soft terrain. *International Journal of Heavy Vehicle Systems* **15** (2/3/4).

Wong, J.Y. and Irwin, G.J. (1992). Measurement and characterization of the pressure–sinkage data for snow obtained using a Rammsonde. *Journal of Terramechanics* **29** (2).

Wong, J.Y. and Kobayashi, T. (2012). Further study of the method of approach to testing the performance of extraterrestrial rovers/rover wheels on earth. *Journal of Terramechanics* **49** (6).

Wong, J.Y. and Preston-Thomas, J. (1983a). On the characterization of the shear stress–displacement relationship of terrain. *Journal of Terramechanics* **19** (4).

Wong, J.Y. and Preston-Thomas, J. (1983b). On the characterization of the pressure–sinkage relationship of snow covers containing an ice layer. *Journal of Terramechanics* **20** (1).

Wong, J.Y. and Preston-Thomas, J. (1986). Parametric analysis of tracked vehicle performance using an advanced computer simulation model. *Proceedings of the Institution of Mechanical Engineers, Part D: Transport Engineering* **200** (D2).

Wong, J.Y. and Preston-Thomas, J. (1988). Investigation into the effects of suspension characteristics and design parameters on the performance of tracked vehicles using an advanced computer simulation model. *Proceedings of the Institution of Mechanical Engineers, Part D: Transport Engineering* **202** (D3).

Wong, J.Y. and Reece, A.R. (1966). Soil failure beneath rigid wheels. *Proc. 2nd Int. Conf. of the International Society for Terrain–Vehicle Systems*, Toronto, Canada.

Wong, J.Y. and Reece, A.R. (1967). Prediction of rigid wheel performance based on the analysis of soil–wheel stresses, part I and part II. *Journal of Terramechanics* **4** (1 and 2).

Wong, J.Y., Radforth, J.R., and Preston-Thomas, J. (1982). Some further studies of the mechanical properties of muskeg. *Journal of Terramechanics* **19** (2).

Wong, J.Y., Garber, M., and Preston-Thomas, J. (1984). Theoretical prediction and experimental substantiation of the ground pressure distribution and tractive performance of tracked vehicles. *Proceedings of the Institution of Mechanical Engineers, Part D: Transport Engineering* **198** (D15).

Wong, J.Y., Jayakumar, P., and Preston-Thomas, J. (2019). Evaluation of the computer simulation model NTVPM for assessing military tracked vehicle cross-country performance. *Proceedings of the Institution of Mechanical Engineers, Part D: Journal of Automobile Engineering* **233** (5).

Wong, J.Y., Garber, M., Radforth, J.R., and Dowell, J.T. (1979). Characterization of the mechanical properties of muskeg with special reference to vehicle mobility. *Journal of Terramechanics* **16** (4).

Wong, J.Y., Jayakumar, P., Toma, E., and Preston-Thomas, J. (2018). Comparison of simulation models NRMM and NTVPM for assessing military tracked vehicle cross-country performance. *Journal of Terramechanics* **80**.

Wong, J.Y., Jayakumar, P., Toma, E., and Preston-Thomas, J. (2020). A review of mobility metrics for next generation vehicle mobility models. *Journal of Terramechanics* **87**.

Wong, J.Y., Senatore, C., Jayakumar, P., and Iagnemma, K. (2015). Predicting mobility performance of a small, lightweight track system using the computer-aided method NTVPM. *Journal of Terramechanics* **61**.

Xia, K. (2011). Finite element modeling of tire/terrain interaction: application to predicting soil compaction and tire mobility. *Journal of Terramechanics* **48** (2).

Yong, R.N. and Fattah, E.A. (1976). Prediction of wheel–soil interaction and performance using the finite element method. *Journal of Terramechanics* **13** (4).

Zhang, T., Lee, J.H., and Liu, Q. (2005). Finite element simulation of tire–snow interaction under combined longitudinal and lateral slip condition. *Proc. 15th Int. Conf. of the International Society for Terrain–Vehicle Systems*, Japan.

Zhou, F., Arvidson, R.E., Bennett, K. et al. (2014). Simulations of Mars rover traverses. *Journal of Field Robotics* **31** (1).

Problems

2.1 The contact area of a tire on a hard and dry soil may be approximated by a circle having a radius of 20 cm (7.9 in.). The contact pressure is assumed to be a uniform 68.95 kPa (10 psi). For this type of soil, the concentration factor v is assumed to be 3. Calculate the resultant vertical stress σ_z in the soil at depths of 20 and 40 cm (7.9 and 15.8 in.) below the center of the contact area. At what depth below the center is the vertical stress one-tenth of the contact pressure on the soil surface?

2.2 A steel cage wheel with 18 lugs on a narrow rim is to be attached to an off-road wheeled vehicle to increase its traction over a wet soil. The outside diameter of the steel wheel across the tips of the lugs is 1.5 m (4.92 ft). The lugs are 25 cm (10 in.) wide and penetrate 12.5 cm (5 in.) into the soil at the vertical position. Estimate the tractive effort that a lug in the vertical position can develop in a soil with $c = 13.79$ kPa (2 psi), $\phi = 5°$, and $\gamma_s = 16$ kN/m^3(102 lb/ft^3). Also calculate the corresponding torque required to drive the lug in the vertical position. The rim of the wheel is narrow, and its effect may be neglected. The surface of the lugs is assumed to be smooth.

2.3 A tracked vehicle with uniform contact pressure weighs 155.68 kN (35 000 lb). Each of its two tracks is 102 cm (40 in.) wide and 305 cm (120 in.) long. Estimate the motion resistance and thrust–slip relationship of the vehicle on a terrain with $n = 0.5$, $k_c = 0.77$ kN/m^{n+1}(0.7 lb/in.$^{n+1}$), $k_\phi = 51.91$ kN/m^{n+2} (1.2 lb/in.$^{n+2}$), $c = 5.17$ kPa (0.75 psi), $\phi = 11°$, and $K = 5$ cm (2 in.). What will be the changes in its performance if the width of the track is reduced by 20% and its length is increased by 25%?

2.4 A four-wheel drive tractor weighs 60 kN (13 489 lb), with equal weight distribution between the axles. All four tires are 11.00 R16XL radial tires with dimensions as given in Example 2.4, and the relationship between the average ground pressure and the inflation pressure for the tires is given in Figure 2.82. The tire inflation pressure is 150 kPa (21.75 psi). Estimate the motion resistance and the thrust of the front axle at 20% slip on a terrain with $n = 0.8$, $k_c = 29.76$ kN/m^{n+1}(9 lb/in.$^{n+1}$), $k_\phi = 2083$ kN/m^{n+2} (16 lb/in.$^{n+2}$), $c = 8.62$ kPa (1.25 psi), $\phi = 22.5°$, and $K = 2.5$ cm (1 in.).

2.5 An inclined bulldozer blade with an angle of 57° to the horizontal is used to remove a layer of soil 15.2 cm (6 in.) deep. The blade has a width of 4.62 m (182 in.), and its surface has an angle of friction with the soil of 24° and no adhesion. The soil has an angle of internal shearing resistance of 35°, a cohesion of 3.79 kPa (0.55 psi), and a weight density of 16 286 N/m^3 (104 lb/ft^3). The values of the parameters characterizing blade–soil interaction $K_{p\gamma}$ and K_{pc} are 10 and 2, respectively. Estimate the resultant force on the blade required to make the initial cut without surcharge. Also identify the magnitudes and directions of the horizontal and vertical components of the resultant force exerted on the blade by the soil.

2.6 A tracked vehicle has a gross weight of 125 kN (28 103 lb). Each of its two tracks has a contact length of 2.65 m (104 in.) and a contact width of 0.38 m (15 in.). The normal pressure distribution under the tracks is of the multipeak sinusoidal form, as shown in Figure 2.52(b). Estimate the compaction resistance and thrust of the vehicle at a slip of 20% on the same terrain as that described in Problem 2.4.

3

Performance Characteristics of Road Vehicles

Performance characteristics of a road vehicle refer to its capability to accelerate, decelerate, and negotiate grades in a straight-line motion. The tractive (or braking) effort developed by the tires and the resisting forces acting on the vehicle determine the performance potential of the vehicle and is discussed in detail in this chapter. Procedures for predicting and evaluating the performance characteristics of road vehicles are also presented.

3.1 Equation of Motion and Maximum Tractive Effort

The major external forces acting on a two-axle vehicle are shown in Figure 3.1. In the longitudinal direction, they include the aerodynamic resistance R_a, rolling resistance of the front and rear tires R_{rf} and R_{rr}, drawbar load R_d, grade resistance $R_g(W \sin \theta_s)$, and tractive effort of the front and rear tires F_f and F_r. For a rear-wheel drive vehicle, $F_f = 0$, whereas for a front-wheel drive vehicle, $F_r = 0$.

The equation of motion along the longitudinal axis x of the vehicle is expressed by

$$m\frac{d^2x}{dt^2} = \frac{W}{g}a = F_f + F_r - R_a - R_{rf} - R_{rr} - R_d - R_g \tag{3.1}$$

where d^2x/dt^2 or a is the linear acceleration of the vehicle along the longitudinal axis, g is acceleration due to gravity, m is vehicle mass, and W is vehicle weight.

By introducing the concept of inertia force, the above equation may be rewritten as

$$F_f + F_r - \left(R_a + R_{rf} + R_{rr} + R_d + R_g + \frac{aW}{g}\right) = 0$$

or

$$F = R_a + R_r + R_d + R_g + \frac{aW}{g} \tag{3.2}$$

where F is the total tractive effort and R_r is the total rolling resistance of the vehicle.

To evaluate the performance potential, the maximum tractive effort that the vehicle can develop has to be determined. There are two limiting factors to the maximum tractive effort of a road vehicle: one is determined by the coefficient of road adhesion and the normal load on the drive axle or axles; the other is determined by the characteristics of the power plant and transmission. The smaller of these two determines the performance potential of the vehicle.

Theory of Ground Vehicles, Fifth Edition. J.Y. Wong.
© 2022 John Wiley & Sons, Inc. Published 2022 by John Wiley & Sons, Inc.
Companion website: www.wiley.com/go/wong/TGV5e

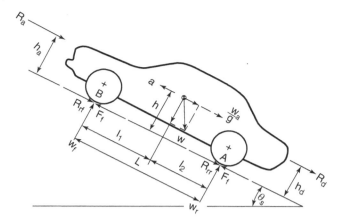

Figure 3.1 Forces acting on a two-axle vehicle.

To predict the maximum tractive effort that the tire–ground contact can support, the normal loads on the axles must be determined. They can be computed readily by summation of the moments about points A and B shown in Figure 3.1. By summing moments about A, the normal load on the front axle W_f can be determined:

$$W_f = \frac{W l_2 \cos \theta_s - R_a h_a - h a W / g - R_d h_d \mp W h \sin \theta_s}{L} \tag{3.3}$$

where l_2 is the distance between the rear axle and the center of gravity of the vehicle; h_a is the height of the point of application of the aerodynamic resistance; h is the height of the center of gravity; h_d is the height of the drawbar hitch; L is the wheelbase; and θ_s is the slope angle. When the vehicle is climbing up a hill, the negative sign is used for the term $W h \sin \theta_s$.

Similarly, the normal load on the rear axle can be determined by summing moments about B:

$$W_r = \frac{W l_1 \cos \theta_s + R_a h_a + h a W / g + R_d h_d \pm W h \sin \theta_s}{L} \tag{3.4}$$

where l_1 is the distance between the front axle and the center of gravity of the vehicle. In the above expression, the positive sign is used for the term $W h \sin \theta_s$ when the vehicle is climbing up a hill.

For small angles of slope, $\cos \theta_s$ is approximately equal to 1. For passenger cars, the height of the point of application of the aerodynamic resistance h_a and that of the drawbar hitch h_d may be assumed to be near the height of the center of gravity h. With these simplifications and assumptions, Eqs. (3.3) and (3.4) may be rewritten as

$$W_f = \frac{l_2}{L} W - \frac{h}{L} \left(R_a + \frac{aW}{g} + R_d \pm W \sin \theta_s \right) \tag{3.5}$$

and

$$W_r = \frac{l_1}{L} W + \frac{h}{L} \left(R_a + \frac{aW}{g} + R_d \pm W \sin \theta_s \right) \tag{3.6}$$

Substituting Eq. (3.2) into the above equations, one obtains

$$W_f = \frac{l_2}{L} W - \frac{h}{L} (F - R_r) \tag{3.7}$$

and

$$W_r = \frac{l_1}{L} W + \frac{h}{L}(F - R_r) \tag{3.8}$$

The first term on the right-hand side of each equation represents the static load on the axle when the vehicle is at rest on level ground. The second term on the right-hand side of each equation represents the dynamic component of the normal load or dynamic load transfer.

The maximum tractive effort that the tire–ground contact can support can be determined in terms of the coefficient of road adhesion μ and vehicle parameters. For a rear-wheel drive vehicle,

$$F_{\max} = \mu W_r = \mu \left[\frac{l_1}{L} W + \frac{h}{L}(F_{\max} - R_r) \right]$$

and

$$F_{\max} = \frac{\mu W (l_1 - f_r h)/L}{1 - \mu h/L} \tag{3.9}$$

where the total rolling resistance R_r is expressed as the product of the coefficient of rolling resistance f_r and the weight of the vehicle W. For a front-wheel drive vehicle,

$$F_{\max} = \mu W_f = \mu \left[\frac{l_2}{L} W - \frac{h}{L}(F_{\max} - R_r) \right]$$

and

$$F_{\max} = \frac{\mu W (l_2 + f_r h)/L}{1 + \mu h/L} \tag{3.10}$$

In deriving the above equations, the transverse load transfer due to engine torque for a longitudinally mounted engine or the longitudinal load transfer due to engine torque for a transversely mounted engine has been neglected, and both the right- and left-hand side tires are assumed to have identical performance.

For a tractor–semitrailer, the calculation of the maximum tractive effort that the tire–ground contact can support is more involved than a two-axle vehicle. The major forces acting on a tractor–semitrailer are shown in Figure 3.2. For most of the tractor–semitrailers, the tractor rear axle is driven. To compute the maximum tractive effort as determined by the nature of tire–road interaction, it is necessary to calculate the normal load on the tractor rear axle under operating conditions. This can be calculated by considering the tractor and the semitrailer as free bodies separately. By taking the semitrailer as a free body, the normal load on the semitrailer axle W_s and the vertical and horizontal loads at the hitch point W_{hi} and F_{hi} can be determined.

The normal load on the semitrailer axle, for small angles of slope, is given by

$$W_s = \frac{W_2 d_2 + R_{a2} h_{a2} + h_2 a W_2 /g \pm W_2 h_2 \sin \theta_s - F_{hi} h_3}{L_2} \tag{3.11}$$

where R_{a2} is the aerodynamic resistance acting on the semitrailer; h_{a2} is the height of the point of application of R_{a2}; and W_2 is the weight of the semitrailer. Other parameters and dimensions are shown in Figure 3.2. When the vehicle is climbing up a hill, the positive sign for the term $W_2 h_2 \sin \theta_s$ in Eq. (3.11) should be used.

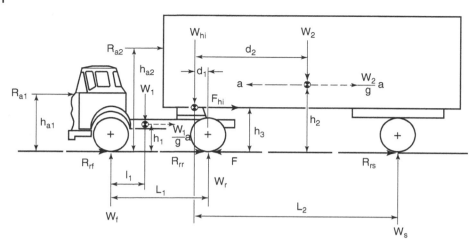

Figure 3.2 Forces acting on a tractor–semitrailer.

If $h_{a2} \cong h_2 \cong h_3$, the expression for W_s may be simplified as

$$W_s = \frac{d_2}{L_2}W_2 + \frac{h_2}{L_2}\left(R_{a2} + \frac{aW_2}{g} \pm W_2\sin\theta_s - F_{hi}\right) \tag{3.12}$$

The longitudinal force at the hitch point is given by

$$F_{hi} = R_{a2} + \frac{aW_2}{g} \pm W_2\sin\theta_s + f_rW_s \tag{3.13}$$

Substitute Eq. (3.13) into Eq. (3.12), and the expression for W_s becomes

$$W_s = \frac{W_2 d_2}{L_2 + f_r h_2}$$

and the load at the hitch point is given by

$$W_{hi} = W_2 - W_s = \left(1 - \frac{d_2}{L_2 + f_r h_2}\right)W_2 \tag{3.14}$$
$$= C_{hi}W_2$$

By taking the tractor as a free body and summing moments about the front tire–ground contact point, the normal load on the tractor rear axle W_r can be determined:

$$W_r = \frac{W_1 l_1 + R_{a1}h_{a1} + h_1 aW_1/g \pm W_1 h_1\sin\theta_s + F_{hi}h_3 + (L_1 - d_1)W_{hi}}{L_1} \tag{3.15}$$

where R_{a1} is the aerodynamic resistance acting on the tractor; h_{a1} is the height of the application of R_{a1}; and W_1 is the weight of the tractor. Other parameters and dimensions are shown in Figure 3.2. When the vehicle is climbing up a hill, the positive sign for the term $W_1 h_1\sin\theta_s$ in Eq. (3.15) should be used.

If $h_{a1} \cong h_1 \cong h_3$, the expression for W_r may be simplified as

$$W_r = \frac{W_1 l_1 + (R_{a1} + aW_1/g \pm W_1\sin\theta_s + F_{hi})h_1 + (L_1 - d_1)W_{hi}}{L_1} \tag{3.16}$$

By equating the forces acting on the tractor in the longitudinal direction, the following expression for the required tractive effort F can be obtained:

$$F = R_{a1} + \frac{aW_1}{g} \pm W_1 \sin \theta_s + f_r(W_1 + W_{hi}) + F_{hi} \tag{3.17}$$

From Eqs. (3.16) and (3.17), the maximum tractive effort that the tire–ground contact can support with the tractor rear axle driven can be expressed by

$$F_{\max} = \mu W_r = \frac{\mu[l_1 W_1 - h_1 f_r(W_1 + W_{hi}) + (L_1 - d_1)W_{hi}]/L_1}{1 - \mu h_1/L_1} \tag{3.18}$$

Substitution of Eq. (3.14) into the above equation yields

$$F_{\max} = \frac{\mu[l_1 W_1 - h_1 f_r(W_1 + C_{hi}W_2) + (L_1 - d_1)C_{hi}W_2]/L_1}{1 - \mu h_1/L_1} \tag{3.19}$$

The maximum tractive effort as determined by the nature of the tire–road interaction imposes a fundamental limit on the vehicle performance characteristics, including maximum speed, acceleration, gradeability, and drawbar pull.

3.2 Aerodynamic Forces and Moments

With growing emphasis on fuel economy and on the reduction of undesirable exhaust emissions, it has become increasingly important to optimize vehicle power requirements. To achieve this, it is necessary to reduce the aerodynamic resistance (drag), rolling resistance, and inertia resistance during acceleration, which is proportional to vehicle weight. For a typical passenger car cruising at a speed higher than approximately 80 km/h (50 mph), the power required to overcome the aerodynamic resistance is greater than that required to overcome the rolling resistance of the tires and the resistance in the transmission, as shown in Figure 3.3 (Kelly and Holcombe 1978). Because of the significant effects of aerodynamic resistance on vehicle power requirements at moderate and higher speeds, continual effort has been expended in improving the aerodynamic performance of road vehicles.

The aerodynamic resistance is generated by two sources: one is the air flow over the exterior of the vehicle body, and the other is the flow through the engine radiator system and the interior of the vehicle for purposes of cooling, heating, and ventilating. Of the two, the former is the dominant one, accounting for more than 90% of the total aerodynamic resistance of a passenger car.

The external air flow generates normal pressure and shear stress on the vehicle body. According to the aerodynamic nature, the external aerodynamic resistance comprises two components, commonly known as the pressure drag and skin friction. The pressure drag arises from the component of the normal pressure on the vehicle body acting against the motion of the vehicle, while the skin friction is due to the shear stress in the boundary layer adjacent to the external surface of the vehicle body. Of the two components, the pressure drag is by far the larger, and constitutes more than 90% of the total external aerodynamic resistance of a passenger car with normal surface finish. The skin friction may become more significant, however, for a long vehicle, such as a bus or a tractor–trailer train. It should be noted that the momentum losses of the air in the wake of the vehicle and the energy imparted to the air by the vortices generated by the vehicle are not additional but are an alternative measure of the pressure drag and skin friction (Carr 1983).

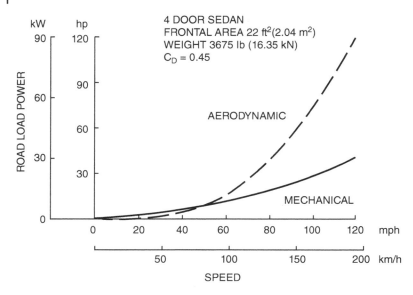

Figure 3.3 Power requirements of a full-size passenger car as a function of speed. Reprinted with permission from SAE Progress in Technology Series Vol. 16 © 1978 Society of Automotive Engineers, Inc.

In practice, the aerodynamic resistance is usually expressed in the following form:

$$R_a = \frac{\rho}{2} C_D A_f V_r^2 \tag{3.20}$$

where ρ is the mass density of the air; C_D is the coefficient of aerodynamic resistance that represents the combined effects of all of the factors described above; A_f is a characteristic area of the vehicle, usually taken as the frontal area, which is the projected area of the vehicle in the direction of travel; and V_r is the speed of the vehicle relative to the wind. Aerodynamic resistance is proportional to the square of speed. Thus, the horsepower required to overcome aerodynamic resistance increases with the cube of speed. When the speed of a vehicle is doubled, the power required for overcoming aerodynamic resistance increases eightfold.

Atmospheric conditions affect air density ρ, and hence aerodynamic resistance. For instance, an increase in ambient temperature from 0 to 38°C (32 to 100°F) will cause a 14% reduction in aerodynamic resistance, and an increase in altitude of 1219 m (4000 ft) will lead to a decrease in aerodynamic resistance by 17%. In view of the significant effects of ambient conditions on the aerodynamic resistance, it is necessary to establish a standard set of conditions to which all aerodynamic test data may be referred. The commonly used standard conditions are temperature 519° Rankine (15°C or 59°F) and barometric pressure 101.32 kPa (14.7 psi, 76 cm or 29.92 in. Hg). In performance calculations, the mass density of the air ρ may be taken as 1.225 kg/m³ (0.002378 slug/ft³, and its equivalent weight density 0.07651 lb/ft³).

The frontal area A_f of the vehicle may be determined from a photograph taken from the front if accurate drawings of the vehicle are not available. For passenger cars, the frontal area varies in the range of 79–84% of the area calculated from the overall vehicle width and height.

The coefficient of aerodynamic resistance C_D may be obtained by wind tunnel testing of full-size vehicles or scale models. Numerous wind tunnels capable of testing full-size passenger cars and commercial vehicles are used in industry and research centers (Hucho 1990a). For instance,

National Research Council Canada has a large wind tunnel with a test section area (nozzle area) 9 m × 9 m (30 ft × 30 ft) and air speed up to 198 km/h (123 mph). It can test a variety of ground vehicles and has a boundary layer control system and a ground effect simulation system (National Research Council Canada 2021). The University of Ontario Institute of Technology (Ontario Tech University), Canada, has a climate wind tunnel with an adjustable test section area from 7.0 to 13.0 m² (75–140 ft²) at a corresponding air speed from 280 to 225 km/h (174–140 mph). It can accommodate vehicles from passenger cars to buses/heavy trucks with length up to 19.1 m (63 ft). The temperature in the tunnel can be varied from 60° to −40° C (140° to −40° F). It has a turntable to facilitate vehicle aerodynamic performance testing at yaw (Ontario Tech University 2021).

While full-scale testing avoids the scaling issues with model testing, it requires a large wind tunnel and is expensive. Consequently, scale model testing, which is comparatively inexpensive and more convenient for shape modifications, is widely used in the development of new products. In the United States, for passenger cars 3/8 scale is widely used, while in Europe 1/4 scale is the most common, though 1/5 scale is also used in small wind tunnels. For commercial vehicles, a scale of 1/2.5 is recommended (Hucho 1990a).

In wind tunnel testing, whether scale models or full-size vehicles are used, two basic issues require special attention: flow field similarity and the modeling of the ground plane. Flow field similarity refers to the similarity between the flow pattern in the wind tunnel and that under actual driving conditions on the road. To ensure flow field similarity in wind tunnel testing of scale models, a basic requirement is that the Reynolds number (RN) for the scale model be equal to that for the full-size vehicle. The RN is the ratio of the product of airstream speed and the characteristic length of the vehicle to the kinematic viscosity of the air. To satisfy this requirement, a 3/8 scale model should, therefore, be tested in the wind tunnel at an airstream speed of 8/3 that of the full-size vehicle. In addition, the blockage ratio, which is the ratio of the frontal area of the model (or the full-size test vehicle) to the test section area of the wind tunnel, should be as small as possible, preferably not exceeding 5% (Hucho 1990a).

The proper modeling of the ground plane in wind tunnel testing is another issue requiring careful consideration. When a vehicle is driven on the road at zero wind speed, the air is at rest relative to the road. In a conventional wind tunnel, the air flows with respect to the tunnel floor and a boundary layer builds up. This may significantly affect the flow pattern under the scale model (or the full-size test vehicle). To alleviate this problem, a moving ground plane has been used.

The deceleration method of road testing, commonly referred to as the coast-down test, may also be used to determine the aerodynamic resistance (Eaker 1988; Rousillon et al. 1973; White and Korst 1972). Using this method, the vehicle is first run up to a certain speed, preferably its top speed, then the drivetrain is disconnected from the engine and the vehicle decelerates. The variations of vehicle speed and/or distance traveled with time are continuously recorded. The deceleration of the vehicle due to the combined effects of the rolling resistance of the tires, drivetrain resistance, and aerodynamic resistance can then be derived from the coast-down test data, such as speed–time or speed–distance relations. From the derived deceleration and taking into account the effects of the rotating inertias of all rotating components in the drivetrain, including the tires, the total resisting force can be deduced. With the effects of the rolling resistance of tires and drivetrain resistance separated from the total resisting force, the aerodynamic resistance can be deduced.

One of the methods that can be used to determine the rolling resistance of the tires and drivetrain resistance is to carry out an additional road test, in which the test vehicle is completely enclosed within a so-called shrouding trailer (Necati 1990). The trailer shrouds the entire test vehicle from any aerodynamic force. However, the tires of the test vehicle maintain full contact with the road and support the entire load of the vehicle. A load cell is placed at the hitch connecting the trailer and the test vehicle to measure the towing force applied to it. During the test, a vehicle is used to tow the trailer together with the test vehicle shrouded by it. The towing force measured by the load cell between the trailer and the test vehicle is then the sum of the rolling resistance of tires and drivetrain resistance of the test vehicle, as the vehicle is shrouded by the trailer and no aerodynamic force applies to it. Alternatively, a procedure recommended by the Society of Automotive Engineers (SAE Recommended Practice J1263) may be followed to deduce the aerodynamic resistance and the combined tire rolling resistance and drivetrain resistance from the coast-down test data.

It has been shown that the coast-down method can yield sufficiently accurate results on the aerodynamic resistance of road vehicles if care is taken to determine the rolling resistance of tires and drivetrain resistance. In comparison with wind tunnel testing, this method does not require expensive facilities. However, it requires a straight and level road (usually not exceeding 0.5% grade) and is subject to the influence of ambient conditions.

The coefficient of aerodynamic resistance (drag) C_D is a function of several vehicle design and operational factors. The shapes of the forebody, afterbody, and underbody, wheels and wheel wells, drip-rails, window recesses, external mirrors, and mud flaps have noticeable effects on the coefficient of aerodynamic resistance. The values of the aerodynamic resistance coefficient for passenger vehicles of various body shapes, ranging from convertible through station wagon to sport utility vehicle (SUV), are shown in Figure 3.4 (Robert Bosch GmbH 2018).

		Aerodynamic drag coefficient C_D	Frontal area	
			m^2	ft^2
	Subcompact	0.29 – 0.37	2.05 – 2.20	22.07 – 23.68
	Compact	0.22 – 0.32	2.18 – 2.28	23.47 – 24.54
	Medium	0.23 – 0.35	2.20 – 2.38	23.68 – 25.62
	Station wagon	0.27 – 0.35	2.20 – 2.38	23.68 – 25.62
	Van	0.25 – 0.35	2.40 – 3.20	25.83 – 34.45
	Convertible -closed -open	0.28 – 0.38 0.35 – 0.50	1.94 – 2.20 1.84 – 2.10	20.88 – 23.68 19.81 – 22.61
	Sport utility	0.29 – 0.55	2.49 – 3.15	26.80 – 33.91
	Sports	0.27 – 0.40	1.65 – 2.20	17.76 – 23.68
	Luxury	0.23 – 0.35	2.28 – 2.65	24.54 – 28.53

Figure 3.4 Aerodynamic resistance coefficients for passenger vehicles. *Source:* Adapted from Robert Bosch GmbH, *Bosch Automotive Handbook*, 10th Edition 2018. Reproduced with permission of Wiley.

Table 3.1 Power required to overcome aerodynamic resistance of various types of passenger vehicle at speeds of 60 km/h (37 mph) and 120 km/h (75 mph).

| Passenger vehicle type[a] | Power required to overcome aerodynamic resistance | | | |
| | Speed at 60 km/h (37 mph) | | Speed at 120 km/h (75 mph) | |
	kW	hp	kW	hp
Subcompact	1.69–2.31	2.27–3.10	13.49–18.47	18.09–24.77
Compact	1.36–2.07	1.82–2.78	10.88–16.55	14.59–22.19
Medium	1.43–2.36	1.92–3.16	11.48–18.90	15.39–25.35
Station wagon	1.68–2.36	2.25–3.16	13.48–18.90	18.08–25.35
Van	1.70–3.18	2.28–4.26	13.61–25.41	18.25–34.08
Convertible:				
closed	1.54–2.37	2.07–3.18	12.32–18.97	16.52–25.44
open	1.83–2.98	2.45–4.00	14.61–23.82	19.59–31.94
Sport utility	2.05–4.91	2.75–6.58	16.38–39.30	21.97–52.70
Sports	1.26–2.50	1.69–3.35	10.11–19.96	13.56–26.77
Luxury	1.49–2.63	2.00–3.53	11.90–21.04	15.96–28.22

[a] Aerodynamic drag coefficients and frontal areas for various passenger vehicle types shown in Figure 3.4 are used in the calculations of the power required to overcome aerodynamic resistance.

Table 3.1 shows the power required to overcome aerodynamic resistance of representative types of passenger vehicle shown in Figure 3.4, at a speed of 60 km/h (37 mph) representative of city driving and at 120 km/h (75 mph) representative of highway driving. The values of aerodynamic drag coefficient and frontal area of various types of passenger vehicle are those given in Figure 3.4. In the calculations, the air mass density of 1.225 kg/m^3 (or air weight density of 0.07651 lb/ft^3) is used. It can be seen that at a speed of 60 km/h (37 mph), the power required to overcome aerodynamic drag is relatively low, while at a speed of 120 km/h (75 mph), the power required increases significantly and is eight times that at the speed of 60 km/h (37 mph).

The influence of body shape details on the aerodynamic resistance coefficient of a passenger car is shown in Figure 3.5 (Janssen and Hucho 1973). The effects of the shapes of the front and rear ends of a passenger car on the values of aerodynamic resistance coefficient are shown in Figures 3.6 and 3.7, respectively (Hucho 1990b; Janssen and Hucho 1973). To improve the aerodynamic performance of vehicles, add-on devices are often used. Figures 3.8 and 3.9 show the effects of front and rear spoilers on the values of aerodynamic resistance coefficient, respectively (Janssen and Hucho 1973).

In addition to the shape of the vehicle body, the attitude of the vehicle defined by the angle of attack (i.e., the angle between the longitudinal axis of the vehicle and the horizontal), ground clearance, loading conditions, and other operational factors, such as radiator open or blanked, and window open or closed, also affect the aerodynamic resistance coefficient. Figure 3.10 shows the effects of the angle of attack on the value of C_D for three types of passenger car, while Figure 3.11 shows the effects of ground clearance on the value of C_D for different types of passenger vehicle (Janssen and

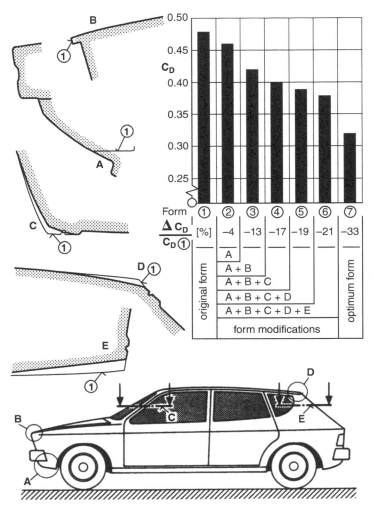

Figure 3.5 Influence of body shape details on aerodynamic resistance coefficient of a passenger car. *Source:* Janssen and Hucho (1973). Reproduced with permission of BHRA Fluid Engineering.

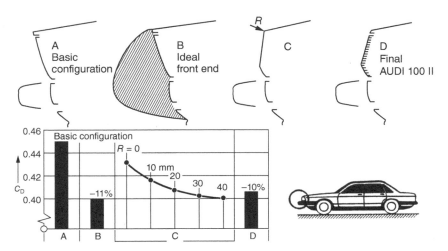

Figure 3.6 Influence of the shape of the front end on aerodynamic resistance coefficient of a passenger car. *Source:* Hucho (1990b). Reproduced with permission of Butterworth-Heinemann.

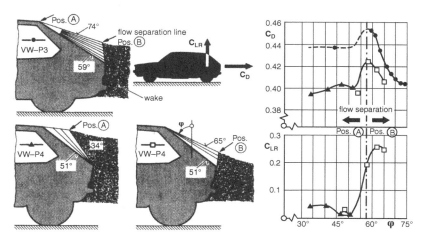

Figure 3.7 Influence of the shape of the rear end on aerodynamic resistance coefficient of a passenger car. *Source:* Janssen and Hucho (1973). Reproduced with permission of BHRA Fluid Engineering.

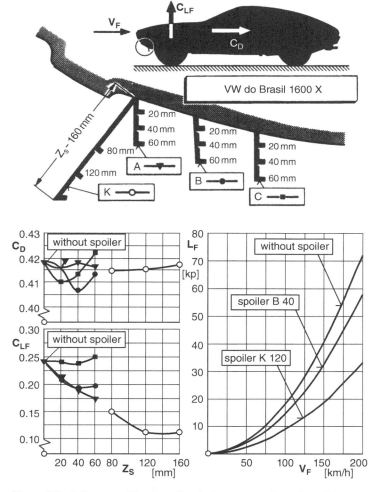

Figure 3.8 Influence of front spoiler design on aerodynamic resistance coefficient and aerodynamic lift coefficient of a passenger car. *Source:* Janssen and Hucho (1973). Reproduced with permission of BHRA Fluid Engineering.

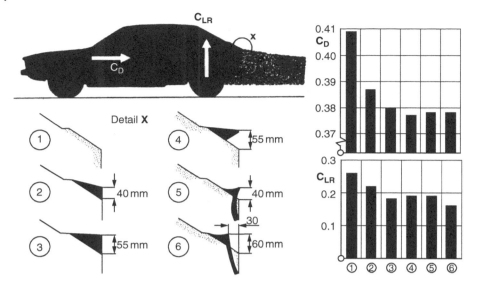

Figure 3.9 Influence of rear spoiler design on aerodynamic resistance coefficient and aerodynamic lift coefficient of a passenger car. *Source:* Janssen and Hucho (1973). Reproduced with permission of BHRA Fluid Engineering.

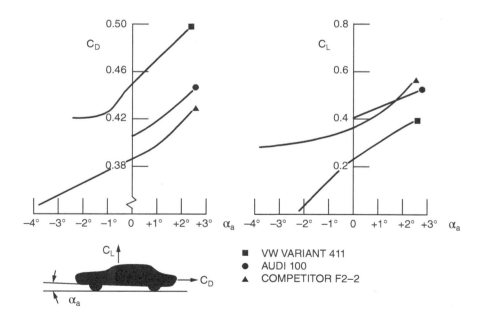

Figure 3.10 Influence of the angle of attack on aerodynamic resistance coefficient and aerodynamic lift coefficient of passenger cars. *Source:* Janssen and Hucho (1973). Reproduced with permission of BHRA Fluid Engineering.

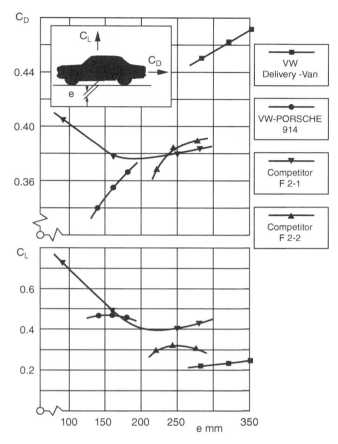

Figure 3.11 Influence of ground clearance on aerodynamic resistance coefficient and aerodynamic lift coefficient of passenger cars. *Source:* Janssen and Hucho (1973). Reproduced with permission of BHRA Fluid Engineering.

Hucho 1973). The loading conditions and the distribution of load among axles may change the attitude (angle of attack) and ground clearance of the vehicle, and hence the aerodynamic resistance coefficient. The influence of loading conditions on the value of C_D for a passenger car is shown in Figure 3.12 (Janssen and Hucho 1973). The influence of operational factors, such as window open or closed, roof open or closed, etc., on the aerodynamic resistance coefficient of a sports car is shown in Figure 3.13 (Janssen and Hucho 1973).

The various components of aerodynamic resistance of a representative passenger car and their minimum feasible values are summarized in Table 3.2 (Carr 1983). The greatest potential for reduction lies in the optimization of the body shape. It is estimated that the component of aerodynamic resistance coefficient due to body shape can be reduced to a practical minimum of 0.1. The estimated minimum feasible value of the coefficient of aerodynamic resistance of a typical car is approximately 0.195, as indicated in Table 3.2.

The effects of the aerodynamic resistance coefficient on the fuel economy under steady-state conditions of a passenger car, with mass 1060 kg (2332 lb), frontal area 1.77 m^2 (19 ft^2), and radial tires, are shown in Figure 3.14 (Hucho et al. 1976). At a steady speed of 97 km/h (60 mph), a reduction of

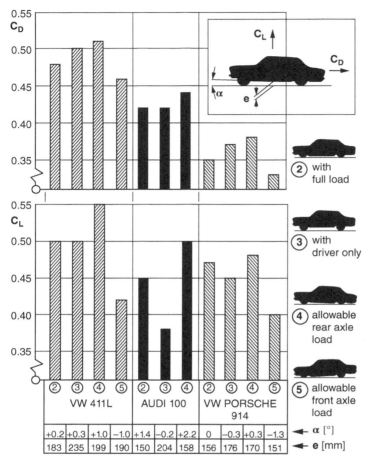

	②	③	④	⑤	②	③	④	②	③	④	⑤	
	VW 411L				AUDI 100			VW PORSCHE 914				
α [°]	+0.2	+0.3	+1.0	−1.0	+1.4	−0.2	+2.2	0	−0.3	+0.3	−1.3	← α [°]
e [mm]	183	235	199	190	150	204	158	156	176	170	151	← e [mm]

Figure 3.12 Influence of load on aerodynamic resistance coefficient and aerodynamic lift coefficient of passenger cars. *Source:* Janssen and Hucho (1973). Reproduced with permission of BHRA Fluid Engineering.

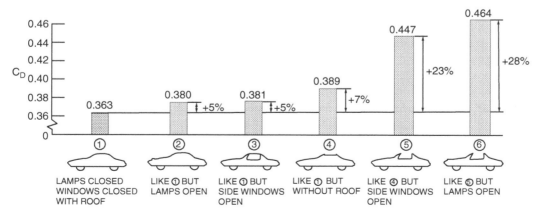

Figure 3.13 Influence of operational factors on aerodynamic resistance coefficient of a sports car. *Source:* Janssen and Hucho (1973). Reproduced with permission of BHRA Fluid Engineering.

Table 3.2 Components of aerodynamic resistance coefficient and potential for reduction.

Components of aerodynamic resistance coefficient	Minimum feasible value
Forebody	−0.015
Afterbody	0.07
Underbody	0.02
Skin friction	0.025
Total body drag	**0.10**
Wheels and wheel wells	0.07
Drip rails	0
Window recesses	0.005
External mirror (one)	0.005
Total protuberance drag	**0.08**
Cooling system	0.015
Total internal drag	**0.015**
Overall total drag	**0.195**

Source: Carr (1983).

Figure 3.14 Effects of reduction in aerodynamic resistance coefficient on fuel economy at different speeds for a midsize passenger car. Reprinted with permission from SAE paper 760185 © 1976 Society of Automotive Engineers, Inc.

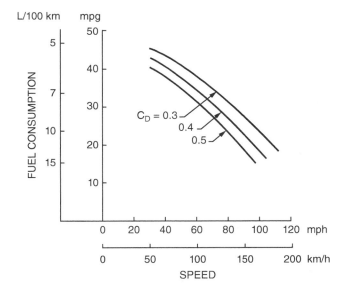

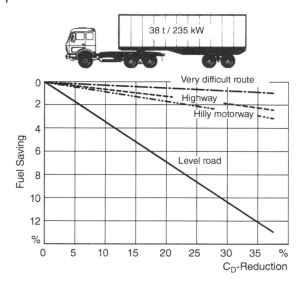

Figure 3.15 Effects of reduction in aerodynamic resistance coefficient on fuel saving for a tractor–semitrailer. *Source:* Gotz (1990). Reproduced with permission of Butterworth-Heinemann.

the aerodynamic resistance coefficient from 0.5 to 0.3 will improve the fuel economy by approximately 23%. Figure 3.15 shows the effects of the reduction in aerodynamic resistance coefficient on fuel saving of a tractor–semitrailer under different operating conditions (Gotz 1990). Operating on a level road at constant speeds, the reduction in aerodynamic resistance coefficient has the most significant effect on fuel saving of a tractor–semitrailer.

In comparison with passenger cars, heavy commercial vehicles, such as trucks, tractor–semitrailers, and truck–trailers, usually have much higher values of aerodynamic resistance coefficient. This is primarily due to their essentially box-shaped body. Figure 3.16 shows the variations of the aerodynamic resistance coefficient of a tractor–semitrailer and a truck–trailer with the yaw angle, which is the angle between the direction of travel of the vehicle and that of the wind (Gotz 1990). It also shows the contributions of the tractor (or truck) and semitrailer (or trailer) to the total aerodynamic resistance coefficient of the combination. It can be seen that for a tractor–semitrailer, the aerodynamic resistance of the tractor is not sensitive to yaw angle, and that the tractor contributes approximately 60% to the total aerodynamic resistance of the tractor–semitrailer combination at 0° yaw angle. For the truck–trailer combination, the truck contributes approximately 62% to the total aerodynamic resistance of the combination at 0° yaw angle. Table 3.3 shows representative values of the aerodynamic resistance coefficient of passenger cars, vans, buses, tractor–semitrailers, and truck–trailers.

To improve the aerodynamic performance of heavy commercial vehicles, add-on devices, such as the air deflector mounted on the roof of the tractor or truck, have been introduced. Figure 3.17 shows the effects of various types of air deflector on the aerodynamic resistance coefficient of a tractor–semitrailer (Berta and Bonis 1980). It can be seen that with the best air deflector among those investigated (type 6 shown in the figure), the aerodynamic resistance coefficient can be reduced by 24%, in comparison with that of the baseline vehicle (type 1). The installation of a gap seal between the tractor and semitrailer (type 8) does not cause a noticeable decrease in the aerodynamic resistance coefficient. With rounded vertical edges in the front of the semitrailer and with smooth, flat panels on the semitrailer body (type 9), the aerodynamic resistance coefficient is reduced by 22%. If this is coupled with the installation of the best air deflector (type 10), a

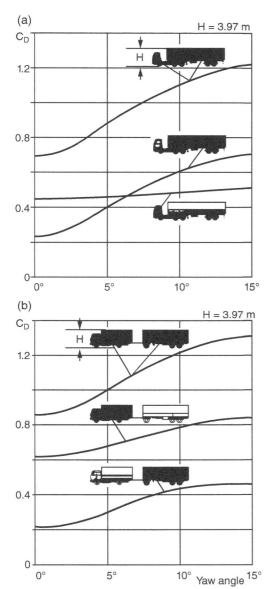

Figure 3.16 Distributions of aerodynamic resistance between (a) tractor and semitrailer and (b) truck and trailer at different yaw angles. *Source:* Gotz (1990). Reproduced with permission of Butterworth-Heinemann.

total reduction of 34% in aerodynamic resistance coefficient can be achieved. Figure 3.18 shows the variations of the aerodynamic resistance coefficient of tractor–semitrailers with various add-on devices and tractor shapes with the yaw angle (Gotz 1990).

In addition to the add-on devices mentioned above, other devices, such as the side skirt (trailer skirt) and boat-tail (trailer tail or rear fairing), for reducing the aerodynamic resistance and for improving fuel economy of tractor–semitrailers have attracted attention. Side skirts are designed to minimize the air turbulence underneath the semitrailer. They are a pair of panels affixed to either side of a semitrailer from its floor down to the ground with a suitable clearance. They usually fill the longitudinal space between the tractor's rear wheels and the semitrailer wheels. A boat-tail is a set of panels which fold out from the rear of the semitrailer, creating a tapered shape to reduce aerodynamic resistance caused by the low-pressure wake behind the semitrailer. Results of a study show that with semitrailer side skirts, fuel savings are between 4% and 5%, in comparison with an unmodified semitrailer (Wood 2012). Side skirts with reduced ground clearance from 16 in. (41 cm) to 8 in. (20 cm) resulted in fuel saving of 4–7%. Boat-tails alone can achieve a fuel saving of between 1% and 5%, and in combination with side skirts, a 9% improvement in fuel economy has been demonstrated. Results of wind tunnel testing of full-scale and 1 : 4 and 1 : 10 scale models of tractor–semitrailers indicate similar effects of side skirts and boat-tails on reduction of aerodynamic resistance coefficient and on improvements in fuel economy (Cooper and Leuschen 2005; Landman et al. 2010).

Aerodynamic lift acting on a vehicle is caused by the pressure differential across the vehicle body from the bottom to the top. It may become significant at moderate speeds. The aerodynamic lift usually causes the reduction of the normal load on the tire–ground contact. Thus, the performance characteristics and directional control and stability of the vehicle may be adversely affected. To improve cornering and traction capabilities of racing cars, externally mounted aerodynamic surfaces that generate a downward aerodynamic force are widely used. This increases the normal load on the tire–ground contact and road holding ability.

Table 3.3 Representative values of aerodynamic resistance coefficient for various types of vehicle.

Vehicle type	Aerodynamic resistance coefficient C_D
Passenger cars (sedans)	0.22–0.35
Vans	0.25–0.35
Buses	0.35–0.6
Tractor–semitrailers	0.55–0.86
Truck–trailers	0.6–0.85

Sources: Robert Bosch GmbH, *Bosch Automotive Handbook*, 10th Edition (2018); Bayindirli et al. (2016); Chowdhury et al. (2013); Patten et al. (2012); Cooper and Leuschen (2005); Gotz (1990); Berta and Bonis (1980).

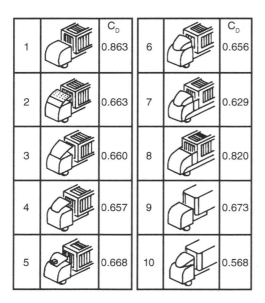

Figure 3.17 Effects of add-on devices and body details on the aerodynamic resistance coefficient of tractor–semitrailers. Reprinted with permission from SAE paper 801402 © 1980 Society of Automotive Engineers, Inc.

The aerodynamic lift R_L acting on a vehicle is usually expressed by

$$R_L = \frac{\rho}{2} C_L A_f V_r^2 \tag{3.21}$$

where C_L is the coefficient of aerodynamic lift usually obtained from wind tunnel testing. Typical values of C_L for passenger cars vary in the range 0.2–0.5 using the frontal area of the vehicle as the characteristic area. Like the coefficient of aerodynamic resistance, it depends not only on the shape of the vehicle but also on a number of operation factors. The effects of the shape of the afterbody on the rear axle aerodynamic lift coefficient C_{LR} are shown in Figure 3.7. Figures 3.8 and 3.9 show the influence of the front and rear spoilers on the front and rear axle aerodynamic lift coefficients, C_{LF} and C_{LR}, respectively. The effects of the angle of attack, ground clearance, and loading conditions on the aerodynamic lift coefficient C_L are illustrated in Figures 3.10, 3.11, and 3.12, respectively.

Figure 3.18 Effects of add-on devices on the aerodynamic resistance coefficient of tractor–semitrailers at different yaw angles. *Source:* Gotz (1990). Reproduced with permission of Butterworth-Heinemann.

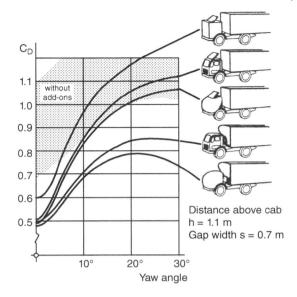

Distance above cab
h = 1.1 m
Gap width s = 0.7 m

The aerodynamic pitching moment may also affect the behavior of a vehicle. This moment is the result of the moments of the aerodynamic resistance and aerodynamic lift about the center of gravity of the vehicle. It may cause significant load transfer from one axle to the other at moderate and higher speeds. Thus, it would affect the performance, as well as the directional control and stability of the vehicle.

The aerodynamic pitching moment M_a is usually expressed by

$$M_a = \frac{\rho}{2} C_M A_f L_c V_r^2 \tag{3.22}$$

where C_M is the coefficient of aerodynamic pitching moment usually obtained from wind tunnel testing and L_c is the characteristic length of the vehicle. The wheelbase or the overall length of the vehicle may be used as the characteristic length in Eq. (3.22). Most passenger cars have a value of C_M between 0.05 and 0.20, using the wheelbase as the characteristic length and the frontal area as the characteristic area.

3.3 Internal Combustion Engines

As mentioned previously, there are two limiting factors to the performance of a road vehicle: one is the maximum tractive effort that can be generated through the interaction between the tire and the ground; the other is the tractive effort that the maximum torque of the power plant with a given transmission can provide. The smaller of these two will determine the performance potential of the vehicle. In lower gears, the tractive effort may be limited by the tire–road adhesion. In higher gears, the tractive effort is usually determined by the power plant and transmission characteristics. To predict the overall performance of a road vehicle, the characteristics of the power plant and transmission must be taken into consideration.

For vehicular applications, the ideal performance characteristics of a power plant are constant power output over a wide operating speed range. Consequently, the output torque varies with speed

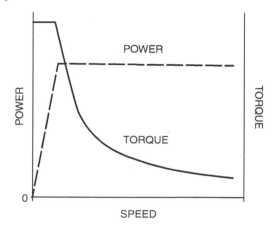

Figure 3.19 Ideal performance characteristics for a vehicular power plant.

hyperbolically in that speed range, as shown in Figure 3.19. This will provide the vehicle with high tractive effort at low speeds where demands for acceleration, drawbar pull, or grade climbing capability are high. There are power plants that have torque/power–speed characteristics close to the ideal for vehicular applications, such as the permanent magnet synchronous AC motor, which will be discussed later.

The internal combustion engine offers less favorable performance characteristics and can only be used in a vehicle with a transmission having a range of appropriate gear reduction ratios. Despite these disadvantages, it is widely used in automotive vehicles to date, because of its relatively high power to weight ratio, reasonable fuel economy, low cost, easiness to start, and a wealth of experience in its production and usage for over a century. It uses high energy density fossil fuels which enable the vehicle to have a relatively long operating range before refueling. Furthermore, the types of fuel used are widely available through an existing infrastructure.

In view of growing concerns over climate change, environmental protection, and energy conservation, increasingly stringent regulations on exhaust emissions and fuel economy have been introduced in many parts of the world. This stimulates a great deal of research and development on reducing greenhouse gas emissions and improving fuel economy of the internal combustion engine, and on alternative power plants and energy sources. Alternatives to the internal combustion engine, such as the electric drive, hybrid electric drive, and fuel cell have received intense worldwide attention and will be discussed later in this chapter.

3.3.1 Performance Characteristics of the Internal Combustion Engine

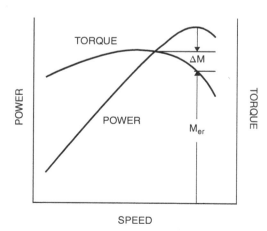

Figure 3.20 Notional performance characteristics of a gasoline engine.

Since the internal combustion engine is currently still widely used in automotive vehicles, the basic features of its characteristics that are essential to the prediction of vehicle performance are presented in this section. Notional performance characteristics of a gasoline engine and a diesel engine with turbocharging are shown in Figures 3.20 and 3.21, respectively. The internal combustion engine starts operating smoothly at a certain speed (the idle speed). Good combustion quality and the maximum engine torque are reached at an intermediate engine speed. As speed increases further, the mean effective pressure decreases because of growing losses in the air-induction manifolds, and the engine torque declines. Power output, however, increases with the increase of speed up to the point of maximum power. Beyond this point, the engine torque decreases more rapidly with the increase of speed. This results in a decline of power output. In vehicular applications, the maximum permissible operating speed of the engine is usually set just above the speed of the maximum power output. Vehicles designed for traction, such as agricultural or industrial tractors, usually operate at much lower engine speeds, since the maximum torque, and not the maximum power, defines their tractive performance. To control the maximum speeds, engines for heavy-duty vehicles are often equipped with a governor.

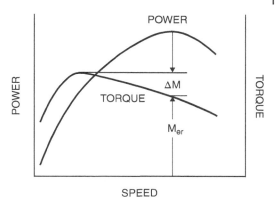

Figure 3.21 Notional performance characteristics of a diesel engine with turbocharging.

Table 3.4 shows a comparison of characteristic values between various types of gasoline and diesel engines (Robert Bosch GmbH 2018). These include naturally aspirated type and supercharged type with intercooling (charge-air cooling) gasoline engines and supercharged type with intercooling diesel engines. The table shows the engine speed range, compression ratio, power output per liter of engine displacement, weight-to-power ratio, specific fuel consumption, and torque increase in percentage (i.e., the ratio of the difference ΔM between the maximum torque and the torque at the rated power (maximum effective power) to the torque at the rated power M_{er} shown in Figures 3.20 and 3.21) for various types of gasoline and diesel engines. As shown in the table, there is a noticeable difference in the operating speed range between the gasoline and diesel engine, with the former having much higher operating speed range than the latter. The diesel engine has much higher compression ratio than the gasoline engine by the nature of the ignition process. The diesel engine is compression ignition while the gasoline engine is spark ignition. As the diesel engine attains lower operating speed, it has lower power output per liter of engine displacement than the gasoline engine. For durability reasons, the diesel engine is heavier than the gasoline engine with the same power output, leading to higher weight-to-power ratio. Due to the lower compression ratio, the gasoline engine has higher fuel consumption than the diesel engine. The difference between the torque at the rated power to the maximum torque for the supercharged gasoline engine is higher than that for the naturally aspirated type. This indicates that the supercharged gasoline engine is more favorable for meeting driving demands than the naturally aspirated gasoline engine.

Engine performance diagrams supplied by manufacturers usually represent the gross engine performance. It is the performance of the engine with only the built-in equipment required for self-sustained operations, and all other installations and accessories not essential to the operation are stripped off. The effective engine power available at the transmission input shaft is therefore reduced by the power consumed by the accessories, such as the fan and water pump for the cooling system, and by losses arising from the air cleaner, exhaust system, etc. There are also auxiliaries, such as the alternator, air conditioning unit, and power-assisted steering and braking, that make a demand on engine power. Figure 3.22 shows the variations of the power consumed by the air conditioning unit, water pump and fan, power steering, and alternator with engine speed for a representative full-size passenger car (Coon and Wood 1974).

In vehicle performance prediction, the power consumption of all accessories over the full engine speed range should be evaluated and subtracted from the gross engine power to obtain the effective power available to the transmission input shaft.

Table 3.4 Typical characteristics of internal combustion engines.

Engine type	Application	Engine speed at rated (maximum) power rpm	Compression ratio	Power output per liter of engine displacement		Weight-to-power ratio		Specific fuel consumption		Torque Increase[a] %
				kW/L	hp/L	kg/kW	lb/hp	g/kWh	lb/hph	
Gasoline	Cars – NAE[b]	5000–8000	9–13	40–80	54–107	2.0–0.8	3.3–1.3	220–270	0.36–0.44	15–20
	Cars – SCE[c]/IC[d]	5000–7500	9–12	60–110	80–148	1.5–0.5	2.5–0.8	220–250	0.36–0.41	20–40
Diesel	Cars and light trucks – SCE[c]/IC[d]	3500–4500	18–22	35–55	47–74	3.0–1.3	4.9–2.1	200–220	0.33–0.36	20–40
	Heavy trucks – SCE[c]/IC[d]	1800–2600	15–18	25–40	36–54	4.0–2.5	6.6–4.1	180–210	0.26–0.35	20–40

[a] Torque increase – the difference between the maximum torque and the toque at the rated power normalized with respect to the torque at the rated power, expressed in percentage.
[b] NAE – Naturally aspirated engine.
[c] SCE – Supercharged engine.
[d] IC – Intercooling (charge-air cooling).
Source: Robert Bosch GmbH, *Bosch Automotive Handbook*, 10th Edition (2018). Reproduced with permission of Wiley.

Figure 3.22 Accessory power requirements for a full-size passenger car. Reprinted with permission from SAE paper 740969 © 1974 Society of Automotive Engineers, Inc.

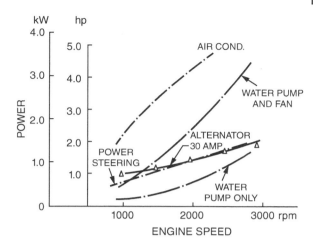

Atmospheric conditions affect engine performance. To allow comparisons of engine power on a common basis, it is necessary to specify the reference atmospheric conditions and to convert engine power measured under various atmospheric conditions to that under the reference atmospheric conditions. There are various standards currently in use for correcting the engine power obtained under the test atmospheric conditions to that under the reference atmospheric conditions. In the following, two standards are briefly outlined: one is the International Organization for Standardization ISO 1585 : 1997-04 and the other is the Society of Automotive Engineers SAE J1349 SEP2011.

1) ISO 1585 : 1997-04

The reference atmospheric conditions specified by the ISO 1585 : 1997-04 are: the reference inlet air temperature is 298 K (25°C), and the reference inlet dry air pressure is 99 kPa. The inlet dry air pressure is based on a total inlet air pressure of 100 kPa and a water vapor pressure of 1 kPa. The methods for correcting the power obtained under the test atmospheric conditions to the power under the reference atmospheric conditions for various types of engine are outlined below.

(A) Naturally aspirated and pressure-charged spark-ignition engines

The test atmospheric conditions shall be within the values given below:

$$288 \text{ K} \leq T_t \leq 308 \text{ K}$$

and

$$80 \text{ kPa} \leq p_t \leq 110 \text{ kPa}$$

where T_t is the temperature at the engine air inlet under test conditions in K (kelvin), which is equal to $273 +$ atmospheric temperature in °C; p_t is the dry atmospheric pressure in kPa (i.e., the total barometric pressure minus the water vapor pressure) under test conditions.

The formula for deriving the engine power at full load under the reference atmospheric conditions from that obtained under the test atmospheric conditions is as follows:

$$P_{re} = \alpha_{ae}P_{te} = \left[\left(\frac{99}{p_t}\right)^{1.2}\left(\frac{T_t}{298}\right)^{0.6}\right]P_{te} \tag{3.23}$$

where P_{re} is the engine power under the reference atmospheric conditions; α_{ae} is the power correction factor for spark-ignition engines; and P_{te} is engine power obtained under the test atmospheric conditions.

Equation (3.23) only applies if $0.93 \leq \alpha_{ae} \leq 1.07$. If these limits are exceeded, the corrected value shall be given, and the test conditions (temperature and pressure) precisely stated in the report.

(B) Compression-ignition engines

The test atmospheric conditions shall be within the values given below:

$$283\,\mathrm{K} \leq T_t \leq 313\,\mathrm{K}$$

and

$$80\,\mathrm{kPa} \leq p_t \leq 110\,\mathrm{kPa}$$

The formula for deriving the engine power under the reference atmospheric conditions from that obtained under the test atmospheric conditions is as follows:

$$P_{re} = \alpha_{ce}P_{te} \tag{3.24}$$

where P_{re} is the engine power under the reference atmospheric conditions; α_{ce} is the power correction factor for compression-ignition engines; P_{te} is engine power obtained under the test atmospheric conditions

The power correction factor α_{ce} for compression-ignition engines at constant fuel delivery setting is obtained by applying the formula given below:

$$\alpha_{ce} = (f_a)^{f_m} \tag{3.25}$$

where f_a is the atmospheric factor and f_m is the engine factor (i.e., the characteristic parameter for each type of engine and adjustment).

The atmospheric factor f_a, which indicates the effects of environmental conditions (pressure, temperature, and humidity) on the air drawn in by the engine shall be calculated from the formulas below, depending on the engine type:

- Naturally aspirated engines, mechanically pressure-charged engines, and turbocharged engines with wastegates operating:

$$f_a = \left(\frac{99}{p_t}\right)\left(\frac{T_t}{298}\right)^{0.7} \tag{3.26}$$

- Turbocharged engines without charge air cooling or with charge air cooling by air/air cooler:

$$f_a = \left(\frac{99}{p_t}\right)^{0.7}\left(\frac{T_t}{298}\right)^{1.2} \tag{3.27}$$

- Turbocharged engines with charge-air cooling by engine coolant:

$$f_a = \left(\frac{99}{p_t}\right)^{0.7} \left(\frac{T_t}{298}\right)^{0.7} \tag{3.28}$$

where T_t and p_t are as defined previously.

The engine factor f_m is a function of the corrected fuel delivery parameter q_c and is calculated from the following formula

$$f_m = 0.036 q_c - 1.14 \tag{3.29}$$

where $q_c = q_f/r_f$, in which q_f is the fuel delivery parameter, in milligrams per cycle per liter of engine swept volume mg/(L · cycle), and is equal to

$$q_f = \frac{(Z) \times (\text{fuel flow in g/s})}{(\text{displacement in L}) \times (\text{engine speed in } \min^{-1})}$$

where $Z= 120{,}000$ for four-stroke cycle engines and $Z= 60{,}000$ for two-stoke cycle engines; r_f is the ratio between the absolute static pressure at the outlet of the pressure charger, or charge-air cooler if fitted, and the ambient pressure; r_f is equal to 1 for naturally aspirated engines.

The formula for the engine factor f_m is only valid for a q_c value between 37.2 mg/ (L · cycle) and 65 mg/(L · cycle). For values less than 37.2 mg/(L · cycle), a constant value of 0.2 shall be taken for f_m.

The correction Eq. (3.25) for α_{ce} only applies if $0.9 \leq \alpha_{ce} \leq 1.1$. If these limits are exceeded, the corrected value obtained shall be given and the test conditions (temperature and pressure) precisely stated in the test report.

For engines not covered by those mentioned above, a correction factor equal to 1 shall be applied when the ambient air density does not vary by more than $\pm 2\%$ from the density at the reference atmospheric conditions (298 K and 99 kPa). When the ambient air density is beyond these limits, no correction shall be applied, but the test conditions shall be stated in the test report.

2) SAE J1349 SEP2011

The reference atmospheric conditions specified in the SAE J1349 SEP2011 are the same as those specified in ISO 1585 : 1997-04 given previously.

The test range limits are temperature 15–35°C and the dry air pressure (absolute) 90–105 kPa. The methods for correcting the power obtained under the test atmospheric conditions to that under the reference atmospheric conditions for various types of engine are outlined below.

(A) Spark-ignition engines

These spark-ignition engine brake power correction formulas are only applicable at full throttle positions.

$$P_{br} = P_{ir} - P_{ft} \tag{3.30}$$

where P_{br} is the engine brake power under the reference atmospheric conditions; P_{ir} is the engine indicated power under the reference atmospheric conditions; P_{ft} is the engine friction power under the test atmospheric conditions.

$$P_{ir} = \alpha_{be} P_{it} = \alpha_{be} (P_{bt} + P_{ft}) \tag{3.31}$$

where α_{be} is the atmospheric correction factor; P_{it} is the engine indicated power under the test atmospheric conditions; P_{bt} is the engine brake power under the test atmospheric conditions.

The atmospheric correction factor α_{be} is given as follows:

$$\alpha_{be} = \left(\frac{99}{p_t}\right)\left(\frac{T_t}{298}\right)^{0.5} \tag{3.32}$$

where p_t is the dry atmospheric pressure in kPa (i.e., the total barometric pressure minus the water vapor pressure) under the test atmospheric conditions; T_t is the temperature at the engine air inlet under the test conditions in K (kelvin), as noted previously.

If the engine friction power is measured, then the engine brake power under the reference atmospheric conditions can be calculated by combining Eqs. (3.30) and (3.31):

$$P_{br} = \alpha_{be}P_{bt} + (\alpha_{be} - 1)\,P_{ft} \tag{3.33}$$

If engine friction power under the test atmospheric conditions P_{ft} is not measured and 85% of engine mechanical efficiency is assumed, then:

$$P_{ft} = (1 - 0.85)P_{it} = (1 - 0.85)\left(P_{bt} + P_{ft}\right)$$

or

$$P_{ft} = \frac{(1 - 0.85)}{0.85}\,P_{bt} = 0.176\,P_{bt} \tag{3.34}$$

By assuming 85% engine mechanical efficiency and combining Eqs. (3.33) and (3.34), one obtains the engine brake power P_{br} under the reference atmospheric conditions as follows:

$$P_{br} = [1 + 1.176(\alpha_{be} - 1)]P_{bt} = \left[1 + 1.176\left\{\left(\frac{99}{p_t}\right)\left(\frac{T_t}{298}\right)^{0.5} - 1\right\}\right]P_{bt}$$

or

$$\frac{P_{br}}{P_{bt}} = 1 + 1.176\left\{\left(\frac{99}{p_t}\right)\left(\frac{T_t}{298}\right)^{0.5} - 1\right\} \tag{3.35}$$

(B) Compression-ignition engines

For turbocharged diesel engines using electronic fuel injection systems, no correction factors shall be applied to engine output for changes in atmospheric density or fuel temperature when the control system will compensate for changes in fuel temperature and air density. For naturally aspirated diesel engines or diesel engines with positive displacement injection systems, please refer to Appendix A of the SAE J1349 SEP2011 for the appropriate correction factors.

3.3.2 Emissions of Internal Combustion Engines

With growing concerns over climate change and air quality, many countries and regions have introduced increasingly stringent regulations governing vehicle exhaust emissions.

For gasoline engines, in addition to unavoidable combustion products of carbon dioxide (CO_2) and water (H_2O), major emissions through the tailpipe of vehicles include the following:

- Carbon monoxide (CO)
- Nitrogen oxides (NO_X)
- Non-combusted hydrocarbons (HC)

For diesel engines, in addition to the emissions from gasoline engines listed above, soot and particulate emissions are of significance and must be considered.

Carbon dioxide is a greenhouse gas that causes climate change. It is a major concern of the global community in recent years. To address the issues of greenhouse effects and climate change, to reduce the dependency on crude oil, and to increase supply reliability, alternative fuels for the internal combustion engine, such as liquefied petroleum gas (LPG), natural gas, and biofuel (such as ethanol and methanol) produced from biomass (such as plants, woods, etc.), have been introduced and have attracted considerable attention. In comparison with gasoline, the combustion of LPG produces approximately 10% less CO_2, and in comparison with diesel fuel, there are virtually no particulates produced by LPG (Robert Bosch GmbH 2018). Vehicles powered by natural gas produce approximately 20–30% less CO_2. In addition to producing no particulates, only very low levels of NO_X, CO, and NMHC(non-methane hydrocarbons) are present in the exhaust, in conjunction with a three-way catalytic converter. Spark-ignition engines running on ethanol or methanol produce lower pollutant emissions and less CO_2, and have reduced ozone and smog formation. Furthermore, these biofuels are completely free of sulfur. It should also be mentioned that gasoline blended with biofuel (such as ethanol) is being used in spark-ignition engines and diesel fuel blended with biofuel, known as biodiesel, is being used in compression-ignition engines in some countries.

There are two basic approaches to the reduction of pollutant emissions: one is to prevent them from forming; the other is to remove them from the exhaust once formed.

For gasoline engines, improving combustion process can reduce undesirable exhaust emissions. These include using direct fuel injection, lean burn combustion with a certain level of excess air, controlled stratification of air/fuel mixture formation in the combustion chamber, variable valve timing, etc. The evaporative emission control system for the gasoline engine intercepts and collects gasoline vapors from the fuel tank. It plays a noticeable role in complying with regulations governing the emission limits of fuel evaporative losses.

Pollutants can be reduced after they have left the combustion chamber by injecting air into the exhaust manifold system for more complete oxidation. An exhaust gas recirculation system, by which the exhaust gas is recirculated to the intake manifold, can also be used to reduce pollutants. This recirculation system plays an important role in reducing emissions of both gasoline and diesel engines.

Exhaust gas of a gasoline engine can also be treated by a three-way catalytic converter, which performs the following three functions:

- Reduction of nitrogen oxides to nitrogen (N_2)

$$2CO + 2NO \rightarrow 2CO_2 + N_2$$

$$\text{Hydrocarbon} + NO \rightarrow CO_2 + H_2O + N_2$$

$$2H_2 + 2NO \rightarrow 2H_2O + N_2$$

- Oxidation of carbon monoxide to carbon dioxide

$$2CO + O_2 \rightarrow 2CO_2$$

- Oxidation of unburnt hydrocarbons (HC) to carbon dioxide and water, in addition to the above NO reaction

$$\text{hydrocarbons} + O_2 \rightarrow H_2O + CO_2.$$

For diesel engines, reducing undesirable exhaust emissions may be achieved by improving the design of the engine and its subsystems. These include, but are not limited to, improving combustion chamber configuration to ensure maximum rate of combustion; using a common rail injection system with high fuel injection pressure to improve air/fuel mixture formation, which leads to more complete combustion; and variable valve actuation and timing.

Treatment of exhaust gas from a diesel engine using a two-way catalytic converter plays a significant role in reducing pollutant emissions. The most used converter is the diesel oxidation catalytic converter, which contains palladium, platinum, and aluminum oxide. It oxidizes carbon monoxide and unburnt hydrocarbons with oxygen to form carbon dioxide and water.

- Oxidation of carbon monoxide to carbon dioxide

$$2CO + O_2 \rightarrow 2CO_2$$

- Oxidation of hydrocarbons (unburnt or partially burnt fuel) to carbon dioxide and water

$$C_xH_{2x+2} + [(3x+1)/2]O_2 \rightarrow xCO_2 + (x+1)H_2O \text{ (a combustion reaction)}$$

A device that combines the features of the catalytic converter and filter technology to reduce the amount of carbon monoxide, unburnt hydrocarbons, and particulates in exhaust gas of diesel engines has been introduced (McKnight 1999).

It should be noted that the two-way catalytic converter for diesel engines does not reduce NO_x. The reduction in NO_x may be achieved by exhaust gas recirculation, as mentioned above, by using a selective catalytic reduction (SCR) device that utilizes a reagent such as ammonia to convert the NO_x into nitrogen and water, or by a NO_x absorber.

Alternatives to the gasoline engine and the diesel engine, such as the gas turbine and the Stirling-cycle engine, have been investigated for use in motor vehicles. The gas turbine has several advantages, which include favorable power-to-weight ratio, ability of operating with a variety of fuels, smooth running, clean exhaust without supplementary emission control devices, good static torque characteristics, and extended maintenance intervals. Its disadvantages include high cost, poor transient response, higher fuel consumption, and less suitability for low-power motor vehicles. The Stirling-cycle engine utilizes alternate heating and cooling of the working medium, such as compressed helium or hydrogen gas, at constant volume to develop useful work. Its advantages include very low concentrations of all the pollutants (CO, NO_x, and unburnt HC), quiet operation without combustion noise, ability of operating with a variety of fuels, and fuel consumption comparable to that of the direct-injection diesel engine. Its disadvantages include poor power-to-weight ratio and bulky and high production cost due to its design complexity.

To address the issue of exhaust emissions from internal combustion engines and to alleviate the dependency on fossil fuels as energy sources, vehicles with electric drive, hybrid electric drive, or fuel cells as energy provider have promising prospects for gradually replacing vehicles powered by internal combustion engines. An introduction to the engineering principles of electrical drive, hybrid electric drive, and fuel cells is presented in the following sections.

3.4 Electric Drives

An electric vehicle denotes a vehicle with an electric drive solely powered by an onboard electrical energy source. Electric vehicles may be powered by batteries, referred to as battery electric vehicles, as well as fuel cells, referred to as fuel cell vehicles. An introduction to the operating principles of the fuel cell will be presented later in Section 3.6.

Figure 3.23 Notional electric motor characteristics. *Source:* Adapted from Toyota Hybrid System II, with permission of Toyota Motor Corporation.

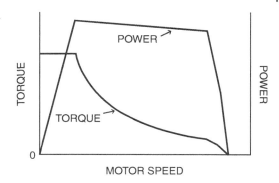

Electric drives are quieter than internal combustion engines and highly efficient. They do not produce tailpipe emissions like internal combustion engines. The processes for generating electricity for charging batteries or for producing hydrogen for fuel cells, however, do not necessarily produce no undesirable emissions. They are dependent upon the type of process and energy source used. Using electricity generated by renewable energy sources, such as wind, solar, or hydro power, for charging the battery or for producing hydrogen would eliminate this concern. In comparison with internal combustion engines, the torque/power–speed characteristics of electric drives are more suited to vehicle driving demands, with high torque at low speed and low torque at high speed over an extended operating range, as shown in Figure 3.23. Furthermore, the use of the electric drive leads to a simpler transmission than that with the internal combustion engine. Between the electric motor and the driven wheel, usually there is only one or two speed reduction stages. The traction motor acts as a generator to convert part of the kinetic energy of the vehicle during braking or of the potential energy of the vehicle in downhill coasting into electrical energy for charging the battery. This feature, commonly referred to as regenerative braking, contributes to the improvement of the operational efficiency and driving range of electric vehicles. The schematic of the regenerative braking process corresponding to the brake pedal depression, with coordination of hydraulic brakes to ensure safety through an electronic control system, is illustrated in Figure 3.24. It should be noted that regenerative braking is applicable only to wheels connected with electric motors. For a two-wheel drive (2WD) electric vehicle, not all the kinetic energy of the vehicle during braking or not all the potential energy during downhill coasting could be utilized for regenerative braking. Results of a study on the regenerative braking of a rear-wheel drive electric vehicle indicate that regenerative braking may increase the driving range of the vehicle by 11–22%, depending on the drive cycle and settings of the regenerative braking system (Wager et al. 2017).

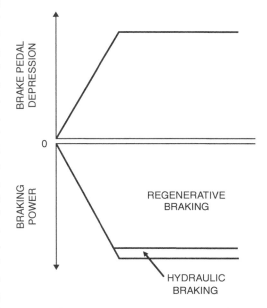

Figure 3.24 Schematic of regenerative braking process. *Source:* Adapted from Toyota Hybrid System II, with permission of Toyota Motor Corporation.

In an electric vehicle, one or more electric motors are used to covert electrical energy into mechanical energy to propel the vehicle. The number of motors used in an electric vehicle is dependent upon vehicle drivetrain design and the number of drive axles or driven wheels.

Based on applications, there are two categories of electric vehicles: one is the electric road vehicle, such as the electric passenger vehicle or electric commercial vehicle; the other is the electric industrial truck, such as the lift truck, also known as a forklift truck, fork truck, or forklift. The industrial truck powered by batteries is used to lift and transport materials for short distances in factory premises or warehouses and is not licensed for use on public roads. Its top speed is usually under 50 km/h. It is estimated that most of all new industrial trucks are powered by batteries. In this section, discussions are focused on the battery electric vehicle for road use.

In the report "*Global EV Outlook 2019*" by the International Energy Agency (IEA), the future development of electric mobility is explored through two scenarios: one aims to illustrate the impact of announced policy ambitions; the other takes into account the pledges of a campaign to reach a 30% market share for electric vehicles in all modes except two-wheelers by 2030 (IEA 2019). With the former scenario, it projects that in 2030, global electric vehicle sales may reach 23 million and the stock exceeds 130 million (excluding two/three-wheelers). For the latter scenario, it estimates that in 2030, sales may reach 43 million and the stock numbering more than 250 million.

Wide acceptance of battery electric passenger vehicles faces noticeable challenges. They include:

- The cost of batteries is relatively high. As a result, battery electric vehicles are currently more expensive than vehicles powered by internal combustion engines, even from a total cost of ownership perspective (considering the price of electricity for charging batteries versus that of gasoline or diesel fuel).
- Charging batteries is time consuming with the usual charging system at 240 V. Battery recharge times for various vehicle models with different operating ranges will be shown later in this section. Charging time can be noticeably reduced using fast/ultra-fast charging systems with much higher charging power, such as 350 kW (Suarez and Martinez 2019). Their use is, however, subject to certain limitations that include: the battery must be designed to accept a fast/ultra-fast charge; the fast/ultra-fast charging only applies during the first charge phase, typically up to approximately 80% of the battery capacity; and the charging may affect the life of the battery and can only be done under moderate temperatures.
- Battery charging infrastructure has yet to be established for interurban travel in many countries or regions. The lack of extensive charging infrastructure causes the so-called range anxiety for the driver on long-distance trips.
- The specific energy of batteries currently available is significantly lower than that of gasoline or diesel fuel. For instance, the specific energy of the lithium-ion battery currently available is in the range of approximately from 75 to 275 Wh/kg (watt-hours per kilogram), whereas that of fossil fuel (gasoline or diesel fuel) is approximately 13,000 Wh/kg. This leads to either the driving ranges on one battery charge being limited or the battery pack being much bigger and heavier than a full fuel tank of a vehicle powered by the internal combustion engine.

The development of battery technology with advancing alternative chemistries to achieve competitive levels of cost, specific energy, specific power (watts per kilogram), and charging time, as well as the infrastructure for supporting electric vehicle operations, is evolving. The challenges facing wide adoption of electric vehicles may gradually be met in the future with further technological innovations, as well as government policy developments and actions.

3.4.1 Elements of an Electric Drive

The basic elements of an electric drive include the battery, electric motor, power control unit, and speed reduction gearset.

1) Batteries

For battery electric vehicles, the battery often accounts for a significant portion of the weight, size, and cost. The challenge is to optimize the battery pack, consisting of many battery cells, in these three aspects. The goal is to reduce the total cost of battery electric vehicles to the level competitive with those powered by internal combustion engines.

In selecting the battery for electric vehicles, consideration should be given to the following:

- Safety, with respect to fire and explosion, as well as electrical safety with voltage higher than 60 V.
- Performance characteristics, in terms of specific energy (Wh/kg), and specific power (W/kg). Specific energy is an indication of the energy storage capability and is related to the operating range of an electric vehicle, while specific power is indicative of its characteristics of utilizing the power from regenerative braking or providing power for vehicle acceleration.
- Calendar life and cycle life, the goal being at least lasting for 10 years (typically 250,000 km or 155,400 miles) and 1000 complete charging cycles, respectively. It is estimated that approximately 3000 complete charging cycles occur throughout the life of a battery electric vehicle, compared with over one million partial charging cycles for the hybrid electric vehicle (Robert Bosch GmbH 2018). The hybrid electric vehicle will be discussed in Section 3.5.

There are three types of battery available for use in various types of electric vehicle: lead-acid, nickel-based, and lithium-ion battery.

(A) Lead-acid batteries

These are less expensive than the nickel-based battery and the lithium-ion battery and are most frequently used in electric industrial trucks. Their specific energy is in the range approximately from 25 to 50 Wh/kg and their specific power is up to 300 W/kg approximately. For industrial trucks, the lead-acid battery can achieve a calendar life of seven to eight years, and a cycle life of 1200–1500 charging cycles. The capability of the lead-acid battery to store electric energy deteriorates with the decrease in temperature. A battery heating system may be required at temperatures below freezing to fully charge the battery within an acceptable period. The amount of power and energy that can be drawn from a lead-acid battery also decreases as temperature drops.

(B) Nickel-metal hydride batteries

These were widely used previously in hybrid electric vehicles, such as the Toyota Prius (2004), Toyota Camry (2007), Toyota Lexus 600H (2008), and Toyota Prius (2010). This type of battery is generally less expensive than the lithium-ion battery. However, the nickel-metal hydride battery is considered as having already reached its best potential, both in terms of cost and performance. Its specific energy is approximately between 60 and 100 Wh/kg, and its specific power is up to 2000 W/kg. The nickel-metal hydride battery is now largely replaced by the lithium-ion battery in electric and hybrid electric passenger cars. For instance, in 2016 the Toyota Prius Prime Plug-In used lithium-ion batteries, replacing the nickel-metal hydride batteries installed in earlier models.

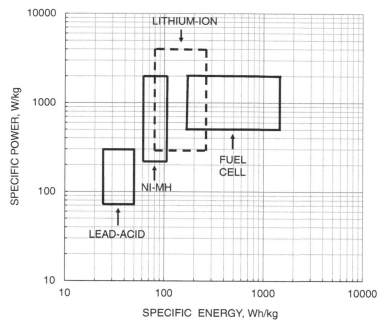

Figure 3.25 Notional performance metrics of batteries and fuel cells. *Sources:* Wang et al. (2020); Beard and Reddy (2019); Robert Bosch GmbH (2018); May et al. (2018); Luo et al. (2015).

(C) Lithium-ion batteries

These are now widely used in electric vehicles, hybrid electric vehicles, and plug-in hybrid electric vehicles. They have specific energy in the range of approximately from 75 to 275 Wh/kg, higher than that of the nickel-metal hydride battery and the lead-acid battery. Their specific power is up to 4000 W/kg approximately. This type of battery, like other rechargeable batteries, can be designed to emphasize energy, power, or some combination of these two capacities. Lithium possesses both high electrochemical potential and low equivalent mass. It has a long lifespan and its potential for further improvement is considered as high. However, it presents a safety challenge, as overcharging can cause fire and explosion. To address this issue, an elaborate monitoring and control system is usually used to keep the battery within strictly defined charge, current, and temperature limits. New materials are in development to improve safety characteristics and to further increase both specific energy and specific power. This includes, for instance, using lithium-ion phosphate in the cathode or lithium-titanate oxide in the anode instead of graphite.

For reference purposes, Figure 3.25. shows a comparison of the approximate ranges of specific energy and specific power of various types of battery, together with the hydrogen fuel cell, for ground vehicles. It should be noted that developments of battery and fuel cell technologies are in a rapid pace and that their ranges of specific energy and specific power shown in the figure will change accordingly with new developments.

2) Electric motors

The drivetrain of an electric vehicle generally consists of three major components, the motor, the power control unit, and the transmission between the motor and the driven wheel. As noted previously, the torque/power–speed characteristics of the electric motor are more suited to vehicle driving demands than those of the internal combustion engine. This leads to a simpler transmission than that with the internal combustion engine as the power plant. Between the electric motor and the driven wheel, usually there is only one or two speed reduction stages.

Three-phase alternating current (AC) asynchronous or synchronous motors have now become the accepted norm for electric vehicles or hybrid electric vehicles, due in part to their low maintenance requirements. In comparison with direct current (DC) motors, AC motors do not require brushes making sliding contact with the successive segments of the commutator. These brushes are made of a soft conductive material, like carbon, and must be regularly replaced, although usually at relatively long intervals.

The types of AC electric motor that are widely used include the following.

(A) Asynchronous motors

Asynchronous motors with squirrel-cage rotors are characterized by simple and robust construction. Their torque density is high and their efficiency improves at high speed and low torque. However, they may require rotor cooling due to ohmic losses in rotor windings. Typical variation of motor efficiency with the speed ratio for an asynchronous motor is shown in Figure 3.26. The efficiency is the ratio of the output power of the motor (i.e. the product of output torque and speed of the motor) to the corresponding input electric power. The speed ratio is the ratio of rotor speed to the maximum rotor speed.

(B) Electrically excited synchronous motors

For this type of motor, the magnetization of its rotor is controlled by electric current. As a result, it can achieve high efficiency with low supply current. This advantage leads to lower costs and reduced power consumption. Its performance characteristics are also shown in Figure 3.26.

(C) Permanent magnet synchronous motors

In this type of motor, permanent magnets, usually made of rare-earth elements, are mounted on the surface of the rotor or in the interior of the rotor. They replace rotor windings (field windings), thus eliminating the excitation current required and the associated ohmic losses. Furthermore, using permanent magnets makes it possible to reduce the overall size of the motor. A compact motor could fit into a small installation space for the electric drivetrain. This feature is of significance to a power-split type of hybrid electric drivetrain, which will be

Figure 3.26 Performance characteristics of various types of motor for electric vehicles: 1 – electrically excited synchronous motor; 2 – permanent magnet synchronous motor; 3 – asynchronous motor. *Source: Robert Bosch GmbH (2018). Reproduced with permission of Wiley.*

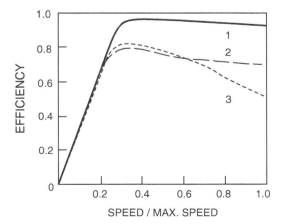

discussed later. As a result, the permanent magnet synchronous motor is widely used in electric and hybrid electric vehicles, such as those produced by Toyota and other manufacturers. The major issues for this type of motor are the relatively expensive rare-earth magnets and the risk of demagnetization. Typical performance characteristics of the permanent magnet synchronous motor are shown in Figure 3.26. It should be noted that the comparison of performance characteristics of the three types of motor shown in Figure 3.26 was conducted for an electric vehicle with installed power of 220 kW (Robert Bosch GmbH 2018).

The selection of a motor for a given application should be made on a detailed evaluation of the operational requirements, installation space available, and costs. For instance, while the synchronous motor with rare-earth magnets is more expensive than the other two types, it is still widely used, particularly in power-split type of hybrid electric vehicles, because of its compact size that enables it to fit into a small installation space, as noted previously.

The operation of an electric motor has certain unique features, such as its operation is limited by the winding or magnet temperatures, in comparison with an internal combustion engine. Figure 3.27 shows representative operating characteristics and limits for torque and power of an electric motor (Robert Bosch GmbH 2018). Curve 1 represents the maximum characteristic torque

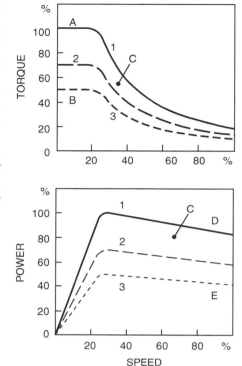

Figure 3.27 Operating limits for an electric motor: 1 – maximum limit characteristic curve; 2 – 60 s characteristic curve; 3 – continuous operation characteristic curve; A – maximum torque; B – continuous torque; C – overload operation; D – maximum power at high speed; E – continuous power. *Source:* Robert Bosch (2018). Reproduced with permission of Wiley.

or power as a function of rotor speed. Curve 2 indicates the permissible characteristic curve for operation not exceeding 60 s. Curve 3 represents the allowable continuous operating characteristics. In the figure, A represents the maximum torque; B indicates the allowable continuous torque; C represents overload operation region; D indicates the maximum power limit; and E represents the allowable power limit for continuous operation.

It should be mentioned that in most electric industrial trucks, series-wound DC motors are widely used. In these motors the field and armature are in series. While the efficiency of series-wound direct current motors is relatively low, it is still being used in electric industrial trucks because of its simple design and low cost.

3) Power control units

The power control unit for an electric vehicle usually includes:

- An inverter for converting DC electricity from the battery to AC for driving AC motors.
- A converter for converting to DC for charging the battery from AC electricity generated by AC motors connected with the driven wheels during regenerating braking. Under these circumstances, the motor functions as a generator.
- A controller for translating the position of the accelerator pedal into corresponding motor current and voltage. The motor driving torque is a function of the accelerator pedal position.

4) Speed reduction gearsets

As noted previously, the electric motor used in the electric drive has favorable torque/power–speed characteristics for meeting driving demands of road vehicles. Consequently, between the electric motor and the driven wheels, usually a one- or two-speed gearset will suffice.

3.4.2 Characteristics of Battery Electric Passenger Vehicles

Based on the data presented in the *2020 Fuel Consumption Guide*, published by the Department of Natural Resources, Government of Canada, Table 3.5 shows the characteristics of various battery electric vehicles. In the table, the following characteristic parameters are shown:

- Power ratings of the electric motor (drive motor) in kW and in hp.
- Energy consumption ratings for city, highway, and combined city and highway driving in kWh/100 km.
- Energy consumption ratings for city, highway, and combined city and highway driving in equivalent miles per US gallon (mpg (US)). One liter of gasoline contains the energy equivalent to 8.9 kW/h of electricity. Equivalent of liters of gasoline L_e = electric energy consumed in kWh/8.9 kWh. Equivalent mpg (US) = 235.21/(L_e/100 km).

Table 3.5 Characteristics of various battery electric vehicles.

Vehicle model	Motor kW (hp)	Consumption City/Highway/ Combined kWh/ 100 km	Consumption City/Highway/ Combined Equivalent mpg (US)	Range km (mile)	Battery recharge time, h
BMW i3 (120Ah)	125 (168)	16.8/20.6/18.5	124.6/101.6/113.2	246 (153)	7
Chevrolet Bolt EV	150 (201)	16.5/19.5/17.8	126.9/107.4/117.6	417 (259)	10
Hyundai Kona Electric	150 (201)	16.2/19.3/17.4	129.2/108.5/120.3	415 (258)	9
Jaguar I-Pace	294 (394)	26.2/29.1/27.5	79.9/71.9/76.1	377 (234)	13
KIA Soul EV (180Ah)	150 (201)	16.8/21.1/18.6	124.6/99.2/112.5	383 (238)	9.5
Nissan Leaf S Plus	160 (215)	17.8/21.5/19.5	117.6/97.4/107.4	363 (226)	11
Porsche Taycan Turbo	170 (228)	30.9/29.5/30.2	67.7/71.0/69.3	323 (201)	10
Tesla Model 3 Standard Range	211 (283)	14.8/16.5/15.6	141.4/126.9/134.2	151 (94)	3.7
Tesla Model 3 Long Range	211 (283)	15.3/17.0/16.1	136.8/123.1/130.0	531 (330)	10
Tesla Model S Standard Range	398 (534)	18.5/20.0/19.2	113.2/104.7/109.0	462 (287)	9
Tesla Model S Long Range	398 (534)	18.2/19.5/18.8	115.0/107.4/111.3	600 (373)	12
Tesla Model X Standard Range	398 (534)	20.0/21.5/20.7	104.7/97.4/101.1	415 (258)	9
Tesla Model X Long Range	398 (534)	21.2/22.5/21.8	98.7/93.0/96.0	528 (328)	12
Volkswagen e-Golf	100 (134)	17.4/19.9/18.6	120.3/105.2/112.5	198 (123)	5.3

Note: City rating represents urban driving in stop-and-go traffic. Highway rating represents a mix of open highway and rural road driving, typical of longer trips. Combined rating reflects 55% city driving and 45% highway driving.
Source: 2020 Fuel Consumption Guide, Department of Natural Resources, Government of Canada.

- Operating range for fully charge battery.
- Battery recharge time is the estimated time (in hours) to fully recharge the battery at 240 V.

Energy consumptions of battery electric vehicles, expressed in equivalent miles per US gallon (mpg (US)) shown in Table 3.5, are lower than corresponding ones powered by internal combustion engines shown later in Table 3.14 in Section 3.9, Operating Fuel Economy of Vehicles with Internal Combustion Engines.

3.5 Hybrid Electric Drives

As discussed in the previous section, electric vehicles do not produce tailpipe emissions. Thus, they contribute noticeably to the overall reduction in greenhouse gases and in air pollution. As noted earlier, however, wide adoption of electric vehicles currently faces noticeable challenges. These include:

- Battery electric vehicles being more expensive than vehicles powered by internal combustion engines, primarily due to the cost of the battery
- Charging the battery of electric vehicles being time consuming using household electricity supply
- The infrastructure for charging electric vehicles being not yet widely established, particularly for long-distance interurban travel

Under these circumstances, it is considered that the hybrid electric vehicle can play a useful role in the transition to the scenario where the electric vehicle would be a prime means of road transportation for a greener global environment.

A hybrid electric vehicle denotes a vehicle propelled by a drive system using two or more types of energy source, such as a combination of the internal combustion engine using fossil fuels and the electric motor using electricity stored in the battery, or a combination of the fuel cell and the battery to power the electric motor for propelling the vehicle. As the fuel cell is still in the developmental stage, this section will focus on the discussion of the hybrid electric vehicle with the combination of the internal combustion engine and the electric motor.

A hybrid electric vehicle whose battery can be recharged by plugging it into an external source of electric power, as well as by its own onboard engine through an electric generator, is called a plug-in hybrid electric vehicle. Charging the battery from the electricity grid can cost less than using the onboard internal combustion engine, thus reducing the operating cost. The plug-in hybrid electric vehicle is gaining increasingly wide acceptance.

The hybrid electric drive, with the combination of the engine and the electric motor as the power unit, has the following merits:

- The provision of an electric motor makes it possible to operate the engine primarily in the range of its optimal efficiency or in the range in which only low emissions generated.
- The combination with an electric motor enables the use of a smaller engine with lower power rating, while retaining the overall power of the combination. This contributes to the reduction in greenhouse gases and overall exhaust emissions.
- Using a planetary gearset as a power-split device in a hybrid electric vehicle, it performs the function as a continuously variable transmission, which is commonly referred to as an electric continuously variable transmission (E-CVT). In addition, the engine operating point may be maintained within the range where its efficiency is high or its emissions are low, by adjusting

the speed of one of the generator/motors, over a wide range of vehicle operating speeds. These topics will be discussed further later where the power-split type of hybrid electric vehicle is discussed.

- The drive motor acts as a generator to convert the kinetic energy of the vehicle during braking or the potential energy of the vehicle in downhill coasting for charging the battery. This regenerative braking feature has been discussed in the previous section on the electric drive.
- For city driving, the engine can often be switched off completely and the vehicle is solely propelled by the electric motor. Under these circumstances, the vehicle is driven with zero emissions.
- With an onboard engine and the infrastructure for refueling gasoline and diesel fuel widely in place, the driver's range anxiety for the battery electric vehicle is essentially eliminated.
- Hybrid electric vehicles with plug-in capability will reduce the operating cost by charging the battery from the electricity grid, in place of the onboard engine.

3.5.1 Types of Hybrid Electric Drive

Dependent on the arrangements for the combined operations of the engine and the electric drive, there are currently three major types of hybrid drive, namely, series, parallel, and power-split.

1) Series type

In this type of hybrid electric drive, the engine drives the electric generator and the electricity generated is used to drive the motor which in turn propels the vehicle. The power flows in series from the engine, through the generator, then the motor, and finally to the driven wheels. Thus, this type of drive is called the series type of hybrid electric drive.

For the series type of hybrid drive, the engine is not coupled to the drive axle of the vehicle and is, therefore, not linked to vehicle speed. This enables the engine to run within its efficient operating region to minimize exhaust emissions, fuel consumption, and noise. For fuel consumption characteristics of the internal combustion engine, please refer to Section 3.9, Operating Fuel Economy of Vehicles with Internal Combustion Engines.

The schematic of energy (power) flow for the series type of hybrid drive is shown in Figure 3.28.

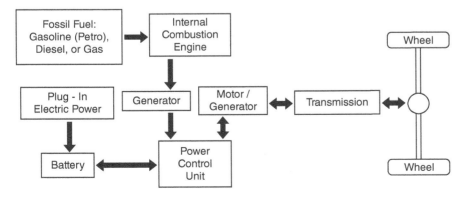

Figure 3.28 Schematic of energy (power) flow for the series type of hybrid electric vehicle.

The series type of hybrid drive has the following operating modes (Ehsani et al. 2019).

- Pure electric drive mode: The vehicle is propelled by the drive motor with the energy stored in the battery and the engine is turned off. In city driving, this operating mode may apply, and the vehicle operates as a zero-emission vehicle.
- Engine operating mode: The vehicle is propelled by the drive motor with electricity generated by the generator driven by the engine. In highway driving, this operating mode may apply.
- Hybrid drive mode: The electricity for the drive motor is drawn from both the battery and engine–generator set. During acceleration or hill climbing, this operating mode may apply.
- Engine operating with battery charging mode: The engine–generator set supplies power to charge the battery and to propel the vehicle.
- Regenerative braking mode: During braking or downhill coasting, the drive motor operates as a generator and the electricity generated is used to charge the battery. The schematic of the regenerative braking process was illustrated previously in Figure 3.24.
- Battery charging mode: The engine–generator set is used to charge the battery while the vehicle is stationary.
- Hybrid battery charging mode: Both the engine–generator set and the drive motor operating as a generator during braking or downhill coasting are used to charge the battery.

The advantages of the series type of hybrid drive include:

- As the engine is not connected to the drive axle of the vehicle, it can be made to operate in the desirable region to improve fuel economy and to minimize emissions.
- Since the electric motor has more favorable torque/power–speed characteristics than the internal combustion engine, as shown in Figure 3.23, the number of gears (or gear ratios) required in the transmission between the drive motor and the drive axle will be less than that with the internal combustion engine as the sole power source.
- As the electric motor is used to directly drive the wheel, the changing of vehicle speed is primarily through regulating motor speeds. Adjusting the motor speed is a simpler procedure and more convenient for the driver than by shifting gears with a mechanical transmission, particularly in a heavy commercial vehicle with a large number of gears. As a result, the series type of hybrid drive has found applications in diesel-electric locomotives, heavy trucks, and buses. Transmissions and their gear ratios for commercial vehicles will be discussed later in Section 3.7.

As noted previously, if the battery is made rechargeable by plugging it into an external source of electric power, as shown in Figure 3.28, then the vehicle will be designated as a plug-in hybrid electric vehicle.

It should be noted that because of the multiple energy conversions involved in this type of hybrid drive (i.e., the energy stored in the fossil fuel is converted into mechanical energy by the internal combustion engine, the mechanical energy developed is converted into electrical energy by the generator, and the electrical energy is then converted back to mechanical energy by the drive motor to propel the vehicle), the overall efficiency of the series type of hybrid vehicle is relatively low.

The series type of hybrid electric vehicle is primarily for reducing emissions. To extend the operating range of some electric vehicles, an internal combustion engine is installed onboard to drive a generator to charge the battery. This arrangement is to use the engine as a range extender.

2) Parallel type

In this type of drive, the power flows in parallel from the internal combustion engine and from the electric motor to propel the vehicle. The parallel type of hybrid drive has the following operating modes:

- Pure electric drive mode: The vehicle is propelled by the drive motor alone, with the energy stored in the battery and the engine is turned off. In city driving, this operating mode may apply, and the vehicle operates as a zero-emission vehicle.
- Pure engine drive mode: The vehicle is propelled by the engine alone. In highway driving, this operating mode may apply.
- Hybrid drive mode: The vehicle is propelled jointly by the engine and the drive motor with electricity drawn from the battery. During acceleration or hill climbing, this operating mode may apply.
- Engine operating with battery charging mode: The engine supplies power to propel the vehicle and to charge the battery, with the drive motor not propelling the vehicle.
- Regenerative braking mode: During braking or downhill coasting, the drive motor operates as a generator and the electricity generated is used to charge the battery.
- Battery charging mode: The engine is used to charge the battery with the drive motor operating as a generator while the vehicle is stationary.
- Hybrid battery charging mode: Both the power from the engine and from regenerative braking or downhill coasting are used to charge the battery with the motor operating as a generator.

In comparison with the series type, the parallel type of hybrid drive has the following merits:

- As the engine and the drive motor directly provide driving torques to the driven wheels, there is no multiple energy conversions involved in the parallel type of hybrid drive, unlike the series type hybrid drive described previously. Consequently, the parallel type of hybrid drive is more energy efficient than the series type.
- The parallel type of hybrid drive is more compact than the series type, as there is no additional generator. Furthermore, the sizes or the power ratings of both the engine and the drive motor for the parallel type will be smaller or lower than those of the series type.

The disadvantages of the parallel type of hybrid drive in comparison with the series type include:

- Since the engine and the drive motor are mechanically coupled, the engine operating point cannot be maintained in the region of optimal efficiency or low emissions over the full range of vehicle operating speeds.
- The control of the combined operations of the engine and the drive motor to achieve optimal efficiency and low emissions is complex.
- When the motor is used either alone or together with the engine to propel the vehicle, charging the battery cannot be realized. This would limit the vehicle operating range with pure electric drive mode.

In operation, when the vehicle speed and the required driving torque is low, such as driving in the city, the engine can be turned off and the vehicle can be driven by the electric motor alone. In this case, the vehicle operates as a zero-emission vehicle. During acceleration or hill climbing, both the engine and the drive motor together drive the wheel to provide the vehicle with the required power, When cruising on highways, the vehicle can be driven by the engine alone, as the efficiency and fuel consumption of the engine are improved when operating at moderate–high loads than that at low loads. If the state of charge of the battery is close to a predetermined minimum level, the battery can be recharged with the motor operating as a generator driven by the engine. During braking or downhill coasting, the drive motor operates as a generator to recharge the battery. By the stop-start operation of the engine in city driving (i.e., when the vehicle stops, the engine is automatically turned off, and when the brake pedal is off, the engine restarts automatically) and by regenerative braking, the parallel hybrid drive can noticeably improve fuel economy and reduce emissions.

For the parallel type of hybrid drive, the power rating of the engine is selected according to the requirement for vehicle operations in the medium power range. When higher power is required, such as during acceleration or hill climbing, the electric drive is automatically engaged to make up for the difference. As an example, the standard version of a compact car has a gasoline engine of 103 kW (138 hp), whereas its hybrid version has a gasoline engine of 70 kW (94 hp) and a drive motor of 15 kW (20 hp). The parallel type of hybrid drive is used for passenger cars, as well as in delivery vans.

There are various configurations for the combined operations of the internal combustion engine and the electric motor for the parallel type of hybrid electric vehicle (Ehsani et al. 2019). Figure 3.29 shows the schematic of the energy (power) flow of a parallel type of hybrid electric vehicle where the engine and drive motor are connected through a mechanical coupler, usually a gearset. This arrangement is referred to as a two-shaft configuration, as there are two input shafts, one from the engine and the other from the electric motor, to the mechanical coupler. For this configuration, the engine and motor torque can be independently controlled, while the speeds of the engine, the motor, and the vehicle are linked, and their relationship is dependent upon the design of the mechanical coupler and the transmission.

A configuration of the parallel hybrid drive that is more compact than the two-shaft one described above is schematically shown in Figure 3.30. In this configuration, commonly known as the single-shaft configuration, the engine output shaft is directly connected to the rotor of the drive motor. Like the two-shaft configuration, the engine and motor torque can be independently controlled. The speeds of the engine, the motor, and the vehicle are, however, linked.

To address the issue that with the parallel type of hybrid electric vehicle, the battery cannot be charged while the motor participates in propelling the vehicle, a separate generator may be installed. The engine can be used to propel the vehicle and to drive the generator for charging the battery at the same time. The schematic of the energy (power) flow of this parallel type of hybrid vehicle is shown in Figure 3.31. This parallel type of hybrid drive can extend the range of using the electric motor for propelling the vehicle.

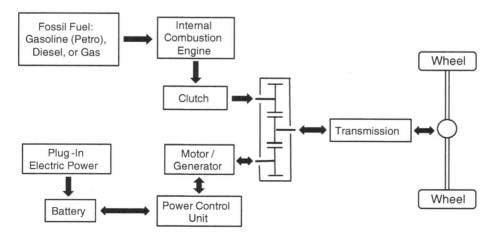

Figure 3.29 Schematic of the energy (power) flow of the two-shaft configuration of the parallel type of hybrid electric drive.

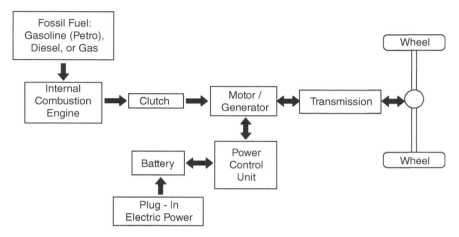

Figure 3.30 Schematic of the energy (power) flow of the single-shaft configuration of the parallel type of hybrid electric drive.

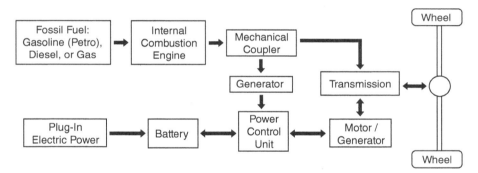

Figure 3.31 Schematic of the energy(power) flow of the parallel type of hybrid electric drive with an additional generator installed.

3) Power-split type

This hybrid electric drive is a combination of the series and parallel type, with a planetary gearset to link the internal combustion engine, the motor, and the generator. Using an appropriate control unit to coordinate their operations, noticeable reduction in emissions, improvement in operational efficiency, and enhancement of driving performance can be achieved. Furthermore, with the planetary gearset, this type of drive performs as an electric continuously variable transmission (E-CVT). It has now become the dominant form of hybrid electric drive.

In comparison with the series or the parallel type, the power-split type has the following features:

- For a series type drive, the engine is used to generate electricity for the motor to propel the vehicle. The engine and the motor work at the same level of power output most of the time. For a parallel type drive, the engine is used as a main power source, with the motor used as an auxiliary power unit during acceleration, hill climbing and the like. This is because the engine power cannot be used for charging the battery when the motor is participating in propelling the vehicle (except the special type of parallel hybrid drive with an additional generator shown in Figure 3.31). This limits the range that the motor can be utilized to propel the vehicle. For the power-split type, as a planetary gearset is used to link the engine, motor, and generator, this enables the engine to use its power partly to drive the vehicle and partly to run the generator for charging the battery. Thus, the motor can be used more often than that of a parallel type.

- The power-split type can maximize the overall efficiency by using the motor to propel the vehicle under conditions in which the efficiency of the engine is low, and by using the engine to run the generator for charging the battery under conditions in which the efficiency of the engine is high.

- As noted previously, the use of the planetary gearset enables the system to perform as an E-CVT, and this eliminates the need for a clutch and a conventional gearbox with a series of gear ratios. Furthermore, over a wide range of vehicle speeds, by regulating the generator speed, the engine speed could be maintained in the operating region where its efficiency is high and fuel consumption and emissions are low.

There are many possible configurations for the power-split type of hybrid electric drive. In the following, three configurations that are used in production vehicles are briefly described. They are the basic configuration, configuration with a speed reduction planetary gearset, and configuration with a speed reduction Ravigneaux gearset.

(A) Basic configuration

The schematic of the energy (power) flow of the basic configuration of power-split type of hybrid electric drive, with a single planetary gear set, is shown in Figure 3.32. It should be noted that the figure is solely for illustrating the operating principles of the drive system and is not to scale. The drive system of an early model of Toyota Prius is of this type.

As shown in the figure, a planetary gear set is used to link the internal combust engine, motor, and generator. The engine is connected to the pinion gear carrier (or carrier) C. Sun gear S is linked to motor/generator 1 (MG1), which functions as a starter for the engine, a generator driven by the engine for generating electricity for charging the battery, or a motor with electricity supplied by the battery to boost the power delivered to the drive axle when needed, such as during acceleration, hill climbing, etc. Ring gear R is connected to motor/generator 2 (MG2), usually referred to as the drive motor, and to the drive axle of the vehicle. During braking or downhill coasting, MG2 driven by the wheels acts as a generator to generate electricity for charging the battery. This is referred to as the regenerative braking discussed previously. The schematic of the electric path connecting MG1, MG2, battery, and power control unit is also shown in Figure 3.32. The power control unit functions in a similar manner as that for electric vehicles discussed in Section 3.4.1. If the battery can be charged by an external electricity source, then it will be a plug-in hybrid electric vehicle, as noted previously. Table 3.6 shows the power ratings of the internal combustion engine, MG1, and MG2 of various power-split types of hybrid electric vehicle. It is shown that in earlier vehicle models the nickel-metal hydride battery was used. As mentioned previously in Section 3.4.1, the nickel-metal hydride battery is considered to have reached its best potential. As a result, in the later vehicle models, the nickel-metal hydride battery is replaced by the lithium-ion battery.

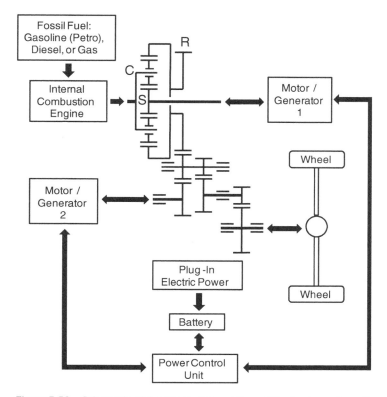

Figure 3.32 Schematic of the energy (power) flow of the basic configuration of the power-split type of hybrid electric drive.

Table 3.6 Power ratings of the engine, MG1, and MG2 of various power-split drives.

Vehicle model	Engine		MG1		MG2		Battery type
	kW	hp	kW	hp	kW	hp	
Toyota Prius (2004)	57	76	33	44	50	67	Nickel-metal hydride
Toyota Lexus RX400h	155	208	109	146	123	165	Nickel-metal hydride
Toyota Prius (2010)	73	98	42	56	60	80.5	Nickel-metal hydride
Toyota Prius Prime Plug-in (2016)	73	98	42	56	60	80.5	Lithium-ion

Source: Oak Ridge National Laboratory (2007) and (2009); manufacturers' published data.

The planetary gearset plays a pivotal role in the power-split hybrid drive and enables the system to function as an E-CVT. Its operating principles, particularly the kinematic relationship of the sun gear, carrier, and ring gear, are briefly reviewed below. To illustrate this relationship, the kinematics of the pinion gear (or pinion) is first analyzed, as shown in Figure 3.33.

As shown in Figure 3.33, when the sun gear is rotating clockwise with angular speed ω_s, the pinion at the contact point with the sun gear has a linear velocity $V_s = \omega_s\, r_s$, where r_s is the pitch radius of the sun gear. When the ring gear is rotating counterclockwise with angular speed ω_r, the pinion at the contact point with the ring gear has a linear velocity $V_r = \omega_r\, r_r$, where r_r is the pitch radius of the ring gear. The instantaneous center of the pinion, which represents the center about which the pinion appears to rotate at that instant, can be located

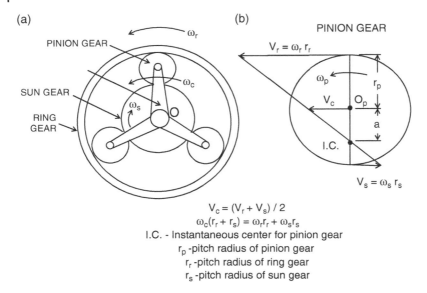

Figure 3.33 Kinematics of the pinion between the ring gear and sun gear in a planetary gearset: (a) schematic of a planetary gearset; (b) kinematic analysis of the pinion gear.

as shown in the figure. With the identification of the location of the instantaneous center for the pinion, the linear velocity V_c of the center O_p of the pinion shown in the figure can be expressed by

$$V_c = (V_r + V_s)/2 \tag{3.36}$$

In the above-noted case, if the linear velocity V_r to the left is denoted as positive, then V_s will be negative in Eq. (3.36).

As V_c is also the linear velocity of the point on the carrier coinciding with the center of the pinion O_p, the angular speed ω_c of the carrier is expressed by

$$\omega_c = \frac{V_c}{r_p + r_s} \tag{3.37}$$

and

$$r_p = \frac{(r_r - r_s)}{2} \tag{3.38}$$

where r_p is the pitch radius of the pinion.

Substituting Eq. (3.38) into Eq. (3.37), one obtains

$$\omega_c = V_c/((r_r + r_s)/2) \tag{3.39}$$

Combining Eqs. (3.39) and (3.36), the relationship of the angular speed of carrier ω_c, the angular speed of the ring gear ω_r, and the angular speed of the sun gear ω_s can be expressed by

$$\omega_c(r_r + r_s) = \omega_r r_r + \omega_s r_s \tag{3.40}$$

Table 3.7 Parameters of various power-split drives and motor speed reduction systems.

	Power-split system		Motor (MG2) speed reduction systems						
	Number of teeth		Planetary system Number of teeth		Ravigneaux gear system Number of teeth				
						Pinion		Sun gear	
Vehicle model	Ring gear R^a	Sun gear S^a	Ring gear $R_1{}^b$	Sun gear $S_1{}^b$	Ring gear $R_2{}^c$	$P_2{}^c$	$P_3{}^c$	$S_2{}^c$	$S_3{}^c$
Toyota Prius (2004)	78	30	–	–	–	–	–	–	–
Toyota Camry (2007)	78	30	57	23	–	–	–	–	–
Toyota Lexus LS 600h (2008)	78	30	–	–	87	28	20	30	27
Toyota Prius (2010)	78	30	58	22	–	–	–	–	–

^{a} For R and S, please refer to Figures 3.32, 3.36 and 3.38.
^{b} For R_1 and S_1, please refer to Figure 3.36.
^{c} For R_2, P_2, P_3, S_2, and S_3, please refer to Figure 3.38.

As the circular pitches of the ring gear and sun gear are the same, in the above equation the pitch radius of ring gear r_r and that of sun gear r_s can be represented by the number of teeth of ring gear n_r and that of sun gear n_s, respectively. As a result, Eq. (3.40) can be rewritten as

$$\omega_c(n_r + n_s) = \omega_r n_r + \omega_s n_s \tag{3.41}$$

Equation (3.41) defines the fundamental kinematic relationship for the planetary gearset.

Table 3.7 shows the number of teeth of the ring gear and that of the sun gear of planetary gearsets for various types of power-split hybrid electric vehicle.

The relationship of the angular speeds of the carrier ω_c, ring gear ω_r, and sun gear ω_s, as described by Eq. (3.41), can be illustrated graphically using the lever diagram (also known as the lever analogy method), as shown in Figure 3.34. In the lever diagram, the three vertical axes represent the angular speeds ω_c, ω_r, and ω_s, respectively. When the horizontal distance between vertical axes for ω_c and ω_r is set to 1 (unity), the distance between the vertical axes for ω_c and ω_s must be set to n_r/n_s, as shown in the figure. If the angular speed of the sun gear ω_s is set to zero and the angular speed of carrier is set to ω_{c1}, then by drawing a straight line from 0 to ω_{c1} and by extending it to the vertical axis for ω_r, the intersection will represent the angular speed of the ring gear ω_{r1} that satisfies Eq. (3.41).

The validity of the lever diagram for illustrating the relationship of ω_c, ω_r, and ω_s can be proved as follows. Considering the triangle in Figure 3.34, which is bounded by the straight line connecting 0, ω_{c1}, and ω_{r1}, the vertical axis for ω_r, and the horizontal line, one can establish the following geometrical relation:

$$\frac{\omega_{r1}}{1 + (n_r/n_s)} = \frac{\omega_{c1}}{n_r/n_s} \tag{3.42}$$

Rearranging Eq. (3.42), one obtains

$$\omega_{c1}(n_r + n_s) = \omega_{r1} n_r \tag{3.43}$$

Equation (3.43) is consistent with Eq. (3.41), when $\omega_s = 0$. This proves that the relationship defined by Eq. (3.41) can be illustrated using the lever diagram shown in Figure 3.34.

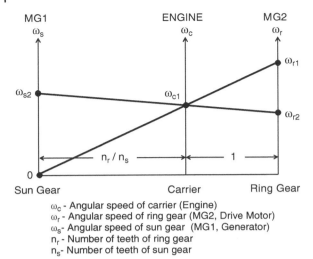

ω_c - Angular speed of carrier (Engine)
ω_r - Angular speed of ring gear (MG2, Drive Motor)
ω_s - Angular speed of sun gear (MG1, Generator)
n_r - Number of teeth of ring gear
n_s - Number of teeth of sun gear

Figure 3.34 Lever diagram illustrating the relationship of the angular speeds of the sun gear, carrier, and ring gear of a planetary gearset.

The lever diagram can be used to demonstrate the special features of the power-split hybrid electric drive for performing like a continuously variable transmission or for maintaining the engine speed at a given operating point over a wide range of vehicle operating speeds. Referring to line $0–\omega_{c1}–\omega_{r1}$ in Figure 3.34, for a given sun gear angular speed ω_s, by regulating the engine speed, which is the carrier angular speed ω_c, the vehicle operating speed, which is related to the ring gear angular speed ω_r, can be continuously varied. This shows that the power-split hybrid electric drive performs as an E-CVT. Referring to line $\omega_{s2}–\omega_{c1}–\omega_{r2}$ and the line $0–\omega_{c1}–\omega_{r1}$ in the figure, it can be seen that by regulating the speed of MG1 (i.e., the angular speed of the sun gear ω_s), the engine speed (i.e., the angular speed of the carrier ω_c) can be maintained at a given operating point over a wide range of vehicle speeds, corresponding to angular speeds of the ring gear ω_r. This means that the power-split hybrid electric drive can maintain the engine operating in an efficient operating region over a wide range of vehicle speeds, by appropriately adjusting the angular speed of the sun gear (i.e., angular speed of MG1).

Various vehicle operations can also be illustrated by the lever diagram, as shown in Figure 3.35. Section (a) in the figure shows the start-up of the engine with MG1 operating as the starter of the engine and the vehicle is stationary with $\omega_r = 0$. Section (b) shows the vehicle in normal driving conditions with the engine and MG2 jointly supplying the power to propel the vehicle, while MG1 is not participating in the operation. Section (c) shows the vehicle in acceleration and the engine and MG2 jointly propel the vehicle, while MG1 driven by the engine generates electricity for supporting the operation of MG2.

(B) Configuration with an additional speed reduction planetary gearset

A configuration of the power-split drive with an additional planetary gearset to provide a speed reduction ratio to the drive axle is schematically shown in Figure 3.36. It shows the energy (power) flow for this configuration. In order not to clutter the figure, the electrical path, like that shown in Figure 3.33, is not included. Ring gear R_1 of the speed reduction planetary gearset, shown on the right of the of the schematic diagram, is connected to ring

Figure 3.35 Various operations of the power-split type of hybrid electric vehicle illustrated using the lever diagram: (a) engine start-up; (b) normal driving; (c) vehicle accelerating. Adapted from Toyota Hybrid System II, with permission of Toyota Motor Corporation.

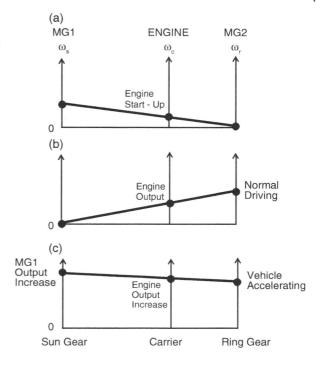

Figure 3.36 Schematic of the energy (power) flow of the power-split type of hybrid electric drive with a speed reduction planetary gearset.

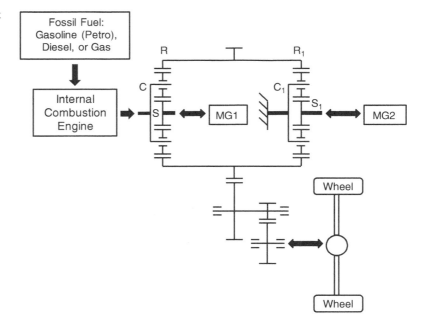

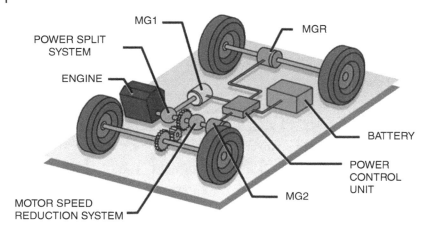

Figure 3.37 Layout of Lexus RX 400h with power-split type of hybrid electric drive having an additional third electric motor for the rear drive axle. Reproduced with permission of Toyota Motor Corporation.

gear R of the power-split unit, shown on the left of the diagram. Ring gear R_1 is linked to the drive axle. Sun gear S_1 of the speed reduction planetary gear set is connected to drive motor MG2. Carrier C_1 of the speed reduction planetary gearset is fixed. This configuration is used in the Toyota Camry (2007) and Toyota Prius (2010) models. As shown in Table 3.7, for the Toyota Camry (2007), ring gear R_1 has 57 teeth and sun gear S_1 has 23 teeth. Using Eq. (3.41), with the carrier C_1 fixed or $\omega_{c1} = 0$, the speed reduction ratio is $\omega_{s1}/\omega_{r1} = 57/23 = 2.48$. For the Toyota Prius (2010), ring gear R_1 has 58 teeth and sun gear S_1 has 22 teeth, so the speed reduction ratio is $\omega_{s1}/\omega_{r1} = 58/22 = 2.64$.

Some hybrid electric vehicles of this configuration, such as the Lexus RX 400h, offer the four-wheel drive feature by adding a third electric motor (MGR) for the rear axle, as shown in Figure 3.37. In this case, the rear axle is purely electrically powered and there is no mechanical link between the engine with the power-split system and the rear driven wheels. The additional planetary gearset to provide the speed reduction ratio for MG2 shown in the figure is linked to the front wheel transaxle only. This arrangement, however, permits regenerative braking on the rear wheels.

(C) Configuration with a Ravigneaux gearset

To provide a compact unit with two speed reduction ratios to the drive axle, a configuration of the power-split drive with a Ravigneaux gearset is used. Figure 3.38 shows the schematic of the energy (power) flow for this configuration. In order not to clutter the figure, the electrical path, like that shown in Figure 3.33, is not included. In the Ravigneaux gearset shown on the right of the schematic diagram Figure 3.38(a), drive motor MG2 is connected to sun gear S_2, which meshes with pinion P_2, as shown in Figure 3.38(a) and (b). In turn, P_2 is in mesh with ring gear R_2. Brake B_1 is connected to R_2. When B_1 is engaged, ring gear R_2 is locked and cannot rotate, whereas when it is released, R_2 can rotate freely. Pinion P_2 meshes with pinion P_3, as shown in Figure 3.38(b). Both P_2 and P_3 are mounted on the same carrier C_2. Pinion P_3 is in mesh with sun gear S_3, which is connected to brake B_2. Like the function of B_1, when B_2 is engaged, sun gear S_3 is locked and cannot rotate, whereas when it is released, S_3 can rotate freely. There are two operating modes for the Ravigneaux gear set: one is B_1 being engaged while B_2 is being released; the other is B_1 being released while B_2 is being engaged. Carrier C_2, on which both P_2 and P_3 are mounted, is linked to the drive axle, as

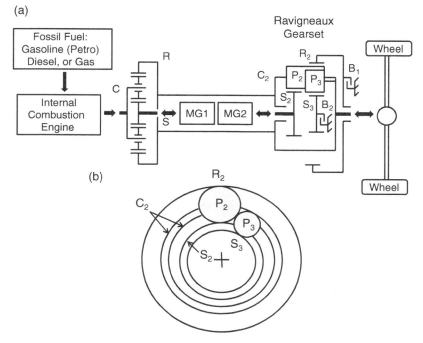

Figure 3.38 The schematic of the energy (power) flow of the power-split type of hybrid electric drive with a Ravigneaux gearset: (a) schematic of a power-split type drive with a Ravigneaux gearset; (b) schematic of a Ravigneaux gearset.

shown in Figure 3.38(a). The power-split drive of Toyota Lexus LS 600h (2008) is of this type. The numbers of teeth of various gears of this configuration are given in Table 3.7.

The operation of the Ravigneaux gearset in the two operating modes are briefly outlined below.

i) For the first operating mode, brake B_1 is engaged and ring gear R_2 is locked, while brake B_2 is released and sun gear S_3 rotates freely. In this mode, sun gear S_2, pinion P_2, and ring gear R_2 constitute a planetary gearset, with the angular speed ω_{r2} of ring gear R_2 equal to zero. Using Eq. (3.41), the following relationship can be established:

$$\omega_{c2}(n_{r2} + n_{s2}) = \omega_{s2}n_{s2} \tag{3.44}$$

Using the Toyota Lexus LS 600h (2008) as an example, with the numbers of teeth of the gears n_{r2} and n_{s2} given in Table 3.7, the speed ratio ω_{s2}/ω_{c2} (i.e., the corresponding gear ratio) is

$$\frac{\omega_{s2}}{\omega_{c2}} = \frac{n_{r2} + n_{s2}}{n_{s2}} = \frac{87 + 30}{30} = 3.9$$

ii) For the second operating mode, brake B_1 is released and ring gear R_2 rotates freely, while brake B_2 is engaged and sun gear S_3 is locked. In this case, sun gears S_2 drives pinion P_2, which in turn drives pinion P_3. As P_3 rotates on the stationary sun gear S_3, it causes carrier C_2 to have planetary motion around the center of sun gear S_3. For this case, the following relationship can be established:

$$\omega_{s2}r_{s2} = \omega_{c2}(r_{p3} + r_{s3}) \tag{3.45}$$

where r_{s2}, r_{p3}, and r_{s3} are the pitch radii of sun gear S_2, pinion P_3, and sun gear S_3, respectively, and ω_{s2} and ω_{c2} are the angular speeds of sun gear S_2 and that of carrier C_2, respectively.

As noted previously, the pitch radii of the gears can be represented by the corresponding numbers of teeth of the gears. Accordingly, Eq. (3.45) can be rewritten as follows:

$$\frac{\omega_{s2}}{\omega_{c2}} = \frac{n_{p3} + n_{s3}}{n_{s2}} \tag{3.46}$$

With the numbers of teeth of gears n_{p3}, n_{s2}, and n_{s3} given in Table 3.7, the speed ratio ω_{s2}/ω_{c2} (i.e., the corresponding gear ratio) is

$$\frac{\omega_{s2}}{\omega_{c2}} = \frac{20 + 27}{30} = 1.57$$

As the power-split type of hybrid drive is more complex than the series or the parallel type, its control system is also more complicated. The development of effective and efficient control strategies for coordinating the operations of various subsystems to achieve improved fuel economy and reduced emissions while maintaining an acceptable level of driving performance is of great importance.

3.5.2 Characteristics of Energy Consumption and Emissions of Hybrid Electric Vehicles

Based on the data from the *2020 Fuel Consumption Guide*, published by the Department of Natural Resources, Government of Canada, Table 3.8 shows a comparison of the fuel consumption and emission ratings of a sample of cars powered by internal combustion engines with that of corresponding hybrid electric versions. It should be noted that the hybrid electric version referred to in the table denotes that the electricity for charging the battery and for driving the motor is solely generated using the power of the internal combustion engine on board the vehicle. The procedures used in obtaining the fuel consumption data presented in the table are outlined in Section 3.9, Operating Fuel Economy of Vehicles with Internal Combustion Engines.

In the table, the following characteristic parameters for various vehicle models are shown:

- Engine size expressed in liters of engine displacement.
- Fuel consumption ratings in city, highway, and combined city and highway driving in liters per 100 kilometers (L/100 km) and in miles per US gallon (mpg (US)). Fuel consumption in mpg (US gallon) = 235.21/(L/100 km).
- Fuel consumption ratio of cars with internal combustion engines in city, highway, and combined city and highway driving to that of corresponding hybrid electric versions: C – fuel consumption ratio for city rating; H – fuel consumption ratio for highway rating; and COM – fuel consumption ratio for combined city (55%) and highway (45%) rating.
- Ratings of vehicle's tailpipe CO_2 and smog emissions on a scale from 1 (worst) to 10 (best).

Table 3.8 shows that in general, the fuel consumptions of cars powered by internal combustion engines in city, highway, and combined city and highway driving are higher than that of corresponding hybrid electric versions. For instance, the fuel consumption ratio of the Ford Fusion powered by the internal combustion engine to that with the hybrid electric drive in city driving is 1.818, in highway driving is 1.228, and in combined city and highway driving is 1.554. This means that the fuel consumptions of the hybrid electric version are 45%, 19%, and 36% less in city, highway, and combined city and highway driving, respectively, than that powered by the internal combustion engine. Emissions of CO_2 and smog-forming pollutants are rated on a scale from 1 (worst) to 10 (best). The rating of 9 for emissions of CO_2 for the hybrid electric version is much better than

Table 3.8 Comparison of fuel consumption and emissions ratings of cars with internal combustion engines versus those of corresponding hybrid electric vehicles.

Vehicle model	Engine size, L	Consumption City/Highway/Combined L/100 km (mpg (U.S.))	Fuel consumption ratio of car with the IC engine to the hybrid	CO_2 rating	Smog rating
Ford Fusion	1.5	10.0/7.0/8.7 (23.5/33.6/27.0)	C = 1.818 H = 1.228 COM = 1.554	6	7
Ford Fusion Hybrid	2.0	5.5/5.7/5.6 (42.8/41.3/42.0)		9	7
Honda Accord	2.0	10.7/7.3/9.2 (22.0/32.2/25.6)	C = 2.14 H = 1.46 COM = 1.84	5	7
Honda Accord Hybrid	2.0	5.0/5.0/5.0 (47.0/47.0/47.0)		10	7
KIA Optima	2.4	9.5/7.1/8.4 (24.8/33.1/28.0)	C = 1.61 H = 1.365 COM = 1.5	6	5
KIA Optima Hybrid	2.0	5.9/5.2/5.6 (39.9/45.2/42.0)		9	7
Lincoln MKZ AWD	2.0	12.1/8.4/10.4 (19.4/28.0/22.6)	C = 2.2 H = 1.474 COM = 1.857	5	5
Lincoln MKZ Hybrid	2.0	5.5/5.7/5.6 (42.8/41.3/42.0)		9	7
Toyota Camry AWD LE/SE	2.5	9.3/6.8/8.2 (25.3/34.6/28.7)	C = 1.898 H = 1.417 COM = 1.673	6	6
Toyota Camry Hybrid LE	2.5	4.9/4.8/4.9 (48.0/49.0/48.0)		10	7
Toyota Camry AWD XLE/XSE	2.5	9.5/7.0/8.4 (24.8/33.6/28.0)	C = 1.792 H = 1.4 COM = 1.647	6	6
Toyota Camry Hybrid XLE/SE	2.5	5.3/5.0/5.1 (44.4/47.0/46.1)		10	7
Toyota Corolla	2.0	8.2/6.5/7.4 (28.7/36.2/31.8)	C = 1.864 H = 1.444 COM = 1.644	7	6
Toyota Corolla Hybrid	1.8	4.4/4.5/4.5 (53.5//52.3/52.3)		10	7

Note: City ratings represent urban driving in stop-and-go traffic. Highway ratings represent a mix of open highway and rural road driving, typical of longer trips. Combined ratings reflect 55% city driving and 45% highway driving.
Source: 2020 Fuel Consumption Guide, Department of Natural Resources, Government of Canada.

that of 6 for the internal combustion engine version. However, for smog emissions, the ratings for both versions are the same. Similar observations may be made for other vehicle models shown in the table.

It should be mentioned that for some vehicle models shown in the table, the engine size, expressed in liters of engine displacement, of the hybrid electric version is different from that of the internal combustion engine version.

Based on the data in the *2020 Fuel Consumption Guide,* Table 3.9 shows the characteristics of various plug-in hybrid electric vehicles. As noted previously, a plug-in hybrid electric vehicle

Table 3.9 Characteristics of various plug-in hybrid electric vehicles.

Vehicle model	Motor kW (hp)	Engine size, L	Electric mode Combined rating kWh/100 km (Equivalent mpg (US))	Hybrid mode City/Highway/Combined L/100 km (mpg (US))	Range Electric mode km (mile)	Range Hybrid mode km (mile)	CO_2 rating	Smog rating	Battery recharge time, h
BMW 530e	83 (111)	2.0	29.1 (71.9)	9.5/7.7/8.7 (24.8/30.5/27.0)	34 (21)	533 (331)	10	7	3
BMW 745Le xDrive	83 (111)	3.0	37.7 (55.5)	12.2/9.1/10.8 (19.3/25.8/21.8)	26 (16)	435 (270)	8	3	3
Chrysler Pacifica Hybrid	89 (119)	3.6	25.8 (81.1)	8.0/7.9/8.0 (29.4/29.8/29.4)	51 (32)	784 (487)	10	7	2
Ford Fusion Energi	68 (91)	2.0	20.5 (102.1)	5.5/5.8/5.6 (42.8/40.6/42.0)	42 (26)	940 (584)	10	7	2.6
Honda Clarity Plug-in Hybrid	135 (181)	1.5	19 (110.2)	5.3/5.9/5.6 (44.4/39.9/42.0)	77 (48)	475 (295)	10	8	2.5
Hyundai IONIQ Electric Plus	45 (60)	1.6	17.7 (118.3)	4.4/4.6/4.5 (53.5/51.1/52.3)	47 (29)	961 (597)	10	7	2.3
KIA Optima Plug-in Hybrid	50 (67)	2.0	20.7 (101.1)	6.0/5.3/5.7 (39.2/44.4/41.3)	45 (28)	962 (598)	10	7	2.7
Mitsubishi Outlander PHEV AWD	60 (80)	2.0	27.7 (75.6)	9.4/9.0/9.2 (25.0/26.1/25.6)	35 (22)	463 (288)	10	7	3.5
Porsche Panamera 4 E-Hybrid	70 (94)	2.9	40.3 (51.9)	11.1/9.7/10.5 (21.2/24.2/22.4)	23 (14)	768 (477)	8	5	3
Subaru Crosstrek Hybrid AWD	100 (134)	2.0	23.5 (89.1)	6.6/6.8/6.7 (35.6/34.6/35.1)	27 (17)	747 (464)	10	6	2
Toyota Prius Prime	71 (95)	1.8	15.8 (132.5)	4.3/4.4/4.3 (54.7/53.5/54.7)	40 (25)	995 (618)	10	7	2
Volvo S60 T8 AWD	65 (87)	2.0	29.0 (72.2)	8.4/7.0/7.8 (28.0/33.6/30.2)	35 (22)	781 (485)	10	7	3
Volvo XC90 T8 AWD	65 (87)	2.0	36.1 (58)	9.1/8.4/8.8 (25.8/28.0/26.7)	29 (18)	813 (505)	10	7	3

Note: For testing during electric mode operation, the vehicles did not use any gasoline.
Source: 2020 Fuel Consumption Guide, published by the Department of Natural Resources, Government of Canada.

denotes that its battery can be recharged by plugging it into an external source of electric power, as well as by its own onboard internal combustion engine.

In the table, the following characteristic parameters for various plug-in hybrid electric vehicle are shown:

- Power rating of the electric motor (drive motor).
- Engine size expressed in liters of engine displacement.
- Energy consumption ratings in pure electric mode for combined city and highway driving in kWh/100 km and in equivalent mpg (US). One liter of gasoline is taken to contain energy equivalent to 8.9 kWh of electricity. Equivalent of liters of gasoline L_e = electric energy consumed in kWh/8.9 kWh. Equivalent mpg (US) = $235.21/(L_e/100\text{ km})$.
- Fuel consumption ratings in hybrid mode for city, highway, and combined city and highway driving in L/100 km and in mpg (US).
- Operating ranges in both electric mode and hybrid mode.
- Ratings of vehicle"s tailpipe CO_2 and smog emissions on a scale from 1 (worst) to 10 (best).
- Battery recharge time to fully recharge of the battery in hours at 240 V.

3.6 Fuel Cells

A fuel cell is an electrochemical device that directly converts chemical energy contained in the fuel (hydrogen) and the oxidant agent (atmospheric oxygen) into electrical energy, in the form of low-voltage direct current electricity. The fuel cell has the following features:

- The electrochemical process of the fuel cell using hydrogen as a fuel produces water only, in contrast with the internal combustion engine using fossil fuels, which produces greenhouse gases and other air pollutants. The process for producing hydrogen, however, does not necessarily produce no emissions. It is dependent upon the type of process and the energy source used. Currently, hydrogen is primarily produced by one of the two processes. One is through steam reforming of natural gas. This process consists of heating the natural gas to between 700 and 1100°C in the presence of steam and a nickel catalyst. The resulting reaction breaks up the methane molecules and produces carbon monoxide (CO) and hydrogen H_2. The byproducts are carbon dioxide (CO_2) and other greenhouse gases. Thus, this process will not meet the general objective of reducing or eliminating greenhouse gases and air pollutants. The other process for producing hydrogen is by electrolysis of water. It uses electricity to split water into hydrogen and oxygen. Water electrolysis can operate between 50 and 80°C. When using renewable energy sources, such as wind, solar or hydro power, to generate electricity, the process for producing hydrogen is without undesirable emissions. The hydrogen so produced is usually referred to as the "clean" (or "green") hydrogen. It meets the environmental objective of zero emissions. However, producing clean hydrogen by electrolysis of water is currently more expensive than by steam reforming of natural gas. Furthermore, to produce clean hydrogen in sufficient quantity to support the wide adoption of fuel cell vehicles would require heavy investment in renewable energy production facilities.
- For passenger and commercial vehicle applications, hydrogen is typically stored on board. To address the issue of lack of infrastructure for distribution of hydrogen, various methods for onboard hydrogen production have been explored. For instance, a process by steam reforming of bio-methane and bio-ethanol in palladium membrane reactors has been proposed (Holgado

and Alique 2019). This process could minimize CO_2 emissions to below the limits proposed by European Union (EU) for the year 2020. It should be noted, however, that adoption of on-board production of hydrogen faces challenges that include: (1) for instance, while the process proposed above could reduce CO_2 emissions, it could not completely eliminate greenhouse gas emissions; (2) carrying an onboard reformer will add weight to the vehicle and increase its capital cost; and (3) the response of the onboard reformer may be sluggish.

- The gasoline engine in a car is less than 20% efficient in converting the chemical energy in gasoline into mechanical energy to propel the vehicle. In comparison, a vehicle powered by the hydrogen fuel cell with an electric motor converts 40–60% of the chemical energy of the fuel into mechanical energy to drive the vehicle. It corresponds to more than twice as efficient as a vehicle powered by the gasoline engine. It should be mentioned that while using hydrogen fuel cells to power vehicles is more efficient than using internal combustion engines, it is more costly to produce the clean hydrogen than gasoline. However, increased demand could reduce the cost of electrolysis of water to produce hydrogen. Coupled with falling renewable energy costs to generate electricity, this would make the clean hydrogen competitive in the future. Furthermore, using fuel cells to power vehicles facilitates the diversification of energy sources away from fossil fuels for the transportation sector.
- It can produce electricity continuously for as long as hydrogen and oxygen are supplied and refueling takes relatively short time. In contrast, the battery in an electric vehicle, when it is exhausted, requires a considerably longer time to recharge.
- Its specific energy is in the range of 150–1500 Wh/kg, as compared with that of approximately 13,000 Wh/kg for fossil fuels (gasoline or diesel fuel). Its specific power is currently up to 2000 W/kg, as shown in Figure 3.25. It should be noted that progresses in fuel cell technologies are continually being made, and the values of specific energy and specific power noted above would change accordingly.

3.6.1 Polymer Electrolyte Membrane Fuel Cells

Many types of fuel cell have been developed, including the polymer electrolyte membrane (PEM, also known as proton exchange membrane) fuel cell, alkaline fuel cell, molten carbonate fuel cell, phosphoric acid fuel cell, solid acid fuel cell, and solid oxide fuel cell. They have different features and are suited for different applications.

Transportation is a prime area of application of the PEM fuel cell. Its noteworthy features include (Wang et al. 2020):

- Compact design with high power density (in watts per liter of fuel cell stack volume or W/L), currently in the range up to 3120 W/L (such as, the fuel cell stack used in the Honda Clarity Fuel Cell car)
- Easy for scaling-up power capacity by packing the required number of fuel cells in a stack to suit a specific need (for instance, the power rating for the fuel cell stack of the Toyota Mirai is 114 kW)
- Low operating temperatures in the range of 50–100°C
- Quick start-up

Other promising areas of application of PEM fuel cells include stationary and portable power generation, auxiliary power units for commercial aircraft, unmanned aerial vehicles, spacecraft, and mobile devices, such as laptop computers, cell phones, and military communication equipment.

Figure 3.39 Basic components in a polymer electrolyte membrane (PEM) fuel cell.

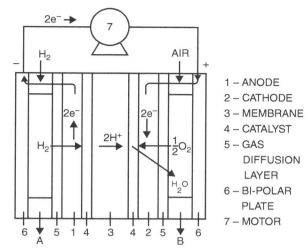

1 – ANODE
2 – CATHODE
3 – MEMBRANE
4 – CATALYST
5 – GAS DIFFUSION LAYER
6 – BI-POLAR PLATE
7 – MOTOR

A schematic diagram of the structure of the PEM fuel cell is shown in Figure 3.39. It consists of the following basic components (Thring 2004):

- *Electrodes* There are a fuel electrode (anode) and an oxidant electrode (cathode). They are made of porous, electrically and thermally conducting material. The anode and cathode are connected with an external electric circuit through bipolar plates, as shown in Figure 3.39.
- *Electrolyte* The PEM fuel cell uses a proton exchange membrane as the electrolyte. The requirements for the membrane are high proton conductivity, high chemical and thermal stability, good durability, and low cost. The typical material for the membrane is called Nafion, which is a brand name for a sulfonated tetrafluoroethylene-based fluoropolymer-copolymer. Its thickness ranges from 0.01 to 0.1 mm and its density is approximately 2 g/cm^3 (Wang et al. 2020).
- *Catalyst Layers* There is one catalyst layer on either side of the membrane. It is on the catalyst surface where the electrochemical reactions take place. The typical material of the catalyst layer is carbon-supported catalyst and ionomer porous composite. The thickness of the catalyst layer is typically in the range from 100 nm to 0.05 mm and its density is approximately 0.4 g/cm^3. The anode, cathode, membrane, and two catalyst layers are usually sealed together to form a single membrane electrode assembly (MEA).
- *Gas Diffusion Layers* The arrangements consist of one gas diffusion layer in contact with the anode and one with the cathode. Their functions are to ensure that the hydrogen gas or air diffuses evenly and efficiently over the catalyst layer; to provide uniform distribution of load on the surface of the MEA from the bipolar plate; and to provide electrical conduction between the bipolar plate and the anode or cathode. It is made of carbon fiber-based porous paper. Its thickness is between 0.1 and 0.4 mm and its density in the range from 0.3 to 0.5 g/cm^3, estimated based on graphite carbon density and component porosity.
- *Bipolar Plates* These are commonly made of carbon-based composites or metals. Their thicknesses are in the range from 0.3 to 2 mm and their density is between 1.7 and 8 g/cm^3, estimated based on graphite and stainless steel. Gas flow channels are machined into the plates to provide inlets for the hydrogen gas and air, as well as outlets for excess hydrogen gas for recirculation (A in Figure 3.39) and for water produced in the fuel cell (B in Figure 3.39).

In operation, hydrogen gas is fed to the fuel cell. It diffuses through the gas diffusion layer and the anode and meets the catalyst surface. On the catalyst surface, electrochemical reactions take place. The hydrogen is split into electrons and hydrogen protons. The flow of electrons from the anode to the cathode, through the bipolar plates, generates electric current, which is used to drive an external load, such as an electric motor shown in Figure 3.39. The hydrogen protons pass through the proton exchange membrane and interact with the electrons and oxygen taken from the ambient air in an electrochemical process at the catalyst surface on the cathode side. As a result of electrochemical reactions, water and heat are produced. Consequently, water and thermal management is one of the major issues that require careful consideration in the design of the fuel cell. The precise control of the flow and pressure of hydrogen gas and air, temperature, and membrane humidity is critical to the efficient operation of the fuel cell (Pukrushpan et al. 2005).

The basic fuel cell chemical reactions are:

$$\text{At the anode :} \quad H_2 \rightarrow 2H^+ + 2e^-$$
$$\text{At the cathode :} \quad \tfrac{1}{2}O_2 + 2H^+ + 2e^- \rightarrow H_2O$$
$$\text{Overall reaction :} \quad H_2 + \tfrac{1}{2}O_2 \rightarrow H_2O$$

As the electrochemical process in the fuel cell with hydrogen as a fuel produces only water and no pollutant emissions, it is a zero-emission power generator.

The hydrogen for PEM fuel cells is typically stored on board the vehicle in a 700-bar (approximately 70,000 kPa or 10,150 psi) high-pressure tank of carbon-fiber casing with high-density polyethylene lining. The hydrogen is expanded to approximately 10 bar (1000 kPa or 145 psi) via a pressure-reducer and fed to the anode through a hydrogen gas injector. The flow rate of hydrogen is dependent on the electrical power of the whole fuel cell stack and may be estimated as follows (Larminie and Dicks 2003):

$$m_{fc} = 1.05 \times 10^{-8} \left(P_{fc}/V_{fc} \right) \tag{3.47}$$

where m_{fc} is hydrogen mass flow rate, kg/s; P_{fc} is the electrical power of the whole fuel cell stack, W (watts); V_{fc} is the mean voltage of fuel cells in the stack, V (volts).

The theoretical electric potential (voltage) generated by the hydrogen/oxygen fuel cell is 1.23 V, at 25°C and atmospheric pressure (Sammes 2006). In operation, because of internal losses in the cell, the available voltage is always lower than the theoretical one. The electric current generated is dependent on the active area of the catalyst layer, with a current density of approximately 1 A/cm^2. With active areas of 200–600 cm^2 (or even higher) for some of the fuel cells currently in use, electric currents of several hundred amperes can be produced. In operation, the voltage decreases with the increase of current. Typical electrical characteristics of a fuel cell are shown in Figure 3.40. It shows that the electric potential in V decreases with the increase of current in mA. The electric power in mW, which is product of V and mA, first increases with the increase of the current and reaches a peak in the mid-range. Beyond that the power decreases with the increase of current. To maintain optimal power output, most fuel cells operate in the range between 0.6 and 0.7 V.

The efficiency of the fuel cell, defined as a ratio of the electricity produced and the hydrogen consumed, is directly proportional to its electric potential. The operating efficiency η_{fc} of the fuel cell stack may be estimated by (Larminie and Dicks 2003):

$$\eta_{fc} = \frac{V_{fc}}{1.48} \tag{3.48}$$

where V_{fc} is the mean voltage of fuel cells in the stack.

Figure 3.40 Notional characteristics of a fuel cell showing the relationships of voltage, current, and power.

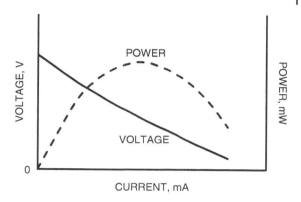

Fuel cells may be connected either in series or in parallel to form a stack to suit specific operational requirements. To achieve high voltages, individual cells are connected in series to form a stack. A stack can be made up of approximately 40–450 cells to achieve maximum voltages between 40 and 450 V. Stacks in the power range up to 120 kW are typically used for vehicle applications. High currents are achieved by appropriately sized PEM surface area. Currents up to 500 A are found in vehicle applications.

3.6.2 Characteristics of Fuel Cell Vehicles

Vehicles powered by fuel cells usually have batteries as additional (or reserved) energy storage devices. Table 3.10 shows the basic characteristics of some production fuel cell vehicles. In the table, the following characteristic parameters are presented:

- Vehicle type: mid-size car, small sport utility vehicle (SUV) with two-wheel drive (2WD), and subcompact car.
- Motor power output in kW and in hp.
- Fuel cell power output in kW and in hp.
- Fuel consumption ratings for city/highway/combined driving in equivalent miles per gallon (US), based on the conversion of energy contained in hydrogen to that in gasoline; conversion of equivalent miles per gallon (US) to liters per 100 km: $L_e/100\,km = 235.21/mpg$ (US).
- Power density of the hydrogen fuel cell stack in kW/L; specific power in kW/kg of fuel cell stack.
- Driving range for one full tank of hydrogen on board the vehicle in km and in miles.
- Battery type: Li-ion (lithium-ion); Ni-MH (nickel-metal hydride).

As of 2020, more than 4000 commercial vehicles and 30 buses powered by fuel cells (FCBs) have been deployed globally (Wang et al. 2020). Three FCBs have been rated at the technically ready stage with full-scale validation in real-world operating environments. A few passenger vehicles powered by fuel cells are on the market, as shown in Table 3.10. For instance, the Toyota Mirai has been on the market since 2017. It has a fuel cell stack with power rating of 114 kW (153 hp), power density of 3.1 kW/L and specific power of 2.0 kW/kg, and has a 122.4 L, at 700 bar, hydrogen tank with an operating range up to 483 km (300 miles).

To achieve wide adoption of fuel cell vehicles by consumers, transit authorities, fleet operators, etc., there are noticeable technical and economic challenges. These include fuel cell durability, adequate and safe onboard hydrogen storage, as well as the all-important cost factor. The U.S. Department of Energy durability tests showed that the Toyota Mirai passed the 3000 hours real-world

Table 3.10 Basic characteristics of some production fuel cell vehicles.

Model	Vehicle type	Motor power, kW (hp)	Fuel cell power, kW (hp)	Fuel consumption rating City/Highway/Combined, Equivalent mpg (US) (Equivalent L_e/100 km)	Fuel cell power density, kW/L	Fuel cell specific power, kW/kg	Range, km (miles)	Battery type
Honda Clarity Fuel Cell 2020	Midsize car	130 (174)	103 (138)	68/67/68 (3.5/3.5/3.5)	3.12	1.98	579 (360)	Li-ion
Hyundai Nexo Blue Fuel Cell 2020	Small SUV (2WD)	120 (161)	95 (127)	65/58/61 (3.6/4.1/3.9)	3.1	–	611 (380)	Li-ion
Hyundai Nexo Limited Fuel Cell 2020	Small SUV (2WD)	120 (161)	95 (127)	59/54/57 (4.0/4.4/4.1)	3.1	–	570 (354)	Li-ion
Toyota Mirai Fuel Cell 2020	Sub-compact	113 (153)	114 (153)	67/67/67 (3.5/3.5/3.5)	3.1	2.0	Up to 483 (300)	Ni-MH

Source: Manufacturers' published data.

testing. The target is, however, over 5000 operating hours by 2025 and ultimately 8000 hours for cars and 25,000 hours for buses (Wang et al. 2020). In addition, developing the infrastructure for production and distribution of hydrogen is another challenge to the successful commercialization of fuel cell vehicles.

With the pressing need to alleviate climate change, intense research and development efforts are being made by the vehicle and related industries. Together with the development of appropriate government policies, challenges facing the wide adaption of fuel cell vehicles would likely be met. Fuel cell vehicles could become competitive and attractive to end users in the future.

3.7 Transmissions for Vehicles with Internal Combustion Engines

To provide a road vehicle with tractive effort–speed characteristics that will satisfy driving demands under various operating conditions, a transmission with appropriate speed reduction ratios between the power plant and the driven wheels is required. As the internal combustion engine is still widely used, transmissions commonly employed in road vehicles with the internal combustion engine as the sole power provider will be discussed. Two major categories of transmission are currently used in road vehicles are: the manual gear transmission and the automatic transmission with a torque converter. Continuously variable transmissions (CVTs) are also in use in some road vehicles, while hydrostatic transmissions are employed in specialized off-road vehicles.

As noted previously, electric motors used in electric vehicles have more favorable torque/power–speed characteristics for meeting driving demands. Consequently, between the electric motor and the driven wheels, usually a one- or two-speed reduction gearset will suffice.

3.7.1 Manual Gear Transmissions

The principal requirements for the transmission between the internal combustion engine and the drive axle are:

- To achieve the desired maximum vehicle speed with an appropriate engine.
- To be able to start, fully loaded, in both forward and reverse directions on a steep gradient, typically 33% (1 in 3), and to be able to maintain a speed of 88–96 km/h (55–60 mph) on a gentle slope, such as 3%, in high gear for passenger cars.
- To properly match the characteristics of an engine with a vehicle to achieve the desired operating fuel economy and acceleration characteristics.

The manual gear transmission usually consists of a clutch, a gearbox, a propeller shaft, and a drive axle with a differential (to allow relative rotation between the left- and right-hand side driven wheels during turning maneuvers). In front-engine and front-wheel drive vehicles or in rear-engine and rear-wheel drive vehicles, the gearbox and differential are usually integrated into a unit, commonly referred to as a transaxle. As a rule, the drive axle has a constant gear reduction ratio, which is determined by the usual practice requiring direct drive (nonreducing drive) in the gearbox in the highest gear, if there is no overdrive gear. For vehicles requiring extremely high torque at low speeds, an additional reduction gear (final drive) may be placed at the driven wheels.

In the past, the number of gears in a gearbox (transmission) for passenger vehicles is usually in the range of 3–5. Figure 3.41 shows the tractive effort–speed relationship, in broken lines, of a vehicle with a three-speed gearbox. The tractive effort–speed relationship of a vehicle with one fixed gear ratio is shown by the chain line in the figure. With only one fixed gear ratio, the tractive effort–speed relationship cannot meet the driving demand. For instance, at low speeds it is not capable of providing sufficiently high tractive effort to accelerate the vehicle. It is noted that with a three-speed gearbox there are relatively large non-utilizable driving ranges which are shown by shaded areas in the figure. They are the areas between the desirable tractive effort–speed curve, represented by the solid line, and the actual tractive effort–speed curves represented by broken lines for the three gears: I, II, and III. The desirable tractive effort–speed relationship is a hyperbola, which represents that when the driving power (the product of the tractive effort and speed) is a constant over the speed range shown. The continuously variable

Figure 3.41 Tractive effort–speed characteristics of a passenger car with a three-speed gearbox. *Source:* Taborek (1957). Reproduced with permission of *Machine Design.*

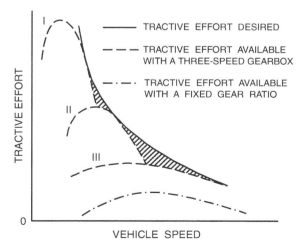

TRACTIVE EFFORT DESIRED

TRACTIVE EFFORT AVAILABLE WITH A THREE-SPEED GEARBOX

TRACTIVE EFFORT AVAILABLE WITH A FIXED GEAR RATIO

TRACTIVE EFFORT

VEHICLE SPEED

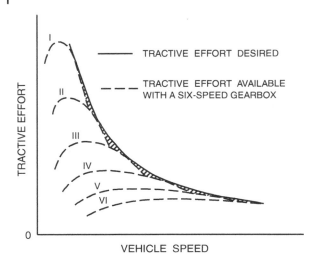

TRACTIVE EFFORT

—— TRACTIVE EFFORT DESIRED

— — — TRACTIVE EFFORT AVAILABLE
WITH A SIX-SPEED GEARBOX

VEHICLE SPEED

Figure 3.42 Tractive effort–speed characteristics of a passenger car with a six-speed gearbox.

transmission can provide a tractive effort–speed relationship approaching that of a hyperbola. The non-utilizable driving ranges can be minimized using more gears. Figure 3.42 shows the tractive effort–speed relationships of a vehicle with a six-speed gearbox. It indicates that the non-utilizable driving ranges are noticeably reduced.

Table 3.11 shows that, for passenger vehicles, the number of gears in the gearboxes for various types of transmission ranges from 6 to 10. In the table, the types of transmission shown include: A – automatic transmissions; AM – automatic manual transmissions; AS – automatic transmission with select shift; AV – continuously variable transmissions; and M – manual transmissions. The number following the transmission designation indicates the number of gears/speeds of the transmission (such as 6–10). In the column K_a of the table, K_a is the average value of the ratios of two consecutive gear ratios. For instance, for a n-speed gearbox, $K_a = (\xi_2/\xi_1 + \xi_3/\xi_2 + \ldots \xi_n/\xi_{n-1})/(n-1)$, where ξ_1, ξ_2, ξ_3,... ξ_n are the gear ratio of the 1st, 2nd, 3rd,... and nth gear, respectively. In the last column of the table, K_g is the geometric progression (or geometric sequence) common ratio. For a n-speed gear box, $K_g = {}^{n-1}\sqrt{(\xi_n/\xi_1)}$. K_a and K_g will be further discussed later in this section.

For commercial vehicles, their engine power to vehicle weight ratios are generally much lower than that of passenger vehicles. Minimizing the non-utilizable driving ranges in the tractive effort–speed relationships has a noticeable impact on improving their driving performance. The number of gears in gearboxes for commercial vehicles, particularly heavy commercial vehicles, is usually in the range of 8 to 18, higher than that for passenger vehicles, as shown in Table. 3.12.

The gear ratio of the highest gear (i.e., the smallest gear reduction ratio) is chosen so that the desired maximum vehicle speed can be achieved with an appropriate engine. The engine should have sufficient power to overcome the internal resistance in the transmission, rolling resistance of the tires, and aerodynamic resistance at the maximum vehicle speed on a level road. The common practice is to select a gear ratio such that at the maximum vehicle speed, the engine speed is slightly higher than that at the maximum engine power, as indicated in Figure 3.43. This ensures sufficient power reserve to maintain a given vehicle speed against a temporary increase in headwind or gradient during operation, or against the possible deterioration in engine performance after

Table 3.11 Gear ratios of transmissions for passenger vehicles.

Vehicle	Transmission type	1st	2nd	3rd	4th	5th	6th	7th	8th	9th	10th	Final drive ratio	K_a	K_g
Acura ILX	AM8	3.08	2.18	1.61	1.22	0.96	0.74	0.62	0.48			3.94	0.768	0.767
Acura TLX	AM8	4.71	2.84	1.91	1.38	1.00	0.81	0.70	0.58			3.52	0.747	0.741
Audi A4	AM7	3.19	2.19	1.52	1.06	0.74	0.51	0.39				4.23	0.705	0.705
Audi A6 Quattro	AM7	3.69	2.15	1.41	1.03	0.79	0.63	0.52				4.09	0.726	0.721
BMW 330i	AS8	5.25	3.36	2.17	1.72	1.32	1.00	0.82	0.64			2.81	0.743	0.740
BMW 540i	AS8	5.00	3.20	2.14	1.72	1.31	1.00	0.82	0.64			2.93	0.748	0.746
Buick Regal Sportback	A9	4.69	3.31	3.01	2.45	1.92	1.45	1.00	0.75	0.62		2.89	0.779	0.777
Cadillac CTS	AS8	4.60	2.72	1.86	1.46	1.23	1.00	0.82	0.69			2.85	0.768	0.763
Chevrolet Malibu	A9	4.69	3.31	3.01	2.45	1.92	1.45	1.00	0.75	0.62		2.89	0.779	0.777
Chrysler 300	A8	4.71	3.14	2.11	1.67	1.28	1.00	0.84	0.67			2.62	0.759	0.757
Ford Fusion	AS6	4.58	2.96	1.91	1.45	1.00	0.75					3.36	0.698	0.696
Ford Taurus	AS6	4.48	2.87	1.84	1.41	1.00	0.74					2.77	0.699	0.698
Honda Accord LX	AV	Continuously variable transmission, ratio range: 2.65–0.41										3.24		
Honda Accord-Sport	M6	3.64	2.08	1.36	1.02	0.83	0.69					4.11	0.724	0.717
Honda Accord-Sport	AS10	5.25	3.27	2.19	1.60	1.30	1.00	0.78	0.65	0.58	0.52	3.55	0.779	0.773
Honda Civic	M6	3.64	2.08	1.36	1.02	0.83	0.69					4.11	0.724	0.717
Hyundai Elanta	AS6	4.40	2.73	1.83	1.39	1.00	0.77					3.06	0.708	0.706
Hyundai Sonata	AS6	4.21	2.64	1.80	1.39	1.00	0.77					2.89	0.714	0.712
Mercedes-Benz C 300	A9	5.35	3.24	2.25	1.64	1.21	1.00	0.86	0.72	0.60		3.07	0.765	0.761
Mercedes-Benz S 560	A9	5.35	3.24	2.25	1.64	1.21	1.00	0.86	0.72	0.60		2.65	0.765	0.761
Nissan Sentra	M6	3.73	2.10	1.52	1.17	0.91	0.77					3.93	0.736	0.729
Toyota Camry	AS8	5.52	3.18	2.05	1.49	1.23	1.00	0.80	0.67			2.56	0.760	0.740
Toyota Corolla	AV	Continuously variable transmission, ratio range: 2.48–0.40										4.76		
Volkswagen Golf	M6	4.11	2.12	1.36	0.97	0.77	0.63					3.39	0.691	0.687
Volkswagen Jetta	A8	5.07	2.97	1.95	1.47	1.23	1.00	0.81	0.67			3.23	0.755	0.757
Volvo S90 T6 AWD	AS8	5.25	3.03	1.95	1.46	1.22	1.00	0.81	0.67			3.08	0.752	0.745

Source: Manufacturers' published data.

Table 3.12 Gear ratios of gearboxes for commercial vehicles.

Gear	ZF-Ecomid 9 S 1110 TD	ZF-Ecosplit 12 S 2130 TD	ZF-Ecosplit 16 S 1820 TO	Eaton Fuller RTO – 15 681
		Transmission type		
1st	8.83	15.57	13.80	14.71
2nd	6.28	12.21	11.54	12.45
3rd	4.64	9.47	9.49	10.20
4th	3.48	7.43	7.93	8.62
5th	2.54	5.82	6.53	7.34
6th	1.81	4.56	5.46	6.21
7th	1.34	3.41	4.57	5,26
8th	1.00	2.68	3.83	4.45
9th		2.07	3.02	3.78
10th		1.63	2.53	3.20
11th		1.27	2.08	2.70
12th		1.00	1.74	2.28
13th			1.43	1.94
14th			1.20	1.64
15th			1.00	1.39
16th			0.84	1.18
17th				1.00
18th				0.85
K_g	0.733	0.779	0.830	0.846

Source: Manufacturers' published data.

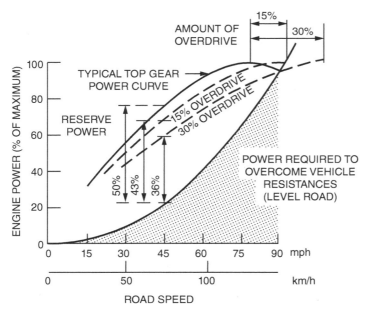

Figure 3.43 Effects of gear ratio of the highest gear on vehicle performance. *Source:* Giles (1969). Reproduced with permission of Butterworth.

extended use. Based on this principle, the gear ratio of the highest gear in the gearbox may be determined as follows:

$$\xi_n = \frac{n_{e1}r(1-i)}{V_{\max}\xi_{ax}} \tag{3.49}$$

where ξ_n is the gear ratio of the highest gear in a gearbox with n-speeds; n_{e1} is the engine speed corresponding to the maximum vehicle speed, which for passenger cars is usually about 10% higher than the speed at the maximum engine power; r is the rolling radius of the tire; i is the tire slip; $V_{\max}$ is the desired maximum vehicle speed; and ξ_{ax} is the gear ratio in the drive axle.

If, in the highest gear, the gearbox is in direct drive (i.e., $\xi_n = 1$), then Eq. (3.49) can be used to determine the gear ratio ξ_{ax} in the drive axle.

The gear ratio of the lowest gear (i.e., the largest gear reduction ratio) is selected on the basis that the vehicle should be able to climb a steep gradient, usually 33% for passenger cars. There is also a suggestion that the gear ratio of the lowest gear in the gearbox should be such that the vehicle can climb the maximum gradient possible without tire spin on a typical road surface. If this approach is followed, then the gear ratio of the lowest gear in the gearbox can be determined following the procedures given below.

For a rear-wheel drive vehicle, from Eq. (3.9), the maximum slope $\theta_{s\max}$ that the vehicle can climb, as determined by the maximum tractive effort that the tire–ground contact can support, is expressed by

$$W \sin\theta_{s\max} = \frac{\mu W(l_1 - f_r h)/L}{1 - \mu h/L} - f_r W \tag{3.50}$$

In the above equation, the aerodynamic resistance is neglected because on a steep slope, the vehicle speed is usually low.

From Eq. (3.50), the gear ratio of the lowest gear ξ_1 in the gearbox is given by

$$\xi_1 = \frac{W(\sin\theta_s\max + f_r)r}{M_{e\max}\xi_{ax}\eta_t} \tag{3.51}$$

where $M_{e\max}$ is the maximum engine torque and η_t is the efficiency of the transmission.

For a front-wheel drive vehicle, a similar expression for the gear ratio of the lowest gear can be derived from Eqs. (3.10) and (3.51).

The method for selecting gear ratios for the intermediate gears between the highest and the lowest is, to a great extent, dependent on the type of vehicle (commercial vehicles or passenger cars). For commercial vehicles, the gear ratios are usually arranged in a geometric progression (geometric sequence). The basis for this is to keep the engine operating within the same speed range in each gear, as shown in Figure 3.44. This would ensure that in each gear, the operating fuel economy is similar. For instance, for a four-speed gearbox, the following relationship can be established (see Figure 3.44):

$$\frac{\xi_2}{\xi_1} = \frac{\xi_3}{\xi_2} = \frac{\xi_4}{\xi_3} = \frac{n_{e2}}{n_{e1}} = K_g$$

and

$$K_g = \sqrt[3]{\frac{\xi_4}{\xi_1}} \tag{3.52}$$

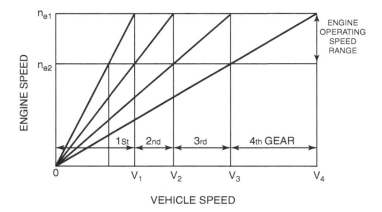

Figure 3.44 Selection of gear ratios based on geometric progression (geometric sequence).

where ξ_1, ξ_2, ξ_3, and ξ_4 are the gear ratios of the first, second, third, and fourth gear, respectively. $\xi_2/\xi_1, \xi_3/\xi_2$, and ξ_4/ξ_3 represent the ratios of two consecutive gear ratios. For a general case, if the ratio of the highest gear ξ_n and that of the lowest gear ξ_1 have been determined, and the number of speeds in the gearbox n_g is known, the common ratio K_g can be determined by

$$K_g = \sqrt[n_g - 1]{\frac{\xi_n}{\xi_1}} \qquad (3.53)$$

and

$$\xi_n = K_g \xi_{n-1} \qquad (3.54)$$

Based on the gear ratios of the gearboxes for commercial vehicles presented in Table 3.12, it can be shown that they are essentially arranged in a geometric progression. It should be mentioned that because the number of teeth of a gear is an integer, it is not possible, in some cases, to arrange gear ratios in an exact geometric progression.

For passenger cars, the gear ratios are not usually arranged in a geometric progression. The ratios of intermediate gears between the highest and the lowest may be chosen to minimize the time required to reach the maximum speed of the vehicle, and to take into consideration that shifting between upper gears (i.e., gears in the range of low gear ratios) happens more frequently than between lower gears (i.e., gears in the range of high gear ratios), particularly in city driving. The ratios of two consecutive gear ratios in the upper gears, such as ξ_n/ξ_{n-1}, are usually higher than those of the lower gears, such as ξ_2/ξ_1. For instance, for the gearbox A9 used in both the Mercedes-Benz C 300 and S 560 presented in Table 3.11, the ratio of $\xi_9/\xi_8 = 0.833$ is higher than $\xi_2/\xi_1 = 0.606$.

The gear ratios of the nine gears of the gearbox A9 for Mercedes-Benz C 300 and S 560 are plotted in Figure 3.45. If the gear ratios are arranged in a geometric progression, the value of the common ratio $K_g = \sqrt[8]{0.6/0.535} = 0.761$, where 0.6 is the gear ratio of the 9th gear and 5.35 is the gear ratio of the 1st gear. The gear ratios for the nine gears from the 1st to the 9th arranged in geometric progression are also plotted in Figure 3.45. It is shown that there are differences between the actual gear ratios for the gearbox A9 used in the Mercedes-Benz C 300 and S 560 and those determined by the geometric progression.

Figure 3.45 Comparisons of the actual gear ratios of a nine-speed gearbox with that determined by geometric progression.

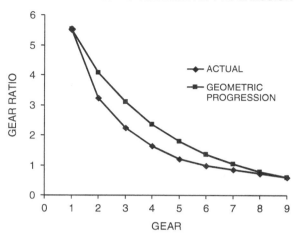

ACTUAL GEAR RATIOS OF A CAR VS. THAT BASED ON GEOMETRIC PROGRESSION

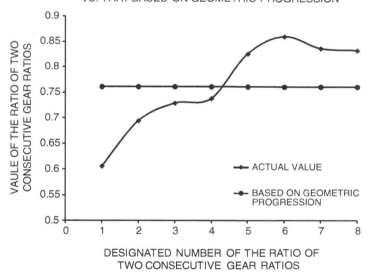

THE RATIO OF TWO CONSECUTIVE GEAR RATIOS OF A CAR VS. THAT BASED ON GEOMETRIC PROGRESSION

Figure 3.46 Variations of the ratio of two consecutive gear ratios (gear step) with the designated number in a sequence of ratios of two consecutive gear ratios (i.e., 1 for ξ_2/ξ_1, 2 for ξ_3/ξ_2, 3 for ξ_4/ξ_3, 4 for ξ_5/ξ_4, 5 for ξ_6/ξ_5, 6 for ξ_7/ξ_6, 7 for ξ_8/ξ_7, and 8 for ξ_9/ξ_8) for the transmission A9 of the Mercedes-Benz C 300 and S 560.

To further illustrate this issue, the values of the ratio of two consecutive gear ratios (gear step, i.e., ξ_2/ξ_1, ξ_3/ξ_2, ξ_4/ξ_3, ... ξ_9/ξ_8) are plotted in Figure 3.46 against the designated number in a sequence of ratios of two consecutive gear ratios (i.e., 1 for ξ_2/ξ_1, 2 for ξ_3/ξ_2, 3 for ξ_4/ξ_3, ...8 for ξ_9/ξ_8). If the gear ratios are arranged in geometric progression, then the ratio of two consecutive gear ratios (i.e., ξ_n/ξ_{n-1}) is equal to the common ratio $K_g = 0.761$ and is represented by a horizontal line

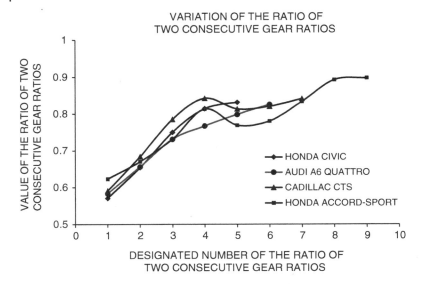

Figure 3.47 Variations of the ratio of two consecutive gear ratios (gear step) with the designated number in a sequence of ratios of two consecutive gear ratios (i.e., 1 for ξ_2/ξ_1, 2 for ξ_3/ξ_2, 3 for ξ_4/ξ_3 ... and n − 1 for ξ_n/ξ_{n-1}) for transmissions of the Audi 6 Quattro, Cadillac CTS, Honda Accord-Sport, and Honda Civic.

in Figure 3.46. It can be seen that the ratio of two consecutive gear ratios gradually increases from 0.606 for ξ_2/ξ_1, which is lower than the common ratio K_g of 0.762, to 0.833 for ξ_9/ξ_8, which is higher than the common ratio K_g. It is interesting to note that the average value $K_a = (\xi_2/\xi_1 + \xi_3/\xi_2 + \xi_4/\xi_3 + \xi_5/\xi_4 + \xi_6/\xi_5 + \xi_7/\xi_6 + \xi_8/\xi_7 + \xi_9/\xi_8)/8 = 0.765$ is very close to the value of $K_g = 0.761$. As shown in Table 3.11, the values of K_a of all gearboxes are very close to their respective values of K_g.

Figure 3.47 shows the variation of the ratio of two consecutive gear ratios (gear step) with the designated number in a sequence of ratios of two consecutive gear ratios (i.e., 1 for ξ_2/ξ_1, 2 for ξ_3/ξ_2, 3 for ξ_4/ξ_3... and n-1 for ξ_n/ξ_{n-1}) for transmissions of the Audi 6 Quattro (AM7), Cadillac CTS (AS8), Honda Accord-Sport (AS10), and Honda Civil (M6). It shows that the ratio of two consecutive gear ratios varies with the designated number in a sequence of ratios ξ_2/ξ_1 to ξ_n/ξ_{n-1}.

In the transmission, there are losses due to friction between gear teeth and in bearings and seals, as well as due to oil churning. The mechanical efficiency of the transmission is a function of load (torque) and speed. Figure 3.48 shows the variations of the mechanical efficiency with the input speed for a three-speed automatic gearbox. The transmission is connected to an engine operating at wide open throttle and developing a maximum torque of 407 N · m (300 lb · ft) (Setz 1961). In vehicle performance predictions, as a first approximation, the following average values for the mechanical efficiency of the major subsystems in the transmission may be used:

Gearbox – direct drive	98%
Gearbox – indirect drive	95%
Drive axle	95%

For a vehicle with a manual gear transmission, the tractive effort of the vehicle is given by

$$F = \frac{M_e \xi_o \eta_t}{r} \tag{3.55}$$

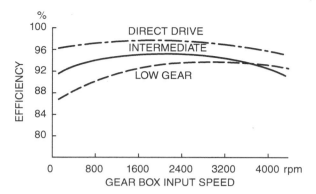

Figure 3.48 Mechanical efficiency of a three-speed automatic gearbox at wide open throttle. Reprinted with permission from SAE paper 610032 © 1961 Society of Automotive Engineers, Inc.

where M_e is the engine output torque, ξ_o is the overall reduction ratio of the transmission (including both the gearbox and drive axle gear ratios), η_t is the overall transmission efficiency, and r is the rolling radius of the tire. The maximum tractive effort that the tire–ground contact can support usually determines the traction capability of the vehicle in low gears.

The relationship between vehicle speed and engine speed is given by

$$V = \frac{n_e r}{\xi_o}(1 - i) \tag{3.56}$$

where n_e is the engine speed and i is the slip of the vehicle running gear. For a road vehicle, the slip is usually assumed to be 2–5% under normal operating conditions. Figure 3.49 shows the variation of the tractive effort with speed for a passenger car equipped with a three-speed manual gear transmission.

3.7.2 Automatic Transmissions

The automatic transmission is widely used in passenger vehicles in North America and elsewhere. It usually comprises a torque converter and an automatic gearbox for which gear shift is performed by a control unit based on vehicle operating conditions and not by the driver.

1) Torque converter characteristics

The torque converter consists of at least three rotary elements known as the pump (impeller), the turbine, and the reactor, as shown in Figure 3.50. The pump is connected to the engine shaft, and the turbine is connected to the output shaft of the converter, which in turn is coupled with the input shaft of a multi-speed automatic gearbox. The reactor is coupled to an external casing to provide a reaction on the fluid circulating in the converter. The function of the reactor is to enable the turbine to develop an output torque higher than the input torque of the converter, thus, to obtain a torque multiplication. The reactor is usually mounted on a free wheel (one-way clutch) so that when the starting period has been completed and the turbine speed is approaching that of the pump, the reactor is in free rotation. At this point, the converter operates as a fluid coupling, with a ratio of output torque to input torque equal to 1.0.

The major advantages of incorporating a torque converter into the transmission may be summarized as follows:

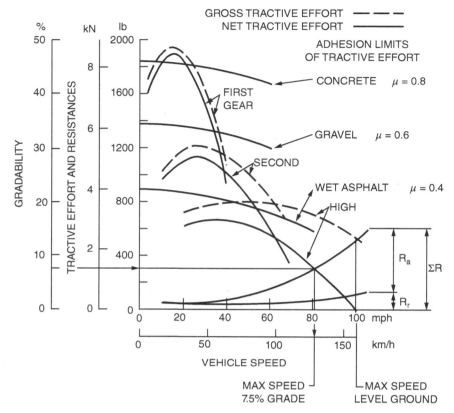

Figure 3.49 Performance characteristics of a passenger car with a three-speed manual transmission. *Source:* Taborek (1957). Reproduced with permission of *Machine Design.*

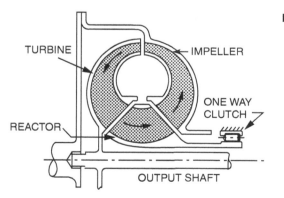

Figure 3.50 Schematic view of a torque converter.

- When properly matched, it will not stall the engine.
- It provides a flexible coupling between the engine and the driven wheels.
- Together with a suitably selected multi-speed automatic gearbox, it provides torque–speed characteristics that approach the ideal.

The performance characteristics of a torque converter are usually described in terms of the following four parameters:

$$\text{Speed ratio } C_{sr} = \text{output speed/input speed}$$

$$\text{Torque ratio } C_{tr} = \text{output torque/input torque}$$

$$\text{Efficiency } \eta_c = \text{output speed} \times \text{output torque/input speed} \times \text{input torque}$$

$$= C_{sr}C_{tr}$$

$$\text{Capacity factor(size factor) } K_{tc} = \text{speed/}\sqrt{\text{torque}}$$

The capacity factor is an indication of the ability of the converter to absorb or to transmit torque.

Notional performance characteristics of the torque converter are shown in Figure 3.51, in which the torque ratio, efficiency, and input capacity factor, which is the ratio of the input speed to the square root of the input torque, are plotted against the speed ratio (Setz 1961). The torque ratio of the converter reaches a maximum at stall condition where the speed ratio is zero. The torque ratio decreases as the speed ratio increases, and the converter eventually acts as a hydraulic coupling with a torque ratio of 1.0. At this point, a small difference between the input and output speed remains because of the slip between the pump (impeller) and the turbine. The efficiency of the converter is zero at stall condition and increases with an increase of the speed ratio. It reaches a maximum when the converter acts as a fluid coupling. The input capacity factor is an important parameter defining the operating conditions of the torque converter and governing the matching between the converter and the engine. The input capacity factor of the converter has a minimum value at stall condition and increases with an increase of the speed ratio.

Since the converter is driven by the engine, to determine the actual operating conditions of the converter, the engine operating point must be specified. To characterize the engine operating

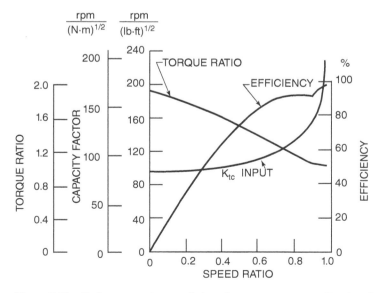

Figure 3.51 Performance characteristics of a torque converter. Reprinted with permission from SAE paper 610032 © 1961 Society of Automotive Engineers, Inc.

conditions for purposes of determining the combined performance of the engine and the converter, an engine capacity factor K_e is introduced and is defined as

$$K_e = \frac{n_e}{\sqrt{M_e}}$$

where n_e and M_e are the engine speed and torque, respectively. The variation of the capacity factor with speed for a particular engine is shown in Figure 3.52 (Setz 1961). To achieve proper matching, the engine and the converter should have a similar range of capacity factor.

As mentioned above, the engine shaft is usually connected to the input shaft of the converter; therefore,

$$K_e = K_{tc}$$

The matching procedure begins with specifying the engine speed and engine torque. Knowing the engine operating point, one can determine the engine capacity factor K_e (Figure 3.52). Since $K_e = K_{tc}$, the input capacity factor of the converter corresponding to the specific engine operating point is then known. For a particular value of the input capacity factor of the converter K_{tc}, the converter speed ratio and torque ratio can be determined from the converter performance curves, as shown in Figure 3.51. The output torque and output speed of the converter are then given by

$$M_{tc} = M_e C_{tr} \tag{3.57}$$

and

$$n_{tc} = n_e C_{sr} \tag{3.58}$$

where M_{tc} and n_{tc} are the output torque and output speed of the converter, respectively.

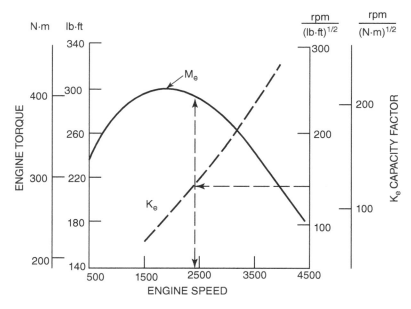

Figure 3.52 Capacity factor of an internal combustion engine. Reprinted with permission from SAE paper 610032 © 1961 Society of Automotive Engineers, Inc.

With the reduction ratios of the gearbox and the drive axle known, the tractive effort and speed of the vehicle can be calculated:

$$F = \frac{M_{tc}\xi_o\eta_t}{r} = \frac{M_e C_{tr}\xi_o\eta_t}{r} \tag{3.59}$$

and

$$V = \frac{n_{tc}r}{\xi_o}(1-i) = \frac{n_e C_{sr}r}{\xi_o}(1-i) \tag{3.60}$$

The efficiency of a torque converter is low over a considerable range of speed ratios, as shown in Figure 3.51. To improve the operating efficiency of the automatic transmission and hence fuel economy, a lockup clutch is incorporated in the torque converter. It is programmed to engage in a predetermined vehicle speed range. When the lockup clutch is engaged, the engine power is directly transmitted to the output shaft of the torque converter, hence power losses in the converter due to the difference in angular speed between the impeller and turbine are eliminated.

Figure 3.53 shows the variation of the tractive effort with speed for a passenger car equipped with a torque converter and a three-speed automatic gearbox (Setz 1961). In comparison with a car with a three-speed manual transmission, shown previously in Figure 3.41, the tractive effort–speed relationship of a car with a three-speed automatic transmission shown in Figure 3.53 is noticeably different, because of the characteristics of the torque converter.

Example 3.1 An engine with torque–speed characteristics shown in Figure 3.52 is coupled with a torque converter with characteristics shown in Figure 3.51. Determine the output speed and output torque of the torque converter when the engine is operating at 2450 rpm with an engine output torque of 393 N · m (290 lb · ft).

Solution

The engine capacity factor K_e

$$K_e = \frac{n_e}{\sqrt{M_e}} = \frac{2450}{\sqrt{393}} = 123 \text{ rpm}/(\text{N} \cdot \text{m})^{1/2}$$

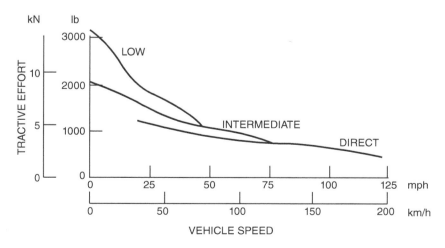

Figure 3.53 Tractive effort–speed characteristics of a passenger car with a three-speed automatic transmission. Reprinted with permission from SAE paper 610032 © 1961 Society of Automotive Engineers, Inc.

Since the capacity factor K_e is equal to that of the torque converter K_{tc},

$$K_{tc} = 123 \, \text{rpm}/(\text{N·m})^{1/2}$$

From Figure 3.51, when $K_{tc} = 123$, the speed ratio $C_{sr} = 0.87$ and the torque ratio $C_{tr} = 1.05$. The output speed of the torque converter n_{tc} is

$$n_{tc} = 0.87 \times 2450 = 2132 \, \text{rpm}$$

The output torque of the torque converter M_{tc} is

$$M_{tc} = 393 \times 1.05 = 413 \, \text{N·m} \, (304 \, \text{lb·ft})$$

The efficiency of the torque converter under this operating condition is

$$\eta_c = 0.87 \times 1.05 = 91.4\%$$

2) Automatic gearbox operations

An automatic gearbox usually comprises a few planetary gearsets and serval shifting components (elements) to facilitate the automatic gear shifting. Brakes and clutches are usually used as shifting components. These brakes and clutches are connected to components of various planetary gearsets, such as sun gears, pinion carriers or ring gears. By activating the appropriate brakes and clutches, gear shifting is accomplished.

Figure 3.54 shows the schematic of an automatic transmission with a torque converter and an eight-speed automatic gearbox (ZF Getriebe GmbH 2020). For this automatic transmission, there is a lockup clutch in the torque converter. As noted previously, it is programmed to be activated in a predetermined vehicle operating range. When the lockup clutch is engaged, the engine power is directly transmitted to the output shaft of the torque converter, hence eliminating power losses in the converter, and improving the operating efficiency of the automatic transmission. The automatic gearbox shown in the figure provides eight forward gear ratios and one reverse gear ratio. It consists of four planetary gearsets (as GS1, GS2, GS3, and GS4, shown in the figure), and five gear shifting components, including two brakes A and B and three clutches C, D, and E. By engaging (closing) three of the five shifting components, while disengaging (opening) the other two, gear shifting to one of the eight forward gears (from the first gear to the eighth gear) or to the reverse gear can be accomplished. By engaging (closing) the brake means that components of the planetary gearsets connected with the brake will be locked up or stationary. By disengaging (opening) the brake means that components of the planetary gearsets connected with the brake will be free and not subject to any constraint. By engaging (closing) the clutch means that torque can be transmitted through it. By disengaging (opening) the clutch means that no torque will be transmitted through it.

For example, as shown in Figure 3.54(a), by engaging brakes A and B and clutch C, and by disengaging clutches D and E, the automatic gear box will be shifted to the first gear. By engaging brakes A and B, sun gear S_1, ring gear R_1, and carrier C_1 of planetary gearset GS1 is locked up. As carrier C_1 is connected with gear R_4 of planetary gearset GS4, R_4 is stationary and not rotating. By engaging clutch C, torque will be transmitted from the output shaft of the torque converter to sun gear S_4 of GS4, while clutches D and E are disengaged. Under these circumstances, for planetary gearset GS4, sun gear S_4 drives carrier C_4, while ring gear R_4 is stationary or its angular speed ω_{R4} is zero. Based on the analysis of the kinematics of planetary gearset discussed in Section 3.5.1, the gear ratio for the first gear can be obtained using Eq. (3.41):

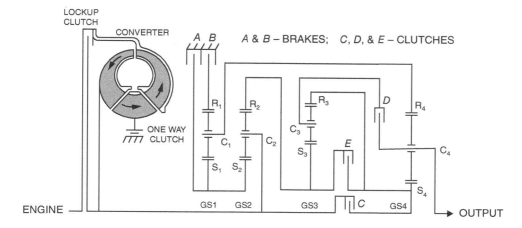

(a) Gear Shifting of the Automatic Transmission

Gear	Brake		Clutch			Gear ratio
	A	B	C	D	E	
1st	•	•	•			4.70
2nd	•	•			•	3.13
3rd		•	•		•	2.10
4th		•		•	•	1.67
5th		•	•	•		1.29
6th			•	•	•	1.00
7th	•		•	•		0.84
8th	•			•	•	0.67
R	•	•		•		−3.30

Note:• Shifting component engaged (closed).

(b) Number of Teeth of Gearset Components

Planetary Gearset	GS1		GS2		GS3		GS4	
Gear	R_1	S_1	R_2	S_2	R_3	S_3	R_4	S_4
Number of teeth	96	48	96	48	111	69	85	23

Figure 3.54 Schematic of a ZF eight-speed automatic transmission: (a) gear shifting scheme; (b) number of teeth of gearset components. *Source:* Adapted from product information released by ZF Getriebe GmbH.

$$\omega_{C4}(n_{R4} + n_{S4}) = \omega_{S4}n_{S4} \tag{3.61}$$

or the gear ratio of the first gear ξ_1 is expressed by

$$\xi_1 = \frac{\omega_{S4}}{\omega_{C4}} = (n_{R4} + n_{S4})/n_{S4} \tag{3.62}$$

where ω_{C4} is the angular speed of carrier C_4 or the angular speed of the output shaft of the automatic gearbox; ω_{S4} is the angular speed of sun gear S_4, or the angular speed of output shaft of the torque converter ω_t, and if the converter lockup clutch is engaged, ω_{S4} will be the angular speed of the engine crankshaft ω_e; n_{R4} and n_{S4} are the number of teeth of ring gear R_4 and that of sun gear S_4 of gearset GS4, respectively.

With the number of teeth of components of planetary gearsets shown in Figure 3.54(b), for gearset GS4, $n_{R4} = 85$ and $n_{S4} = 23$, from Eq. (3.62) the gear ratio of the first gear $\xi_1 = (85 + 23)/23 = 4.70$, as shown in Figure 3.54(a).

To illustrate the feature of the automatic gearbox in providing direct drive or shifting into the sixth gear with a gear ratio of 1, shown in Figure 3.54(a), let us consider the operations of the shifting components. As indicated in Figure 3.54(a), to shift into the sixth gear, clutches C, D, and E are engaged (closed), while brakes A and B are disengaged (opened). Under these circumstances, sun gear S_4, sun gear S_3, carrier C_3, and ring gear R_3 are all rotating with the angular speed of the output shaft of the torque converter ω_t. If the lockup clutch of the converter is engaged, they are rotating with the angular speed of the engine crankshaft ω_e. With brakes A and B being disengaged (opened), sun gear S_1, ring gear R_1, and carrier C_1 are all free (not subject to any constraint). As ring gear R_4 is connected with carrier C_1, ring gear R_4 is also free. Since Carrier C_3 is connected with carrier C_4 through clutch D, carrier C_4 and sun gear S_4 are rotating with the same speed as that of the torque converter output shaft ω_t. If the converter lockup clutch is engaged, sun gear S_4 and carrier C_4 are rotating with the engine crankshaft angular speed ω_e. This indicates that with clutches C, D, and E engaged (closed), and with brakes A and B disengaged (opened), the angular speeds of the input and output shaft of the automatic gearbox, ω_t (or ω_e) and ω_{C4}, are the same, and therefore the gear ratio of the sixth gear ξ_6 is 1, as shown in Figure 3.54(a).

For automatic transmissions currently in production, gear shifting is electronically controlled, based on a number of inputs from sensors for monitoring vehicle and engine operating conditions, such as inputs from the vehicle speed sensor, throttle position sensor, and torque converter turbine sensor, as well as inputs from the engine control system, traction control system, cruise control module, etc. The aim of the control system is to optimize gear shifting processes for achieving improved vehicle fuel economy and performance.

As shown in Table 3.11, for manual transmissions of passenger cars, the number of gears (gear ratios) is usually limited to six, taking into account human factors of an ordinary driver in handling gear shifting, whereas the number of gears for automatic transmissions can be much higher, as gear shifting is not performed by the driver. For instance, as shown in Table 3.11, for the automatic transmission AS10 of the Honda Accord-Sport, the number of gears is 10. As mentioned previously, with the increase in the number of gears, vehicle non-utilizable driving range is reduced, and the tractive effort–speed relationship approaches to that of a continuously variable transmission. This enables the engine to operate in the optimal efficiency range, leading to improved vehicle fuel economy and driving performance. Vehicle operating fuel economy and a comparison of fuel consumption of vehicles with manual transmissions and that with automatic transmissions will be discussed in Section 3.9.

3.7.3 Continuously Variable Transmissions

With growing interest in improving fuel economy of automotive vehicles, continuously variable transmissions have attracted a great deal of interest. This type of transmission provides a continuously variable reduction ratio that enables the engine to operate under the optimal fuel economy conditions over a wide range of vehicle speeds. The operating fuel economy of automotive vehicles will be discussed later in this chapter.

Two representative continuously variable transmissions of the mechanical type are the Van Doorne belt system and the Perbury system. The Van Doorne system has a pair of conically faced pulleys, as shown in Figure 3.55. The effective radius of the pulleys, and hence the reduction ratio, can be varied by adjusting the distance between the two sides of the pulleys. On the original system, the reduction ratio was controlled by mechanical means through centrifugal weights on the driving pulley and an engine vacuum actuator. Later, a microprocessor-based control system was

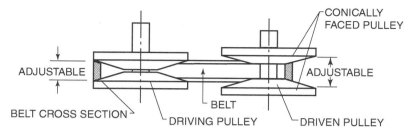

Figure 3.55 A Van Doorne type of continuously variable transmission.

Figure 3.56 Variations of mechanical efficiency with input torque at a constant reduction ratio for a Van Doorne type of continuously variable transmission. Reprinted with permission from SAE paper 850569 © 1985 Society of Automotive Engineers, Inc.

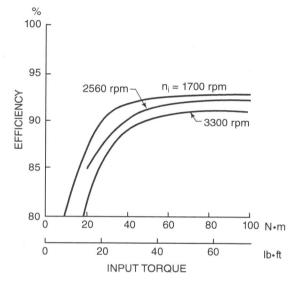

developed (Bonthron 1985; Hahne 1984). This type of continuously variable transmission can achieve a reduction ratio ranging from 4 to 6. The mechanical efficiency of this transmission varies with the load and speed. The variations of the efficiency with input torque and speed at a reduction ratio of 1 for a system designed for a lightweight passenger car are shown in Figure 3.56 (Bonthron 1985). To improve the efficiency of the system and to reduce noise and wear, a "segmented steel belt" or a "push belt system" has been developed (Newton et al. 1983). It comprises a set of belt elements about 2 mm (0.078 in.) thick, with slots on each side to fit two high-tensile steel bands, which hold them together. Unlike the conventional V-belt, it transmits power by the compressive force between the belt elements, instead of tension. The Van Doorne system is most suited to low-power applications and has been used in small-size passenger cars and snowmobiles.

The Perbury system is shown schematically in Figure 3.57 (Greenwood 1984). The key component of this system is the variator, which consists of three disks, with the outer pair connected to the input shaft and the inner one connected to the output shaft. The inner surfaces of the disks are of a toroidal shape, upon which the spherical rollers roll. The rollers can rotate about their own axes. By

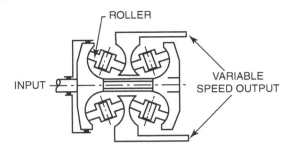

ROLLER

INPUT

VARIABLE
SPEED OUTPUT

Figure 3.57 A Perbury type of continuously variable transmission. *Source:* Greenwood (1984). Reproduced with permission of the Council of the Institution of Mechanical Engineers.

varying the inclination of the roller axes, a continuously variable reduction ratio can be achieved. The carriers for the rollers are fixed. To minimize wear, lubrication is provided between the rollers and the disks. To transmit adequate torque, high normal forces across the contact points are required, as the coefficient of traction between the roller and the disk surfaces is low, typically less than 0.1. The relative slip between the two is 1–2% under normal operating conditions. This system can provide a reduction ratio of about 5. Common with the Van Doorne belt system, it cannot provide zero output and a reverse ratio. Consequently, separate devices are required for starting and for reverse. The Perbury system is suited to higher power applications than the Van Doorne belt system and has been designed for buses, delivery trucks, etc. with a rated power up to 375 kW (502 hp). The average value of the mechanical efficiency of this system is approximately 90%.

In addition to continuously variable transmissions of the mechanical type, there are electric-continuously variable transmissions (E-CVT) used in hybrid electric vehicles, as noted in Section 3.5.

3.7.4 Hydrostatic Transmissions

Hydrostatic transmissions are primarily used in off-road vehicles and have enjoyed a degree of success in those of a specialized nature (Price and Beasley 1964). Hydrostatic drives may be divided into three categories.

1) Constant displacement pump with fixed displacement motor

This type of hydrostatic drive usually consists of a gear or vane pump driving a gear, vane, or piston motor through control valves. The maximum working pressure of the fluid in this type of system is usually about 20,685 kPa (3000 psi). This simple hydrostatic transmission has been quite widely used in construction machinery, such as excavators, to drive the tracks. Each gear pump drives its own hydraulic motor, which allows the two tracks to be operated individually, thus providing a mechanism for steering. The tractive effort–speed characteristics of this system with a multispeed gearbox are illustrated in Figure 3.58(a).

2) Variable displacement pump with fixed displacement motor

This type of hydrostatic drive has certain advantages over the fixed displacement pump and motor system. Variable displacement pumps are piston pumps that permit higher pressure to be used. They also permit stepless speed control from zero to maximum. Closed-loop fluid circuits can be employed to provide both forward and reverse motions and braking functions. To extend the tractive effort and vehicle speed range, a gearbox is frequently employed. The performance characteristics of this type of system coupled with a two-speed gearbox are shown in Figure 3.58(b).

Figure 3.58 Tractive effort–speed characteristics of a vehicle equipped with various types of hydrostatic transmission. *Source:* Wardill (1974). Reproduced with permission of the Council of the Institution of Mechanical Engineers.

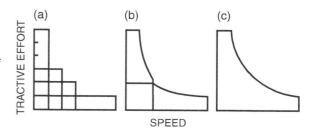

3) Variable displacement pump and motor

In this system, the displacement of both the pump and the motor can be varied continuously. The performance characteristics of the vehicle are approaching the ideal ones, as shown in Figure 3.58(c). Although the performance and control of this type of transmission are beyond question, the problems of cost, reliability, maintenance, and service remain to be addressed.

In comparison with the automatic transmission with a torque converter, the hydrostatic drive can provide a more positive speed control and flexibility in vehicle layout. Figure 3.59 shows the efficiencies of a hydrostatic drive and a comparable automatic transmission (a torque converter with a two-speed gearbox) (Price and Beasley 1964). It appears that there is relatively little difference between the two types of transmission from the point of view of efficiency over the operating range. However, for vehicles designed for traction, operating with high tractive efforts at low speeds, the hydrostatic transmission seems to be more suitable. Figure 3.60 shows the fuel consumption and power output of a vehicle equipped with these two types of transmission (Price and Beasley 1964). It is apparent that the hydrostatic transmission permits full power to be developed by the engine once the output speed of the transmission is high enough. Thus, a faster rate of work can be achieved with the hydrostatic drive. Tests of off-road vehicles equipped with hydrostatic transmissions have shown improved productivity as compared with those equipped with manual gear transmissions, even though manual gear transmissions have higher efficiency (Price and Beasley 1964).

Figure 3.59 Variations of transmission efficiency with output speed of an automatic transmission with a torque converter and of a hydrostatic transmission. *Source:* Price and Beasley (1964). Reproduced with permission of the Council of the Institution of Mechanical Engineers.

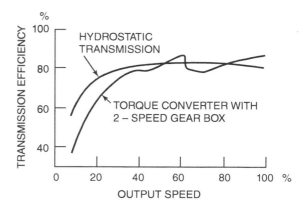

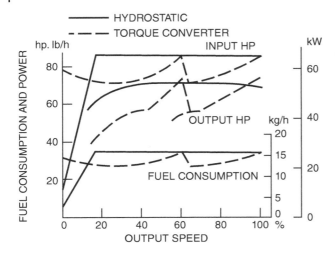

Figure 3.60 Variations of fuel consumption and output power with output speed of an automatic transmission with a torque converter and of a hydrostatic transmission. *Source:* Price and Beasley (1964). Reproduced with permission of the Council of the Institution of Mechanical Engineers.

3.8 Prediction of Vehicle Performance

The passenger car with an internal combustion engine and having tractive effort–speed character-istics shown in Figure 3.49 will be used as an example to illustrate the procedures for predicting acceleration characteristics and gradeability. A similar approach can be followed to predicting the performance of an electric vehicle, with given torque/power–speed characteristics of the electric motor and the gear ratios of the one- or two-speed transmission.

To fully describe the performance of a vehicle, in addition to the relationship between the trac-tive effort and vehicle speed, the resistance of the vehicle as a function of speed must also be deter-mined. On level ground without a drawbar load, the major resisting forces include the rolling resistance R_r and the aerodynamic resistance R_a, and they can be predicted using methods dis-cussed previously. The variation of R_r and R_a with speed for the passenger car is shown in Figure 3.49. The difference between the tractive effort (thrust) and the resultant resisting force is the net thrust (net tractive effort) F_{net} available for accelerating the vehicle or for overcoming grade resistance. The intersection of the vehicle tractive effort and the resultant resistance curve determines the maximum speed that the vehicle can achieve, as shown in Figure 3.49. Note that the nature of tire–road interaction imposes a fundamental limit on the maximum tractive effort. The maximum tractive effort of the passenger car that the tire–ground contact can support on various surfaces including concrete, gravel, and wet asphalt are shown in the figure. They are determined using the method discussed in Section 3.1. For instance, with the second gear engaged, the maximum tractive effort as determined by the engine torque and transmission char-acteristics is about 5.5 kN (1240 lb), whereas the maximum tractive effort on wet asphalt that the tire–road contact can support is only 4 kN (900 lb). This indicates that the maximum tractive effort on wet asphalt that the car can develop with the second gear engaged is 4 kN. Figure 3.49 also shows that with the second gear engaged, when the vehicle speed is below 112 km/h (70 mph), the tractive effort of the vehicle on wet asphalt is limited by the tire–road adhesion, and not by the engine torque.

3.8.1 Acceleration Time and Distance

Having determined the net thrust of the vehicle as a function of speed, one can then compute the acceleration of the vehicle using Newton's second law. However, the translational motion of the vehicle is coupled to the rotational motion of the components connected with the wheels, including the engine and the drivetrain. Any change of translational speed of the vehicle will, therefore, be accompanied by a corresponding change of the rotational speed of the components coupled with the wheels. To consider the effects of the inertia of the rotating parts on vehicle acceleration characteristics, a mass factor γ_m is introduced into the following equation for calculating vehicle acceleration a:

$$F - \sum R = F_{\text{net}} = \gamma_m m a \tag{3.63}$$

where m is the vehicle mass.

γ_m can be determined from the moments of inertia of the rotating parts by

$$\gamma_m = 1 + \frac{\sum I_w}{mr^2} + \frac{\sum I_1 \xi_1^2}{mr^2} + \frac{\sum I_2 \xi_2^2}{mr^2} + \cdots + \frac{\sum I_n \xi_n^2}{mr^2} \tag{3.64}$$

where I_w is the mass moment of inertia of the wheel, $I_1, I_2, ..., I_n$ are the mass moments of inertia of the rotating components connected with the drivetrain having gear ratios $\xi_1, \xi_2, ..., \xi_n$, respectively, with reference to the driven wheel, and r is the rolling radius of the tire. For passenger cars, the mass factor γ_m may be calculated using the following empirical relation (Taborek 1957):

$$\gamma_m = 1.04 + 0.0025 \xi_o^2 \tag{3.65}$$

The first term on the right-hand side of the above equation represents the contribution of the rotating inertia of the tires, while the second term represents the contribution of the inertia of the components rotating at the equivalent engine speed with the overall gear reduction ratio ξ_o with respect to the driven wheel.

In the evaluation of vehicle acceleration characteristics, time–speed and time–distance relationships are of prime interest. These relationships can be derived using the equation of motion of the vehicle in a differential form:

$$\gamma_m m \frac{dV}{dt} = F - \sum R = F_{\text{net}}$$

and

$$dt = \frac{\gamma_m m dV}{F_{\text{net}}} \tag{3.66}$$

As can be seen from Figure 3.49, the net tractive effort F_{net} available for accelerating the vehicle is a function of vehicle speed:

$$F_{\text{net}} = f(V) \tag{3.67}$$

This makes the expression relating the time and speed of the following form not integrable by analytic methods:

$$t = \gamma_m m \int_{V_1}^{V_2} \frac{dV}{f(V)} \tag{3.68}$$

To predict the time required to accelerate the vehicle from speed V_1 to V_2, the integration is best handled by numerical methods.

The distance S that the vehicle travels during an acceleration period from speed V_1 to V_2 can be calculated by integrating the following equation:

$$S = \int_{v_1}^{v_2} \frac{V\, dV}{F_{net}/\gamma_m m} = \gamma_m m \int_{V_1}^{V_2} \frac{V\, dV}{f(V)} \qquad (3.69)$$

When a vehicle having a manual gear transmission starts from rest, in the initial period, slip occurs between the driving and driven parts of the clutch, and the vehicle speed is not directly related to the engine speed. There are mathematical models available for analyzing the dynamics of the clutch engagement process. In a first approximation, however, it can be assumed that during the clutch engagement period, the maximum engine torque is transmitted to the input shaft of the gearbox. The acceleration time and distance from zero vehicle speed to the next speed increment can then be calculated using the procedures outlined above.

In the evaluation of acceleration time and distance, the engine is usually assumed to be operating at wide open throttle. It should be noted that a certain amount of time is required for gear shift during acceleration. For manual transmissions, gear shift causes a time delay of 1–2 s; for automatic transmissions, the delay is typically 0.5–1 s. To obtain a more accurate estimate of acceleration time and distance, this delay should be taken into consideration.

Figure 3.61 shows the acceleration time–distance curve and the acceleration time–speed curve for a passenger car with a gross weight of 17.79 kN (4000 lb) and with the thrust–speed characteristics as shown in Figure 3.49 (Taborek 1957). The kinks in the time–speed curve represent the delays caused by gear shift.

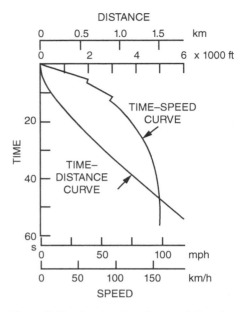

Figure 3.61 Acceleration characteristics of a passenger car with a three-speed manual transmission. *Source:* Taborek (1957). Reproduced with permission of *Machine Design.*

Example 3.2 A vehicle weighs 21.24 kN (4775 lb), including the four road wheels. Each of the wheels has a rolling radius of 33 cm (13 in.) and a radius of gyration of 25.4 cm (10 in.), and weighs 244.6 N (55 lb). The engine develops a torque of 325 N · m (240 lb · ft) at 3500 rpm. The equivalent mass of moment of inertia of the parts rotating at engine speed is 0.733 kg · m²(0.54 slug · ft²). The transmission efficiency is 85% and the total reduction ratio of the drivetrain in the third gear is 4.28:1. The vehicle has a frontal area of 1.86 m² (20 ft²), and the aerodynamic drag coefficient is 0.38. The coefficient of rolling resistance is 0.02. Determine the acceleration of the vehicle on a level road under these conditions.

Solution

(a) The mass factor γ_m for the vehicle in the third gear can be calculated using Eq. (3.64):

$$\gamma_m = 1 + \frac{\sum I_w + \sum I\xi^2}{mr^2}$$

$$= 1 + \frac{4 \times 1.61 + 0.733 \times 4.28^2}{2165 \times 0.33^2} = 1.084$$

(b) The thrust of the vehicle F is determined using Eq. (3.55):

$$F = \frac{M_e \xi_0 \eta_t}{r} = 3583 \text{ N } (806 \text{ lb})$$

(c) The vehicle speed V can be calculated using Eq. (3.56):

$$V = \frac{n_e r}{\xi_0} (1 - i)$$

Assume that $i = 3\%$; the vehicle speed V is

$$V = 98.7 \text{ km/h } (61.3 \text{ mph})$$

(d) The total resistance of the vehicle is the sum of the aerodynamic resistance R_a and the rolling resistance R_r:

$$\sum R = R_a + R_r = 752 \text{ N } (169 \text{ lb})$$

(e) The acceleration a of the vehicle can be determined using Eq. (3.63):

$$a = \frac{F - \sum R}{\gamma_m m} = 1.2 \text{ m/s}^2 \left(3.94 \text{ ft/s}^2\right)$$

3.8.2 Gradeability

Gradeability is usually defined as the maximum grade a vehicle can negotiate at a given steady speed. This parameter is primarily intended for the evaluation of the performance of heavy commercial vehicles and off-road vehicles. On a slope at a constant speed, the tractive effort must overcome grade resistance, rolling resistance, and aerodynamic resistance:

$$F = W \sin \theta_s + R_r + R_a$$

For a relatively small angle of θ_s, $\tan\theta_s \cong \sin \theta_s$. Therefore, the grade resistance may be approximated by $W \tan \theta_s$ or WG, where G is the grade in percent.

The maximum grade a vehicle can negotiate at a constant speed, therefore, is determined by the net tractive effort available at that speed:

$$G = \frac{1}{W} (F - R_r - R_a) = \frac{F_{\text{net}}}{W} \tag{3.70}$$

Use can be made of the performance curves of a vehicle, such as those shown in Figure 3.49, to determine the speed obtainable on each grade. For instance, the grade resistance of the passenger car with a weight of 17.79 kN (4000 lb) on a grade of 7.5% is 1.34 kN (300 lb). A horizontal line representing this grade resistance can be drawn on the diagram, which intersects the net tractive effort curve at a speed of 133 km/h (82 mph). This indicates that for the passenger car under consideration, the maximum speed obtainable at a grade of 7.5% is 133 km/h (82 mph). It should be noted that the tractive effort limited by the nature of tire–road adhesion usually determines the maximum gradeability of the vehicle. For instance, it can be seen from Figure 3.49 that the maximum grade the vehicle can negotiate at low speeds on a gravel surface with $\mu = 0.6$ will be approximately 35%.

3.9 Operating Fuel Economy of Vehicles with Internal Comustion Engines

The operating fuel economy of an automotive vehicle with an internal combustion engine depends on a number of factors, including the fuel consumption characteristics of the engine, transmission characteristics, weight of the vehicle, aerodynamic resistance, rolling resistance of tires, driving conditions, and driver behavior.

Notional fuel economy characteristics of a gasoline and a diesel engine are shown in Figures 3.62 and 3.63, respectively (Giles 1969). Internal combustion engines usually have reduced fuel economy at low throttle and low torque settings. Operations at low engine speeds and high torques are always more economical than at high speed and low torque settings with the same power output. For instance, it can be seen from Figure 3.62 that for the gasoline engine to develop 22 kW (30 hp) of power, it can run at a speed of 2500 or 4000 rpm. At 2500 rpm, the specific fuel consumption is approximately 0.29 kg/kW · h (0.48 lb/hp · h), whereas at 4000 rpm, it is 0.37 kg/kW · h (0.60 lb/hp · h). By connecting the engine operating points with the lowest specific fuel consumption for each power setting, an optimum fuel economy line (maximum efficiency line) of the engine can be drawn, as shown in Figure 3.62 (Giles 1969).

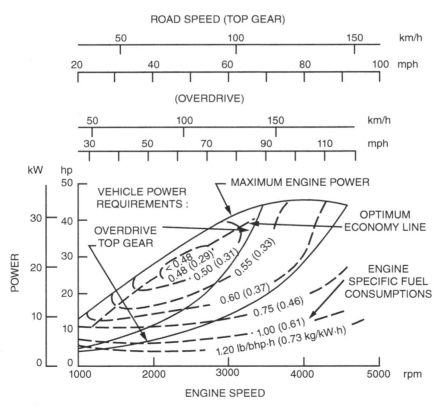

Figure 3.62 Fuel economy characteristics of a gasoline engine. *Source:* Giles (1969). Reproduced with permission of Butterworth.

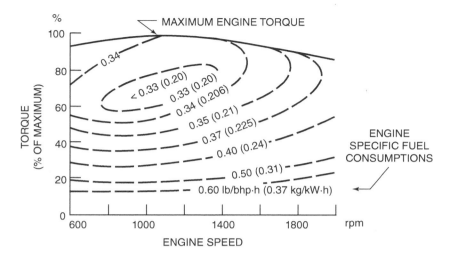

Figure 3.63 Fuel economy characteristics of a diesel engine. *Source:* Giles (1969). Reproduced with permission of Butterworth.

For a given power requirement at a specific vehicle speed, the engine operating point is determined by the gear ratio of the transmission. Ideally, the gear ratio of the transmission can be continuously varied to any desired value so that the engine operating point will follow the optimum fuel economy line for all power settings. This has stimulated the development of a variety of continuously variable transmissions, as described in Section 3.7.3. The potential gain in fuel economy using a continuously variable transmission may be illustrated using the example given above. As shown in Figure 3.62, when the vehicle is operating in the top gear at 128 km/h (80 mph), the power required is 22 kW (30 hp), and the engine is running at 4000 rpm with a specific fuel consumption of 0.37 kg/kW · h (0.60 lb/hp · h). However, if a continuously variable transmission (CVT) is used, the transmission gear ratio can be varied so that at the same vehicle speed with the same power output, the engine is running at 2500 rpm with a specific fuel consumption of approximately 0.29 kg/kW · h (0.48 lb/hp · h). This represents a potential fuel saving of 21.6%. However, the mechanical efficiency of the current generation of continuously variable transmissions is generally lower than that of the manual gear transmission. The actual saving in fuel using the CVT may not be as high as that given in the above example. Figure 3.64 shows a comparison of fuel consumption of a small passenger car with a 1.6-liter engine equipped with a Van Doorne type CVT with that with a manual five-speed transmission (Hahne 1984). It shows that under steady operating conditions in the speed range 60 – 150 km/h (37 – 93 mph), the vehicle with the Van Doorne transmission achieves better fuel economy than the manual five-speed transmission, despite its lower mechanical efficiency (approximately 86–90%).

To further illustrate the effects of the gear ratio of the transmission on the operating fuel economy of road vehicles, the overdrive gear of a passenger vehicle may be used as an example. As shown in Figure 3.43, the gear ratio of the top gear in the transmission is usually selected in such a way that the curve representing the power available at the driven wheels meets the resultant resistance curve at a speed slightly higher than that of maximum power (Giles 1969). This typical choice of gear ratio provides the vehicle with sufficient power reserve to maintain a given vehicle speed against a temporary increase in resistance due to headwind or gradient. The vehicle power requirements at various vehicle speeds in top gear can be plotted in the engine performance diagram as shown in

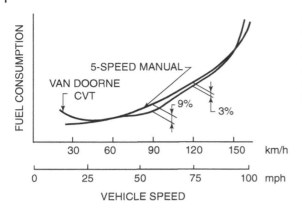

Figure 3.64 Comparisons of the fuel consumption of a small car equipped with a continuously variable transmission versus that with a manual transmission. *Source:* Hahne (1984). Reproduced with permission of the Council of the Institution of Mechanical Engineers.

Figure 3.62. A maximum vehicle speed of 145 km/h (90 mph) is equivalent to an engine speed of 4500 rpm in top gear in the example shown, and the power required to overcome the resultant resistance at that speed is about 32.8 kW (44 hp). When the engine is running at 4500 rpm and developing 32.8 kW (44 hp), the specific fuel consumption will be 0.40 kg/kW · h (0.65 lb/hp · h), as shown in Figure 3.62. Thus, in one hour at 145 km/h (90 mph), the vehicle will consume 13.1 kg (28.8 lb) of fuel in top gear.

If, however, an overdrive gear with a gear ratio approximately 30% less than that of the top gear is used in the transmission, the vehicle can still achieve a maximum speed of 145 km/h (90 mph), as shown in Figure 3.62, but with reduced power reserve over the entire speed range. Owing to the lower gear ratio that an overdrive gear introduces, for the same vehicle speed the engine speed will be lower than that when using the top gear, as shown in Figure 3.62. For instance, at a speed of 145 km/h (90 mph), with the overdrive gear, the engine is running at 3400 rpm as compared with 4500 rpm when using the top gear. Accordingly, using the overdrive gear, the engine-specific fuel consumption is reduced to 0.32 kg/kW · h (0.53 lb/hp · h), as compared with 0.40 kg/kW · h (0.65 lb/ hp · h) when using the top gear. Thus, with the overdrive gear, the vehicle will consume only 10.5 kg (23.1 lb) of fuel per hour at a speed of 145 km/h (90 mph). This represents a saving of fuel of approximately 20%. Although this example is for a specific engine and vehicle, most cars will show similar gains with an overdrive. The improvement in fuel economy obtained by an overdrive gear under steady-state cruising conditions is an exploitation of the fact that for the same power output, the internal combustion engine is always more economical to operate at lower speeds and higher torques than at higher speeds and lower torques. It is interesting to point out that all the passenger vehicles shown in Table 3.11 have two or more overdrive gears. For the Acura ILX with AM8 transmission and Honda Accord-Sport with AS10 transmission, they have four overdrive gears with gear ratios less than one.

The reduction of vehicle weight is also one of the important measures for achieving improved fuel economy. This is because the propelling force, and hence power, required to accelerate a vehicle is proportional to its weight. In stop-and-go driving conditions in an urban environment, the frequent acceleration leads to higher fuel consumption for a heavier vehicle than a lighter vehicle. To reduce vehicle weight, unibody construction has largely replaced the separate body-frame construction, and computer-aided techniques have been introduced to optimize the design of vehicle structures. Furthermore, lightweight materials, such as composites, high-strength low-alloy steel, plastics, aluminum, and metal–plastic laminates, have found increasing use in vehicle components. The reconfiguration of the vehicle from front engine–rear wheel drive to front engine–front wheel drive also leads to a considerable reduction in vehicle weight. It is estimated that a reduction of 1 kg in vehicle mass is equivalent to a reduction in fuel consumption of 7.24×10^{-5} L/km (or each pound of weight saved will result in a reduction in fuel consumption of 1.4×10^{-5} US gal/mile (Cole 1984). It is observed that a 10% change in tire rolling resistance will result in an approximately 2% change in fuel economy for passenger cars. The effects of aerodynamic resistance on vehicle fuel consumption is also noticeable, which has been discussed in Section 3.2.

Driving conditions significantly affect fuel consumption. It is obvious that the fuel consumption for driving in the city with slow speeds and frequent stop-and-go traffic is substantially higher than that for driving on the highway with steadier and higher speeds. To provide a common basis for comparing the fuel economy of different vehicles, the Environmental Protection Agency (EPA) of the United States has devised a series of laboratory tests that follow specific driving cycles (schedules). For 2007 and earlier model year vehicles, only the city and highway driving cycles were used. Beginning with 2008 models, three additional tests are used to adjust the city and highway fuel consumption ratings to account for higher speeds, air conditioning use, and colder temperatures. Attributes of these five test cycles specified by EPA are given in Table 3.13 (Environmental Protection Agency 2021).

Table 3.13 EPA fuel economy test cycles.

Test cycle attributes	Test cycle				
	City	Highway	High speed	Air conditioning	Cold Temperature
Trip type	Low speeds in stop-and -go urban traffic	Free-flow traffic at highway speeds	Higher speeds; harder acceleration and braking	A/C use under hot ambient conditions	City test with colder outside temperatures
Top speed	90 km/h (56 mph)	97 km/h (60 mph)	129 km/h (80 mph)	88 km/h (54.8 mph)	90 km/h (56 mph)
Average speed	34 km/h (21.2 mph)	78 km/h (48.3 mph)	78 km/h (48.4 mph)	34 km/h (21.2 mph)	34 km/h (21.2 mph)
Maximum Acceleration	5.3 km/h/s (3.3 mph/s)	5.3 km/h/s (3.2 mph/s)	13.6 km/h/s (8.46 mph/s)	8.2 km/h/s (5.1 mph/s)	5.3 km/h/s (3.3 mph/s)
Simulated distance	17.8 km (11 miles)	16.5 km (10.3 miles)	12.9 km (8 miles)	5.8 km (3.6 miles)	17.8 km (11 miles)
Time	31.2 min	12.75 min	9.9 min	9.9 min	31.2 min
Stops	23	None	4	5	23
Idling time	18% of time	None	7% of time	19% of time	18% of time
Engine startup	Cold	Warm	Warm	Warm	Cold
Laboratory Temperature	20–30°C (68–86°F)			35°C (95°F)	−7°C (20°F)
Vehicle air conditioning	Off	Off	Off	On	Off

Source: U.S. Environmental Protection Agency (2021).

Based on the results of the EPA city and highway tests, a combined fuel consumption rating in miles per US gallon (mpg (US)) is calculated by weighting the city value by 55% and the highway value by 45%.

In the United States, vehicle fuel economy is measured under controlled conditions in a laboratory dynamometer, following the series of five tests specified by EPA noted above. Manufacturers test their own vehicles – usually pre-production prototypes – and report the results to EPA. EPA reviews the results and confirms about 15–20% of them through its own tests.

In Canada, manufacturers use controlled laboratory testing that follows the above-noted series of five tests specified by EPA to generate fuel consumption data. Department of the Environment and Climate Change, Government of Canada (Environment and Climate Change Canada) collects the data from vehicle manufacturers. Department of Natural Resources, Government of Canada (Natural Resources Canada) put the data and other information together to publish the *Fuel Consumption Guide*.

Table 3.14 shows the fuel consumption and emissions (CO_2 and smog) rating for a sample of cars given in the *Fuel Consumption Guide*. In the table, the following are presented: model name; engine size (displacement) in liters (L); transmission type: A – automatic, AM – automated manual, AS – automatic with select shift, AV – continuously variable, M – manual; number of gears/speeds (1–10); fuel type: X – regular gasoline, Z – premium gasoline; fuel consumptions ratings in city/highway/combined in L/100 km and in mpg (US); and ratings on vehicle tailpipe emissions of CO_2 and smog-forming pollutants on a scale from 1 (worst) to 10 (best).

In recent years, SUVs have become increasingly popular in North America and elsewhere. Table 3.15 shows the fuel consumption and emissions (CO_2 and smog) rating for a sample of SUVs given in the *Fuel Consumption Guide*, Department of Natural Resources, Government of Canada. In the table, the following are presented: model name; engine size (displacement) in liters (L); transmission type: A – automatic, AM – automated manual, AS – automatic with select shift, AV – continuously variable, M – manual; number of gears/speeds (1–10); fuel type: X – regular gasoline, Z – premium gasoline; fuel consumption ratings in city/highway/combined in L/100 km and in mpg (US); and vehicle tailpipe emissions of CO_2 and smog-forming pollutants ratings on a scale from 1 (worst) to 10 (best).

SUVs are generally larger in size and heavier in weight than cars with engines of similar displacements. Furthermore, most of them are all-wheel drive. As a result, their fuel consumptions shown in Table 3.15 are generally higher in comparison with that in Figure 3.14.

Several operating factors, including engine start-up and warm-up processes, ambient conditions, road surface conditions, vehicle maintenance, and driver behavior (habits), affect fuel consumption and emissions. So, the fuel consumption and emissions under actual driving conditions may be different from those presented in Tables 3.14 and 3.15.

In the past, fuel consumptions of cars with manual transmissions were generally lower than those with automatic transmissions. Low efficiency of the torque converter over a range of operating conditions is a factor, as discussed in Section 3.7.2. With new developments and innovations in automatic transmissions in recent years, fuel consumption of a vehicle with automatic transmission is now comparable to or lower than that of the same vehicle with manual transmission. Some of the salient points are outlined below:

- The use of electronic control systems to optimize automatic transmission gear shift operations, based on inputs from sensors monitoring vehicle and engine operating conditions, leads to improvements in vehicle fuel economy and performance, as noted previously in Section 3.7.2.

Table 3.14 Fuel consumption and emissions ratings for a sample of cars with internal combustion engines.

Model	Engine size, L	Transmission type	Fuel type	Fuel consumption City/Highway/Combined, L/100 km (mpg (US))	CO_2 rating	Smog rating
Acura ILX	2.4	AM8	Z	9.9/7.0/8.6 (23.8/33.6/27.4)	6	3
Acura TLX	2.4	AM8	Z	10.0/7.1/8.7 (23.5/33.1/27.0)	6	3
Audi A4	2.0	AM7	Z	8.6/6.9/7.8 (27.4/34.1/30.2)	7	5
Audi A6 Quattro	3.0	AM7	Z	10.7/8.2/9.6 (22.0/28.7/24.5)	5	5
BMW 330i xDrive Touring	2.0	AS8	Z	10.2/7.2/8.8 (23.1/32.7/26.7)	6	7
BMW 540i xDrive	3.0	AS8	Z	11.2/8.1/9.8 (21.0/29.0/24.0)	5	5
Buick Regal AWD	2.0	AS8	Z	11.0/8.0/9.6 (21.4/29.4/24.5)	5	5
Cadillac CTS	3.6	AS8	X	12.3/8.2/10.5 (19.1/28.7/22.4)	4	5
Chevrolet Malibu	2.0	A9	Z	10.5/7.4/9.1 (22.4/31.8/25.8)	5	5
Chrysler 300 AWD	3.6	A8	X	12.8/8.7/11.0 (18.4/27.0/21.4)	4	3
Ford Fusion	1.5	AS6	X	10.0/7.0/8.7 (23.5/33.6/27.0)	6	7
Ford Taurus AWD	3.5	AS6	X	14.6/10.0/12.5 (16.1/23.5/18.8)	3	3
Honda Accord	2.0	M6	X	10.7/7.3/9.2 (22.0/32.2/25.6)	5	6
Honda Civic Sedan	2.0	AV	X	7.9/6.1/7.1 (29.8/38.6/33.1)	8	3
Hyundai Ioniq	1.6	AM6	X	4.3/4.4/4.3 (54.7/53.5/54.7)	10	7
Hyundai Sonata	2.4	AS6	X	9.3/7.1/8.3 (25.3/33.1/28.3)	6	7
KIA Forte	2.0	AV	X	7.7/5.9/6.9 (30.5/39.9/34.1)	8	5
KIA Optima	2.4	AS6	X	9.5/7.1/8.4 (24.8/33.1/28.0)	6	5
Mercedes-Benz C 300 4MATIC	2.0	A9	Z	11.0/7.3/9.4 (21.4/32.2/25.0)	5	5
Mercedes-Benz E 450 4MATIC	3.0	A9	Z	11.8/8.5/10.3 (19.9/27.7/22.8)	5	5
Mercedes-Benz S 560 4MATIC	4.0	A9	Z	13.5/8.6/11.3 (17.4/27.4/20.8)	4	5
Nissan Altima AWD	2.5	AV	X	9.1/6.5/7.9 (25.8/36.2/29.8)	7	7

(Continued)

Table 3.14 (Continued)

Model	Engine size, L	Transmission type	Fuel type	Fuel consumption City/Highway/Combined, L/100 km (mpg (US))	CO$_2$ rating	Smog rating
Nissan Sentra	1.8	AV	X	8.1/6.3/7.3 (29.0/37.3/32.2)	7	7
Porsche 911 Carrera	3.0	M7	Z	11.8/8.1/10.1 (19.9/29.0/23.3)	5	1
Toyota Camry	2.5	AS8	X	8.1/5.7/6.9 (29.0/41.3/34.1)	8	7
Toyota Crolla	1.8	AV	X	8.3/6.5/7.5 (28.3/36.2/31.4)	7	3
Volkswagen Golf	1.4	M6	X	8.2/6.3/7.4 (28.7/37.3/31.8)	7	7
Volkswagen Passat	2.0	AS6	X	9.3/6.5/8.1 (25.3/36.2/29.0)	6	7
Volvo S60 T5	2.0	AS8	Z	9.9/6.6/8.4 (23.8/35.6/28.0)	6	5
Volvo S60 T6 AWD	2.0	AS8	Z	11.1/7.3/9.4 (21.2/32.2/25.0)	5	7
Volvo S90 T6 AWD	2.0	AS8	Z	11.1/7.3/9.4 (21.2/32.2/25.0)	5	7

Source: *2019 Fuel Consumption Guide*, Department of Natural Resources, Government of Canada.

- With automatic transmissions, more gears, such as up to 10 shown in Table 3.11, can be used. This would reduce the non-utilizable operating ranges of the vehicle tractive effort–speed relation and enable the transmission to function close to a continuously variable transmission, as discussed previously in Section 3.7.2. This would allow the engine to operate in the optimal fuel economy range, as described earlier.
- In city driving, the often stop-and-go traffic causes frequent engine idling when the vehicle stops, while the torque converter operates at zero speed ratio with high slippage between the impeller and turbine, as shown in Figure 3.51. This would adversely affect vehicle fuel economy. An innovative control system is introduced into some current automatic transmissions. It senses once the car is completely stopped and it shuts down the engine. Once the driver's foot is off the brake pedal and the engine starts up again automatically. This means that engine idling is essentially eliminated, which leads to improved fuel economy.
- As noted previously in Section 3.7.2, a lockup clutch is incorporated in the current design of torque converters. It is programmed to engage in a predetermined vehicle operating range that allows the torque directly transmitted from the engine to the output shaft of the torque converter. This improves the efficiency of the torque converter in a certain operating range leading to improved vehicle economy.

Figures 3.65 and 3.66 show the variations of the ratio of fuel consumption of a vehicle with manual transmission to that of the same vehicle with automatic transmission as a function of engine size (displacement) in liters for city rating and highway rating, respectively. It is shown that, in most cases, the fuel consumption of a vehicle with manual transmission is higher than that of the same

Table 3.15 Fuel consumption and emissions ratings for a sample of sport utility vehicles.

Model	Engine size, L	Transmission type	Fuel type	Fuel consumption City/Highway/Combined, L/100 km (mpg (US))	CO_2 rating	Smog rating
Acura MDX SH-AWD	3.5	AS9	Z	12.2/9.0/10.8 (19.3/26.1/21.8)	4	3
Acura RDX AWD	2.0	AS10	Z	11.0/8.6/9.9 (21.4/27.4/23.8)	5	6
Audi Q3 Quattro	2.0	AS8	X	12.3/8.6/10.6 (19.1/27.4/22.2)	4	7
Audi Q5	2.0	AM7	Z	10.9/8.7/9.9 (21.6/27.0/23.8)	5	3
BMW X3 xDrive30i	2.0	AS8	Z	10.8/8.0/9.6 (21.8/29.4/24.5)	5	7
BMW X5 xDrive40i	3.0	AS8	Z	11.7/9.1/10.5 (20.1/25.8/22.4)	5	3
Buick Enclave	3.6	A9	X	13.0/9.1/11.2 (18.1/25.8/21.0)	4	6
Buick Encore AWD	1.4	AS6	X	9.9/8.1/9.1 (23.8/29.0/25.8)	5	5
Cadillac XT5	3.6	AS8	X	12.1/8.9/10.6 (19.4/26.4/22.2)	4	6
Chevrolet Blazer AWD	3.6	A9	X	12.7/9.5/11.3 (18.5/24.8/20.8)	4	6
Chevrolet Equinox	1.5	A6	X	9.2/7.3/8.3 (25.6/32.2.28.3)	6	5
Dodge Journey	2.4	A4	X	12.7/9.2/11.1 (18.5/25.6/21.2)	4	3
Ford Escape	1.5	AS6	X	10.2/7.8/9.1 (23.1/30.2/25.8)	5	7
Ford Explorer AWD	2.3	AS6	X	13.1/9.2/11.4 (18.0/25.6/20.6)	4	3
GMC Acadia	2.5	A6	X	11.1/9.2/10.2 (21.2/25.6/23.1)	5	5
GMC Terrain AWD	1.5	A9	X	9.6/8.3/9.0 (24.5/28.3/26.1)	5	5
Honda CR-V	1.5	AV	X	8.4/7.0/7.8 (28.0/33.6/30.2)	7	5
Honda Pilot AWD	3.5	A6	X	13.0/9.3/11.3 (18.1/25.3/20.8)	4	3
Hyundai Santa Fe	2.4	AS8	X	10.8/8.0/9.6 (21.8/29.4/24.5)	5	5
Hyundai Tucson AWD	2.0	AS6	X	10.8/9.2/10.1 (21.8/25.6/23.3)	5	5
Jeep Cherokee	3.2	A9	X	11.9/8.2/10.2 (19.8/28.7/23.1)	5	3
Jeep Compass	2.4	A6	X	10.6/7.6/9.3 (22.2/30.9/25.3)	5	5
Ranger Rover 3.0	3.0	AS8	Z	14.1/10.3/12.4 (16.7/22.8/19.0)	3	7
Lexus NX 300 AWD	2.0	AS6	Z	10.7/8.5/9.7 (22.0/27.7/24.2)	5	3

(Continued)

Table 3.15 (Continued)

Model	Engine size, L	Transmission type	Fuel type	Fuel consumption City/Highway/Combined, L/100 km (mpg (US))	CO$_2$ rating	Smog rating
Lexus RX 350 AWD	3.5	AS8	X	12.2/9.0/10.8 (19.3/26.1/21.8)	4	5
Mercedes-Benz GLC 300 4MATIC	2.0	A9	Z	11.0/8.7/10.0 (21.4/27.0/23.5)	4	5
Mercedes-Benz GLS 450 4MATIC	3.0	A9	Z	14.9/11.2/13.2 (15.8/21.0/17.8)	3	3
Nissan Pathfinder 4WD	3.5	AV	X	12.1/8.9/10.7 (19.4/26.4/22.0)	4	5
Nissan Rogue 4WD	2.5	AV	X	9.6/7.5/8.7 (24.5/31.4/27.0)	6	7
Toyota Highlander AWD	3.5	AS8	X	12.1/9.0/10.6 (19.4/26.1/22.2)	4	5
Toyota RAV4 AWD	2.5	AS8	X	9.2/7.1/8.3 (25.6/33.1/28.3)	6	6
Volkswagen Tiguan	2.0	AS8	X	10.7/8.0/9.5 (22.0/29.4/24.8)	5	7
Volkswagen Tiguan 4Motion	2.0	AS8	X	11.1/8.1/9.8 (21.2/29.0/24.0)	5	7
Volvo XC60 T5 AWD	2.0	AS8	Z	11.3/8.5/10.0 (20.8/27.7/23.5)	5	5
Volvo XC90 T5 AWD	2.0	AS8	Z	11.3/8.5/10.0 (20.8/27.7/23.5)	5	5

Source: 2019 Fuel Consumption Guide, Department of Natural Resources, Government of Canada.

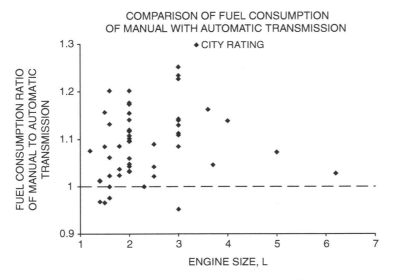

Figure 3.65 Variations of the ratio of fuel consumption of a vehicle with manual transmission to that of the same vehicle with automatic transmission as a function of engine size (displacement) in liters for city rating. *Source: 2019 Fuel Consumption Guide*, Department of Natural Resources, Government of Canada.

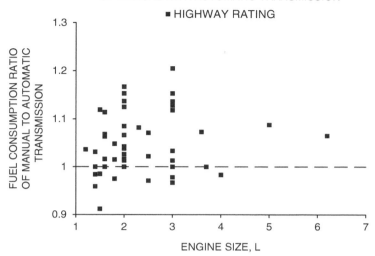

COMPARISON OF FUEL CONSUMPTION
OF MANUAL WITH AUTOMATIC TRANSMISSION
■ HIGHWAY RATING

Figure 3.66 Variations of the ratio of fuel consumption of a vehicle with manual transmission to that of the same vehicle with automatic transmission as a function of engine size (displacement) in liters for highway rating. *Source: 2019 Fuel Consumption Guide*, Department of Natural Resources, Government of Canada.

vehicle with automatic transmission for both city and highway ratings. For some vehicles, the fuel consumptions with manual transmissions are as much as 25% and 20% higher than that with automatic transmissions for city and highway rating, respectively.

Table 3.16 shows a sample of the data presented in Figures 3.65 and 3.66. In the table, the following are presented: model name; engine size (displacement) in liters; fuel type: X – regular gasoline, Z – premium gasoline; transmission type: A – automatic, AM – automated manual, AS – automatic with select shift, AV – continuously variable, M – manual; number of gears/speeds (1–10); fuel consumption ratings in city/highway/combined in L/100 km and in mpg (US); and the ratio of fuel consumption of a vehicle with manual transmission to that of the same vehicle but with automatic transmission for city (C)/highway (H)/combined city and highway (COM) rating. As shown in the table, for the Jaguar F-Type Convertible, the ratio of the fuel consumption with manual transmission M6 to that with automatic transmission AS8 for city rating is 1.252. For the BMW 440i Coupe, the ratio of the fuel consumption with manual transmission M6 to that with automatic transmission AS8 for highway rating is 1.205. In other words, for these two cases, the fuel consumption with manual transmissions is more than 20% higher than that with automatic transmissions.

In recent years, all-wheel drive (AWD) vehicles have become increasingly popular in North America and elsewhere. In the *Fuel Consumption Guide* published by the Department of Natural Resources, Canada, an AWD vehicle refers to a vehicle designed to operate with all wheels powered, while a four-wheel drive (4WD or 4×4) vehicle refers a vehicle designed to operate with either two wheels or four wheels powered upon demand. With AWD, vehicle mobility on slippery or snow-covered surfaces or in off-road environments is better than 2WD (front-wheel drive or rear-wheel drive) vehicles. For AWD, the entire vehicle weight W can be used to generate tractive effort. For instance, on a road with a coefficient of road adhesion μ, the maximum tractive effort F_{max} that an AWD vehicle can develop is given by $F_{max} = \mu W$, whereas for a front-wheel drive vehicle, the maximum tractive effort $F_{fmax} = \mu W_f$, where W_f is the normal load supported by the front

Table 3.16 Comparison of fuel consumptions of vehicles with manual transmissions versus those of the same vehicles with automatic transmissions.

Model	Engine size, L	Fuel type	Transmission type	Fuel consumption City/Highway/Combined, L/100 km (mpg (US))	Fuel consumption ratio of manual to automatic transmission
BMW 440i Coupe	3.0	Z	M6	12.8/8.8/11.0 (18.4/26.7/21.4)	C = 1.143 H = 1.205 COM = 1.170
BMW 440i Coupe	3.0	Z	AS8	11.2/7.3/9.4 (21.0/32.2/25.0)	
Chevrolet Camaro	2.0	Z	M6	11.9/7.9/10.1 (19.8/29.8/23.3)	C = 1.102 H = 1.026 COM = 1.074
Chevrolet Camaro	2.0	Z	AS8	10.8/7.7/9.4 (21.8/30.5/25.0)	
Dodge Challenger SRT Hellcat	6.2	Z	M6	18.1/11.4/15.1 (13.0/20.6/15.6)	C = 1.028 H = 1.065 COM = 1.041
Dodge Challenger SRT Hellcat	6.2	Z	A8	17.6/10.7/14.5 (13.4/22.0/16.2)	
Fiat 500	1.4	X	M5	8.5/7.1/7.9 (27.7/33.1/29.8)	C = 0.876 H = 0.959 COM = 0.908
Fiat 500	1.4	X	A6	9.7/7.4/8.7 (24.2/31.8/27.0)	
Ford Mustang	2.3	X	M6	11.2/7.9/9.7 (21.0/29.8/24.2)	C = 1.0 H = 1.082 COM = 1.032
Ford Mustang	2.3	X	AS10	11.2/7.3/9.4 (21.0/32.2/25.0)	
Honda Accord	1.5	X	M6	8.9/6.7/7.9 (26.4/35.1/29.8)	C = 1.085 H = 0.985 COM = 1.039
Honda Accord	1.5	X	AV7	8.2/6.8/7.6 (28.7/34.6/30.9)	
Hyundai Elantra	1.6	X	M6	10.7/7.8/9.4 (22.0/30.2/25.0)	C = 1.202 H = 1.114 COM = 1.160
Hyundai Elantra	1.6	X	AM7	8.9/7.0/8.1 (26.4/33.6/29.0)	
Jaguar F-Type Convertible	3.0	Z	M6	14.9/9.8/12.6 (15.8/24.0/18.7)	C = 1.252 H = 1.153 COM = 1.212
Jaguar F-Type Convertible	3.0	Z	AS8	11.9/8.5/10.4 (19.8/27.7/22.6)	
KIA Forte	2.0	X	M6	8.6/6.4/7.6 (27.4/36.8/30.9)	C = 1.117 H = 1.085 COM = 1.101
KIA Forte	2.0	X	AV	7.7/5.9/6.9 (30.5/39.9/34.1)	
Mazda CX-3	2.0	X	M6	8.8/7.0/8.0 (26.7/33.6/29.4)	C = 1.060 H = 1.014 COM = 1.039
Mazda CX-3	2.0	X	AS6	8.3/6.9/7.7 (28.3/34.1/30.5)	
Mini Cooper 3 Door	1.5	Z	M6	8.5/6.2/7.5 (27.7/37.9/31.4)	C = 0.966 H = 0.912 COM = 0.949
Mini Cooper 3 Door	1.5	Z	AS6	8.8/6.8/7.9 (26.7/34.6/29.8)	

Table 3.16 (Continued)

Model	Engine size, L	Fuel type	Transmission type	Fuel consumption City/Highway/Combined, L/100 km (mpg (US))	Fuel consumption ratio of manual to automatic transmission
Nissan Sentra	1.8	X	M6	8.8/6.6/7.8 (26.7/35.6/30.2)	C = 1.086 H = 1.048
Nissan Sentra	1.8	X	AV	8.1/6.3/7.3 (29.0/37.3/32.2)	COM = 1.068
Porsche 911 Carrera	3.0	Z	M7	11.8/8.1/10.1 (19.9/29.0/23.3)	C = 1.113 H = 1.013
Porsche 911 Carrera	3.0	Z	AM7	10.6/8.0/9.4 (22.2/29.4/25.0)	COM = 1.074
Subaru Impreza 5-Door AWD	2.0	X	M5	10.1/7.7/9.0 (23.3/30.5/26.1)	C = 1.202 H = 1.167
Subaru Impreza 5-Door AWD	2.0	X	AV7	8.4/6.6/7.6 (28.0/35.6/30.9)	COM = 1.184
Toyota Corolla	1.8	X	M6	8.5/6.6/7.6 (27.7/35.6/30.9)	C = 1.024 H = 1.015
Toyota Corolla	1.8	X	AV	8.3/6.5/7.5 (28.3/36.2/31.4)	COM = 1.013
Volkswagen Golf	1.4	X	M6	8.2/6.3/7.4 (28.7/37.3/31.8)	C = 1.012 H = 0.984
Volkswagen Golf	1.4	X	AS8	8.1/6.4/7.4 (29.0/36.8/31.8)	COM = 1.0
Volkswagen Jetta	1.4	X	M6	7.9/5.9/7.0 (29.8/39.9/33.6)	C = 1.013 H = 1.0
Volkswagen Jetta	1.4	X	AM8	7.8/5.9/7.0 (30.2/39.9/33.6)	COM = 1.0

Source: 2019 Fuel Consumption Guide, Department of Natural Resources, Government of Canada.

wheels, and for a rear-wheel drive vehicle, the maximum tractive effort $F_{rmax} = \mu W_r$, where W_r is the normal load supported by the rear wheels. It is apparent that an AWD vehicle has better mobility, as well as acceleration capability (as limited by μW) than a two-wheel drive vehicle. The mobility of an AWD off-road vehicle on unprepared terrain will be discussed in detail in Chapter 4, Section 4.1.3.

An AWD vehicle will also have better handling capability than a two-wheel drive vehicle. This is because with AWD, vehicle tractive effort is distributed between the front and rear wheels. Consequently, the cornering force that the front tires can support for an AWD vehicle would be higher than that for the font tires of a front-wheel drive vehicle. For a front-wheel drive vehicle, the vehicle tractive effort is entirely developed by the front tires, which lowers the cornering force that the front tires can support. As a result, the propensity to the loss of directional control for an AWD vehicle is less than that of a front-wheel drive vehicle. Similarly, the propensity to the loss of directional stability for an AWD vehicle is less than that of a rear-wheel drive vehicle. The loss of vehicle directional control or stability will be discussed in Section 3.11.1.

While an AWD vehicle has better mobility and handling capability than a two-wheel drive vehicle, the additional drive for the front or rear wheels will increase mechanical losses in the drivetrain. This leads to higher fuel consumptions of AWD vehicles in comparison with that of two-wheel drive vehicles.

Figures 3.67 and 3.68 show the variations of the ratio of fuel consumption of a vehicle with AWD to that of the same vehicle with two-wheel drive, as a function of engine size (displacement) in liters

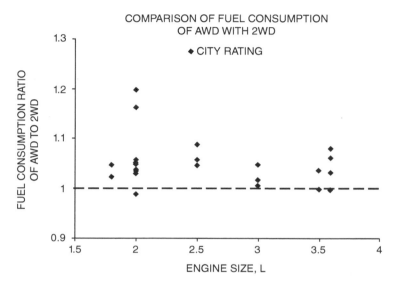

Figure 3.67 Variations of the ratio of fuel consumption of a vehicle with all-wheel drive to that of the same vehicle with two-wheel drive as a function of engine (size) displacement in liters for city rating. *Source: 2019 Fuel Consumption Guide*, Department of Natural Resources, Government of Canada.

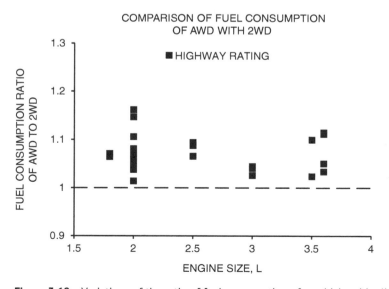

Figure 3.68 Variations of the ratio of fuel consumption of a vehicle with all-wheel drive to that of the same vehicle with two-wheel drive as a function of engine (size) displacement in liters for highway rating. *Source: 2019 Fuel Consumption Guide*, Department of Natural Resources, Government of Canada.

for city and highway rating, respectively. It is shown that for city rating, the fuel consumption of a vehicle with AWD is either close to or higher than the same vehicle with two-wheel drive. For highway rating, the fuel consumption of a vehicle with AWD is higher than the same vehicle with two-wheel drive for all cases shown.

Table 3.17 shows a sample of the data presented in Figures 3.67 and 3.68. In the table, the following are presented: model name; engine size (displacement) in liters; fuel type: X – regular gasoline, Z – premium gasoline; transmission type: A – automatic, AM – automated manual, AS – automatic with select shift, AV – continuously variable, M – manual; number of gears/speeds (1–10): fuel consumption ratings in city/highway/combined in L/100 km and in mpg (US); and the ratio of fuel consumption of a vehicle with AWD to that of the same vehicle but with two-wheel drive for city (C)/highway (H)/combined city and highway (COM) rating. As shown in the table, for instance, the fuel consumption of Audi A3 Quattro (AWD) is approximately 20% and 16% higher than that of Audi A3 (2WD) for city and highway rating, respectively.

For tradeoffs between higher mobility and lower fuel economy of all-wheel drive vehicles, it appears that the use of AWD is justified in regions with extended cold and snowy winters. For other regions, it appears that the often called "automatic AWD," which operates most of the time in two-wheel drive mode, with power automatically delivered to all four wheels only when additional traction is needed, may offer a compromise between mobility and fuel economy. This issue, however, may require further detailed evaluation of all factors involved, prior to an informed decision being made.

Table 3.17 Comparison of fuel consumptions of vehicles of all-wheel drive with that of two-wheel drive.

Model	Engine size, L	Fuel type	Drive type	Transmission	Fuel consumption City/Highway/Combined, L/100 km (mpg (US))	Fuel consumption ratio of AWD to 2WD
Alfa Romeo Giulia AWD	2.0	Z	AWD	A8	10.5/7.7/9.2 (22.4/30.5/25.6)	C = 1.05 H = 1.069 COM = 1.057
Alfa Romeo Giulia	2.0	Z	2WD	A8	10.0/7.2/8.7 (23.5/32.7/27.0)	
Audi A3 Quattro	2.0	Z	AWD	AM7	10.7/7.9/9.6 (21.6/29.8/24.5)	C = 1.198 H = 1.162 COM = 1.85
Audi A3	2.0	Z	2WD	AM7	9.1/6.8/8.1 (25.8/34.6/29.0)	
BMW 440i xDrive Coupe	3.0	Z	AWD	AS8	11.4/7.6/9.7 (20.6/30.9/24.2)	C = 1.018 H = 1.041 COM = 1.032
BMW 440i Coupe	3.0	Z	2WD	AS8	11.2/7.3/9.4 (21.0/32.2/25.0)	
Buick Lacrosse AWD	3.6	X	AWD	AS9	11.7/8.2/10.1 (20.1/28.7/23.3)	C = 1.0 H = 1.051 COM = 1.020
Buick Lacrosse	3.6	X	2WD	AS9	11.7/7.8/9.9 (20.1/30.2/23.8)	
Cadillac CTS AWD	2.0	Z	AWD	AS8	11.4/8.1/9.9 (20.6/29.0/23.8)	C = 1.036 H = 1.038 COM = 1.042
Cadillac CTS	2.0	Z	2WD	AS8	11.0/7.8/9.5 (21.4/30.2/24.8)	

(Continued)

Table 3.17 (Continued)

Model	Engine size, L	Fuel type	Drive type	Transmission	Fuel consumption City/Highway/Combined, L/100 km (mpg (US))	Fuel consumption ratio of AWD to 2WD
Chrysler 300 AWD	3.6	X	AWD	A8	12.8/8.7/11.0 (18.4/27.0/21.4)	C = 1.032 H = 1.115 COM = 1.068
Chrysler 300	3.6	X	2WD	A8	12.4/7.8/10.3 (19.0/30.2/22.8)	
Ford Taurus FFV AWD	3.5	X	AWD	AS6	14.0/9.9/12.4 (16.8/23.8/19.0)	C = 1.037 H = 1.1 COM = 1.078
Ford Taurus FFV	3.5	X	2WD	AS6	13.5/9.0/11.5 (17.4/26.1/20.5)	
Honda HR-V AWD	1.8	X	AWD	AV	8.8/7.5/8.2 (26.7/31.4/28.7)	C = 1.048 H = 1.071 COM = 1.051
Honda HR-V	1.8	X	2WD	AV	8.4/7.0/7.8 (28.0/33.6/30.2)	
Jaguar F-Type Coupe R-Dynamic AWD	3.0	Z	AWD	AS8	13.0/9.2/11.3 (18.1/25.6/20.8)	C = 1.048 H = 1.045 COM = 1.046
Jaguar F-Type Coupe R-Dynamic	3.0	Z	2WD	AS8	12.4/8.8/10.8 (19.0/26.7/21.8)	
Mazda CX-3 4WD	2.0	X	AWD	AS6	8.6/7.4/8.1 (27.4/31.8/29.0)	C = 1.036 H = 1.072 COM = 1.052
Mazda CX-3	2.0	X	2WD	AS6	8.3/6.9/7.7 (28.3/34.1/30.5)	
Mercedes-Benz B 250 4MATIC	2.0	Z	AWD	AM7	10.3/7.8/9.1 (22.8/30.2/25.8)	C = 1.051 H = 1.147 COM = 1.083
Mercedes-Benz B 250	2.0	Z	2WD	AM7	9.8/6.8/8.4 (24.0/34.6/28.0)	
Nissan Altima AWD	2.5	X	AWD	AV	9.1/6.5/7.9 (25.8/36.2/29.8)	C = 1.058 H = 1.066 COM = 1.068
Nissan Altima	2.5	X	2WD	AV	8.6/6.1/7.4 (27.4/38.6/31.8)	
Toyota Prius AWD	1.8	X	AWD	AV	4.5/4.9/4.7 (52.3/48.0/50.0)	C = 1.023 H = 1.065 COM = 1.068
Toyota Prius	1.8	X	2WD	AV	4.4/4.6/4.4 (53.5/51.1/53.5)	

Source: 2019 Fuel Consumption Guide, Department of Natural Resources, Government of Canada.

3.10 Internal Combustion Engine and Transmission Matching

From the discussions presented previously, the engine and transmission characteristics are two of the most significant design factors that affect the performance and fuel economy of a vehicle. For a given vehicle to achieve a desired level of performance and fuel economy, proper matching of the

Figure 3.69 Effects of drive axle ratio on performance and fuel economy of a passenger car. Reprinted with permission from SAE paper 770418 © 1977 Society of Automotive Engineers, Inc.

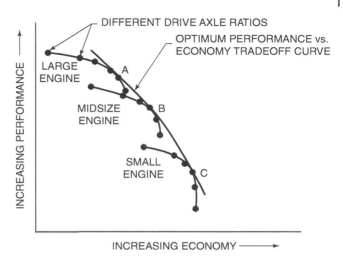

transmission to the engine is important. The performance of a road vehicle may be characterized by its acceleration time from standstill to a given speed, usually 100 km/h, or 60 mph, or the time required to travel a given distance, such as 1/4 mile or 0.4 km. This can be predicted using the method described in Section 3.3. The fuel economy may be characterized by the fuel consumed for a given distance traveled under city and highway driving cycle, as described in the previous section. The combined city and highway fuel consumption in liters per 100 kilometer (L/100 km) or miles per US gallon (mpg (US)), calculated by weighting the value of fuel consumption for city driving by 55% and that for highway driving by 45%, is generally used as a fuel economy indicator, as discussed in Section 3.9.

For a given vehicle with a particular engine and gearbox having a specific gear ratio span (i.e., the ratio of the gear ratio of the lowest gear to that of the highest gear), the acceleration time and fuel economy are a function of the drive axle gear ratio. Figure 3.69 shows the effects of the gear ratio of the drive axle on the performance and fuel economy of a vehicle with different engine sizes (small, midsize, and large) (Chana et al. 1977). The enveloping curve shown in Figure 3.69 represents the optimum performance versus fuel economy tradeoff curve. For instance, points A, B, and C represent the gear ratios of the drive axle with a specific gearbox that achieve the optimum tradeoff between performance and fuel economy for the large, midsize, and small engine, respectively. Figure 3.70 shows the optimum performance versus economy tradeoff curves for various four-speed gearboxes having different gear ratio spans from 3.56 to 5.94. If the vehicle is to achieve an acceleration time of 13.5 s from standstill to 60 mph (97 km/h) with a given engine, then using gearbox unit C with a gear ratio span of 5.94 and with an optimum drive axle ratio will improve the fuel economy by 4.4%, in comparison with that using gearbox unit A with a gear ratio span of 3.56. The fuel economy of the vehicle is measured by the combined city and highway fuel consumption in L/100 km or mpg (US).

Figure 3.71 shows the effects of the gear ratio span of a three-speed automatic transmission with an optimum drive axle gear ratio and of the engine displacement (in liters) on the performance and fuel economy of a General Motors front-wheel drive car (Porter 1979). It shows the gain in fuel

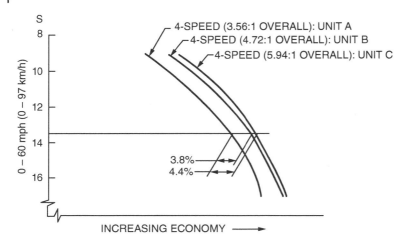

Figure 3.70 Effects of gear ratio span of a four-speed manual transmission on performance and fuel economy of a passenger car. Reprinted with permission from SAE paper 770418 © 1977 Society of Automotive Engineers, Inc.

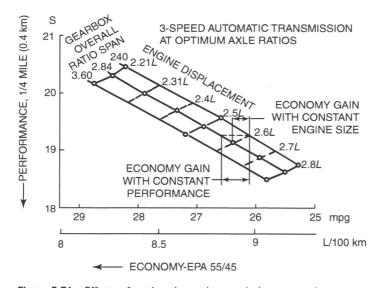

Figure 3.71 Effects of engine size and transmission gear ratio span on performance and fuel economy of a passenger car. Reprinted with permission from SAE paper 790721 © 1979 Society of Automotive Engineers, Inc.

economy that can be obtained by changing the transmission gear ratio span or engine size for a given performance level, expressed in terms of the time taken to travel 1/4 mile (0.4 km). This provides the vehicle engineer with quantitative information to use in selecting the proper combination of engine and transmission to achieve a desired level of performance and fuel economy.

3.11 Braking Performance

Braking performance of motor vehicles is undoubtedly one of the most important characteristics that affect vehicle safety. With increasing emphasis on traffic safety, intensive efforts have been directed toward improving braking performance. Safety standards that specify performance requirements of various types of brake systems have been introduced in many countries.

In this section, the method of approach to the analysis of the braking performance of road vehicles is presented. Criteria for the evaluation of braking capability and approaches to improving braking performance are discussed.

3.11.1 Braking Characteristics of a Two-Axle Vehicle

The major external forces acting on a decelerating two-axle vehicle are shown in Figure 3.72. The braking force F_b originating from the brake system and developed on the tire–road interface is the primary retarding force. When the braking force is below the limit of tire–road adhesion, the braking force F_b is given by

$$F_b = \frac{T_b - \sum I\alpha_{an}}{r} \tag{3.71}$$

where T_b is the applied brake torque, I is the rotating inertia connected with the wheel being decelerated, α_{an} is the corresponding angular deceleration, and r is the rolling radius of the tire.

In addition to the braking force, the rolling resistance of tires, aerodynamic resistance, transmission resistance, and grade resistance (when traveling on a slope) also affect vehicle motion during braking. Thus, the resultant retarding force F_{res} can be expressed by

$$F_{\text{res}} = F_b + f_r W \cos\theta_s + R_a \pm W \sin\theta_s + R_t \tag{3.72}$$

where f_r is the rolling resistance coefficient, W is the vehicle weight, θ_s is the angle of the slope with the horizontal, R_a is the aerodynamic resistance, and R_t is the transmission resistance. When the vehicle is moving uphill, the positive sign for the term $W \sin\theta_s$ should be used. On a downhill grade, the negative sign should, however, be used. Normally, the magnitude of the transmission resistance is small and can be neglected in braking performance calculations.

Figure 3.72 Forces acting on a two-axle vehicle during braking.

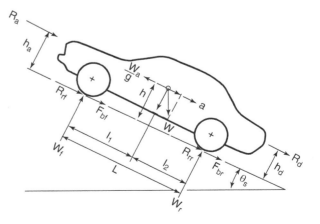

During braking, there is a load transfer from the rear axle to the front axle. By considering the equilibrium of the moments about the front and rear tire–ground contact points, the normal loads on the front and rear axles, W_f and W_r, can be expressed as

$$W_f = \frac{1}{L}\left[Wl_2 + h\left(\frac{W}{g}a - R_a \pm W\sin\theta_s\right)\right] \tag{3.73}$$

and

$$W_r = \frac{1}{L}\left[Wl_1 - h\left(\frac{W}{g}a - R_a \pm W\sin\theta_s\right)\right] \tag{3.74}$$

where a is the deceleration. When the vehicle is moving uphill, the negative sign for the term $W\sin\theta_s$ should be used. In the above expression, it is assumed that the aerodynamic resistance is applied at the center of gravity of the vehicle, and that there is no drawbar load.

By considering the force equilibrium along the slope, the following relationship can be established:

$$F_b + f_r W = F_{bf} + F_{br} + f_r W = \frac{W}{g}a - R_a \pm W\sin\theta_s \tag{3.75}$$

where F_{bf} and F_{br} are the braking forces of the front and rear axles, respectively. Substituting Eq. (3.75) into Eqs. (3.73) and (3.74), the normal loads on the axles become

$$W_f = \frac{1}{L}[Wl_2 + h(F_b + f_r W)] \tag{3.76}$$

and

$$W_r = \frac{1}{L}[Wl_1 - h(F_b + f_r W)] \tag{3.77}$$

The maximum braking force that the tire–ground contact can support is determined by the normal load and the coefficient of road adhesion. With four-wheel brakes, the maximum braking forces on the front and rear axles are given by (assuming the maximum braking force of the vehicle $F_{b\,max} = \mu W$)

$$F_{bf\,max} = \mu W_f = \frac{\mu W[l_2 + h(\mu + f_r)]}{L} \tag{3.78}$$

$$F_{br\,max} = \mu W_r = \frac{\mu W[l_1 - h(\mu + f_r)]}{L} \tag{3.79}$$

where μ is the coefficient of road adhesion. When the braking forces reach the values determined by Eqs. (3.78) and (3.79), tires are at the point of sliding. Any further increase in the braking force would cause the tires to lock up.

The distribution of the braking forces between the front and rear axles is a function of the design of the brake system when no wheels are locked. For conventional brake systems, the distribution of the braking forces is primarily dependent on the hydraulic (or air) pressures and brake cylinder (or chamber) areas in the front and rear brakes. From Eqs. (3.78) and (3.79), it can be seen that only when the distribution of the braking forces between the front and rear axles is in exactly the same proportion as that of normal loads on the front and rear axles will the maximum braking forces of the front and rear tires be developed at the same time:

$$\frac{K_{bf}}{K_{br}} = \frac{F_{bf\,max}}{F_{br\,max}} = \frac{l_2 + h(\mu + f_r)}{l_1 - h(\mu + f_r)} \tag{3.80}$$

where K_{bf} and K_{br} are the proportions of the total braking force on the front and rear axles, respectively, and are determined by the brake system design. They are also referred to as the installed braking force (effort) distribution on the front and rear axles, respectively.

For instance, for a light truck with 68% of the static load on the rear axle ($l_2/L = 0.32$ and $l_1/L = 0.68$), $h/L = 0.18$, $\mu = 0.85$, and $f_r = 0.01$, the maximum braking forces of the front and rear tires that the tire–ground contact can support will be developed at the same time only if the installed braking force distribution between the front and rear brakes satisfies the following condition:

$$\frac{K_{bf}}{K_{br}} = \frac{0.32 + 0.18(0.85 + 0.01)}{0.68 - 0.18(0.85 + 0.01)} = \frac{47}{53}$$

In other words, 47% of the total braking force must be placed on the front axle and 53% on the rear axle to achieve optimum utilization of the potential braking capability of the vehicle. The braking force distribution that can ensure the maximum braking forces of the front and rear tires develop at the same time is referred to as the ideal braking force distribution. If the braking force distribution is not ideal, then either the front or the rear tires will lock up first.

When the rear tires lock up first, the vehicle will lose directional stability. This can be visualized with the aid of Figure 3.73. The figure shows the top view of a two-axle vehicle acted upon by the braking force and the inertia force. When the rear tires lock, the capability of the rear tires to resist lateral force is reduced to zero. If some slight lateral movement of the rear tires is initiated by side wind, road camber, or centrifugal force, a yawing moment due to the inertia force about the yaw center of the front axle will be developed. As the yaw motion progresses, the moment arm of the inertia force increases, resulting in an increase in yaw acceleration. As the rear end of the vehicle swings around 90°, the moment arm gradually decreases, and eventually the vehicle rotates 180°, with the rear end leading the front end. Figure 3.74 shows the measured angular deviation of a vehicle when the front and rear tires do not lock at the same instant (Lister 1965).

The lockup of front tires will cause a loss of directional control, and the driver will no longer be able to exercise effective steering. However, front tire lockup does not cause directional instability. This is because whenever lateral movement of the front tires occurs, a self-correcting moment due to the inertia force of the vehicle about the yaw center of the rear axle will be developed, which tends to bring the vehicle back to a straight line path.

Loss of steering control may be detected more readily by the driver, and control may be regained by release or partial release of the brakes. Contrary to the case of front tire lockup, when rear tires lock and the angular deviation of the vehicle exceeds a certain level, control cannot be regained, even by complete release of the brakes and by the most skillful driver. This suggests that rear tire

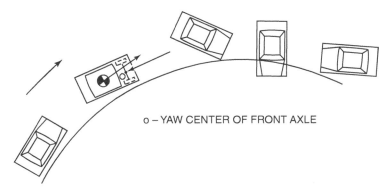

o – YAW CENTER OF FRONT AXLE

Figure 3.73 Loss of directional stability due to lockup of rear tires.

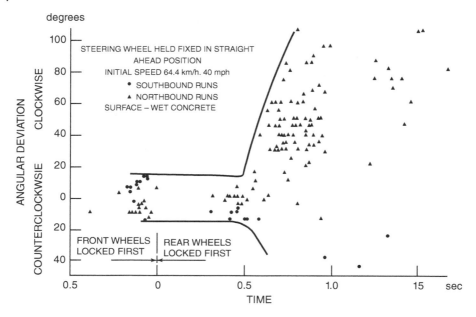

Figure 3.74 Angular deviation of a car when the front and rear tires do not lock at the same instant. Reprinted with permission from SAE paper 650092 © 1965 Society of Automotive Engineers, Inc.

lockup is a more critical situation, particularly on a road surface with a low coefficient of adhesion. Since on slippery surfaces the value of the available braking force is low, the kinetic energy of the vehicle will dissipate at a slow rate, and the vehicle will experience a serious loss of directional stability over a considerable distance. Because of the importance of the sequence of locking of the tires to vehicle behavior during braking, braking performance standards in some countries require that the installed braking effort distribution be such that the front tires lock before the rear tires, up to a critical acceleration, such as 0.85g.

The conditions under which the front or the rear tires will lock first can be quantitatively determined. To facilitate the understanding of the problem, only the braking force and rolling resistance will be considered in the following analysis. Thus,

$$F_b + f_r W = F_{bf} + F_{br} + f_r W = \frac{W}{g} a \tag{3.81}$$

Substituting Eq. (3.81) into Eqs. (3.76) and (3.77) yields

$$W_f = \frac{W}{L}\left(l_2 + \frac{a}{g}h\right) \tag{3.82}$$

$$W_r = \frac{W}{L}\left(l_1 - \frac{a}{g}h\right) \tag{3.83}$$

The braking forces of the front and rear axles as determined by the brake system design are expressed by

$$F_{bf} = K_{bf}F_b = K_{bf}W\left(\frac{a}{g} - f_r\right) \tag{3.84}$$

and

$$F_{br} = K_{br}F_b = \left(1 - K_{bf}\right)F_b = \left(1 - K_{bf}\right)W\left(\frac{a}{g} - f_r\right) \qquad (3.85)$$

The front tires approach lockup when

$$F_{bf} = \mu W_f \qquad (3.86)$$

Substituting Eqs. (3.82) and (3.84) into Eq. (3.86) yields

$$K_{bf}W\left(\frac{a}{g} - f_r\right) = \mu W\left(\frac{l_2}{L} + \frac{a}{g}\frac{h}{L}\right) \qquad (3.87)$$

From Eq. (3.87), the vehicle deceleration rate (in g units) associated with the impending lockup of the front tires can be defined by

$$\left(\frac{a}{g}\right)_f = \frac{\mu l_2/L + K_{bf}f_r}{K_{bf} - \mu h/L} \qquad (3.88)$$

Similarly, it can be shown that the rear tires approach lockup when the deceleration rate is

$$\left(\frac{a}{g}\right)_r = \frac{\mu l_1/L + \left(1 - K_{bf}\right)f_r}{1 - K_{bf} + \mu h/L} \qquad (3.89)$$

For a given vehicle with a particular installed braking force distribution on a given road surface, the front tires will lock first if

$$\left(\frac{a}{g}\right)_f < \left(\frac{a}{g}\right)_r \qquad (3.90)$$

On the other hand, the rear tires will lock first if

$$\left(\frac{a}{g}\right)_r < \left(\frac{a}{g}\right)_f \qquad (3.91)$$

From the above analysis, for a given vehicle with a fixed braking force distribution, both the front and rear tires will lock at the same deceleration rate only on a particular road surface. Under this condition, the maximum braking forces of the front and rear axles that the tire–ground contact can support are developed at the same time, which indicates an optimum utilization of the potential braking capability of the vehicle. Under all other conditions, either the front or rear tires will lock first, resulting in a loss of either steering control or directional stability. This suggests that, ideally, the braking force distribution should be adjustable to ensure optimum braking performance under various operating conditions.

Based on the analysis described above, the interrelationships among the sequence of locking of tires, the deceleration achievable prior to any tire lockup, the design parameters of the vehicle, and operating conditions can be quantitatively defined. As an illustrative example, Figure 3.75 shows the braking characteristics of a light truck as a function of the braking effort distribution on the front axle under loaded and unloaded conditions (Bickerstaff and Hartley 1974). In this analysis, the effects of tire rolling resistance are neglected. For the loaded condition, the gross vehicle weight is 44.48 kN (10 000 lb), and for the unloaded case, it is 26.69 kN (6000 lb). The ratio of the height of the center of gravity to the wheelbase is 0.18 for both loaded and unloaded conditions. The coefficient of road adhesion is 0.85.

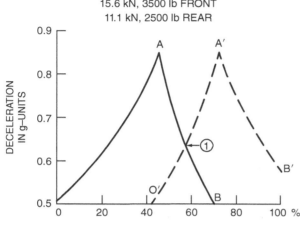

LIGHT TRUCK
ROAD ADHESION COEFFICIENT : 0.85

———— LOADED :
14.2 kN, 3200 lb FRONT
30.2 kN, 6800 lb REAR

—— —·UNLOADED :
15.6 kN, 3500 lb FRONT
11.1 kN, 2500 lb REAR

FRONT BRAKING FORCE / TOTAL BRAKING FORCE

Figure 3.75 Effects of braking effort distribution on braking performance of a light truck. Reprinted with permission from SAE paper 741137 © 1974 Society of Automotive Engineers, Inc.

In Figure 3.75, the solid line and the dashed line represent the boundaries of the deceleration rate that the vehicle can achieve prior to the locking of any tires under loaded and unloaded condition, respectively. Lines OA and $O'A'$ represent the limiting values of the deceleration rate the vehicle can achieve without locking the rear tires under loaded and unloaded condition, respectively. Lines AB and $A'B'$ represent the limiting values of the deceleration rate the vehicle can achieve without locking the front tires under loaded and unloaded condition, respectively. Use can be made of Figure 3.75 to determine the braking characteristics of the light truck under various operating conditions. For instance, if the brake system is designed to have 40% of the total braking force placed on the front axle, then, for the loaded vehicle, the lockup of the rear tires will take place prior to the lockup of the front tires and the highest deceleration rate the vehicle can achieve just prior to rear tire lockup will be approximately 0.77 g. Conversely, if 60% of the total braking force is placed on the front, then, for the loaded case, the lockup of the front tires will take place prior to that of the rear tires, and the highest deceleration rate the vehicle can achieve without the locking of any tires will be approximately 0.61 g. To achieve the maximum deceleration rate of 0.85 g, which indicates the optimum utilization of the potential braking capability on a surface with a coefficient of road adhesion of 0.85, 47% of the total braking force on the front is required for the loaded case as compared to 72% for the unloaded case. Therefore, there is a difference of 25% in the optimum braking force distribution between the loaded and unloaded cases. A compromise in the selection of the installed braking force distribution must be made. Usually, the value of the braking force distribution on the front axle corresponding to the intersection of lines AB and $O'A'$, point 1 in Figure 3.75, is selected as a compromise. Under these circumstances, the maximum deceleration that the truck can achieve

Figure 3.76 Effects of braking effort distribution on braking performance of a passenger car. Reprinted with permission from SAE paper 741137 © 1974 Society of Automotive Engineers, Inc.

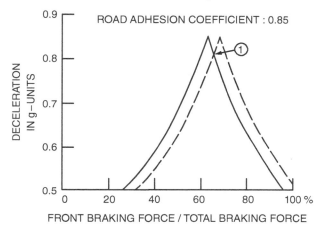

without locking any tires under both loaded and unloaded conditions is 0.64 g on a surface with a coefficient of road adhesion of 0.85.

Figure 3.76 illustrates the braking characteristics of a passenger car (Bickerstaff and Hartley 1974). In this analysis, the effects of tire rolling resistance are neglected. Because the difference in vehicle weight between the loaded and unloaded cases for a passenger car is much smaller than that for a truck, the braking characteristics under these two conditions are very close, which can readily be seen from Figure 3.76. To achieve the maximum deceleration rate of 0.85 g, 62% of the total braking force on the front is required for the loaded case as compared to 67% for the unloaded case, a difference of 5%. An installed braking force distribution with 64.5% of the total braking force on the front, corresponding to point 1 in Figure 3.76, may be selected as a compromise under these circumstances. The maximum deceleration that the vehicle can achieve prior to any tire lockup under both loaded and unloaded conditions is therefore approximately 0.82 g.

The analysis and examples given above indicate the complex nature of the braking process. It is shown that the optimum braking force distribution, which ensures the maximum deceleration rate, varies with the loading conditions of the vehicle, vehicle design parameters, and road surface conditions. In practice, the operating conditions vary in a wide range; thus, for a given vehicle with a fixed braking force distribution, only under a specific set of loading and road conditions will the maximum braking forces on the front and rear axles be developed at the same time and will the maximum deceleration rate be achieved. Under all other conditions, the achievable deceleration rate without causing a loss of steering control or directional stability will be reduced. To improve braking performance, pressure proportioning valves or load-sensing proportioning valves have been used. Pressure proportioning valves commonly in use provide equal pressure to both front and rear brakes up to a certain pressure level, and then reduce the rate of pressure rise to the rear brakes. Load-sensing proportioning valves have been used on trucks. These valves adjust the braking effort distribution as a function of load distribution between the axles.

An electronic braking force distribution system (EBD) has been introduced. This system utilizes the information from wheel speed sensors and monitors the differences in skid between the front and rear wheels during braking. If the system detects over braking of the rear wheels, the system reduces the rise in brake pressure at the rear wheels and thereby reduces the risks of vehicle directional instability.

To ensure steering control and directional stability of the vehicle under all operating conditions, the antilock brake system has been introduced. Its prime function is to prevent tires from locking; thus, the capability of the tire to sustain a side force is maintained. The operating principles of the antilock brake system are briefly discussed in Section 3.11.4.

Example 3.3 A passenger car weighs 21.24 kN(4775 lb) and has a wheelbase of 2.87 m (113 in.). The center of gravity is 1.27 m (50 in.) behind the front axle and 0.508 m (20 in.) above ground level. The installed braking effort distribution on the front axle is 60%. The coefficient of rolling resistance is 0.02. Determine which set of the tires will lock first on two road surfaces: one with a coefficient of road adhesion $\mu = 0.8$; the other with $\mu = 0.2$.

Solution

(a) On the road surface with $\mu = 0.8$, the vehicle deceleration associated with the impending lockup of the front tires is determined by Eq. (3.88):

$$\left(\frac{a}{g}\right)_f = \frac{\mu l_2/L + K_{bf} f_r}{K_{bf} - \mu h/L} = \frac{0.8 \times 0.558 + 0.6 \times 0.02}{0.6 - 0.8 \times 0.177} = 1.0$$

The vehicle deceleration associated with the impending lockup of the rear tires is determined by Eq. (3.89):

$$\left(\frac{a}{g}\right)_r = \frac{\mu l_1/L + (1 - K_{bf}) f_r}{1 - K_{bf} + \mu h/L} = \frac{0.8 \times 0.442 + 0.4 \times 0.02}{0.4 + 0.8 \times 0.177} = 0.67$$

Since $(a/g)_f > (a/g)_r$, the rear tires will lock first on the road surface with $\mu = 0.8$.

(b) On the road surface with $\mu = 0.2$,

$$\left(\frac{a}{g}\right)_f = \frac{0.2 \times 0.558 + 0.6 \times 0.02}{0.6 - 0.2 \times 0.177} = 0.219$$

$$\left(\frac{a}{g}\right)_r = \frac{0.2 \times 0.442 + 0.4 \times 0.02}{0.4 + 0.2 \times 0.177} = 0.221$$

Since $(a/g)_f < (a/g)_r$, the front tires will lock first on the road surface with $\mu = 0.2$.

3.11.2 Braking Efficiency and Stopping Distance

To characterize the braking performance of a road vehicle, braking efficiency can be used. Braking efficiency η_b is defined as the ratio of the maximum deceleration rate in g units (a/g) achievable prior to any tire lockup to the sum of the coefficient of road adhesion μ and the tire rolling resistance coefficient f_r. It is expressed by

$$\eta_b = \frac{a/g}{\mu + f_r} \tag{3.92}$$

The braking efficiency indicates the extent to which the vehicle utilizes the available coefficient of road adhesion and the tire rolling coefficient for braking. Thus, when $a/g < \mu + f_r$, hence $\eta_b < 1.0$, the deceleration is less than the maximum achievable, resulting in an unnecessarily long stopping distance. Referring to Figure 3.75, if 57% of the total braking force is placed on the front axle, corresponding to point 1, the maximum deceleration achievable prior to any tire lockup is 0.64 g. This indicates that on a surface with a coefficient of road adhesion of 0.85 and with the tire rolling resistance coefficient being neglected as noted previously, the braking efficiency is 75.3%.

Stopping distance is another parameter widely used for evaluating the overall braking performance of a road vehicle. It is made up of the following factors.

1) Reaction distance

This is the distance traveled by the vehicle during the reaction time of the driver and the delay time of the brake system from the actuation of the brake pedal to the initiation of braking action. It is estimated that the sum of the reaction time of the driver and the delay time of the brake system t_r varies from 0.75 to 2 s. During this period, the vehicle essentially travels at the initial vehicle speed V_0 just prior to applying brakes. The reaction distance S_r may be estimated by

$$S_r = V_0 t_r \tag{3.93}$$

2) Pressure buildup distance

This is the distance traveled by the vehicle from the initiation of braking action to the full development of brake pressure in the brake system. Pressure buildup time t_p is the time required by the driver and brake system to overcome the internal resistances in the brake system and to allow brake pressure to reach the cylinders of the wheel brakes, thus building up braking forces between the wheels and the road. The pressure buildup time is dependent on the actuating behavior of the driver and the design of the brake system. Based on available data, t_p varies in the range 0.3–0.75 s. During this period, the average deceleration of the vehicle a_p may be taken as approximately one-half of the deceleration when brake pressure is fully developed, and can be expressed by

$$a_p = \frac{1}{2}\left(\frac{F_b + \Sigma R}{\gamma_b W/g}\right)$$

where F_b is the total braking force of the vehicle when brake pressure is fully developed; g is acceleration due to gravity; W is the vehicle weight; ΣR is the resultant resistance of the vehicle; and γ_b is an equivalent mass factor taking into account the mass moments of inertia of all rotating components involved during braking. Since the clutch is usually disengaged during braking, the value of γ_b is not necessarily the same as that of γ_m used in the prediction of vehicle acceleration. For passenger cars or light trucks, γ_b is in the range 1.03–1.05.

The pressure buildup distance S_p may be estimated by

$$S_p = V_0 t_p - \frac{1}{2} a_p t_p^2 \tag{3.94}$$

3) Braking distance

This is the distance traveled by the vehicle during the period when brake pressure is fully developed. The braking distance S_b from an initial speed $V_1 = V_0 - a_p t_p$ to a final speed V_2 ($V_2 = 0$ for a complete stop) can be expressed by

$$S_b = \int_{V_2}^{V_1} \frac{\gamma_b W}{g} \frac{V \, dV}{F_b + \sum R}$$

Substituting Eq. (3.72) into the above equation and neglecting the transmission resistance R_t, one obtains

$$S_b = \frac{\gamma_b W}{g} \int_{V_2}^{V_1} \frac{V \, dV}{F_b + f_r W \cos \theta_s \pm W \sin \theta_s + R_a} \tag{3.95}$$

The aerodynamic resistance is proportional to the square of speed, and it may be expressed as

$$R_a = \frac{\rho}{2} C_D A_f V^2 = C_{ae} V^2 \tag{3.96}$$

With substitution of $C_{ae} V^2$ for R_a and integration, the braking distance can be expressed by (Taborek 1957)

$$S_b = \frac{\gamma_b W}{2g C_{ae}} \ln \left(\frac{F_b + f_r W \cos \theta_s \pm W \sin \theta_s + C_{ae} V_1^2}{F_b + f_r W \cos \theta_s \pm W \sin \theta_s + C_{ae} V_2^2} \right) \tag{3.97}$$

For final speed $V_2 = 0$, Eq. (3.97) reduces to the form

$$S_b = \frac{\gamma_b W}{2g C_{ae}} \ln \left(1 + \frac{C_{ae} V_1^2}{F_b + f_r W \cos \theta_s \pm W \sin \theta_s} \right) \tag{3.98}$$

For a given vehicle, if the braking force distribution and road conditions are such that the maximum braking forces of the front and rear tires that the tire–ground contact can support are developed at the same time – that is, the braking efficiency $\eta_b = 100\%$ – the minimum braking distance will be achieved. In this case, the braking torque generated by the brakes has already overcome the inertia of the rotating parts connected with the wheels; the maximum braking forces developed at the tire–ground contact are retarding only the translational inertia. The mass factor γ_b is therefore 1 (unity). The minimum braking distance can be expressed as

$$S_{b\,min} = \frac{W}{2g C_{ae}} \ln \left(1 + \frac{C_{ae} V_1^2}{\mu W \cos \theta_s + f_r W \cos \theta_s \pm W \sin \theta_s} \right) \tag{3.99}$$

If the braking efficiency η_b is less than 100% (i.e., the maximum deceleration rate in g units achievable prior to tire lockup is less than the sum of the coefficient of road adhesion and the tire rolling resistance coefficient available), then the braking distance will be longer than that determined using Eq. (3.99). In this case, the braking distance may be calculated from

$$S_b = \frac{W}{2g C_{ae}} \ln \left(1 + \frac{C_{ae} V_1^2}{\eta_b (\mu W + f_r W) \cos \theta_s \pm W \sin \theta_s} \right) \tag{3.100}$$

The stopping distance S_s is the sum of the reaction distance, pressure buildup distance, and braking distance, and is expressed by

$$S_s = S_r + S_p + S_b$$

3.11.3 Braking Characteristics of a Tractor–Semitrailer

In comparison with a two-axle vehicle, the braking characteristics of a tractor–semitrailer are more complex. For a given two-axle vehicle, the load transfer is only a function of the deceleration rate, whereas for a tractor–semitrailer, the load transfer during braking is dependent not only on the deceleration rate, but also on the braking force of the semitrailer. Consequently, the optimum braking for a tractor–semitrailer is even more difficult to achieve than for a two-axle vehicle. A tractor–semitrailer during emergency braking could exhibit behavior of a more complex nature than that of a two-axle vehicle. In addition to the possibility of loss of directional control due to the lockup of tractor front tires, directional instability of a tractor–semitrailer may be caused by the locking of either the tractor rear tires or the semitrailer tires. The locking of the tractor rear tires first usually causes jackknifing, which puts the vehicle completely out of control and often causes considerable damage to both the vehicle itself and other road users. On the other hand, the lockup of the semitrailer tires causes trailer swing. Although trailer swing has little effect on the stability of the tractor, it could be very dangerous to other road users, particularly to the oncoming traffic (Wong and Guntur 1978).

To reach a better understanding of the braking characteristics of a tractor–semitrailer, it is necessary to review its mechanics of braking. Figure 3.77 shows the major forces acting on a tractor–semitrailer during braking. To simplify the analysis, the aerodynamic drag and rolling resistance will be neglected.

The equilibrium equations are as follows.

- For the tractor

$$W_f + W_r = W_1 + W_{hi} \tag{3.101}$$

$$C_f W_f + C_r W_r = a/g W_1 + F_{hi} \tag{3.102}$$

$$(a/g)W_1 h_1 + F_{hi} h_3 + W_1(L_1 - l_1 - d_1) + W_r d_1 = W_f(L_1 - d_1) \tag{3.103}$$

- For the semitrailer

$$W_{hi} + W_s = W_2 \tag{3.104}$$

$$F_{hi} + C_{se} W_s = (a/g) W_2 \tag{3.105}$$

$$W_2 d_2 + F_{hi} h_3 = (a/g) W_2 h_2 + W_s L_2 \tag{3.106}$$

- For the tractor–semitrailer combination

$$W_f + W_r + W_s = W_1 + W_2 \tag{3.107}$$

$$C_f W_f + C_r W_r + C_{se} W_s = (a/g)(W_1 + W_2) \tag{3.108}$$

$$(a/g)W_1 h_1 + (a/g)W_2 h_2 + W_r L_1 + W_s[L_1 - d_1 + L_2] = W_1 l_1 + W_2[L_1 - d_1 + d_2] \tag{3.109}$$

where W_{hi} is the vertical load on the fifth wheel, F_{hi} is the horizontal load on the fifth wheel, a is the deceleration of the vehicle, and C_f, C_r, and C_{se} are the ratios of the braking force to the normal load of the tractor front axle, rear axle, and semitrailer axle, respectively. Other parameters are shown in Figure 3.77. The equations described above are applicable to tractors with a single rear axle and semitrailers with a single axle. For tractors and semitrailers having tandem axles without equalization, the equations must be modified, because of interaxle load transfer.

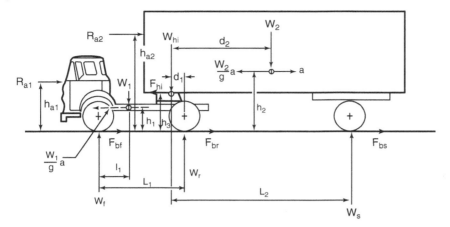

Figure 3.77 Forces acting on a tractor–semitrailer during braking.

From the above equations, the normal loads on various axles can be expressed by the following.

- Tractor front axle

$$W_f = \frac{W_1[L_1 - l_1 + (a/g)h_1 + (C_r - a/g)h_3]}{L_1 + (C_r - C_f)h_3} + \frac{W_2[L_2 - d_2 + (C_{se} - a/g)h_3 + (a/g)h_2](d_1 + C_r h_3)}{(L_2 + C_{se}h_3)[L_1 + (C_r - C_f)h_3]}$$

(3.110)

- Tractor rear axle

$$W_r = \frac{W_1[l_1 - (a/g)h_1 + (a/g - C_f)h_3]}{L_1 + (C_r - C_f)h_3} + \frac{W_2[(L_2 - d_2) + (C_{se} - a/g)h_3 + (a/g)h_2][(L_1 - d_1) - C_f h_3]}{(L_2 + C_{se}h_3)[L_1 + (C_r - C_f)h_3]}$$

(3.111)

- Semitrailer axle

$$W_s = W_2 \frac{d_2 + (h_3 - h_2)a/g}{C_{se}h_3 + L_2}$$

(3.112)

It shows that to determine the normal loads on various axles of a given tractor–semitrailer, the deceleration rate and the braking force coefficient of the semitrailer axle C_{se} must be specified. When the deceleration and the braking force of the semitrailer axle are known, the normal load on the semitrailer axle can be determined from Eq. (3.112), and the vertical and horizontal loads on the fifth wheel, W_{hi} and F_{hi}, can be calculated from Eqs. (3.104) and (3.105). With the values of W_{hi} and F_{hi} known, from Eqs. (3.101) and (3.103), the normal loads on the front and rear axles of the tractor can be calculated.

For the optimum braking condition where the maximum braking forces of all axles that the tire–ground contact can support are developed at the same time, the braking force coefficients for all axles and the deceleration rate in g units are equal to the coefficient of road adhesion, $C_f = C_r = C_{se} = a/g = \mu$. The expressions for the axle loads given above can be simplified as follows.

- Tractor front axle

$$W_f = \frac{W_1(L_1 - l_1 + \mu h_1)}{L_1}$$
$$+ \frac{W_2(L_2 - d_2 + \mu h_2)(d_1 + \mu h_3)}{L_1(L_2 + \mu h_3)} \tag{3.113}$$

- Tractor rear axle

$$W_r = \frac{W_1(l_1 - \mu h_1)}{L_1}$$
$$+ \frac{W_2(L_2 - d_2 + \mu h_2)(L_1 - d_1 - \mu h_3)}{L_1(L_2 + \mu h_3)} \tag{3.114}$$

- Semitrailer axle

$$W_s = \frac{W_2[d_2 + \mu(h_3 - h_2)]}{\mu h_3 + L_2} \tag{3.115}$$

Under the optimum braking condition, the braking forces on the axles are proportional to the corresponding normal loads. The required braking force distribution among the axles, therefore, can be determined from Eqs. (3.113) to (3.115). Figure 3.78 shows the variation of the optimum braking force distribution with the coefficient of road adhesion for a particular tractor–semitrailer under various loading conditions (Wong and Guntur 1978). The parameters of the vehicle used in

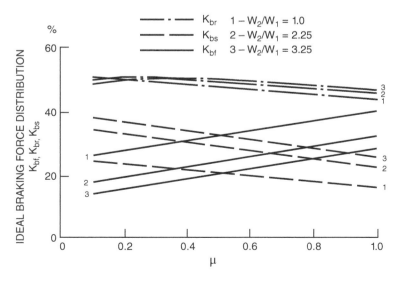

Figure 3.78 Variations of ideal braking force distribution with road adhesion coefficient and loading conditions for a tractor–semitrailer.

the analysis are as follows: $W_1 = 75.62$ kN (17,000 lb), $W_2 = 75.62$ kN (17,000 lb) (semitrailer empty), $W_2 = 170.14$ kN (38,250 lb) (semitrailer partially loaded), $W_2 = 245.75$ kN (55,250 lb) (semitrailer fully loaded), $L_1 = 5.0$ m (16.5 ft), $L_2 = 9.75$ m (32 ft), $l_1 = 2.75$ m (9 ft), $d_1 = 0.3$ m (1 ft), $d_2 = 4.88$ m (16 ft), $h_1 = 0.84$ m (2.75 ft), $h_2 = 2.44$ m (8 ft), and $h_3 = 0.98$ m (3.20 ft). It shows that the optimum value of the braking force distribution on the tractor rear axle K_{br} varies very little over a wide range of road and loading conditions. On the other hand, the optimum value of the braking force distribution on the tractor front axle K_{bf} and that on the semitrailer axle K_{bs} vary considerably with the coefficient of road adhesion and with the loading conditions of the semitrailer. This indicates that for a tractor–semitrailer combination with a fixed braking force distribution, the optimum braking condition can be achieved only with a particular loading configuration over a specific road surface. Under all other conditions, one of the axles will lock up first. As mentioned previously, the locking of tractor front tires results in a loss of steering control, the locking of tractor rear tires first leads to jackknifing, and the locking of semitrailer tires causes trailer swing. This indicates that the locking sequence of the tires is of importance to the behavior of the tractor–semitrailer during braking. As jackknifing is the most critical situation, the preferred locking sequence, therefore, appears to be tractor front tires locking up first, then semitrailer tires, and then tractor rear tires. A procedure for predicting the locking sequence of tires of tractor–semitrailers has been developed (Wong and Guntur 1978). It has been shown that by careful selection of the braking force distribution among the axles coupled with the proper control of loading conditions, the preferred locking sequence may be achieved over a certain range of road conditions, thus minimizing the undesirable directional response. However, a loss of braking efficiency will result under certain operating conditions (Wong and Guntur 1978).

The dynamic behavior and directional response of tractor–semitrailers during braking is of practical importance to traffic safety. Extensive study has been made, and the results have been reported in the literature (Mikulcik 1971; Murphy et al. 1972; Winkler et al. 1976; Wong et al. 1977; Wong and Guntur 1978; Guntur and Wong 1978; van Zanten and Krauter 1978; Lam et al. 1979; Guntur and Wong 1980; Verma et al. 1980; Ellis 1994).

3.11.4 Antilock Brake Systems

As mentioned previously, when a tire is locked (i.e., 100% skid), the coefficient of road adhesion falls to its sliding value, and its ability to sustain side force is reduced to almost null. As a result, the vehicle will lose directional control or stability, dependent upon whether the front tires or the rear tires lock up first, as discussed previously. Also, the stopping distance will be longer than the minimum achievable. Figure 3.79 shows the general characteristics of the braking effort coefficient (i.e., the ratio of the braking effort to the normal load of the tire) and the cornering force coefficient (i.e., the ratio of the cornering force to the normal load of the tire) as a function of skid at a given slip angle for a pneumatic tire. When the braking effort coefficient is the region beyond its maximum value, the tire operates in an unstable condition and the skid of the tire tends to increase rapidly toward being locked up.

The prime function of an antilock brake system is to prevent the tire from locking, and ideally to keep the skid of the tire within a desirable range, such as that shown in Figure 3.79. This will ensure that the tire can develop a sufficiently high braking force for stopping the vehicle, and at the same time it can retain an adequate cornering force for directional control or stability. Data collected in Germany and other countries have shown that the introduction of antilock brake systems has reduced a noticeable number of traffic accidents involving passenger cars and has also mitigated the consequences of a number of accidents (Klein 1986).

Figure 3.79 Effects of skid on cornering force coefficient of a tire.

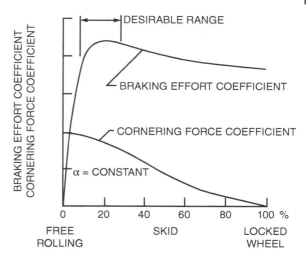

To appreciate the operation of an antilock brake system, it would be useful to briefly review the dynamics of the tire during braking, with the tire rolling resistance being neglected. When a braking torque T_b is applied to the tire, a corresponding braking effort F_b is developed on the tire–ground contact patch, as shown in Figure 3.80. This braking effort F_b has a moment about the tire center, which is equal to $F_b r$ and acts in the opposite direction of the applied braking torque T_b, as shown in Figure 3.80. The difference between $F_b r$ and T_b causes an angular acceleration (or deceleration) $\dot{\omega}$ of the tire:

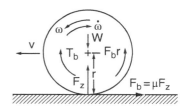

Figure 3.80 Forces and moments acting on a tire during braking.

$$\dot{\omega} = (F_b r - T_b)/I_w \tag{3.116}$$

where I_w is the mass of moment of inertia of the tire assembly about its center and r is the radius of the tire. When the difference between $F_b r$ and T_b is positive, the tire rotation accelerates, and when it is negative, the tire rotation decelerates.

The braking effort F_b also causes a linear deceleration a_c of the tire center:

$$a_c = F_b/(W/g) \tag{3.117}$$

where W is the load carried by the tire and g is the acceleration due to gravity.

Because of the skid of the tire during braking, the linear deceleration of the tire center a_c is not equal to $r\dot{\omega}$. As noted previously, the skid i_s of the tire is defined by Eq. (1.30) as follows:

$$i_s = \left(1 - \frac{r\omega}{V}\right) \times 100\%$$

where ω and V are the angular speed and linear speed of the center of the tire, respectively.

If the applied braking torque T_b is large and the torque T_t due to the braking effort F_b on the tire–ground contact patch is small, then according to Eq. (3.116), the angular deceleration $\dot{\omega}$ will be high. This indicates that the tire will tend to lock (i.e., its angular speed ω becomes zero, while the linear speed of the tire center V is not zero) within a short period of time. The basic function of an antilock device is to monitor the operating conditions of the tire and to control the applied braking torque by modulating the brake pressure to prevent the tire from becoming locked.

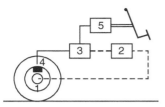

1. SENSOR
2. CONTROL UNIT
3. MODULATOR
4. WHEEL CYLINDER
5. MASTER CYLINDER

Figure 3.81 Elements of an antilock braking system.

An antilock brake system is an electronic feedback control system. It consists of sensors, an electronic control unit, and a brake pressure modulator, as schematically shown in Figure 3.81.

Sensors with electromagnetic pulse pickups and toothed wheels are usually used to monitor the rotation of individual tires or the propeller shaft for determining the average angular speed of the tires on a drive axle. To monitor the rotation of individual tires, sensors are mounted directly to the wheeled hubs. They usually generate 90–100 pulses per wheel revolution. The angular speed and angular deceleration (or angular acceleration) of the tire are derived from these digital pulse signals by differentiation with respect to time. A linear accelerometer is used in some antilock devices to monitor the longitudinal deceleration of the vehicle. The signals generated by the sensors are transmitted to the electronic control unit for processing.

The control unit usually consists of four modules: a signal processing module, a module for predicting whether the tire is at the point of locking, a module for determining whether the danger of locking the tire is averted, and a module for generating a command signal for activating the pressure modulator. In the control unit, after signals generated by the sensors have been processed, the measured parameters and/or those derived from them are compared with the corresponding predetermined threshold values. When certain conditions that indicate the impending lockup of the tire are met by the measured parameters and/or their derivatives, a command signal is sent to the modulator to release the brake (Guntur and Ouwerkerk 1972; Wong et al. 1977; Leiber and Czinczel 1979, 1983; Guntur and Wong 1980; Satoh and Shiraishi 1983a, b; van Zanten 2014).

In practice, the actual skid of the tire is difficult to determine, primarily due to the lack of a cost-effective means to directly measure the linear speed V of the tire center during braking. For antilock brake systems currently available, tire skid can neither be directly determined nor controlled. The control logic of an antilock device, therefore, is usually formulated based on some easily measurable parameters, such as the angular deceleration (or angular acceleration) of the tire.

A high angular deceleration of the tire will generally indicate that tire skid is higher than the desirable value at the maximum braking effort coefficient and that the tire will tend to lock, as noted previously. The methods for predicting the locking of the tire used in some antilock brake systems are described below.

- In some of the existing antilock devices, the locking of the tire is predicted, and a command signal is transmitted to the modulator to release the brake whenever the product of the angular deceleration $\dot{\omega}$ of the tire and its rolling radius r exceeds a predetermined value. In some systems, the threshold value used is in the range 1–1.6 g.
- In an antilock system designed for passenger cars, the angular speed signal of the tire is tracked by a track-and-hold circuit in the control unit, and when the value of $r\dot{\omega}$ is greater than 1.6g, the tracked signal is held in a memory circuit for about 140 ms. During this period of time, if the measured angular speed of the tire decreases by 5% of the already held value, and at the same time if the linear deceleration of the vehicle measured by a linear accelerometer is not higher

than 0.5 g, it is predicted that the tire is at the point of locking, and a command signal for releasing the brake is sent to the modulator. On the other hand, if the linear deceleration of the vehicle is higher than 0.5 g, locking of the tire is predicted, and the brake is released whenever the decrease in angular speed of the tire is 15% of the already stored value.

- In many current antilock brake systems, the brake pressure will be reduced if the following two conditions are met: the estimated tire skid $i_s > i_{s0}$ and $r\dot\omega = r\dot\omega_0$, where i_{s0} is the threshold value for tire skid, typically 10%, and $r\dot\omega_0$ is the threshold value for the circumferential deceleration of the tire, typically $1 - 1.6g$ (Leiber and Czinczel 1979; Satoh and Shiraishi 1983a). As mentioned previously, the actual skid of the tire during braking is difficult to determine. Therefore, in many cases, the tire skid i_s is calculated from the estimated linear speed of the tire center based on the radius r and the measured angular speed ω of the tire, using various estimation methods. To avoid excessive operation of the antilock device at low vehicle speeds due to errors in estimating tire skid, the threshold value for skid i_{s0} is increased with a decrease in the vehicle speed.

During the braking process, the operating conditions of the tire and the vehicle are continuously monitored by the sensors and the control unit. After the danger of locking the tire is predicted and the brake is released, another module in the control unit will determine at what point the brake should be reapplied. A variety of criteria are employed in existing antilock systems and some of them are described below.

- In some systems, a command signal will be sent to the modulator to reapply the brake whenever the criteria for releasing the brake discussed previously are no longer satisfied.
- In certain devices, a fixed time delay is introduced to ensure that the brake is reapplied only when a fixed time has elapsed after the release of the brake.
- When the brake is released, the forward momentum of the vehicle causes the tire to have an angular acceleration. In some systems, the brake is reapplied as soon as the product of the angular acceleration $\dot\omega$ of the tire and the rolling radius r exceeds a predetermined value. Threshold values for $r\dot\omega$ in the range $2.2 - 3g$ have been used. In some devices, the brake pressure buildup rate is made dependent on the angular acceleration of the tire.

As an example, Figure 3.82 shows the characteristics of an antilock system designed for heavy commercial vehicles (Srinivasa et al. 1980). The variations with time of the brake pressure, tire skid,

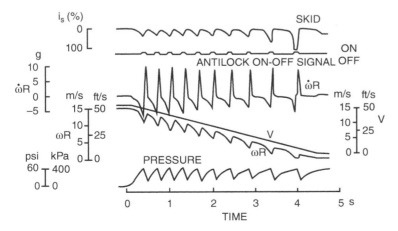

Figure 3.82 Operating characteristics of an antilock system for heavy commercial vehicles with pneumatic braking systems.

vehicle speed, circumferential speed $r\omega$, and circumferential acceleration $r\dot\omega$ of the tire during a simulated braking maneuver over a wet pavement are shown. As can be seen, the brake pressure fluctuates during the operation of an antilock device. The cycle of reducing, holding, and restoring the brake pressure is repeated several times, typically from 5 to 16 times per second, until the vehicle has slowed to a speed of approximately $3 - 5$ km/h ($2 - 3$ mph), at which point the antilock device is usually deactivated.

Various layouts for antilock devices on road vehicles have been used. The primary consideration is to ensure directional control and stability of the vehicle, not only when braking in a straight line, but also when braking in a turn and on an asymmetrical road surface having different values of coefficient of road adhesion for the left- and right-hand-side tires.

The common layouts for passenger cars are the four-channel and four-sensor, three-channel and three-sensor, and three-channel and four-sensor configurations, as schematically shown in Figure 3.83 (Leiber and Czinczel 1983). A channel refers to the portion of the brake system that the control unit/modulator controls independently of the rest of the brake system. For instance, the four-channel and four-sensor configuration shown in the figure has four hydraulic brake circuits with sensors for monitoring the operating conditions of the four tires separately. The two front tires are controlled individually, based on the information obtained by the respective sensors. However, the two tires on the rear axle are jointly controlled in the "select-low" operating mode. "Select-

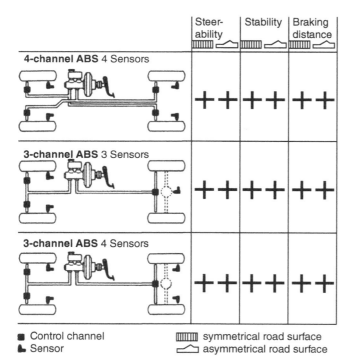

Figure 3.83 Various layouts of antilock brake system for passenger cars. Reprinted with permission from SAE paper 830481 © 1983 Society of Automotive Engineers, Inc.

low" means that the control unit will use the information from the slower of the two tires to jointly control both tires with the same brake pressure, whereas "select-high" means that the control unit will use the information from the faster of the two tires to control the brake pressure applied to both tires. Using the "select-low" operating mode for controlling the two tires on the rear axle will ensure vehicle directional stability when braking on an asymmetrical road or on a turn, in contrast to using the "select-high" operating mode. This is because, with the "select-high" mode, the tire on the low friction side of an asymmetrical road or on the inside of a turn will be locked up, while the other tire on the high friction side of the road or on the outside of a turn develops a higher braking force. This results in a reduction of the cornering force available for the axle, and a large yawing moment, which will have adverse effects on the directional stability of the vehicle. However, using the "select-low" operating mode will have a lower braking efficiency and a longer stopping distance than using the "select-high" mode. Also, with the four tires individually controlled, different braking forces are acting on the left- and right-hand-side tires on the front axle, as well as on the rear axle on an asymmetrical road. This results in a large yawing moment, which will adversely affect the directional stability of the vehicle. The three-channel and three-sensor configuration has three hydraulic circuits for the control of the two front tires individually and the two rear tires jointly, based on the information obtained by the two sensors mounted on the two front tires and by the one sensor for monitoring the average operating conditions of the two rear tires. The operation of the three-channel and four-sensor configuration with the "select-low" operating mode for the rear tires is similar to that of the four-channel and four-sensor configuration described above.

A two-channel and four-sensor configuration in which the two front tires are jointly controlled by the "select-high" operating mode and the two rear tires by the "select-low" operating mode has also been developed (Satoh and Shiraishi 1983b). It is claimed that this system can substantially curb the excessive operation of the antilock brake system on rough roads, and that the combination of the "select-high" for the front axle and the "select-low" for the rear axle offers a reasonable compromise between achieving a sufficiently short stopping distance and retaining adequate directional stability.

3.11.5 Traction Control Systems

Similar to the functions of antilock devices for improving vehicle braking performance, traction control systems have been developed for improving vehicle tractive performance and for maintaining directional control and stability during acceleration (Bleckmann et al. 1986; Demel and Hemming 1989). The prime functions of a traction control system are as follows:

- To improve traction on an asymmetrical road surface with different values of coefficient of road adhesion for the left- and right-hand side tires
- To prevent the tire from spinning during acceleration or on slippery surfaces, and ideally to keep the slip of the tire within a desired range to retain an adequate cornering force for direction control or stability

Like an antilock brake system, a typical traction control system includes sensors, an electronic control unit, and a brake pressure modulator. In addition, it has an engine control device, which controls the throttle, fuel injection system, or ignition. Because of the similarity between the two systems, the traction control system is usually integrated with the antilock brake system, sharing a large number of common components, such as the sensor, electronic control unit, and brake pressure modulator.

For a drive axle with a simple differential, the driving torque is evenly distributed between the left- and right-hand-side half-shafts. As a result, when operating on an asymmetrical road, the tire on the side of the road having a lower value of coefficient of road adhesion will slip excessively, and will impose a limit on the tractive effort that the other tire with a higher value of coefficient of road adhesion can develop. While a differential lock, limited-slip differential or viscous coupling would provide a solution to this problem, a traction control system is a viable alternative. When the slip of a tire on one side of a drive axle is determined to be excessive, a braking torque is applied to that tire through brake pressure modulation so as to increase the driving torque available to the other tire for improving traction. For a vehicle with only one drive axle, the slip of the driven tires can be determined directly using the angular speed of the free-rolling tires on the nondriven axle as a reference.

When the vehicle operates on a slippery surface with a low coefficient of road adhesion or during start-up, both tires on the drive axle may slip excessively. Under these circumstances, the traction control system will apply braking torques to both driven tires and/or decrease the engine output torque to reduce the tractive effort and slip. This ensures adequate directional control for a front-wheel drive vehicle or directional stability for a rear-wheel drive vehicle.

When operating on a long stretch of slippery road, to avoid brake overheating, a combined brake pressure modulation and engine control is necessary. The response of traction control to brake pressure modulation is usually faster than that to engine control because of the actuation time of the engine control device, engine rotating inertia, and the elasticity of the drivetrain.

References

Bayindirli, C., Akansu, Y., and Salman, M.S. (2016). The determination of aerodynamic drag coefficient of truck and trailer model by wind tunnel tests. *International Journal of Automotive Engineering and Technologies* **5** (2).

Beard, K.W. and Reddy, T.B. (eds.) (2019). *Linden's Handbook of Batteries*, 5e. New York: McGraw-Hill.

Berta, C. and Bonis, B. (1980). Experimental shape research of ideal aerodynamic characteristics for industrial vehicles. Society of Automotive Engineers paper 801402.

Bickerstaff, D.J. and Hartley, G. (1974). Light truck tire traction properties and their effect on braking performance. *SAE Transactions* **83**, paper 741137.

Bleckmann H.W., Fennel, H., Graber, J., and Selbert, W.W. (1986). Traction control system with Teves ABS Mark II. Society of Automotive Engineers paper 860506.

Bonthron, A. (1985). CVT-efficiency measured under dynamic running conditions. Society of Automotive Engineers paper 850569.

Carr, C.W. (1983). Potential for aerodynamic drag reduction in car design. In: *Impact of Aerodynamics on Vehicle Design: Technological Advances in Vehicle Design Series SP3* (ed. M.A. Dorgham). Jersey, Channel Islands, UK: Inderscience Enterprises.

Chana, H.E., Fedewa, W.L., and Mahoney, J.E. (1977). An analytical study of transmission modifications as related to vehicle performance and economy. Society of Automotive Engineers paper 770418.

Chowdhury, H., Moria, H., Ali, A. et al. (2013). A study on aerodynamic drag of a semi–trailer truck. *Procedia Engineering* **56**.

Cole, D. (1984). Automotive fuel economy. In: *Fuel Economy in Road Vehicles Powered by Spark Ignition Engines* (eds. J.C. Hilliard and G.S. Springer). New York: Plenum Press.

Coon, C.W. and Wood, C.D. (1974). Improvement of automobile fuel economy. Society of Automotive Engineers paper 740969.

Cooper, K.R. and Leuschen, J. (2005). Model and full-scale wind tunnel tests of second-generation aerodynamic fuel saving devices for tractor–trailers. Society of Automotive Engineers paper 2005-01-3512.

Demel, H. and Hemming, H. (1989). ABS and ASR for passenger cars: Goals and limits. Society of Automotive Engineers paper 890834.

Eaker, G.W. (1988). Wind tunnel-to-road aerodynamic drag correlation. Society of Automotive Engineers paper 880250. doi:https://doi.org/10.4271/880250.

Ehsani, M., Gao, Y., Longo, S., and Ebrahimi, K. (2019). *Modern Electric, Hybrid Electric, and Fuel Cell Vehicles*, 3e. Boca Raton, FL: CRC Press.

Ellis, J.R. (1994). *Vehicle Handling Dynamics*. London: Mechanical Engineering Publications.

Environmental Protection Agency (2021). Fuel Economy. https://www.fueleconomy.gov/feg/fe_test_schedules.shtml (accessed 10 March 2021).

Giles, J.G. (1969). *Gears and Transmissions: Automotive Technology Series 4*. London: Butterworth.

Gotz, H. (1990). Commercial vehicles. In: *Aerodynamics of Road Vehicles* (ed. W.-H. Hucho). London: Butterworth-Heinemann.

Greenwood, C.J. (1984). The design, construction and operation of a commercial vehicle continuously variable transmission. *Driveline '84*. London: Institution of Mechanical Engineers.

Guntur, R.R. and Ouwerkerk, H. (1972). Adaptive brake control systems. *Procedings of the Institution of Mechanical Engineers* **186**.

Guntur, R.R. and Wong, J.Y. (1978). Application of the parameter plane method to the analysis of directional stability of tractor–semitrailers. *ASME Transactions, Journal of Dynamic Systems, Measurement and Control* **100** (1).

Guntur, R.R. and Wong, J.Y. (1980). Some design aspects of anti-lock brake systems for commercial vehicles. *Vehicle System Dynamics* **9** (3).

Hahne, D. (1984). A continuously variable automatic transmission for small front wheel drive cars. *Driveline '84*. London: Institution of Mechanical Engineers.

Holgado, M. and Alique, D. (2019). Preliminary equipment design for onboard hydrogen production by steam reforming in palladium membrane reactors. *ChemEngineering* **3** (1).

Hucho, W.-H. (1990a). Wind tunnels for automobile aerodynamics. In: *Aerodynamics of Road Vehicles* (ed. W.-H. Hucho). London: Butterworth-Heinemann.

Hucho, W.-H. (1990b). Aerodynamic drag of passenger cars. In: *Aerodynamics of Road Vehicles* (ed. W.-H. Hucho). London: Butterworth-Heinemann.

Hucho, W.-H, Janssen, L.J., and Emmelmann, H.J. (1976). The optimization of body details - A method for reducing the aerodynamic drag of road vehicles. Society of Automotive Engineers paper 760185.

IEA (2019). Global EV Outlook 2019: Scaling-up the transition to electric mobility. Technology Report, May 2019. Paris: International Energy Agency.

International Organization for Standardization (1997). Road Vehicles: Engine Test Code – Net Power. ISO 1585:1997-04.

Janssen, L.J. and Hucho, W.-H. (1973). The effect of various parameters on the aerodynamic drag of passenger cars. In: *Advances in Road Vehicle Aerodynamics* (ed. H.S. Stephens). Cranfield, England: BHRA Fluid Engineering.

Kelly, K.B. and Holcombe, H.J. (1978). Aerodynamics for body engineers. In: *Automotive Aerodynamics: Progress in Technology Series* **16**, Society of Automotive Engineers.

Klein, H.-C. (1986). Anti-lock brake systems for passenger cars: State of the art 1985. *Proc. XXI FISITA Congress*, paper 865139, Belgrade, Yugoslavia.

Lam, C.P., Guntur, R.R., and Wong, J.Y. (1979). Evaluation of the braking performance of a tractor–semitrailer equipped with two different types of antilock System. *SAE Transactions* **88**, Society of Automotive Engineers paper 791046.

Landman, D., Wood, R., Seay, W., and Bledsoe, J. (2010). Understanding practical limits to heavy truck drag reduction. *SAE International Journal of Commercial Vehicles* **2** (2) https://doi.org/10.4271/2009-01-2890.

Larminie, J. and Dicks, A. (2003). *Fuel Cell Systems Explained*, 2e. Hoboken, NJ: Wiley.

Leiber, H. and Czinczel, A. (1979). Antiskid system for passenger cars with a digital electronic control unit. Society of Automotive Engineers paper 790458.

Leiber, H. and Czinczel, A. (1983). Four years of experience with 4-wheel antiskid brake systems (ABS). Society of Automotive Engineers paper 830481.

Lister, R.D. (1965). Retention of directional control when Braking. SAE Transactions **74**, Society of Automotive Engineers paper 650092.

Luo, X., Wang, J., Dooner, M., and Clarke, J. (2015). Overview of current development in electrical energy storage technologies and the application potential in power system operation. *Applied Energy* **137**.

May, G.J., Davidson, A., and Monahov, B. (2018). Lead batteries for utility energy storage: a review. *Journal of Energy Storage* **15**.

McKnight, J. (1999). Transportation in the 21st century: Our challenges and opportunities. The 11th George Stephenson Lecture, Institution of Mechanical Engineers.

Mikulcik, E.C. (1971). The dynamics of tractor–semitrailer vehicles: The jackknifing problem. *SAE Transactions* **80**, Society of Automotive Engineers paper 710045.

Murphy, R.W., Bernard, J.E., and Winkler, C.B. (1972). A computer based mathematical method for predicting the braking performance of trucks and tractor–trailers. Report of the Highway Safety Research Institute, University of Michigan, Ann Arbor, MI.

National Research Council Canada (2021). Research facilities navigator. https://navigator.innovation.ca/en/facility/national-research-council-canada/9-m-wind-tunnel (accessed 6 March 2021).

Necati, G.A. (1990). Measurement and test techniques. In: *Aerodynamics of Road Vehicles* (ed. W.-H. Hucho). London: Butterworth-Heinemann.

Newton, K., Steeds, W., and Garrett, T.K. (1983). *The Motor Vehicle*, 10e. London: Butterworth.

Oak Ridge National Laboratory (2007). Evaluation of 2004 Toyota Prius hybrid electric drive system. Interim Report: Revised.

Oak Ridge National Laboratory (2009). Evaluation of the 2008 Lexus LS 600H Hybrid Synergy Drive System.

Ontario Tech University (2021). ACE: Climatic aerodynamic wind tunnel. https://ace.ontariotechu.ca (accessed 6 March 2021).

Patten, J., McAuliffe, B., Mayda, W., and Tanguay B. (2012). Review of aerodynamic drag reduction devices for heavy trucks and buses. Technical Report CSTT-HVC-TR-205. National Research Council Canada.

Porter, F.C. (1979). Design for fuel economy: The new GM front drive cars. Society of Automotive Engineers paper 790721.

Price, C.K.J. and Beasley, S.A. (1964). Aspects of hydraulic transmissions for vehicles of specialized nature. *Proceeding of the Institution of Mechanical Engineers* **178** (Part 3C).

Pukrushpan, J.T., Stefanopoulou, A.G., and Peng, H. (2005). *Control of Fuel Cell Power Systems*. London: Springer-Verlag.

Robert Bosch GmbH (2018). *Bosch Automotive Handbook*, 10e. Hoboken, NJ: Wiley.

Rousillon, G., Marzin, J., and Bourhis, J. (1973). Contribution to the accurate measurement of aerodynamic drag by the deceleration method. In: *Advances in Road Vehicle Aerodynamics*. Cranfield, England: BHRA Fluid Engineering.

Sammes, N. (ed.) (2006). *Fuel Cell Technology: Reaching Towards Commercialization*. London: Springer-Verlag.

Satoh, M. and Shiraishi, S. (1983a). Excess operation of antilock brake system on a rough road. In: *Braking of Road Vehicles 1983*. London: Institution of Mechanical Engineers.

Satoh, M and Shiraishi, S. (1983b). Performance of antilock brakes with simplified control technique. Society of Automotive Engineers paper 830484.

Setz, H.I. (1961). Computer predicts car acceleration. *SAE Transactions* **69** Society of Automotive Engineers paper 610032.

Society of Automotive Engineers (2011). Engine Power Test Code: Spark Ignition and Compression Ignition – As Installed Net Power Rating. SAE J1349 SEPT2011.

Srinivasa, R., Guntur, R.R., and Wong, J.Y. (1980). Evaluation of the performance of anti-lock brake systems using laboratory simulation techniques. *International Journal of Vehicle Design* **1** (5).

Suarez, C. and Martinez, W.H. (2019). Fast and ultra-fast charging for battery electric vehicles: A Review. *2019 IEEE Energy Conversion Conference and Exposition*, Baltimore, MD.

Taborek, J.J. (1957). Mechanics of vehicles. *Machine Design*, May 30–December 26.

Thring, R.H. (ed.) (2004). *Fuel Cells for Automotive Applications*. New York: American Society of Mechanical Engineers.

van Zanten (2014). Control of horizontal vehicle motion. In: *Road and Off-Road Vehicle Dynamics Handbook* (eds. G. Mastinu and M. Ploechl). Boca Raton, FL: CRC Press.

van Zanten, A. and Krauter, A. (1978). Optimal control of the tractor–semitrailer. *Vehicle System Dynamics* **7**.

Verma, V.S., Guntur, R.R., and Wong, J.Y. (1980). Directional behavior during braking of a tractor-semitrailer fitted with antilock devices. *International Journal of Vehicle Design* **1** (3).

Wager, G., Braunl, T., and Whale, J. (2017). Performance evaluation of regenerative braking systems. *Proceedings of the Institution of Mechanical Engineers, Part D: Journal of Automobile Engineering* **232** (10) https://doi.org/10.1177/0954407017728651.

Wang, Y., Diaz, D.F.R., Chen, K.S. et al. (2020). Materials, technological status, and fundamental of PEM fuel cells. *Materials Today* **32**.

Wardill, R. (1974). Why has the British manufacturer been hesitant to adopt hydrostatic drives? *Proc. Institution of Mechanical Engineers Conference on Making Technology Profitable – Hydrostatic Drives*, London, 19–20 March. London: Mechanical Engineering Publications.

White, R.A. and Korst, H.H. (1972). The determination of vehicle drag contributions from coastdown tests. *SAE Transactions* **81**, Society of Automotive Engineers paper 720099.

Winkler, C.B., Bernard, J.E., Fancher, P.S. et al. (1976). *Predicting the Braking Performance of Trucks and Tractor–Trailers*. Report of the Highway Safety Research Institute. Ann Arbor, MI: University of Michigan.

Wong, J.Y. and Guntur, R.R. (1978). Effects of operational and design parameters on the sequence of locking of the wheels of tractor–semitrailers. *Vehicle System Dynamics* **7** (1).

Wong, J.Y., Ellis, J.R., and Guntur, R.R. (1977). *Braking and Handling of Heavy Commercial Vehicle*. Monograph. Department of Mechanical and Aeronautical Engineering, Carleton University, Ottawa, Canada.

Wood, R. (2012). EPA SmartWay verification of trailer undercarriage advanced aerodynamic drag reduction technology. Society of Automotive Engineers paper 2012-01-2043.

ZF Getriebe GmbH (2020). The freedom to exceed limits (online). https://brainfunkers.co/blog/files/file_008971%20-%20The%20freedom%20to%20exceed%20limits.pdf (accessed 24 November 2020).

Problems

3.1 A vehicle weighs 20.02 kN (4500 lb) and has a wheelbase of 279.4 cm (110 in.). The center of gravity is 127 cm (50 in.) behind the front axle and 50.8 cm (20 in.) above ground level. The frontal area of the vehicle is 2.32 m²(25 ft²) and the aerodynamic drag coefficient is 0.45. The coefficient of rolling resistance is given by $f_r = 0.008 + 0.4 \times 10^{-7}V^2$, where V is the speed of the vehicle in kilometers per hour. The rolling radius of the tires is 33 cm (13 in.). The coefficient of road adhesion is 0.8. Estimate the possible maximum speed of the vehicle on level ground and on a grade of 25% as determined by the maximum tractive effort that the tire–road contact can support if the vehicle is (a) rear-wheel drive and (b) front-wheel drive. Plot the resultant resistance versus vehicle speed and show the maximum thrust of the vehicle with the two types of drive.

3.2 The vehicle described in Problem 3.1 is equipped with an engine having torque–speed characteristics as shown in the following table. The gear ratios of the gearbox are first, 4.03; second, 2.16; third, 1.37; and fourth, 1.0. The gear ratio of the drive axle is 3.54. The transmission efficiency is 88%. Estimate the maximum speed of the vehicle on level ground and on a grade of 25% as determined by the tractive effort that the engine torque with the given transmission can provide if the vehicle is rear-wheel drive. Plot the vehicle thrust in various gears versus vehicle speed.

Engine Characteristics

Engine speed, rpm	500	1000	1750	2500	3000	3500	4000	4500	5000
Engine torque, N-m	339	379.7	406.8	393.2	363.4	325.4	284.8	233.2	189.8

3.3 A vehicle is equipped with an automatic transmission consisting of a torque converter and a three-speed gearbox. The torque converter and the engine characteristics are shown in Figures 3.51 and 3.52, respectively. The total gear reduction ratio of the gearbox and the drive axle is 2.91 when the third gear is engaged. The combined efficiency of the gearbox, propeller shaft, and the drive axle is 0.90. The rolling radius of the tire is 33.5 cm (1.1 ft). Determine the tractive effort and speed of the vehicle when the third gear is engaged and the engine is running at 2000 rpm with an engine torque of 407 N · m (300 lb · ft). Also determine the overall efficiency of the transmission, including the torque converter.

3.4 A passenger car weighs 12.45 kN (2800 lb), including the four tires. Each of the tires has an effective diameter of 67 cm (2.2 ft) and a radius of gyration of 27.9 cm (11 in.), and weighs 222.4 N (50 lb). The engine develops 44.8 kW (60 hp) at 4000 rpm, and the equivalent weight of the rotating parts of the drivetrain at that engine speed is 444.8 N (100 lb) with a radius of gyration of 10 cm (4 in.). The transmission efficiency is 88% and the total reduction ratio of the drivetrain in the second gear is 7.56 to 1. The vehicle has a frontal area of 1.67 m²(18 ft²) and the aerodynamic drag coefficient is 0.45. The average coefficient of rolling resistance is 0.008. Estimate the acceleration of the vehicle on a level road under these conditions.

3.5 A passenger car weighs 20.02 kN (4500 lb) and has a wheelbase of 279.4 cm (110 in.). The center of gravity is 127 cm (50 in.) behind the front axle and 50.8 cm (20 in.) above ground level. In practice, the vehicle encounters a variety of surfaces, with the coefficient of road adhesion ranging from 0.2 to 0.8 and the coefficient of rolling resistance of 0.015. To prevent the loss of directional stability on surfaces with a low coefficient of adhesion under emergency braking conditions, what would you recommend regarding the braking effort distribution between the front and rear axles?

3.6 For a tractor–semitrailer combination, the tractor weighs 66.72 kN (15 000 lb) and the semitrailer weighs 266.88 kN (60 000 lb). The wheelbase of the tractor is 381 cm (150 in.), and the semitrailer axle is 1016 cm (400 in.) behind the rear axle of the tractor. The hitch point is 25 cm (10 in.) in front of the tractor rear axle and 122 cm (48 in.) above the ground level. The center of gravity of the tractor is 203.2 cm (80 in.) behind the tractor front axle and 96.5 cm (38 in.) above the ground. The center of gravity of the semitrailer is 508 cm (200 in.) in front of the semitrailer axle and 177.8 cm (70 in.) above the ground. What is the ideal braking effort distribution between the axles that ensures all the tires being locked up at the same time on a surface with a coefficient of road adhesion $\mu = 0.6$? Also determine the normal loads on the axles and the forces acting at the hitch point.

3.7 A "coastdown" test was performed to estimate the aerodynamic resistance coefficient C_D and the rolling coefficient f_r of a road vehicle. The test was conducted on a level road with a tail wind of 8 km/h (5 mph). The vehicle was first run up to a speed of 97 km/h (60 mph) and then the gear was shifted to neutral. The vehicle decelerated under the action of the aerodynamic resistance, the rolling resistance of the tires, and the internal resistance of the drivetrain. The vehicle slowed down from 97 km/h (60 mph) to 88.5 km/h (55 mph) in a distance of 160 m (525.5 ft), and from 80 km/h (50 mph) to 72.4 km/h (45 mph) in a distance of 162.6 m (533.5 ft). The vehicle weighs 15.568 kN (3500 lb) and has a frontal area of 2.32 m^2(25 ft^2). Assuming that the rolling resistance of the tires is independent of speed and that the internal resistance of the drivetrain may be neglected, estimate the values of the aerodynamic resistance coefficient C_D and the rolling resistance coefficient of the tires f_r.

3.8 For an internal combustion engine, an optimum fuel economy line can be drawn on the engine performance diagram, as shown in Figure 3.62. This line identifies the engine operating point where the specific fuel consumption is the lowest for a given engine output power. A gasoline engine with performance shown in Figure 3.62 is to be installed in a subcompact car with a weight of 11.28 kN (2536 lb). It has a frontal area of 1.83 m^2(19.7 ft^2) and an aerodynamic drag coefficient of 0.4. The rolling radius and rolling resistance coefficient of the tire are 0.28 m (11 in.) and 0.008, respectively. The drive axle gear ratio is 2.64.
(a) If the car is equipped with a continuously variable transmission (CVT) with a mechanical efficiency of 0.88, determine its gear ratio that provides the maximum fuel economy at a vehicle speed of 100 km/h (62 mph).
(b) Estimate the fuel consumption of the car in liters per 100 km (or in miles per US gallon) under the above-noted operating conditions. The mass density of the gasoline is taken to be 0.75 kg/L (or weight density of 6.26 lb per US gallon).

3.9 A passenger car weighs 20.75 kN (4665 lb). Its frontal area is 2.02 m^2(21.74 ft^2) and the aerodynamic drag coefficient is 0.36. The coefficient of rolling resistance is 0.008.

(a) Estimate the maximum engine power required to accelerate the car at a constant rate from standstill to a speed of 97 km/h (60 mph) in 8 s, on a level road and with zero wind speed. The mass factor γ_m at the highest gear of the transmission is 1.08 and the average transmission efficiency is 0.9.

(b) Estimate the engine power required for cruising at a constant speed of 97 km/h (60 mph). Compare the power requirement for cruising with that for acceleration under the condition described in (a).

(c) If the weight of the car is reduced by 20% (by using high-strength, lightweight materials and advanced design and production techniques), estimate the maximum engine power required for acceleration under the same condition as that described in (a). Compare the power requirement for weight reduction of 20% with that for the original weight of the car.

4

Performance Characteristics of Off-Road Vehicles

Performance of an off-road vehicle refers to its ability to overcome motion resistance, to develop net force (usually represented by its drawbar pull), to negotiate grades, or to accelerate in straight-line motion. Depending on their functions, different criteria are employed to evaluate the performance characteristics of various types of off-road vehicle.

- For tractors, such as those used in farming and construction, their main function is to provide adequate drawbar pull to pull (or push) various types of implement or machinery. The drawbar performance under steady-state conditions is, therefore, of prime interest. It can be characterized by the drawbar pull, drawbar power (the product of drawbar pull and actual forward speed of the tractor), and drawbar (tractive) efficiency (the ratio of the drawbar power to the corresponding engine (power plant) output power or the ratio of the drawbar power to the corresponding power input to the driven tires of wheeled vehicles or that to the drive sprockets of tracked vehicles). To provide a common basis for comparing the performance of different tractors, a nondimensional parameter known as the drawbar pull coefficient is commonly used. It is the ratio of the drawbar pull to vehicle weight, as noted previously in Chapter 2.
- For cross-country transport vehicles, the transport productivity and transport efficiency are often used as basic criteria for evaluating their performance. The transport productivity is defined as the product of payload and the average steady cross-country speed of the vehicle. The transport efficiency is defined as the ratio of transport productivity to the corresponding engine (power plant) output power. If transport productivity and corresponding engine output power are expressed in the same units, transport efficiency is a nondimensional parameter.
- For military vehicles, on the other hand, the maximum feasible operating speed under steady-state conditions between two specific points in an area of operation may be employed as a criterion for evaluating their mobility.

Although differing criteria are used to assess the performance of different kinds of off-road vehicle, there is a basic requirement common to all off-road vehicles, that is the mobility over unprepared terrain. Mobility in the broad sense refers to the cross-country performance of the vehicle on unprepared terrain, ability in obstacle negotiation and avoidance, ride quality over rough terrain, and ability in water crossing. The tractive performance on unprepared terrain constitutes a central issue in vehicle mobility, and a detailed analysis of the relationship among vehicle tractive performance, vehicle design parameters, and the terrain is, therefore, of prime importance, as discussed in Chapter 2.

In this chapter, methods for evaluating and predicting the tractive performance of cross-country vehicles are discussed. Performance criteria for various types of off-road vehicle are also examined in detail.

4.1 Drawbar Performance

4.1.1 Drawbar Pull and Drawbar Power

The major external forces acting on a tracked vehicle and on an off-road wheeled vehicle in the longitudinal direction and on level ground are shown in Figure 4.1(a) and (b), respectively. As shown in Figure 4.1(a), the major external forces acting on a tracked vehicle include the thrust (propelling force) F, the motion resisting force of the running gear R_v, the aerodynamic resistance (drag) R_a, and the drawbar pull F_d (net force for pulling or pushing implements or machinery). The major external forces acting on a four-wheel drive vehicle are shown in Figure 4.1(b). They include the thrust of the front wheel F_f, the thrust of the rear wheel F_r, the motion resisting force on the front wheel R_{vf}, the motion resisting force on the rear wheel R_{vr}, the aerodynamic resistance R_a, and the drawbar pull F_d. When an off-road vehicle is climbing a slope, the gravitational force component along the slope constitutes a resisting force. During acceleration or deceleration, a corresponding vehicle inertia force is also present.

For off-road vehicles designed for traction (such as tractors), the drawbar performance under steady-state conditions is of prime importance, as it represents the ability of the vehicle to pull or push various types of implement or machinery, such as plow, bulldozer blade, or earth-moving equipment, as noted previously. Drawbar pull F_d is the force available at the drawbar hitch, and is equal to the difference between the thrust (tractive effort) F developed by the running gear and the resultant resisting force ΣR acting on the vehicle:

$$F_d = F - \sum R \tag{4.1}$$

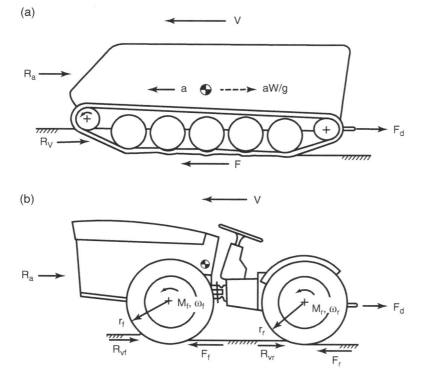

Figure 4.1 Major external forces acting on (a) a tracked vehicle and (b) on an off-road wheeled vehicle.

To provide a common basis for comparing the drawbar performance of off-road vehicles of different weights, a nondimensional parameter known as the drawbar pull coefficient is frequently used. As noted previously, it is ratio of the drawbar pull to total vehicle weight. The drawbar pull coefficient is not only an indication of its capability for developing drawbar pull, but also its ability for slope climbing or for acceleration. For instance, if the drawbar pull coefficient is 30% under a given operating condition, then it indicates the vehicle has the potential to climb up a grade of 30% or to accelerate at a rate of 0.3 g on level ground under a similar operating condition.

For a vehicle with known power plant and transmission characteristics, the tractive effort and vehicle theoretical speed can be determined using methods like those described in Chapter 3. However, in cross-country operations, the maximum tractive effort is often limited by the characteristics of vehicle–terrain interaction, as described in Chapter 2. Furthermore, the development of thrust often results in considerable slip over unprepared terrain. Thus, the drawbar pull and vehicle actual forward speed are functions of slip.

The resultant resisting force acting on an off-road vehicle includes the internal resistance of the running gear, resistance due to vehicle–terrain interaction (such as the resisting force caused by terrain compaction or bulldozing effects on the front of the running gear), obstacle resistance, aerodynamic drag, as well as grade resistance if the vehicle is operating on a longitudinal slope. The internal resistance of the running gear is taken into account, when the net tractive effort (thrust) of the vehicle is calculated from the driving torque at the sprocket of a tracked vehicle or at the driven wheel of a wheeled vehicle.

1) Internal resistance of the running gear

For wheeled vehicles, the internal resistance of the running gear (tire) is mainly due to hysteresis losses in the tire, which has been discussed in Chapter 1. For tracked vehicles, the internal resistance of the track system may be substantial. Frictional losses in track pins, between the driving sprocket teeth and track links, in the bearings of sprocket hubs and roadwheels, as well as the resistance of roadwheels rolling on the inner surface of the track, constitute the major portion of the internal resistance of the track system. Experimental results show that of the total power consumed in the track–suspension system, 63–75% is due to losses in the track itself. Among the operational parameters, track tension and vehicle speed have noticeable effects on the internal resistance, as shown in Figures 4.2 and 4.3, respectively (Cleare 1963–1964; Bekker 1969).

Because of the complex nature of the internal resistance in the track system, it is difficult, if not impossible, to establish an analytical framework for predicting it with sufficient accuracy. Usually, the internal resistance of the track system is measured on hard, smooth road surfaces using the following methods:

- The coast-down method. This method has been discussed in Section 3.2. The vehicle is run up to a maximum speed, the transmission is shifted to neutral, and the vehicle is allowed to decelerate to a standstill by external resisting forces acting on the vehicle. The vehicle speed is measured against time, from which the deceleration and hence the resultant resisting force can be calculated at various speeds. Corrections may be made for aerodynamic resistance in deriving the internal resistance of the track system.
- The towing method. The vehicle is towed at various steady speeds and the towing force representing the resultant resisting force is measured. Corrections may be made for aerodynamic resistance.
- The method of measuring the sprocket torque. The sprocket torque is measured under self-propelled conditions at various steady speeds. With the measured sprocket torque and the known sprocket pitch radius, the resultant resisting force can be determined at various speeds. Corrections may be made for aerodynamic resistance.

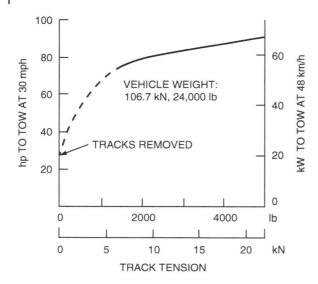

Figure 4.2 Effects of track tension on power consumption. *Source:* Cleare (1963–1964). Reproduced with permission of the Council of the Institution of Mechanical Engineers.

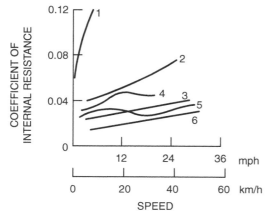

Figure 4.3 Effects of speed on the coefficient of internal resistance of various types of track system. *Source: Introduction to Terrain–Vehicle Systems* by M.G. Bekker, copyright © 1969 by the University of Michigan. Reproduced with permission of the University of Michigan Press.

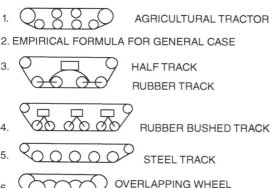

1. AGRICULTURAL TRACTOR

2. EMPIRICAL FORMULA FOR GENERAL CASE

3. HALF TRACK
   RUBBER TRACK

4. RUBBER BUSHED TRACK

5. STEEL TRACK

6. OVERLAPPING WHEEL
   LUBRICATED TRACK

Figure 4.3 shows the variations of the coefficient of internal resistance (i.e., the ratio of the internal resistance to normal load) of various track systems with speed, including track systems for agricultural tractors, half track systems with rubber tracks, track systems with rubber-bushed pins, steel track systems with dry pins of modern versions of tank, and tracks with sealed, lubricated pins supported by needle bearings used in some World War II German military tracked vehicles with overlapping wheels (Bekker 1969). For the general case represented by Line 2 in Fig. 4.3, the variation of the coefficient of internal resistance f_r with speed V up to 40 km/h may be approximately expressed by:

$$f_r = 0.033 + 0.00105\ V \tag{4.2}$$

where V is track speed in kilometers per hour

From test data of a group of primarily U.S. and German military tracked vehicles, the coefficient of internal resistance f_r can be approximately represented by the following equation, similar in form to Eq. (4.2) (Ogorkiewicz 1991):

$$f_r = f_o + f_s V \tag{4.3}$$

where f_o and f_s are empirical coefficients and their values depend on the type of track system, and V is track speed in kilometers per hour. For tanks with tracks having double, rubber-bushed pins and rubber pads, f_o may be taken as 0.03 and f_s as 0.0009. With all-steel, single-pin tracks, f_o is typically 0.025 and with tracks having sealed, lubricated pins with needle bearings for some World War II German military tracked vehicles with overlapping wheels noted previously, f_o can be as low as 0.015.

From test data of a group of British military tracked vehicles, the variation of the coefficient of internal resistance with track speed in kilometers per hour for some track systems may also be expressed by Eq. (4.3) (Maclaurin 2018). For instance, for track systems with rubber-bushed pins and roadwheels running on metal path (inner surface) of the track, f_o is 0.0267 and f_s is 0.00018, and for track systems with rubber-bushed pins and roadwheels running on rubber path (inner surface) of the track, f_o is 0.03 and f_s is 0.00011. However, for track systems with dry pins and roadwheels running on metal path (inner surface) of the track, the variation of the coefficient of internal resistance with speed may be approximately expressed by

$$f_r = f_o + f_s V + f_{s1} V^2 \tag{4.4}$$

where f_o is 0.0304, f_s is 0.0001, f_{s1} is 0.00001, and V is track speed in kilometers per hour.

2) Resistance due to vehicle–terrain interaction

This type of resistance is the most significant one for off-road vehicles, and determines, to a great extent, the mobility of the vehicle over unprepared terrain. It includes the resistance due to compacting the terrain and the bulldozing effect, and it may be predicted using the methods described in Chapter 2 or determined experimentally.

3) Ground obstacle resistance

In off-road operations, obstacles such as stumps and stones may be encountered. The obstacle resistance may be considered a resisting force, usually variable in magnitude, acting parallel to the ground at a certain effective height. When the line of action of this resisting force is high above ground level, it produces a moment that would cause significant load transfer, and it should be taken into consideration in formulating the equations of motion for the vehicle. In general, the value of the obstacle resistance is obtained from tests.

4) Aerodynamic resistance

Aerodynamic resistance is usually not a significant factor for off-road vehicles operating at speeds below 48 km/h (30 mph). For vehicles designed for higher-speed operations, such as military vehicles, aerodynamic resistance may have to be taken into consideration in performance predictions. The aerodynamic resistance can be predicted using the methods described in Chapter 3.

The aerodynamic resistance coefficient mainly depends on the shape of the vehicle, as mentioned previously in Chapter 3. For heavy fighting vehicles, such as battle tanks, the aerodynamic resistance coefficient C_D is approximately 1.0, and the frontal area is of the order of $6 - 8 \, \text{m}^2 \, (65 - 86 \, \text{ft}^2)$ (Ogorkiewicz 1991). For a tank weighing 50 t and having an aerodynamic resistance coefficient of 1.17 with a frontal area of $6.5 \, \text{m}^2 (70 \, \text{ft}^2)$, the power required to overcome aerodynamic resistance at 48 km/h (30 mph) amounts to about 11.2 kW (15 hp). A light tracked vehicle weighing 10 t and having a frontal area of $3.7 \, \text{m}^2 \, (40 \, \text{ft}^2)$ may require 10.5 kW (14 hp) to overcome the aerodynamic resistance at a speed of 56 km/h (35 mph).

In addition to the resisting forces described above, the grade resistance must be taken into consideration when the vehicle is climbing up a slope. For heavy fighting vehicles, the usual requirement is that they should be able to climb up a gradient of 30° (58%).

To characterize the drawbar performance, the slip of the running gear and the vehicle speed in each gear are usually plotted against tractive effort and drawbar pull, as shown in Figure 4.4. The product of drawbar pull and vehicle speed is usually referred to as the drawbar power that represents the potential productivity of the vehicle, that is, the rate at which productive work may be done. The drawbar power P_d is given by

$$P_d = F_d V = \left(F - \sum R\right) V_t (1 - i) \tag{4.5}$$

where V and V_t are the actual forward speed and the theoretical speed of the vehicle, respectively. The theoretical speed is the speed of the vehicle if there is no slip (or skid), and is determined by the engine speed, reduction ratio of the transmission, and the radius of the tire (or sprocket).

Usually, the variation of drawbar power with drawbar pull in each gear is also shown in the drawbar performance diagram. As an example, Figure 4.5 shows the measured drawbar performance of an MF-165 tractor on tarmacadam (Kolozsi and McCarthy 1974). The drawbar performance diagram provides a basis for comparing and evaluating the tractive performance of tractors. It also provides the user with the required information to achieve proper matching of the tractor with implements or working machinery.

4.1.2 Drawbar (Tractive) Efficiency

To characterize the efficiency of an off-road vehicle in transforming the engine (power plant) power to the power available at the drawbar hitch, the overall drawbar (tractive) efficiency is often used.

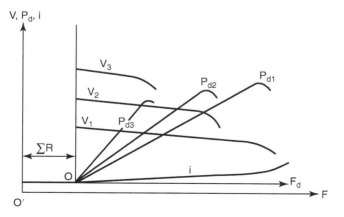

Figure 4.4 Drawbar performance diagram for tractors.

Figure 4.5 Drawbar performance of an MF-165 tractor on tarmacadam. *Source:* Kolozsi and McCarthy (1974). Reproduced with permission of the *Journal of Agricultural Engineering Research.*

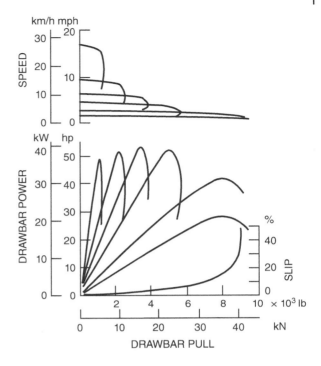

The overall drawbar efficiency η_{do} is defined as the ratio of drawbar power P_d to the corresponding output power of the engine P:

$$\eta_{do} = \frac{P_d}{P} = \frac{F_dV}{P} = \frac{(F - \Sigma R)V_t(1 - i)}{P} \tag{4.6}$$

The output power of the engine may be expressed in terms of the power available at the driven wheel (or sprocket) and the transmission efficiency η_t:

$$P = \frac{FV_t}{\eta_t} \tag{4.7}$$

Substituting Eq. (4.7) into Eq. (4.6), the expression for overall tractive efficiency η_{do} becomes

$$\eta_{do} = \frac{(F - \Sigma R)}{F}(1 - i)\eta_t = \eta_m\eta_s\eta_t \tag{4.8}$$

where η_m is the efficiency of motion, which is equal to F_d/F, and η_s is the efficiency of slip, which is equal to $(1 - i)$.

The variation of the overall efficiency η_{do} with drawbar pull is shown in Figure 4.6. The efficiency of motion η_m indicates the losses in transforming the tractive effort at the driven wheels to the pull at the drawbar. For motion resistance having a constant value, the efficiency of motion η_m increases with an increase of drawbar pull, as shown in the figure. The efficiency of slip η_s characterizes the power losses and the reduction in forward speed of the vehicle due to slip of the running gear. Since slip increases with an increase of tractive effort and drawbar pull, the efficiency of slip η_s decreases as the drawbar pull increases. When the tire slip is 100%, the vehicle does not move forward and the efficiency of slip η_s is zero, and hence the overall drawbar efficiency η_{do} is zero as well, as shown in the figure.

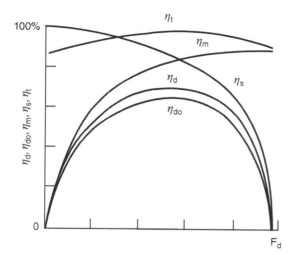

100%

η_t

η_m

η_d　η_s

η_{do}

0

$\eta_d, \eta_{do}, \eta_m, \eta_s, \eta_t$

F_d

Figure 4.6 Variations of the overall drawbar efficiency η_{do} and drawbar efficiency η_d with drawbar pull.

As can be seen from Eq. (4.8), the overall drawbar efficiency η_{do} is the product of the efficiency of motion η_m, efficiency of slip η_s, and efficiency of transmission η_t. In general, it exhibits a peak at an intermediate value of drawbar pull, as shown in Figure 4.6. To increase the overall tractive efficiency, optimization of the form and size of the vehicle running gear is of importance. In this respect, terramechanics, described in Chapter 2, plays an important role.

In addition to the overall drawbar efficiency η_{do} given by Eq. (4.8), another frequently used drawbar efficiency η_d is defined as the ratio of the drawbar power to the corresponding power input to the driven tires of wheeled vehicles or to the drive sprockets of tracked vehicles. It is expressed by

$$\eta_d = \frac{F_d V}{F V_t} = \frac{(F - \Sigma R) V_t (1 - i)}{F V_t} = \frac{(F - \Sigma R)}{F} (1 - i) = \eta_m \eta_s \tag{4.9}$$

In this case, the transmission efficiency η_t is not included in the drawbar efficiency η_d.

Example 4.1 An off-road wheeled vehicle is equipped with an engine having torque–speed characteristics given in Table 4.1. The vehicle is to operate on a soil with thrust–slip characteristics given in Table 4.2. The total motion resistance is 2.23 kN (500 lb). The transmission efficiency is 0.85, and the rolling radius of the tire is 0.76 m (2.5 ft). Determine the drawbar power and the overall drawbar (tractive) efficiency of the vehicle when the fourth gear with a total reduction ratio $\xi_o = 20.5$ is engaged.

Table 4.1 Engine characteristics.

	Engine torque M_e	
Engine speed n_e (rpm)	**N · m**	**lb · ft**
800	393	290
1200	650	479
1600	732	540
2000	746	550
2400	705	520
2800	610	450

Table 4.2 Thrust–slip characteristics.

Slip (%)	Thrust F	
	kN	lb
5	10.24	2303
10	16.0	3597
15	20.46	4600
20	24.0	5396
25	26.68	5998
30	28.46	6398
40	32.02	7199

Solution

The thrust can be calculated using Eq. (3.55) in Chapter 3:

$$F = \frac{M_e \xi_o \eta_t}{r}$$

From Table 4.2, the slip at a particular thrust can be determined. The vehicle speed can be determined using Eq. (3.56) in Chapter 3:

$$V = \frac{n_e r}{\xi_o}(1 - i)$$

For a given engine operating point, the vehicle speed and thrust are related. Therefore, the slip at a particular theoretical speed can be determined, and the actual speed of the vehicle can be calculated. The results of the calculations for F and V are tabulated in Table 4.3. The drawbar pull is given by

$$F_d = F - \sum R$$

Table 4.3 Drawbar performance.

Engine speed n_e(rpm)	Thrust F		Slip i (%)	Vehicle speed V		Drawbar pull F_d		Drawbar power P_d		Transmission efficiency η_t (%)	Efficiency of slip η_s (%)	Overall tractive efficiency η_{do} (%)
	kN	lb		km/h	mph	kN	lb	kW	hp			
800	9.01	2025	4.4	10.7	6.7	6.78	1525	20.1	27	85	95.6	61.1
1200	14.90	3350	9.0	15.3	9.5	12.67	2850	53.8	72.2	85	91.0	65.7
1600	16.78	3772	10.9	19.9	12.4	14.55	3272	80.4	107.8	85	89.1	65.6
2000	17.10	3844	11.23	24.9	15.5	14.87	3344	102.8	137.9	85	88.7	65.6
2400	16.16	3633	10.2	30.2	18.8	13.93	3133	116.8	156.7	85	89.8	65.8
2800	13.98	3143	8.3	36.0	22.4	11.75	2643	117.5	157.5	85	91.7	65.5

Since the total motion resistance is given, the drawbar pull can be calculated and the results are shown in Table 4.3. The drawbar power can be determined using Eq. (4.5):

$$P_d = F_d V$$

and the overall drawbar efficiency can be calculated using Eq. (4.6):

$$\eta_{do} = \frac{F_d V}{P}$$

The results of the calculations for P_d and η_{do} are tabulated in Table 4.3. The maximum drawbar efficiency of the vehicle under the operating conditions specified is approximately 66%, which indicates that 34% of the corresponding engine power is lost in the transmission, in overcoming the motion resistance, and in vehicle slip.

Engines for off-road vehicles are often equipped with a governor to limit maximum operating speed. When the engine characteristics over the operating range of the governor (i.e., between the full load and no-load settings of the governor) are known, the drawbar performance of the vehicle in that range can be predicted in the same way as that described above.

4.1.3 All-Wheel Drive

To enhance the mobility of wheeled vehicles, all-wheel drive has been gaining popularity, as they can fully utilize the loads and contact areas of all the wheels for generating propelling force (tractive effort or thrust). Thus, in the defense sector, fighting and logistics vehicles are mostly all-wheel drive. In the civilian sector, such as farming, construction, logging, and cross-country transport, there is a growing demand for higher productivity. Consequently, there has been a steady increase in the engine (power plant) power for vehicles used in these fields. For instance, for wheeled tractors to fully utilize the high engine power and to maintain high tractive efficiency, four-wheel drive has gained increasingly wide acceptance. As shown in Figure 4.1(b), the weights and contact areas of the wheels on both the front and rear axles of a four-wheel drive tractor are utilized for the development of thrust. On the other hand, for a rear-wheel drive tractor only about 60–70% of the total weight of the tractor is applied on the driven wheels and utilized for developing thrust. Consequently, over soft terrain, a four-wheel drive tractor has the capability of developing higher thrust than an equivalent rear-wheel drive tractor at the same slip, leading to higher drawbar power and productivity. Furthermore, a four-wheel drive vehicle with the same size tires on both the front and rear axles usually has a lower overall coefficient of motion resistance than an equivalent rear-wheel drive vehicle (Osborne 1969–1970). This is because its rear tires run entirely in the ruts formed by the front tires in straight-line motion, thus reducing the overall motion resistance. Figure 4.7(a) and (b) show a comparison of the drawbar performance of a rear-wheel drive and a four-wheel drive tractor on a dry loam, stubble field and on a wet clayey loam, respectively (Söhne 1968). On the dry loam, stubble field, the drawbar pull of the four-wheel drive tractor at 20% slip is 27% higher and at 50% slip is 20% higher than the corresponding values of the rear-wheel drive tractor. The peak tractive efficiency of the four-wheel drive tractor and that of the rear-wheel drive tractor are 77% and 70%, respectively. On the wet clayey loam, the drawbar pull of the four-wheel drive tractor at 30% slip is as much as 57% higher, and at 50% slip is 44% higher, than the corresponding values of the rear-wheel drive tractor. The values of the peak tractive efficiency of the four-wheel drive and rear-wheel drive tractor are 51% and 40%, respectively.

Figure 4.7 Comparisons of drawbar performance of a four-wheel drive tractor with a corresponding rear-wheel drive tractor on (a) dry loam, stubble field and (b) wet clayey loam. *Source:* Söhne (1968). Reproduced with permission of the International Society for Terrain–Vehicle Systems.

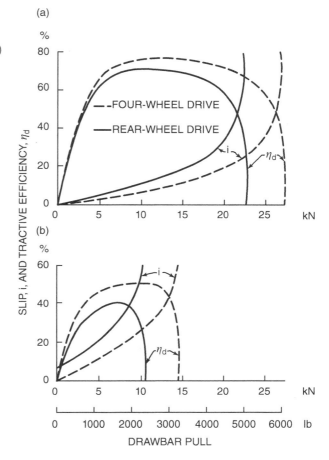

For an all-wheel drive off-road vehicle to achieve the maximum tractive efficiency, certain requirements must be met. To define these requirements quantitatively, it is necessary to examine the tractive efficiency, particularly the efficiency of slip, of an all-wheel drive off-road vehicle. To illustrate the approach to analyzing and optimizing the tractive efficiency of an all-wheel drive vehicle, a four-wheel drive vehicle is used as an example.

For a four-wheel drive vehicle, power losses due to slip occur at both the front and rear driven wheels. Assuming that the performances of the two front tires are identical and so are those of the two rear tires, the slip efficiency η_{s4} of a four-wheel drive vehicle is determined by (Wong 1970)

$$\eta_{s4} = 1 - \frac{i_f M_f \, \omega_f + i_r \, M_r \omega_r}{M_f \, \omega_f + M_r \, \omega_r} = 1 - \frac{i_f \, V_{tf} F_f + i_r \, V_{tr} \, F_r}{V_{tf} F_f + V_{tr} \, F_r} \qquad (4.10)$$

where M_f and M_r are the driving torques at the front and rear wheels, respectively, as shown in Figure 4.1(b); ω_f and ω_r are the angular speeds of the front and rear wheels, respectively; V_{tf} and V_{tr} are the theoretical speeds of the front and rear wheels, respectively; F_f and F_r are the tractive efforts of the front and rear wheels, respectively; and i_f and i_r are the slips of the front and rear wheels, respectively.

As the front and rear axles are connected to the same vehicle frame, the actual forward speed of the front wheels must be equal to that of the rear wheels for a four-wheel drive vehicle in straight-line motion. This kinematic relationship can be expressed by

$$V_{tf}(1-i_f) = V_{tr}(1-i_r) = V$$

or

$$K_v = \frac{V_{tf}}{V_{tr}} = \frac{\omega_f r_f}{\omega_r r_r} = \frac{(1-i_r)}{(1-i_f)} \tag{4.11}$$

where K_v is the ratio of the theoretical speed of the front wheels V_{tf} to that of the rear wheels V_{tr}, which is usually referred to as the theoretical speed ratio; V is the actual forward speed of the vehicle as a whole; r_f and r_r are the free-rolling radius of the front and rear wheels, respectively.

Therefore,

$$\eta_{s4} = 1 - \frac{[(1-i_r)/(1-i_f)]i_f V_{tr} F_f + i_r V_{tr} F_r}{[(1-i_r)/(1-i_f)]V_{tr} F_f + V_{tr} F_r}$$

$$= 1 - \frac{i_f(1-i_r) - (i_f - i_r)K_d}{(1-i_r) - (i_f - i_r)K_d} \tag{4.12}$$

where K_d is the coefficient of thrust distribution and is equal to $F_r/(F_f + F_r)$.

Equation (4.12) shows that, in general, the efficiency of slip of a four-wheel drive vehicle depends not only on the slips of the front and rear wheels, but also on the distribution of the thrust between them. As an example, Figure 4.8 shows the variations of the efficiency of slip η_{s4} with the coefficient of thrust distribution K_d at various values of the vehicle total thrust-to-weight ratio F/W for a four-wheel drive vehicle on a farm soil (Wong 1970). The results shown in the figure are based on the assumptions that the vehicle weight is equally distributed between the front and rear wheels; the front and rear wheels run on the same soil; and the relationship between the thrust coefficient

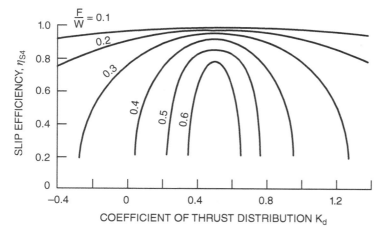

Figure 4.8 Effects of thrust distribution between driven axles on the slip efficiency of a four-wheel drive off-road vehicle.

(i.e., the ratio of thrust to normal load) and slip for the front wheels and that for the rear wheels are identical.

Since the thrust distribution affects the tractive efficiency of a four-wheel drive vehicle considerably, it is of importance to analyze the factors that in practice affect the thrust distribution. Generally speaking, there are two basic factors: firstly, the type of coupling between the front and rear axles, which may be rigid coupling, interaxle differential, overrunning clutch, viscous coupling, etc.; and secondly, the difference in theoretical speed (wheel speed when no slip or skid occurs) between the front and rear wheels.

A difference in theoretical speed often exists under operating conditions and is usually caused by the variation of the radii of the front or rear tires, owing to differences in tire inflation pressure, uneven wear of tires, or load transfer. It has been shown that the difference in theoretical speed between the drive axles of an all-wheel drive off-road vehicle could be as much as 10% in practice. When the sizes of the front and rear tires are not the same, sometimes it is difficult to provide the right gear ratio that exactly matches the sizes of the tires, and this also causes the difference in theoretical speed between them.

The most common configuration of four-wheel drive tractors has rigid coupling between the front and rear drive axles. For this type of vehicle, the ratio of the angular speed of the front wheel to that of the rear wheel is fixed. The relationship between the slip of the front wheel and that of the rear wheel in straight-line motion is therefore a function of the theoretical speed ratio, K_v:

$$i_r = 1 - \frac{V_{tf}}{V_{tr}}(1 - i_f) = 1 - K_v(1 - i_f) \tag{4.13}$$

The variation of i_r with i_f for different values of K_v is shown in Figure 4.9. When the theoretical speed ratio K_v is equal to 0.85 (i.e., the theoretical speed of the front wheel is 85% of that of the rear wheel) and i_r is less than 15%, the front wheel skids and develops negative thrust (braking force). On the other hand, when K_v is equal to 1.15 (i.e., the theoretical speed of the front wheel is 15% higher than that of the rear wheel) and i_f is less than 13%, the rear wheel skids and develops negative thrust. In both cases, the maximum forward thrust of the vehicle is reduced, and torsional wind-up in the transmission inevitably occurs. This results in an increase of stress in the components of the driveline and a reduction of transmission efficiency. Figure 4.10 shows the torques on the front and rear axles of a four-wheel drive vehicle, with rigid coupling between the front and rear axles, on a dry concrete surface in forward motion (shown as period 1) as well as in reverse (period 2) (Dudzinski 1986). It shows that when the radius of the front wheel is smaller than that of the rear wheel, then in

Figure 4.9 Effects of theoretical speed ratio on the slips of the front and rear tires of a four-wheel drive tractor with rigid interaxial coupling.

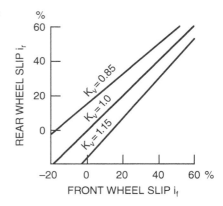

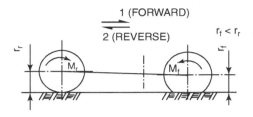

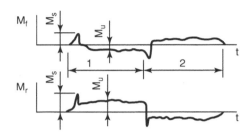

Figure 4.10 Torque distributions between the front and rear axles of a four-wheel drive tractor with rigid interaxial coupling when the radius of the front tire is smaller than that of the rear tire. *Source:* Dudzinski (1986). Reproduced with permission of the International Society for Terrain–Vehicle Systems.

forward motion, after the initial start-up period with torque M_s shown in the figure, the torque on the front axle M_u is negative, while that on the rear axle is positive. This indicates that the front wheel skids and develops a braking force, while the rear wheel slips and develops a forward thrust. Under these circumstances, torsional wind-up occurs. When the vehicle is in reverse, a similar situation can be observed.

In a turning maneuver, the wheels on the front and rear axles usually follow different paths with different turning radii, which requires the front and rear wheels to have different translatory speeds. If the front and rear axles are rigidly coupled, the front wheel will skid and develop a braking force. Figure 4.11 shows the torques on the front and rear axles of a four-wheel drive vehicle during a turning maneuver on dry concrete (Dudzinski 1986). When the vehicle is in forward

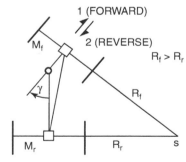

Figure 4.11 Torque distributions between the front and rear axles of a four-wheel drive tractor with rigid interaxial coupling during a turn. *Source:* Dudzinski (1986). Reproduced with permission of the International Society for Terrain–Vehicle Systems.

Figure 4.12 Variations of coefficient of thrust distribution with theoretical speed ratio of a four-wheel drive tractor with rigid interaxial coupling.

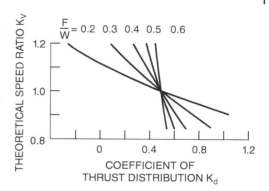

motion, after the initial start-up period, the torque on the front axle is negative, while that on the rear axle is positive. This indicates that the front wheel skids and develops a braking force, while the rear wheel slips and develops a forward thrust. Under these circumstances, torsional wind-up again occurs.

Since wheel slip is related to thrust, as described in Chapter 2, the thrust distribution between the front and rear drive axles depends on the theoretical speed ratio. Figure 4.12 shows the relationships between the coefficient of thrust distribution K_d and the theoretical speed ratio K_v in straight-line motion, at various values of vehicle total thrust-to-weight ratio F/W for the same four-wheel drive vehicle as that for Figure 4.8 (Wong 1970). When the vehicle total thrust-to-weight ratio is high, such as when the vehicle is pulling a heavy load, the difference in the theoretical speed between the front and rear wheels has a less significant effect on the thrust distribution. It is interesting to note that when the vehicle total thrust-to-weight ratio F/W is 0.2 and the theoretical speed ratio K_v is 0.9, the coefficient of thrust distribution K_d is equal to 1.0. This indicates that the vehicle is essentially a rear-wheel drive vehicle and that the potential advantages of four-wheel drive are not realized.

Since the theoretical speed ratio affects the relationship between the slip of the front wheel and that of the rear wheel, and hence the thrust distribution between the drive axles, the slip efficiency is a function of the theoretical speed ratio. Figure 4.13 shows the variations of the slip efficiency η_{s4}

Figure 4.13 Variations of slip efficiency with thrust/weight ratio at various theoretical speed ratios of a four-wheel drive tractor with rigid interaxial coupling.

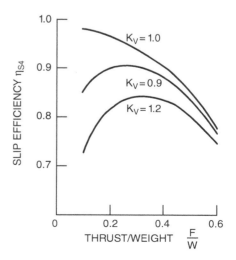

with the vehicle total thrust-to-weight ratio F/W in straight-line motion, at various values of theoretical speed ratio K_v for the same four-wheel drive vehicle as that for Figure 4.8.

As Eq. (4.12) shows that the efficiency of slip of a four-wheel drive vehicle is a function of thrust distribution between the front and rear wheels, it is of interest to explore whether there is an optimum coefficient of thrust distribution that enables the efficiency of slip to reach its maximum under a given operating condition. To find this optimum thrust distribution, the first partial derivative of η_{s4} with respect to K_d is taken and set equal to zero:

$$\frac{\partial \eta_{s4}}{\partial K_d} = \frac{(1-i_f)(1-i_r)(i_f-i_r)}{\left[(1-i_r)-(i_f-i_r)K_d\right]^2} = 0 \qquad (4.14)$$

This condition can be satisfied only if the slip of the front wheel or that of the rear wheel is 100%, or the slip of the front wheel is equal to that of the rear wheel. When the slip of either the front or the rear wheel is 100%, the vehicle cannot move forward, and the efficiency of slip becomes zero. Therefore, under normal operating conditions, only when the slip of the front wheel equals that of the rear wheel will the first partial derivative be zero. In other words, the slip of the front wheel being equal to that of the rear wheel is the necessary condition for achieving the optimal slip efficiency of a four-wheel drive vehicle (Wong 1970).

The efficiency of slip η_{s4} may be expressed in an alternate form by substituting the theoretical speed ratio K_v for $(1-i_r)/(1-i_f)$ in Eq. (4.12):

$$\eta_{s4} = 1 - \frac{i_f K_v - (K_v - 1)K_d}{K_v - (K_v - 1)K_d} \qquad (4.15)$$

Taking the first partial derivative of η_{s4} with respect to K_d and setting it to zero, one obtains

$$\frac{\partial \eta_{s4}}{\partial K_d} = \frac{(K_v - 1)K_v (1 - i_f)}{\left[K_v - (K_v - 1)K_d\right]^2} = 0 \qquad (4.16)$$

This leads to the same conclusion that when the value of the theoretical speed ratio K_v is equal to 1.0, which is equivalent to the slips of the front and rear wheels being equal in straight-line motion, the first partial derivative of η_{s4} with respect to K_d will be equal to zero.

The results of a detailed analysis indicate that the slips of the front and rear wheels being equal is the necessary, as well as the sufficient condition, for achieving the maximum efficiency of slip under a given operating condition (such as operating on a particular terrain at a given vehicle total thrust to weight ratio), if one of the following three conditions is satisfied (Huang et al. 2014):

- The relationship between the thrust coefficient and slip for the front wheel and that for the rear wheel are the same.
- The relationship between the thrust coefficient and slip for the front wheel and that for the rear wheel are different, but their relationships are linear.
- Their relationships are nonlinear (such as that described by an exponential function), but they have the same characteristics, such as

$$F_f = A\left(1 - e^{-i_f/B}\right)$$

and

$$F_r = A\left(1 - e^{-i_r/B}\right)$$

Under normal operating conditions of off-road vehicles, such as farm tractors, the wheel slip would normally be kept to less than 20%. In the range between 0 and 20% slip, usually the relationship between the thrust coefficient and slip is approximately linear on a variety of terrains. Thus, according to the finding noted in the second bullet point above, even the relationship between the thrust coefficient and slip for the front wheel is different from that for the rear wheel, but with slip less than 20% their thrust–slip relationships are approximately linear, the conclusion that when the slips of the front and rear wheels are equal (or the theoretical speed ratio K_v is equal to 1.0), the efficiency of slip will be an optimum is valid for practical purposes.

The results of a detailed analysis also indicate that only when the relationship between the thrust coefficient and slip for the front wheel and that for the rear wheel are nonlinear and have different characteristics, such as $F_f = A_1\left(1 - e^{-i_f/B_1}\right)$ and $F_r = A_2\left(1 - e^{-i_f/B_2}\right)$, where $A_1 \neq A_2$ and $B_1 \neq B_2$, will the condition for achieving the optimum slip efficiency deviate from that defined by $i_f = i_r$ or $K_v = 1$ (Huang et al. 2014). It should be noted, however, that even under extremely unusual operating scenario, such as the front wheel operating on a surface like farm soil and the rear wheel operating on a surface as hard as tarmac, the optimal theoretical speed ratio K_v deviates only slightly from 1.0, when the vehicle total thrust-to-weight ratio F/W is within the normal operating range up to 0.6 (Besselink 2003). For instance, under the above-noted extremely unlikely operating conditions, when F/W is 0.6 (corresponding to a wheel slip of approximately 20% on the farm soil), the optimal theoretical speed ratio K_v is 1.0406, which is only 4.06% different from $K_v = 1.0$. The efficiency of slip at the optimal theoretical speed ratio $K_v = 1.0406$ is 90.84%, as against that of 90.44% at theoretical speed ratio $K_v = 1.0$. The difference in efficiency of slip between these two cases amounts to merely 0.4%, which is negligible for practical purposes.

The approach to the analysis of the efficiency of slip of four-wheel drive off-road vehicles presented above can equally be applied to any all-wheel drive off-road vehicle with any number of driven wheels, such as 6×6 and 8×8.

A series of field tests was performed to examine the effects of theoretical speed ratio on slip efficiency, tractive efficiency, and fuel efficiency of a four-wheel drive vehicle (Wong et al. 1998, 1999, 2000). The test vehicle was an instrumented four-wheel drive tractor with front-wheel-assist, Case-IH Magnum. When the front-wheel drive is engaged, the two drive axles were rigidly coupled with a gear ratio of 0.752:1. The static load on the front axle with ballast was 43.26 kN (9726 lb) and that on the rear axle was 64.16 kN (14,424 lb). The drawbar performance of the test vehicle, with seven combinations of front and rear tires of different sizes and at various inflation pressures, was measured in a farm field. The seven front and rear wheel combinations were: 13.6R28 and 20.8R38; 14.9R28 and 20.8R38; 16.9-26 bias and 20.8R38; 16.9R26 and 20.8R38; 16.9R28 and 20.8R38; 13.6R28 and 18.4R38; and 14.9R28 and 18.4R38. The inflation pressure of the tires varied in the range of 82 kPa (12 psi) to 193 kPa (28 psi). The various combinations of front and rear tires produce a variety of values of theoretical speed ratio ranging from 0.908 to 1.054. The theoretical speed ratios were calculated from the free-rolling radii of the front and rear tires in the field and taking into account the fixed gear ratio between the front and rear drive axles. During field tests, the driving torques on the front and rear axles, slips of the front and rear tires, drawbar pull, vehicle forward speed, and fuel consumption of the tractor were monitored.

Field test data show that the thrust–slip relations for all front tires under various inflation pressures are quite similar and so are those for the two rear tires. Thus, all performance data of approximately 350 sets obtained with the seven sets of tires under various inflation pressures are combined in the evaluation of the effects of theoretical speed ratio K_v on the efficiency of slip, tractive efficiency, and fuel efficiency of the tractor.

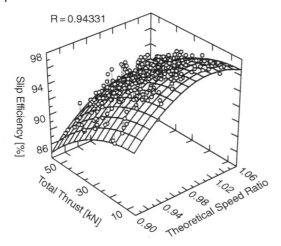

R = 0.94331

Figure 4.14 Measured relationships of slip efficiency, total thrust, and theoretical speed ratio of a four-wheel drive tractor on a clayey loam.

Figure 4.14 shows the relationship of the measured efficiency of slip (defined by Eq. (4.12) or Eq. (4.15)), total thrust (i.e., the sum of the thrusts of the front and rear tires), and theoretical speed ratio. A three-dimensional curved surface, representing the efficiency of slip as a quadratic function of total thrust and theoretical speed ratio, was fitted to the measured data, using the method of least-squares, as shown in Figure 4.14. By slicing the curved surface along a constant total thrust plane, the relationships between the efficiency of slip and theoretical speed ratio at various values of total thrust can be obtained, as shown in Figure 4.15. It shows that for the total thrust-to-weight ratio F/W up to 0.45 or the total thrust up to 48 kN (10,791 lb), when the theoretical speed ratio is close to or equal to 1.0, the efficiency of slip of the four-wheel drive tractor is indeed at a maximum. Thus, the analytical finding that when the theoretical speed ratio is equal to 1.0, the efficiency of slip will reach its peak, is experimentally substantiated. It is also shown in Figure 4.15 that if the theoretical speed ratio is either higher or lower than 1.0, the efficiency of slip will be lower than the maximum.

Figure 4.16 shows the relationship of the measured tractive efficiency η_d (defined by Eq. (4.9)), total thrust, and theoretical speed ratio. Again, by slicing the fitted curved surface shown in Figure 4.16 along a constant total thrust plane, the relationships between the tractive efficiency and theoretical speed ratio at various values of

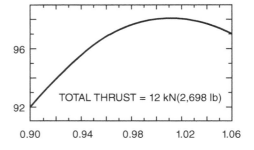

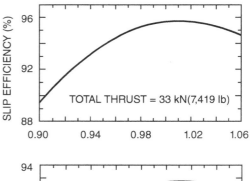

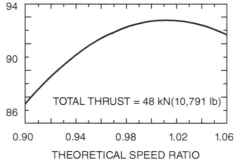

Figure 4.15 Relationships between slip efficiency and theoretical speed ratio at various values of total thrust.

Figure 4.16 Measured relationships of tractive efficiency, total thrust, and theoretical speed ratio of a four-wheel drive tractor on a clayey loam.

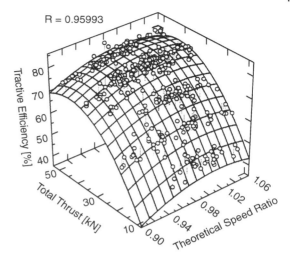

total thrust can be obtained, as shown in Figure 4.17. It shows that when the theoretical speed ratio is close to or equal to 1.0, the tractive efficiency of the four-wheel drive tractor is also at its peak. This is mainly because the efficiency of slip is a major component of the tractive efficiency.

Figure 4.18 shows the relationship of the measured fuel efficiency (defined as the drawbar power per unit volume of diesel fuel consumed per hour), total thrust, and theoretical speed ratio. Again, by slicing the fitted curved surface shown in Figure 4.18 along a constant total thrust plane, the relationships between the fuel efficiency and theoretical speed ratio at various values of total thrust can be obtained, as shown in Figure 4.19. When the theoretical speed ratio is close to or equal to 1.0, the fuel efficiency of the four-wheel drive tractor is also at a maximum.

The test results show that under the operating conditions described, the slip efficiency, tractive efficiency, and fuel efficiency of the test vehicle do reach their respective maximums when the theoretical speed ratio is close to or equal to 1.0. This indicates that maintaining equal slips for all driven tires or theoretical speed ratio $K_v = 1.0$ is a useful and practical guide for achieving high operating efficiency of all-wheel drive off-road vehicles.

Analytical and experimental studies of tractive and fuel efficiency of multi-wheel drive vehicles following other approaches have also been made. Like the findings presented above, it is shown from these

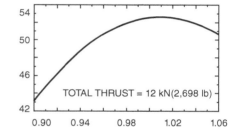

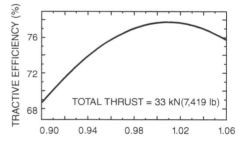

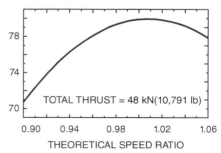

Figure 4.17 Relationships between tractive efficiency and theoretical speed ratio at various values of total thrust.

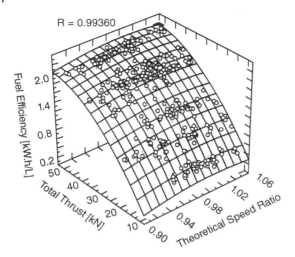

Figure 4.18 Measured relationships of fuel efficiency, total thrust, and theoretical speed ratio of a four-wheel drive tractor on a clayey loam.

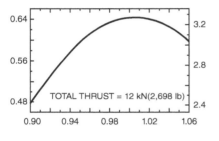

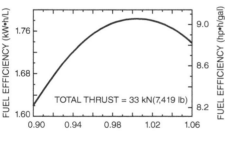

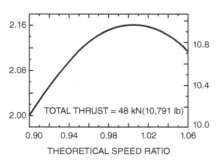

Figure 4.19 Relationships between fuel efficiency and theoretical speed ratio at various values of total thrust.

studies that optimum driveline systems for multi-wheel drive vehicles are expected to be designed in such a way that leads to the same slips for all the driven tires (Vantsevich 2007, 2008).

4.1.4 Coefficient of Traction

In the evaluation of the drawbar performance of off-road vehicles, the ratio of the drawbar pull to the normal load on the driven wheels W_d, which is usually referred to as the coefficient of traction μ_{tr}, is a widely used parameter. It is expressed by

$$\mu_{tr} = \frac{F_d}{W_d} = \frac{F - \sum R}{W_d} \qquad (4.17)$$

Since drawbar pull is a function of slip, the coefficient of traction of different vehicles should be compared at the same slip. Figure 4.20 shows a comparison of the coefficient of traction of a two-wheel drive and a comparable four-wheel drive tractor on a farm soil (Osborne 1969–1970).

4.1.5 Weight-to-Power Ratio for Off-Road Vehicles

For off-road vehicles designed for traction, the desirable weight-to-engine-power ratio is determined by the necessity for the optimal utilization of the engine power to produce the required drawbar pull. It is, therefore, a function of the operating speed. From

Figure 4.20 Variations of coefficient of traction with slip for a two-wheel drive and a four-wheel drive tractor. *Source:* Osborne (1969–1970). Reproduced with permission of the Council of the Institution of Mechanical Engineers.

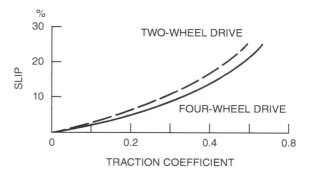

Eq. (4.7), the relationship between the vehicle-weight-to-engine-power ratio and the operating speed can be expressed by (Reece 1969–1970)

$$P = \frac{FV_t}{\eta_t} = \frac{(F_d + \Sigma R)V_t}{\eta_t} = \frac{(\mu_{tr} W_d + f_r W)V}{(1-i)\eta_t} \tag{4.18}$$

and

$$\frac{W}{P} = \frac{(1-i)\eta_t}{(\mu_{tr} W_d/W + f_r)V} = \frac{(1-i)\eta_t}{(\mu_{tr}K_{we} + f_r)V} \tag{4.19}$$

where W is the total vehicle weight; f_r is the coefficient of motion resistance; and K_{we} is called the weight utilization factor and is the ratio of W_d to W. The weight utilization factor K_{we} is less than unity for a two-wheel drive tractor and is equal to unity for a four-wheel drive vehicle or a tracked vehicle. If there is load transfer from the implement to the vehicle, the value of K_{we} may be greater than unity.

Equation 4.19 indicates that for a vehicle designed to operate in a given speed range, the weight-to-engine-power ratio should be within a particular boundary, so that a specific level of tractive efficiency can be maintained. Figure 4.21 shows the variations of the desirable weight-to-engine-power ratio with operating speed for a two-wheel drive and a four-wheel drive tractor under a particular operating environment.

In agriculture, attempts are being made to achieve higher productivity in the field by increasing the operating speed of the tractor–implement system. This would require the development of appropriate implements as well as tractors, so that full advantage of high-speed operation can be realized. Equation 4.19 provides guiding principles for the selection of the design and performance parameters of tractors designed for operation at increased speeds. It shows that to achieve optimum utilization of engine power and to maintain a high level of tractive efficiency, an increase of the operating speed must be accompanied by a corresponding reduction of the tractor weight-to-engine-power ratio.

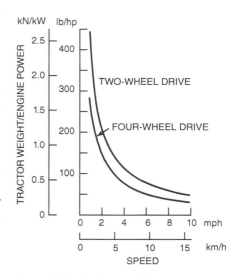

Figure 4.21 Variations of the optimum weight-to-power ratio with operating speed for a two-wheel drive and a four-wheel drive tractor. *Source:* Reece (1969–1970). Reproduced with permission of the Council of the Institution of Mechanical Engineers.

4.2 Fuel Economy of Cross-Country Operations

The fuel economy of off-road vehicles depends not only on the fuel consumption characteristics of the engine but also on the transmission characteristics, internal resistance of the running gear, external resisting forces, drawbar pull, and operating speed. When the resultant resisting force, drawbar pull, and operating speed are known, the required engine output power P is determined by

$$P = \frac{(\sum R + F_d)V}{(1-i)\eta_t} \tag{4.20}$$

The fuel consumed per hour of operation u_h can then be calculated by

$$u_h = P u_s \tag{4.21}$$

where u_s is the specific fuel consumption of the engine in kg/kW · h (or lb/hp · h). For changing operating conditions, the basic equations given above can still be used to compute, step by step, the changing power requirements and fuel consumption.

The motion resistance and slip of an off-road vehicle over a given terrain affect the power requirements and are dependent, to a great extent, on the design of the running gear and vehicle configuration, as discussed in Chapter 2. Consequently, the tractive performance of the vehicle has a considerable impact on the fuel economy of cross-country operations. This may be illustrated by the following example.

Example 4.2   Referring to Example 2.3 in Chapter 2, if vehicles A and B are to pull a load that requires a drawbar pull of 45 kN (10 117 lb), estimate the difference in fuel consumption of the two vehicles when traveling at a speed of 10 km/h (6.2 mph). In the calculations, an average specific fuel consumption of 0.25 kg/kW · h (0.41 lb/hp · h) for a diesel engine may be assumed.

Solution

Referring to Example 2.3, the compaction resistance of vehicle A is 3.28 kN (738 lb) and that of vehicle B is 2.6 kN (585 lb). Both vehicles have a weight of 135 kN (30 350 lb). The internal resistance of the two vehicles traveling at 10 km/h may be estimated using Eq. (4.2):

$$R_{in} = f_r W = (0.033 + 0.00105V)W = 0.0435 \times 135 = 5.87 \text{ kN } (1320 \text{ lb})$$

(a) To pull the load specified, vehicle A should develop a thrust F equal to the sum of the resultant motion resistance and drawbar pull:

$$F = \sum R + F_d = 3.28 + 5.87 + 45 = 54.15 \text{ kN } (12, 174 \text{ lb})$$

From Table 2.14, to develop this thrust, the slip of vehicle A is approximately 56.8%. The transmission efficiency η_t is taken as 85%. The engine power required for vehicle A traveling at 10 km/h (6.2 mph) can be calculated using Eq. (4.20):

$$P = \frac{(\sum R + F_d)}{(1-i)\eta_t} = 410 \text{ kW } (550 \text{ hp})$$

The fuel consumption per hour of operation is

$$u_h = Pu_s = 410 \times 0.25 = 102.5 \text{ kg/h } (226 \text{ lb/h})$$

(b) To pull the load specified, the thrust that vehicle B should develop is

$$F = 2.6 + 5.87 + 45 = 53.47 \text{ kN } (12{,}021 \text{ lb})$$

From Table 2.14, to develop this thrust, the slip of vehicle B is approximately 33.1%. The engine power required for vehicle B traveling at 10 km/h (6.2 mph) is

$$P = \frac{(\sum R + F_d)}{(1-i)\eta_t} = 261.4 \text{ kW } (350.5 \text{ hp})$$

The fuel consumption per hour of operation is

$$u_h = Pu_s = 65.4 \text{ kg/h } (144 \text{ lb/h})$$

The results indicate that vehicle A consumes approximately 56.7% more fuel than vehicle B, because vehicle A has higher motion resistance and slip than vehicle B under the circumstances specified. The difference in tractive performance and fuel consumption between the two vehicles is due to the difference in the dimensions of the tracks.

The operational fuel economy of various types of cross-country vehicle may also be evaluated using parameters reflecting the productive work performed by the vehicle. For instance, for an agricultural tractor, the operating fuel economy may be expressed in terms of fuel consumed for work performed in unit area, u_a:

$$u_a = \frac{Pu_s}{B_m V_m} \tag{4.22}$$

where B_m is the working width of the implement or machinery that the tractor pulls and V_m is the average operating speed.

To evaluate the fuel economy of a tractor in developing drawbar power, the fuel consumption per unit drawbar power per hour u_d may be used as a criterion:

$$u_d = \frac{u_h}{P_d} = \frac{Pu_s}{P\eta_m\eta_s\eta_t} = \frac{u_s}{\eta_m\eta_s\eta_t} \tag{4.23}$$

where u_h is the fuel consumed per hour of operation; P is the engine power; P_d is the drawbar power; u_s is the specific fuel consumption of the engine; η_m is the efficiency of motion; η_s is the efficiency of slip; and η_t is the transmission efficiency.

In the University of Nebraska's test programs, the energy obtained at the drawbar E_d per unit volume of fuel consumed u_e is used as an index for evaluating fuel economy:

$$u_e = \frac{E_d}{u_t} = \frac{F_d V t}{u_t} = \frac{P_d t}{u_t} \tag{4.24}$$

where u_t is the fuel consumed during time t.

For cross-country transporters, the fuel consumption per unit payload transported over a unit distance u_{tr} may be used as a criterion for evaluating fuel economy:

$$u_{tr} = \frac{Pu_s}{W_p V_m} \tag{4.25}$$

where W_p is the payload and u_{tr} may be expressed in liters per tonne · kilometer or gallons per ton · mile.

When operating in areas where fuel is not readily available, special fuel carriers must be used to supply the payload carriers with the required fuel. Thus, the total fuel consumption per unit payload transported should include the consumption of the fuel carriers (Bekker 1969).

4.3 Transport Productivity and Transport Efficiency

The absolute criterion for comparing one commercial off-road transporter with another is the relative cost of transporting a unit payload on a particular route. This involves not only the performance and fuel consumption characteristics of the vehicle but also factors not known before a vehicle has been operated, such as load factor and customer's preference. However, certain basic performance criteria exist that enable some assessment and comparison to be made in the preliminary stage of development. Some of these are discussed below.

Transport productivity, which is defined as the product of payload and the average cross-country speed through a specific region, may be used as a criterion for evaluating the performance of off-road transporters. For an existing vehicle, the average speed may be measured experimentally. However, for a vehicle under development, the prediction of its average operating speed through a particular region may be quite complex, as the terrain conditions may vary considerably from one patch to another.

In addition to vehicle tractive performance, a number of other factors, such as the ability in obstacle negotiation, mobility in a riverine environment, and vehicle vibrations excited by ground roughness, also affect the cross-country speed of the vehicle.

To characterize the efficiency of a transport system, the transport efficiency η_{tr}, which is defined as the ratio of the transport productivity to the corresponding power input to the system, may also be used (Wong 1972, 1975):

$$\eta_{tr} = \frac{W_p V}{P} \tag{4.26}$$

where W_p is the payload; V is the average cross-country speed; and P is the power input to the system.

The transport efficiency as defined in Eq. (4.26) has three basic components, namely, the lift/drag ratio C_{ld} (the ratio of the vehicle total weight to the resultant motion resistance), structural efficiency η_{st} (the ratio of the payload to the vehicle total weight), and propulsive efficiency η_p:

$$\eta_{tr} = \frac{W_p V}{P} = \frac{W_p V}{(\Sigma R) V / \eta_p} = \frac{W}{\Sigma R} \frac{W_p}{W} \eta_p \tag{4.27}$$
$$= C_{ld} \eta_{st} \eta_p$$

The propulsive efficiency includes the transmission efficiency and slip efficiency of the vehicle.

The reciprocal of transport efficiency expressed in terms of power consumption per unit transport productivity may also be used to characterize the performance of a transport system.

4.4 Mobility Map and Mobility Profile

The maximum possible speed in straight-line motion under steady-state operating conditions from one location to another is used as a metric for characterizing military ground vehicle mobility in the NATO Reference Mobility Model (NRMM) and its predecessors, such as the AMM-75 and AMC-71 (Ahlvin and Haley 1992; Jurkat et al. 1975; Nuttall Jr. et al. 1974). This maximum possible speed, also known as the speed-made-good, is a highly aggregated parameter representing the results of numerous interactions between the vehicle and the operating environment. In view of the variation of environmental conditions in the field, in these simulation models, the area of interest is first divided into patches, within each of which the terrain is considered sufficiently uniform to permit the use of the maximum possible speed in straight-line motion to define military ground vehicle mobility.

For these models, the characteristics of the terrain, the vehicle, and the driver are required as inputs. Terrain surface composition, surface geometry, vegetation, and linear geometry, such as stream cross-section and water speed and depth, must be specified. Vehicle geometric characteristics, inertia properties, and mechanical characteristics together with the driver's reaction time, recognition distance, and ride comfort limits must also be defined.

The terrain is classified into three categories: areal patch, linear feature segment, such as a stream, ditch, or embankment, and road or trail segment.

When a vehicle is crossing an areal terrain unit, the maximum speed may be limited by one or a combination of the following factors:

- The tractive effort available for overcoming the resisting forces due to sinkage, slope, obstacles, vegetation, etc.
- The driver's tolerance to ride discomfort when traversing rough terrain and to obstacle impacts.
- The driver's reluctance to proceed faster than the speed at which the vehicle would be able to decelerate to a stop, within the limited visibility distance prevailing in that patch.
- Vehicle maneuverability to avoid obstacles.
- The acceleration and deceleration between obstacles, and speed reduction due to maneuvering to avoid obstacles.

The speed limited by each of the above factors is calculated and compared, and the maximum possible speed within a particular terrain patch is determined. When the vehicle is traversing a linear feature segment, such as a stream, man-made ditch, canal, escarpment, railroad, or highway embankment, appropriate models are used to determine the maximum feasible speed. In the models, the time required to enter and cross the segment, and that required to egress from it, are taken into consideration. Both include allowance for engineering effort, such as winching and excavating, whenever required.

To predict the maximum possible speed of a vehicle on roads or trails, in addition to the speed limited by the motion resistance, the speed limited by ride discomfort, visibility, tire characteristics, or road curvature must be taken into consideration. The least of them is taken as the maximum feasible speed for the road or trail segment.

The results obtained from the analysis may be conveniently shown in a mobility map, as illustrated in Figure 4.22 (Nuttall Jr. et al. 1974). The numbers in the map indicate the speed (in miles per hour) of which a particular vehicle is capable in each patch throughout the region under consideration. This provides the basis for selecting the optimum route for the vehicle to maximize the average speed through a given area. The information contained in the mobility map may be

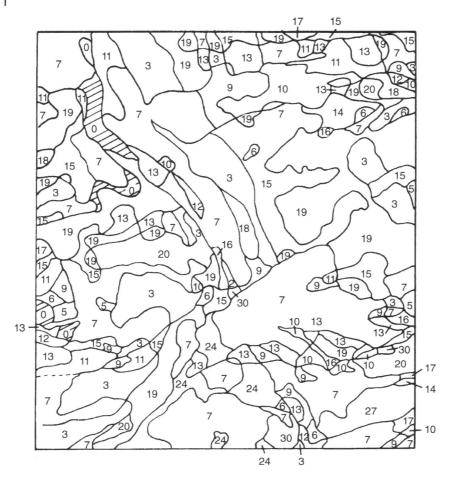

Figure 4.22 Mobility map of a 2.5 ton truck. Numbers in the map indicate the maximum achievable speed in miles per hour in each patch; cross-hatched areas indicate where the vehicle is immobile. Reprinted with permission from SAE paper 740426 © 1974 Society of Automotive Engineers, Inc.

generalized in a mobility profile shown in Figure 4.23 (Nuttall Jr. et al. 1974), which conveys a complete statistical description of vehicle mobility in a particular area. It indicates the speed at which the vehicle can sustain as a function of the percentage of the total area under consideration. For instance, the intercept of 90% point (point *A*) in Figure 4.23 indicates that the vehicle can achieve an average speed of 13.7 km/h (8.5 mph) over 90% of the area. The mobility map and mobility profile are suitable formats for characterizing vehicle mobility for many purposes, such as operational planning and effectiveness analysis. However, they are not directly suitable for parametric analysis of vehicle design.

The mobility of a fighting vehicle, such as a tank, may be described in terms of its operational mobility and battlefield mobility (Ogorkiewicz 1991).

The operational mobility is the ability of the tank to move in the zone of operations. It is related to the power-to-weight ratio, vehicle weight, operating range, and reliability. The higher the power-to-weight ratio of the vehicle, the higher is the potential speed with which it can move from one area to another. Vehicle weight affects tractive performance over soft terrain, as discussed in Chapter 2,

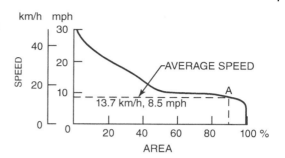

Figure 4.23   Mobility profile of a 2.5 ton truck. Reprinted with permission from SAE paper 740426 © 1974 Society of Automotive Engineers, Inc.

and restricts the type of road bridge that the vehicle can cross. The operating range of the vehicle affects the frequency of refueling stops from origin to destination, and hence its average speed. Vehicle reliability also influences operational mobility, as the higher it is, the greater is the probability that the tank will arrive at the destination on schedule.

The battlefield mobility is the ability of the tank to move when engaging enemy forces in the battlefield. It should be able to move over various types of terrain, ranging from soft soil to hard, rough ground, and to negotiate obstacles at the highest possible speed to minimize its exposure to enemy fire. The weight and the design of the track–suspension system of a fighting vehicle greatly affect its performance over soft terrain and its speed over rough ground. The power-to-weight ratio, to a great extent, determines the acceleration and agility of the vehicle, and hence its ability to take evasive maneuvers under battlefield conditions. Figure 4.24 shows the relationships of the

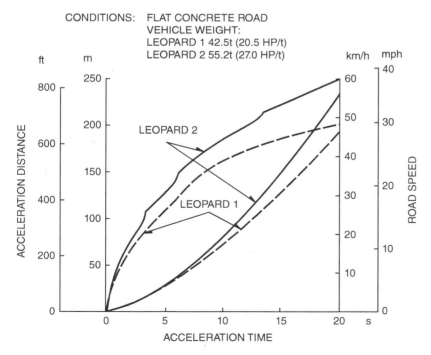

Figure 4.24   Speed (top curves) and distance (bottom curves) versus acceleration time of two main battle tanks: Leopard 1 and 2. Reproduced with permission of MTU Motoren-und Turbinen-Union Friedrichshafen, Germany.

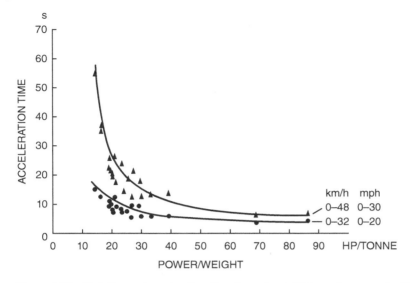

Figure 4.25 Variations of acceleration time from standstill to a given speed with vehicle power-to-weight ratio. Reproduced from *Technology of Tanks* by R.M. Ogorkiewicz (1991), with permission of Jane's Information Group.

acceleration distance and time for two tanks, Leopard 1 and 2, with different power-to-weight ratios. Leopard 2, with a power-to-weight ratio of 27 hp/t (20 kW/t), can attain a given speed or travel a specific distance faster than Leopard 1 with 20.5 hp/t (15.3 kW/t) (Ogorkiewicz 1991). Figure 4.25 shows the time taken by several tanks with different power-to-weight ratios to accelerate from standstill to 32 or 48 km/h (20 or 30 mph) on hard road surfaces. The time required to accelerate the vehicle to a speed of 32 km/h (20 mph) approaches a more or less constant value when the power-to-weight ratio is up to approximately 40 hp/t (30 kW/t). This indicates that it is not effective to increase the power-to-weight ratio beyond a certain level for a given operating condition. Armor protection also affects the battlefield mobility of a tank. With better armor protection, it can move more freely under battlefield conditions, and it has improved battlefield survivability.

4.5 Selection of Vehicle Configurations for Off-Road Operations

Vehicle configuration can generally be defined in terms of its form, size, weight, and power (Bekker 1969). Selection of vehicle configuration is primarily based on its mission and operational requirements and on the environment in which the vehicle is expected to operate. In addition, fuel economy, safety, cost, impact on the environment, reliability, maintainability, and other factors must be taken into consideration. To define an optimum vehicle configuration for a given mission and environment, a systems analysis approach should therefore be adopted.

The analysis of terrain–vehicle systems usually begins with defining mission requirements, such as the type of work to be performed, the kind of payload to be transported, and the operational characteristics of the vehicle system, including output rates, cost, and economy. The physical and geometric properties of the terrain over which the vehicle is expected to operate are collected as inputs. Competitive vehicle concepts with the probability of accomplishing the specified mission requirements are chosen, based on experience and future development trends. The operational characteristics and performance of the vehicle candidates are then analyzed and compared. In the evaluations, the methods and techniques discussed in Chapter 2 and in the preceding sections of this chapter may be employed. As a result of systems analysis, an order of merit for the vehicle candidates is established, from which an optimal vehicle configuration is selected (Bekker 1969).

Thus, selection of vehicle configuration for a given mission and environment is a complex process, and it is not possible to define the optimal configuration without detailed analysis. However, based on the current state of the art of off-road transport technology, some generalization of the merits and limitations of existing vehicle configurations may be made (Wong 2010). There are currently two major types of ground vehicle capable of operating over a specific range of unprepared terrains: wheeled vehicles and tracked vehicles.

4.5.1 Wheeled Vehicles

Referring to the analysis of the tractive performance of off-road vehicles given in Chapter 2 and in the preceding sections of this chapter, the maximum drawbar pull coefficient of a vehicle may be expressed by

$$
\frac{F_d}{W} = \frac{F - \sum R}{W} = \frac{cA + W \tan \phi - f_r W}{W}
$$
$$
= \frac{c}{p} + \tan \phi - f_r
$$
(4.28)

This equation indicates that for a given terrain with specific values of cohesion and angle of internal shearing resistance, c and ϕ, the maximum drawbar pull coefficient is a function of the contact pressure p and the coefficient of motion resistance f_r. The lower the contact pressure and the coefficient of motion resistance, the higher is the maximum drawbar pull coefficient. Since the contact pressure and the motion resistance are dependent on the design of the vehicle, the proper selection of vehicle configuration and design parameters is of utmost importance.

For given overall dimensions and gross weight, a tracked vehicle will have a larger ground contact area than a wheeled vehicle. Consequently, the ground contact pressure, and hence the sinkage and external motion resistance of the tracked vehicle, would generally be lower than that of an equivalent wheeled vehicle. Furthermore, a tracked vehicle has a longer contact length than a wheeled vehicle of the same overall dimensions. Thus, the slip of a tracked vehicle is usually lower than that of an equivalent wheeled vehicle for the same thrust. As a result, the mobility of the tracked vehicle is generally superior to that of the wheeled vehicle in soft terrain.

The wheeled vehicle is, however, a more suitable choice than the tracked one when frequent on-road travel and high road speeds are required.

4.5.2 Tracked Vehicles

Although the tracked vehicle has the capability of operating over a wide range of unprepared terrain, to fully realize its potential, careful attention must be given to the design of the track–roadwheel system. The nominal ground pressure of the tracked vehicle (i.e., the ratio of the vehicle gross weight to the nominal ground contact area) has been quite widely used in the past as a design parameter of relevance to soft ground mobility. However, the shortcomings of its general use are now evident, both in its neglect of the actual pressure variation under the track and in its inability to distinguish between track–roadwheel–suspension system designs with the same nominal ground pressure but having different soft ground performance. It has been shown in Chapter 2 that vehicle sinkage, and hence motion resistance, depends on the maximum pressure exerted by the vehicle on the ground and not the nominal ground pressure. Therefore, it is of prime importance that the design of the track–roadwheel–suspension system provides ground contact pressure as uniform as possible under operating conditions.

For low-speed tracked vehicles, such as those used in farming and construction, uniform ground contact pressure could be achieved by using a relatively long-pitch link track with a large number of small-diameter roadwheels. To achieve optimal tractive performance, the ratio of roadwheel spacing to track pitch should, however, be within a certain range, as discussed in Section 2.7.

For high-speed tracked vehicles, such as those used in cross-country transport or in defense operations, to minimize the vibration of the vehicle and to provide adequate obstacle-crossing capability, relatively large-diameter roadwheels with considerable suspension travel are used. The track in these vehicles is usually either short-pitch link tracks or rubber belt (band) tracks. This would generally result in a rather non-uniform pressure distribution under the track. The overlapping roadwheel system shown in Figure 2.58(c) provides a possible compromise in meeting the conflicting requirements for soft ground mobility and high-speed operations. It is also shown in Section 2.6 that the initial track tension generally has a noticeable effect on ground pressure distribution. To improve tractive performance, careful selection of the initial track tension is of importance. The installation of an initial track tension regulating system controlled by the driver or automatically based on track slip would be a cost-effective means to improve vehicle mobility over soft terrain. The driver-controlled system enables the driver to conveniently increase the initial track tension when traversing soft terrain is anticipated. It also enables the driver to conveniently reduce the initial track tension to the regular level when the vehicle has passed over the soft patch, to minimize the wear and tear of the track system. The role of the initial track tension regulating system in improving tracked vehicle mobility on soft ground is analogous to that of the central tire inflation system in improving wheeled vehicle mobility on weak terrain.

The rubber belt (band) track is gaining increasingly wide applications to agricultural and industrial tractors. This is primarily due to its ability to operate over paved roads without damaging the surfaces, its low operating noise, and its light weight in comparison with the metal track. The rubber belt (band) track has also been proposed for use in future combat vehicles.

Experience and analysis have shown that the method of steering is also of importance to the mobility of tracked vehicles in soft terrain. Articulated steering provides the vehicle with better mobility and maneuverability than skid-steering over weak terrain. Articulated steering also makes it possible for the vehicle to achieve a more rational form since a long, narrow vehicle encounters less external resistance over soft ground than does a short, wide vehicle with the same ground contact area. From an environmental point of view, articulated steering causes less damage to the terrain during maneuvering than skid-steering. A detailed analysis of the characteristics of various steering methods for tracked vehicles is given in Chapter 6.

The characteristics of the transmission also play a noticeable role in vehicle mobility over soft ground. Generally speaking, the automatic transmission is preferred as it allows gear changing with less interruption of power flow to the running gear.

4.5.3 Wheeled Vehicles versus Tracked Vehicles

While the merits and limitations of wheeled and tracked vehicles for off-road operations have been generally outlined above, it is useful to provide more detailed guidance for the selection of vehicle configuration for a given mission and environment. In the following, the issue of wheeled vehicles versus tracked vehicles is examined in some detail from a traction perspective (Wong and Huang 2006).

For a wheeled vehicle, to simplify the analysis it is assumed that all tires have an essentially flat and rectangular contact patch with the same contact length L_{ti} and uniform normal pressure, and that the vehicle weight is uniformly distributed among the tires, then the total thrust (or tractive effort) F_{ti} developed by n_{ti} number of tires of the vehicle at a given slip i can be expressed by (assuming the shear stress–shear displacement relationship is described by Eq. (2.58))

$$F_{ti} = n_{ti}[cb_{ti}L_{ti} + (W/n_{ti})\tan\phi][1 - (K/iL_{ti})(1 - \exp(-iL_{ti}/K))] \qquad (4.29)$$

where b_{ti} is tire contact width and is assumed to be the same for all tires; W is the total weight of the vehicle; i is the slip and is assumed to be the same for all tires; and K is the shear deformation parameter.

For a tracked vehicle, to simplify the analysis it is assumed that the tracks have an essentially flat and rectangular contact patch with the same contact length L_{tr} and uniform normal pressure, and that the vehicle weight is uniformly distributed among the tracks, then the total thrust F_{tr} developed by n_{tr} number of tracks of the vehicle at a given slip i can be expressed by

$$F_{tr} = n_{tr}[cb_{tr}L_{tr} + (W/n_{tr})\tan\phi][1 - (K/iL_{tr})(1 - \exp(-iL_{tr}/K))] \qquad (4.30)$$

where b_{tr} is track contact width and is assumed to be the same for all tracks; W is the total weight of the vehicle; and i is the slip and is assumed to be the same for all tracks.

Based on Eqs. 4.29 and 4.30, the ratio of the thrust of a wheeled vehicle to that of a tracked vehicle F_{ti}/F_{tr}, hereinafter called the thrust ratio, is expressed by

$$\frac{F_{ti}}{F_{tr}} = \frac{n_{ti}[cb_{ti}L_{ti} + (W/n_{ti})\tan\phi][1 - (K/iL_{ti})(1 - \exp(-iL_{ti}/K))]}{n_{tr}[cb_{tr}L_{tr} + (W/n_{tr})\tan\phi][1 - (K/iL_{tr})(1 - \exp(-iL_{tr}/K))]} \qquad (4.31)$$

To illustrate the difference between a wheeled vehicle and a tracked vehicle from the traction perspective, the thrust developed by an off-road wheeled vehicle and that of a tracked vehicle are compared. Figure 4.26 is a schematic diagram of the four tires on one side of an 8 × 8 wheeled vehicle and that of a track on one side of a comparable tracked vehicle with two tracks. In the comparison, it is assumed that $b_{ti} = b_{tr} = 0.38$ m (15 in.), B (the wheelbase of the wheeled vehicle) $= L_{tr}$ (the ground contact length of the track) $= 3.3$ m(130 in.), tire outside diameter $= 0.984$ m (38.7 in.) (corresponding to that of a 325/85R16 tire), and both the wheeled and tracked vehicle weight $W = 110.57$ kN (24,858 lb).

Based on Eq. (4.31), the variations of the thrust ratio F_{ti}/F_{tr} with the contact length ratio L_{ti}/L_{tr} for various values of shear deformation parameter K on two types of soil, sand and clay with high

Figure 4.26 Schematic of four tires on one side of an 8 × 8 wheeled vehicle and that of a track on one side of a comparable tracked vehicle used in the study.

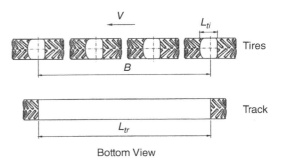

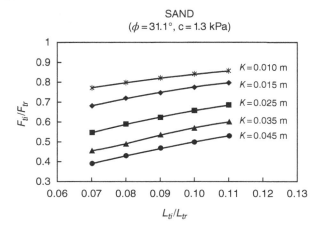

Figure 4.27 Variations of the ratio of thrust of a wheeled vehicle to that of a tracked vehicle at 20% slip with contact length ratio on sand.

moisture content (HMC), are shown in Figures 4.27 and 4.28, respectively. On the two types of terrain, the thrust ratio F_{ti}/F_{tr} is always lower than one, which indicates that the thrust developed by the wheeled vehicle is always lower than that of the comparable tracked vehicle. This is primarily because, for a tire, which has a much shorter contact length, the shear displacement at the rear of the contact patch will be much smaller than that of a track at the same slip. This may limit the full development of shear stress on the tire contact patch in many cases, particularly when the value of K is high. Since the thrust developed by a vehicle running gear is the integration of the shear stress over the contact area, with less developed shear stress and lower contact area, the thrust developed by a wheeled vehicle will generally be lower than that developed by a comparable tracked vehicle. This is illustrated by Figure 4.29, in which the development of shear stress under the tires of the 8 × 8 wheeled vehicle is compared with that of the comparable tracked vehicle. However, with the same weight but having a smaller contact area, the average normal pressure under the tire of a wheeled

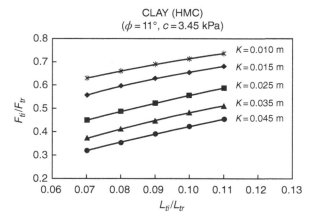

Figure 4.28 Variations of the ratio of thrust of a wheeled vehicle to that of a tracked vehicle at 20% slip with contact length ratio on a clayey soil with high moisture content.

Figure 4.29 Comparisons of the idealized shear stress distributions under (a) a track and (b) tires.

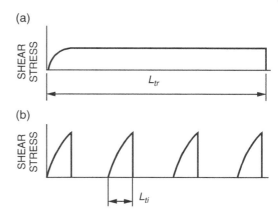

(a)

(b)

vehicle will be higher than that under a comparable track. This factor has been taken into account in the results shown in Figures 4.27 and 4.28.

In the figures the thrust ratio F_{ti}/F_{tr} generally increases with the contact length ratio L_{ti}/L_{tr}. It should be mentioned that tire contact length L_{ti} increases with the lowering of tire inflation pressure. This indicates that over soft terrain, reducing tire inflation pressure generally improves the thrust of a wheeled vehicle, and hence the thrust ratio F_{ti}/F_{tr}.

Figures 4.27 and 4.28 also show that when the value of K increases, the thrust ratio F_{ti}/F_{tr} decreases noticeably on both types of soil examined. This is because when the value of K increases, the shear displacement required for the development of the maximum shear stress also increases (please refer to Eq. (2.58) and Figure 2.43 in Section 2.4.3). This means that because a tire has a much shorter contact length than a track, the increase of the value of K has a much more noticeable effect on the thrust of a wheeled vehicle than that of a tracked vehicle, as noted previously and illustrated in Figure 4.29.

While the basic difference in traction between a wheeled vehicle and a tracked vehicle is illustrated by the analysis given above, it is based on simplifying assumptions, such as the normal pressure under the tire or the track being uniform and the contact patch being of rectangular shape, which are not necessarily realistic in many cases. Furthermore, the tractive performance of a vehicle is dependent on not only the thrust but also on the motion resistance, as discussed in Section 4.1. To realistically compare the performance of a wheeled vehicle with that of a tracked vehicle, computer-aided methods, such as those described in Sections 2.6, 2.7, and 2.9, should be used. Figure 4.30 shows a comparison of the drawbar performance at 20% slip of an 8 × 8 off-road wheeled vehicle (resembling a widely used light armored vehicle) at various tire inflation pressures with that of a vehicle with flexible tracks (similar to a widely used armored personnel carrier) on four types of terrain (sand, loam, clay with HMC, and clay with medium moisture content (MMC)) (Wong and Huang 2006).

The basic parameters of these two vehicles are given in Table 4.4. In the figure, $(DP/W)_{ti}$ and $(DP/W)_{tr}$ represent the drawbar-pull-to-vehicle-weight ratio of the wheeled vehicle and that of the tracked vehicle, predicted using NWVPM and NTVPM outlined in Sections 2.9 and 2.6, respectively (Wong and Huang 2006). The ratio of $(DP/W)_{ti}$ to $(DP/W)_{tr}$ on the four types of terrain examined is always less than 1 and the ratio decreases with the increase of tire inflation pressure. On the clay with HMC, the ratio becomes negative when the tire inflation pressure exceeds approximately 180 kPa (26 psi), which indicates that the wheeled vehicle will be unable to propel itself and will become

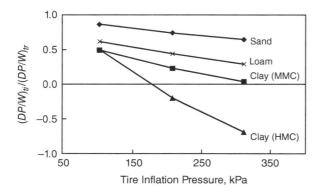

Figure 4.30 Comparisons of the drawbar pull to weight ratio of the wheeled vehicle at different tire inflation pressures to that of the tracked vehicle at 20% slip on various types of soil.

Table 4.4 Basic parameters of the 8 × 8 wheeled vehicle and tracked vehicle.

Vehicle parameters	Wheeled vehicle	Tracked vehicle
Total weight, kN	127.48	110.57
Wheelbase (wheeled vehicle) or nominal track contact length (tracked vehicle), m	3.475	2.67
Number of tires or roadwheels on one side	4	5
Tire or roadwheel outside diameter, m	0.984	0.61
Tire or track width, m	0.393	0.38

immobile. This further demonstrates that the computer simulation models presented in Chapter 2 are useful tools for evaluating the performance and design of off-road vehicles from the traction perspective.

References

Ahlvin, R.B. and Haley, P.W. (1992). NATO Reference Mobility Model Edition II, NRMM II Users Guide. Technical Report GL-92-19. U.S. Army Corps of Engineers Waterways Experiment Station, Vicksburg, MS.

Bekker, M.G. (1969). *Introduction to Terrain–Vehicle Systems*. Ann Arbor, MI: University of Michigan Press.

Besselink, B.C. (2003). Tractive efficiency of four-wheel drive vehicles: An analysis for non-uniform traction conditions. *Proceedings of the Institution of Mechanical Engineers, Part D: Journal of Automobile Engineering* **217** (5).

Cleare, G.V. (1963–1964). Factors affecting the performance of high-speed track layers. *Proceedings of the Institution of Mechanical Engineers* **178** (2A): 2.

Dudzinski, P.A. (1986). The problems of multi-axle vehicle drives. *Journal of Terramechanics* **23** (2).

Huang, W., Wong, J.Y., and Knezevic, Z. (2014). Further study of the optimization of the tractive efficiency of all-wheel-drive vehicles. *International Journal of Heavy Vehicle Systems* 21 (2).

Jurkat, M.P., Nuttall, C.J., and Haley, P.W. (1975). The U.S. Army Mobility Model (AMM-75). *Proc. 5th Int. Conf. of the International Society for Terrain–Vehicle Systems* 4, Detroit, MI.

Kolozsi, Z. and McCarthy, T.T. (1974). The prediction of tractor field performance. *Journal of Agricultural Engineering Research* 19: 167–172.

Maclaurin, B. (2018). *High Speed Off-Road Vehicles: Suspensions, Tracks, Wheels and Dynamics*. Chichester, England: Wiley

Nuttall, C.J. Jr., Rula, A.A., and Dugoff, H.J. (1974). Computer model for comprehensive evaluation of cross-country vehicle mobility. *SAE Transactions*, 83, Society of Automotive Engineers paper 740426.

Ogorkiewicz, R.M. (1991). *Technology of Tanks*. Surrey, UK: Jane's Information Group.

Osborne, L.E. (1969–1970). Ground-drive systems for high-powered tractors. *Proceedings of the Institution of Mechanical Engineers* 184 (3Q).

Reece, A.R. (1969–1970). The shape of the farm tractor. *Proceedings of the Institution of Mechanical Engineers* 184 (3Q).

Söhne, W. (1968). Four-wheel drive or rear-wheel drive for high power farm tractors. *Journal of Terramechanics* 5 (3).

Vantsevich, V.V. (2007). Multi-wheel-drive vehicle energy/fuel efficiency and traction performance: objective function analysis. *Journal of Terramechanics* 44 (3).

Vantsevich, V.V. (2008). Power losses and energy efficiency of multi-wheel drive vehicles: A method for evaluation. *Journal of Terramechanics* 45 (3).

Wong, J.Y. (1970). Optimization of the tractive performance of four-wheel drive off-road vehicles. *SAE Transactions* 79, Society of Automotive Engineers paper 700723.

Wong, J.Y. (1972). On the application of air cushion technology to overland transport. *High Speed Ground Transportation Journal* 6 (3).

Wong, J.Y. (1975). System energy in high speed ground transportation. *High Speed Ground Transportation Journal* 9 (1).

Wong, J.Y. (2010). *Terramechanics and Off-Road Vehicle Engineering*, 2e. Oxford, England: Elsevier.

Wong, J.Y. and Huang, W. (2006). Wheels vs. tracks: a fundamental evaluation from the traction perspective. *Journal of Terramechanics* 43 (1).

Wong, J.Y., McLaughlin, N.B., Knezevic, Z., and Burtt, S. (1998). Optimization of the tractive performance of four-wheel-drive tractors: theoretical analysis and experimental substantiation. *Proceedings of the Institution of Mechanical Engineers, Part D: Journal of Automobile Engineering* 212 (D4).

Wong, J.Y., McLaughlin, N.B., Zhao, Zhiwen, Li, Jianqiao, and Burtt, S. (1999). Optimizing tractive performance of four-wheel-drive tractor—Theory and practice. *Proc. 13th Int. Conf. of the International Society for Terrain–Vehicle Systems* 2, Munich, Germany.

Wong, J.Y., Zhao, Zhiwen, Li, Jianqiao, McLaughlin, N.B., and Burtt, S. (2000). Optimization of the performance of four-wheel-drive tractors: Correlation between analytical predictions and experimental data. *SAE Transactions, Journal of Commercial Vehicles*, Society of Automotive Engineers paper 2000-01-2596; and *SAE Journal of Off-highway Engineering*, February 2001: 46–50.

Problems

4.1 Calculate the drawbar power and tractive efficiency of the off-road vehicle described in Example 4.1 at various operating speeds when the third gear with a total reduction ratio of 33.8 is engaged.

4.2 An off-road vehicle pulls an implement having a resistance of 17.792 kN (4000 lb). The motion resistance of the vehicle is 6.672 kN (1500 lb). Under these circumstances, the slip of the running gear is 35%. The transmission efficiency is 0.80. What percentage of the power is lost in converting engine power into drawbar power?

4.3 A four-wheel drive off-road vehicle has a rigid coupling between the front and rear drive axles. The thrust–slip characteristics of the front and rear axles are assumed to be identical and are given in the following table. Owing to differences in tire inflation pressure and uneven wear of tires, the theoretical speed of the front tires is 6% higher than that of the rear tires. The motion resistance of the vehicle is 1.67 kN (375 lb). The vehicle is to pull an implement having a resistance of 16.51 kN (3712 lb). Determine whether torsional wind-up in the transmission will occur. Also determine the thrust distribution between the front and rear axles and the slip efficiency of the vehicle.

Thrust-Slip Relationship for the Drive Axle (Front or Rear)

Slip (%)	5	10	15	20	25	30	40
Thrust (kN)	5.12	8.0	10.23	12.0	13.34	14.23	16.01

4.4 If the four-wheel drive off-road vehicle described in Problem 4.3 is equipped with an overrunning clutch, instead of a rigid coupling, between the front and rear drive axles, so that the front axle will not be driven until the slip of the rear tires is up to 10%, determine the thrust distribution between the drive axles and the slip efficiency of the vehicle when it pulls an implement having a resistance of 14.06 kN (3160 lb). The motion resistance of the vehicle is 1.67 kN (375 lb).

4.5 A two-wheel drive tractor with a weight utilization factor of 75% is to be designed mainly for operation in the speed range 10–15 km/h (6.2–9.3 mph). Both the transmission efficiency and the slip efficiency are assumed to be 85%. The average value of the coefficient of motion resistance is 0.1, and that of the traction coefficient is 0.4. Determine the appropriate range of the power-to-weight ratio for the tractor.

4.6 An off-road transporter with a gross weight of 44.48 kN (10,000 lb) carries a payload of 17.79 kN (4000 lb). The coefficient of motion resistance of the vehicle is 0.15. Both the transmission efficiency and the slip efficiency are 0.85. If the transporter travels at a speed of 15 km/h (9.3 mph) and the average specific fuel consumption of the engine is 0.25 kg/kW · h (0.41 lb/hp · h), determine the fuel consumed in transporting 1 t (1000 kg) of payload for 1 km. Also calculate the transport productivity, the power consumption per unit productivity, and the transport efficiency of the vehicle system.

5

Handling Characteristics of Road Vehicles

Handling characteristics of a road vehicle refer to its response to steering commands and to environmental inputs, such as wind gust and road disturbances, that affect its direction of motion. There are two basic issues in vehicle handling: one is the control of the direction of motion of the vehicle; the other is its ability to stabilize its direction of motion against external disturbances.

The vehicle as a rigid body has six degrees of freedom, translations along the x, y, and z axes, and rotations about these axes, as shown in Figure 5.1. The primary motions associated with the handling behavior of a vehicle are longitudinal, lateral, and yaw motions (i.e., translation along the x axis, translation along the y axis, and rotation about the z axis, respectively). In practice, during a turning maneuver, the vehicle body rolls (i.e., rotating about the x axis). This roll motion may cause the wheels to steer, thus affecting the handling behavior of the vehicle. Furthermore, bounce and pitch motions of the vehicle body (i.e., translation along the z axis and rotation about the y axis, respectively) may also affect the steering response of the vehicle. However, the inclusion of these motions in the analysis becomes necessary only when considering the limits of handling characteristics.

This chapter is intended to serve as an introduction to the study of the handling characteristics of road vehicles. Simplified linear models for the handling behavior of passenger cars and tractor–semitrailers in which suspension characteristics are not taken into account are presented. The models demonstrate the effects on handling behavior of major vehicle design and operational parameters, such as tire properties, location of the center of gravity, and forward speed, and lead to conclusions of practical significance concerning vehicle directional control and stability. The response of the vehicle to steering input and its directional stability associated with a fixed steering wheel, which are usually referred to as fixed-control characteristics, are analyzed.

5.1 Steering Geometry

In examining the handling characteristics of a road vehicle, it is convenient to begin with a discussion of the cornering behavior of the vehicle at low speeds, with the effect of the centrifugal force being neglected. For road vehicles, steering is normally effected by changing the heading of the front wheels through the steering system, although four-wheel steering has been introduced to some passenger cars. At low speeds, there is a simple relationship between the direction of motion of the vehicle and the steering wheel angle. The prime consideration in the design of the steering system is minimizing tire scrub during cornering. This requires that during the turn, all tires should be in pure rolling without lateral sliding. To satisfy this requirement, the wheels should follow curved paths with different radii originating from a common turn center, as shown in

Theory of Ground Vehicles, Fifth Edition. J.Y. Wong.
© 2022 John Wiley & Sons, Inc. Published 2022 by John Wiley & Sons, Inc.
Companion website: www.wiley.com/go/wong/TGV5e

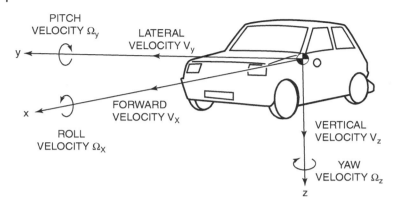

Figure 5.1 A vehicle axis system.

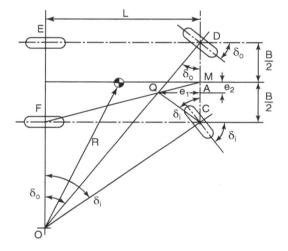

Figure 5.2 Steering geometry.

Figure 5.2. This establishes the proper relationship between the steer angle of the inside front wheel δ_i and that of the outside front wheel δ_o. From Figure 5.2, it can be readily seen that the steer angles δ_i and δ_o should satisfy the following relationship:

$$\cot \delta_o - \cot \delta_i = B/L \tag{5.1}$$

where B and L are the tread (or track) and wheelbase of the vehicle, respectively.

The steering geometry that satisfies Eq. (5.1) is usually referred to as the Ackermann steering geometry.

The relationship between δ_i and δ_o that satisfies Eq. (5.1) can be illustrated graphically. Referring to Figure 5.2, first connect the midpoint of the front axle M with the center of the inside rear wheel F. Then lay out the steer angle of the outside front wheel δ_o from the front axle. Line DO intersects line MF at Q. Connect point Q with the center of the inside front wheel C; then angle $\angle QCM$ is the steer angle of the inside front wheel δ_i that satisfies Eq. (5.1). This can be proved from the geometric relations shown in Figure 5.2:

$$\cot \delta_o = (B/2 + e_2)/e_1$$
$$\cot \delta_i = (B/2 - e_2)/e_1$$

and

$$\cot \delta_o - \cot \delta_i = 2e_2/e_1 \qquad (5.2)$$

Since triangle ΔMAQ is similar to triangle ΔMCF,

$$\frac{e_2}{e_1} = \frac{B/2}{L}$$

Eq. (5.2) can then be rewritten as

$$\cot \delta_o - \cot \delta_i = B/L$$

The results of the above analysis indicate that if the steer angles of the front wheels δ_i and δ_o satisfy Eq. (5.1), then by laying out the steer angles δ_i and δ_o from the front axle, the intersection of the noncommon sides of δ_i and δ_o (i.e., point Q in Figure 5.2) will lie on the straight line connecting the midpoint of the front axle and the center of the inside rear wheel (i.e., line MF in Figure 5.2).

Figure 5.3 shows the relationship between δ_o and δ_i that satisfies Eq. (5.1) for a vehicle with B/L = 0.56, as compared to a parallel steer curve ($\delta_i = \delta_o$) and a typical steering geometry used in practice (Bunker 1968).

To evaluate the characteristics of a particular steering linkage with respect to the Ackermann steering geometry, a graphic method may be employed. First, the steer angles of the inside front wheel δ_i with suitable increment are laid out from the initial position of the steer arm *CH*, as shown in Figure 5.4. Then, from the pivot of the inside steer arm *H*, an arc is struck with a radius equal to the length of the tie rod *HI*. This intersects the arc generated by the steer arm of the outside front wheel *DI*. The intersection then defines the corresponding steer angle of the outside front wheel δ_o. By laying out the steer angles of the inside front wheel δ_i and the corresponding steer angles of the outside front wheel δ_o from the front axle, the noncommon sides of δ_i and δ_o will intersect at points

Figure 5.3 Characteristics of various types of steering linkage. *Source:* Bunker (1968). Reproduced with permission of Butterworths.

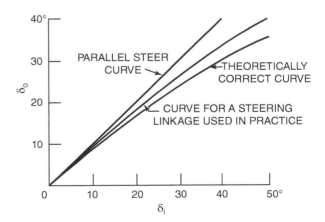

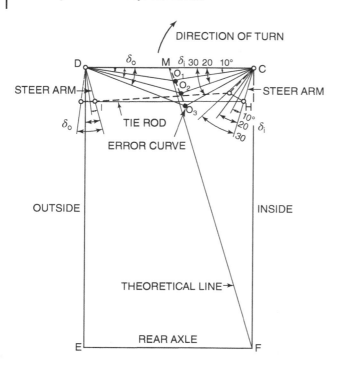

Figure 5.4 Error curve of a steering linkage.

O_1, O_2, and O_3, as shown in Figure 5.4. If the steering geometry satisfies Eq. (5.1), the intersections of the noncommon sides of δ_i and δ_o will lie on the straight line *MF*, as mentioned previously. The deviation of the curve connecting O_1, O_2, and O_3 from line *MF* is therefore an indication of the error of the steering geometry with respect to the Ackermann steering geometry. Steering geometry with an error curve that deviates excessively from line *MF* shown in Figure 5.4 will exhibit considerable tire scrub during cornering. This results in excessive tire wear and increased steering effort.

The graphic method described above is applicable only to the type of coplanar steering linkage shown in Figure 5.4, which is commonly used in vehicles with a front beam axle. For vehicles with front independent suspensions, the steering linkage will be more complex. Dependent on the type of independent suspension used, the front wheels may be steered via a three-piece tie rod or by a rack and pinion with outer tire rods. The approach for constructing steering error curves for these linkages is similar to that described above. The procedure, however, is more involved.

5.2 Steady-State Handling Characteristics of a Two-Axle Vehicle

Steady-state handling performance is concerned with the directional behavior of a vehicle during a turn under non-time-varying conditions. An example of a steady-state turn is a vehicle negotiating a curve with constant radius at a constant forward speed. In the analysis of steady-state handling behavior, the inertia properties of the vehicle are not involved.

When a vehicle is negotiating a turn at moderate or higher speeds, the effect of the centrifugal force (an inertia force arising from the normal component of acceleration toward the center of the

Figure 5.5 Simplified steady-state handling model for a two-axle vehicle.

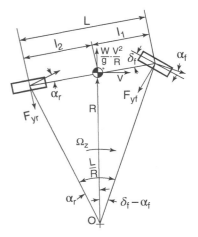

turn) acting at the vehicle center of gravity can no longer be neglected. To balance the centrifugal force, the tires must develop appropriate cornering forces. As discussed in Chapter 1, a side force acting on a tire produces a sideslip angle. Thus, when a vehicle is negotiating a turn at moderate or higher speeds, the four tires will develop appropriate slip angles. To simplify the analysis, the pair of tires on an axle are represented by a single tire with double the cornering stiffness, as shown in Figure 5.5. The handling characteristics of the vehicle depend, to a great extent, on the relationship between the slip angles of the front and rear tires, α_f and α_r.

The steady-state response to steering input of a vehicle at moderate and higher speeds is more complex than that at low speeds. From the geometry shown in Figure 5.5, the relationship of the steer angle of the front tire δ_f, the turning radius R, the wheelbase L, and the slip angles of the front and rear tires α_f and α_r is approximately given by (Bundorf 1968)

$$\delta_f - \alpha_f + \alpha_r = L/R$$

or

$$\delta_f = L/R + \alpha_f - \alpha_r \tag{5.3}$$

This indicates that the steer angle δ_f required to negotiate a given curve is function of not only the turning radius R but also the front and rear slip angles α_f and α_r. The slip angles α_f and α_r are dependent on the side forces acting on the tires and their cornering stiffness. The cornering forces on the front and rear tires F_{yf} and F_{yr} can be determined from the dynamic equilibrium of the vehicle in the lateral direction. For small steer angles, the cornering forces acting at the front and rear tires are approximately given by

$$F_{yf} = \frac{W}{g} \frac{V^2}{R} \frac{l_2}{L} \tag{5.4}$$

$$F_{yr} = \frac{W}{g} \frac{V^2}{R} \frac{l_1}{L} \tag{5.5}$$

where W is the total weight of the vehicle; g is the acceleration due to gravity; V is the vehicle forward speed; and other parameters are shown in Figure 5.5.

The normal load on each of the front tires W_f and that on each of the rear tires W_r under static conditions are expressed by

$$W_f = Wl_2/2L$$

$$W_r = Wl_1/2L$$

Equations (5.4) and (5.5) can be rewritten as

$$F_{yf} = 2W_f \frac{V^2}{gR} \tag{5.6}$$

$$F_{yr} = 2W_r \frac{V^2}{gR} \tag{5.7}$$

Within a certain range, the slip angle and cornering force may be considered linearly related with a constant cornering stiffness, as discussed in Section 1.4.1. The slip angles α_f and α_r, therefore, are given by

$$\alpha_f = \frac{F_{yf}}{2C_{af}} = \frac{W_f}{C_{af}} \frac{V^2}{gR} \tag{5.8}$$

$$\alpha_r = \frac{F_{yr}}{2C_{ar}} = \frac{W_r}{C_{ar}} \frac{V^2}{gR} \tag{5.9}$$

where C_{af} and C_{ar} are the cornering stiffness of each of the front and rear tires, respectively. As described in Chapter 1, the cornering stiffness of a given tire varies with a number of operational parameters, including inflation pressure, normal load, tractive (or braking) effort, and lateral force. It may be regarded as a constant only within a limited range of operating conditions.

Substituting Eqs. (5.8) and (5.9) into Eq. (5.3), the expression for the steer angle δ_f required to negotiate a given curve becomes (Bundorf 1968)

$$\delta_f = \frac{L}{R} + \left(\frac{W_f}{C_{af}} - \frac{W_r}{C_{ar}} \right) \frac{V^2}{gR}$$

$$= \frac{L}{R} + K_{us} \frac{V^2}{gR} \tag{5.10}$$

$$= \frac{L}{R} + K_{us} \frac{a_y}{g}$$

where K_{us} is usually referred to as the understeer coefficient and is expressed in radians, and a_y is the lateral acceleration.

Equation (5.10) is the fundamental equation governing the steady-state handling behavior of a two-axle road vehicle. It indicates that the steer angle required to negotiate a given curve depends on the wheelbase, turning radius, forward speed (or lateral acceleration), and understeer coefficient of the vehicle, which is a function of the weight distribution and front and rear tire cornering stiffnesses.

Dependent on the values of the understeer coefficient K_{us} or the relationship between the slip angles of the front and rear tires, the steady-state handling characteristics may be classified into three categories: neutral steer, understeer, and oversteer (Bundorf 1968).

5.2.1 Neutral Steer

When the understeer coefficient $K_{us} = 0$, which is equivalent to the slip angles of the front and rear tires being equal (i.e., $\alpha_f = \alpha_r$ and $W_f/C_{\alpha f} = W_r/C_{\alpha r}$), the steer angle δ_f required to negotiate a given curve is independent of forward speed and is given by

$$\delta_f = L/R \tag{5.11}$$

A vehicle having this handling property is said to be "neutral steer." Its handling characteristics for a constant radius turn are represented by a horizontal line in the steer angle–speed diagram shown in Figure 5.6.

When a neutral steer vehicle is accelerated in a constant radius turn, the driver should maintain the same steering wheel position. In other words, when it is accelerated with the steering wheel fixed, the turning radius remains the same, as illustrated in Figure 5.7. When a neutral steer vehicle originally moving along a straight line is subjected to a side force acting at the center of gravity, equal slip angles will be developed at the front and rear tires (i.e., $\alpha_f = \alpha_r$). As a result, the vehicle follows a straight path at an angle to the original, as shown in Figure 5.8.

5.2.2 Understeer

When the understeer coefficient $K_{us} > 0$, which is equivalent to the slip angle of the front tire α_f being greater than that of the rear tire α_r (i.e., $\alpha_f > \alpha_r$ and $W_f/C_{\alpha f} > W_r/C_{\alpha r}$), the steer angle δ_f required to negotiate a given curve increases with the square of vehicle forward speed (or lateral acceleration). A vehicle with this handling property is said to be "understeer." Its handling characteristics for a constant radius turn are represented by a parabola in the steer angle–speed diagram shown in Figure 5.6.

When an understeer vehicle is accelerated in a constant radius turn, the driver must increase the steer angle. In other words, when it is accelerated with the steering wheel fixed, the turning radius increases, as illustrated in Figure 5.7. At the same steering wheel position and vehicle forward

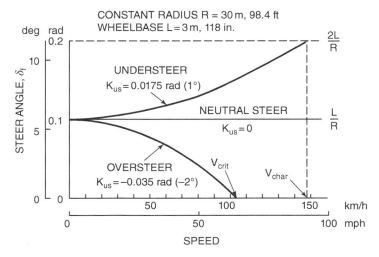

Figure 5.6 Relationships between steer angle and speed of neutral steer, understeer, and oversteer vehicles.

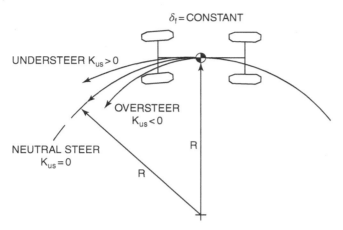

Figure 5.7 Curvature responses of neutral steer, understeer, and oversteer vehicles at a fixed steering angle.

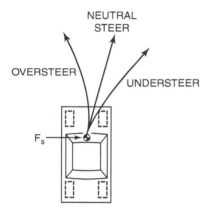

Figure 5.8 Directional responses of neutral steer, understeer, and oversteer vehicles to a side force at the center of gravity.

speed, the turning radius of an understeer vehicle is larger than that of a neutral steer vehicle. When a side force acts at the center of gravity of an understeer vehicle originally moving along a straight line, the front tires will develop a slip angle greater than that of the rear tires (i.e., $\alpha_f > \alpha_r$). As a result, a yaw motion is initiated, and the vehicle turns away from the side force, as shown in Figure 5.8.

For an understeer vehicle, a characteristic speed V_{char} may be identified. It is the speed at which the steer angle required to negotiate a turn is equal to $2L/R$, as shown in Figure 5.6. From Eq. (5.10),

$$V_{char} = \sqrt{\frac{gL}{K_{us}}} \tag{5.12}$$

5.2.3 Oversteer

When the understeer coefficient $K_{us} < 0$, which is equivalent to the slip angle of the front tire α_f being less than that of the rear tire α_r (i.e., $\alpha_f < \alpha_r$ and $W_f/C_{\alpha f} < W_r/C_{\alpha r}$), the steer angle δ_f required to negotiate a given curve decreases with an increase of vehicle forward speed (or lateral acceleration). A vehicle with this handling property is said to be "oversteer." The relationship between the required steer angle and forward speed for this kind of vehicle at a constant radius turn is illustrated in Figure 5.6.

When an oversteer vehicle is accelerated in a constant radius turn, the driver must decrease the steer angle. In other words, when it is accelerated with the steering wheel fixed, the turning radius decreases, as illustrated in Figure 5.7. For the same steering wheel position and vehicle forward speed, the turning radius of an oversteer vehicle is smaller than that of a neutral steer vehicle. When a side force acts at the center of gravity of an oversteer vehicle originally moving along a straight line, the front tires will develop a slip angle less than that of the rear tires (i.e., $\alpha_f < \alpha_r$). As a result, a yaw motion is initiated, and the vehicle turns into the side force, as illustrated in Figure 5.8.

For an oversteer vehicle, a critical speed V_{crit} can be identified. It is the speed at which the steer angle required to negotiate any turn is zero, as shown in Figure 5.6. From Eq. (5.10),

$$V_{crit} = \sqrt{\frac{gL}{-K_{us}}} \tag{5.13}$$

For an oversteer vehicle, the understeer coefficient K_{us} in the above equation has a negative sign. It will be shown later that the critical speed also represents the speed above which an oversteer vehicle exhibits directional instability.

The prime factors controlling the steady-state handling characteristics of a vehicle are the weight distribution of the vehicle and the cornering stiffness of the tires. A front-engine, front-wheel drive vehicle with a large proportion of the vehicle weight on the front tires may tend to exhibit understeer behavior. A rear-engine, rear-wheel drive car with a large proportion of the vehicle weight on the rear tires, on the other hand, may tend to have oversteer characteristics (Bunker 1968). Changes in load distribution will alter the handling behavior of a vehicle. For instance, when a vehicle is accelerating during a turn, due to longitudinal load transfer from the front to the rear, the slip angles of the front tires increase while those of the rear tires decrease, as discussed in Section 1.4.1. Consequently, the vehicle tends to exhibit understeer characteristics. On the other hand, when it is decelerating, due to load transfer from the rear to the front, the slip angles of the front tires decrease while those of the rear tires increase. As a result, the vehicle tends to exhibit oversteer behavior. Figure 5.9 shows the measured turning behavior of a four-wheel drive car, with an equal distribution of driving torque between the front and rear axles, at a fixed steering wheel position during acceleration and deceleration (Shibahata et al. 1993). It shows that the car exhibits increased understeer behavior during acceleration. On the other hand, during deceleration when the lateral acceleration is up to approximately 0.7g, the vehicle demonstrates oversteer characteristics.

A number of design and operational parameters affect the cornering stiffness of tires, and thus the handling performance of the vehicle. Mixing of radial-ply with bias-ply tires in a vehicle may have serious consequences in its handling characteristics. Installing laterally stiff radial-ply tires on the front and relatively flexible bias-ply tires on the rear may change an otherwise understeer vehicle to an oversteer one. Lowering the inflation pressure in the rear tires can have similar effects, as the cornering stiffness of a tire usually decreases with a decrease of inflation pressure. The lateral load transfer from the inside tire to the outside tire on an axle during a turn will increase the slip angle required to generate a given cornering force, as discussed in Chapter 1. Thus, lateral load transfer will affect the handling behavior of the vehicle. The application of a driving or braking torque to the tire during a turn will also affect the cornering behavior of the vehicle, as a driving or braking torque modifies the cornering properties of the tire, as mentioned in Chapter 1. For a rear-wheel drive vehicle, the application of tractive effort during a turn reduces the effective cornering stiffness

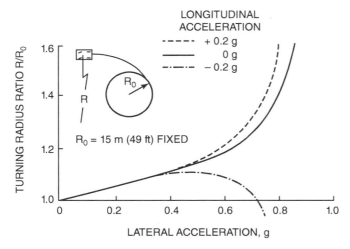

Figure 5.9 Variations of turning radius with lateral acceleration of a four-wheel drive car at various longitudinal accelerations. *Source:* Shibahata et al. (1993). Reproduced with permission of Swets & Zeitlinger.

of the rear tires, producing an oversteering effect. On the other hand, for a front-wheel drive car, the application of tractive effort during a turn reduces the effective cornering stiffness of the front tires, thus introducing an understeering effect.

Effects of roll steer (the steering motion of the front or rear wheels due to the relative roll motion of the sprung mass with respect to the unsprung mass), roll camber (the change in camber of the wheels due to the relative motion of the sprung mass with respect to the unsprung mass), and compliance steer (the steering motion of the wheels with respect to the sprung mass resulting from compliance in, and forces on, the suspension and steering linkages) would be significant under certain circumstances, and they should be taken into account in a more comprehensive analysis of vehicle handling. The effects of these factors can, however, be included in a modified form of the understeer coefficient K_{us}, and Eq. (5.10) that describes the steady-state handling behavior still holds.

In summary, there are a number of design and operational factors that would affect the understeer coefficient of a vehicle, and hence its handling characteristics. For a practical vehicle, the understeer coefficient would vary with operating conditions. Figure 5.10 shows the changes in

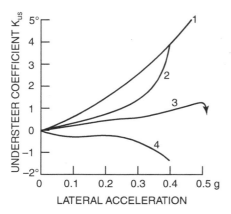

Figure 5.10 Variations of understeer coefficient with lateral acceleration of various types of car: 1 – a conventional front-engine/rear-wheel drive car; 2 – a European front-engine/front-wheel drive car; 3 – a European rear-engine/rear-wheel drive car; 4 – an American rear-engine/rear-wheel drive car. *Source:* Fenton (1996). Reproduced with permission of Butterworths.

the understeer coefficient, expressed in degrees, with lateral acceleration for four different types of passenger car (Fenton 1996). Curve 1 represents the characteristics of a conventional front-engine/rear-wheel drive car. It shows that the understeer coefficient increases sharply with an increase of lateral acceleration. Curve 2 represents the behavior of a European front-engine/front-wheel drive car. It exhibits similar understeer characteristics. The characteristics of a European rear-engine/rear-wheel drive car are represented by curve 3. It indicates that the vehicle exhibits understeer behavior up to lateral acceleration of approximately 0.5g, above which it tends to become oversteer. The behavior of an American rear-engine/rear-wheel drive compact car is represented by curve 4. It shows that the vehicle exhibits oversteer characteristics in the operating range shown.

Among the three types of steady-state handling behavior, oversteer is not desirable from a directional stability point of view, which will be discussed later in this chapter. It is considered desirable for a road vehicle to have a small degree of understeer up to a certain level of lateral acceleration, such as 0.4g, with increasing understeer beyond this point (Fenton 1996). This would have the advantages of sensitive steering response associated with a small degree of understeer during the majority of turning maneuvers. The increased understeer at higher lateral accelerations, on the other hand, would provide greater stability during tight turns.

To illustrate the changes in the handling behavior of road vehicles with operating conditions, a handling diagram is often used. In this diagram, the vehicle lateral acceleration in g units, a_y/g (V^2/gR), is plotted as a function of the parameter ($L/R - \delta_f$), where L is the wheelbase; R is the turning radius; and δ_f is the average front tire steer angle. During a turning maneuver, the turning radius R may be difficult to measure directly. However, it can be readily determined from the yaw velocity Ω_z (measured using a rate-gyro) and the forward speed V of the vehicle ($R = V/\Omega_z$). Therefore, in the handling diagram, the lateral acceleration a_y/g (in g units) is often plotted as a function of ($\Omega_z L/V - \delta_f$), as shown in Figure 5.11. From Eq. (5.10), the relationship between a_y/g and ($\Omega_z L/V - \delta_f$) is expressed by

$$K_{us}\frac{V^2}{gR} = K_{us}\frac{a_y}{g} = -\left(L/R - \delta_f\right) = -\left(\Omega_z L/V - \delta_f\right)$$

The slope of the curve shown in the handling diagram (Figure 5.11) is given by

$$\frac{d\left(a_y/g\right)}{d\left(\Omega_z L/V - \delta_f\right)} = -\frac{1}{K_{us}} \tag{5.14}$$

Figure 5.11 Handling diagram.

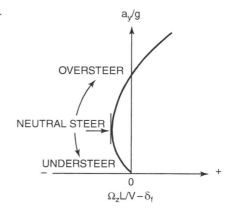

This indicates that the handling behavior of a road vehicle can be identified by the slope of the curve shown in the handling diagram. If the slope is negative, then it implies that the understeer coefficient K_{us} is positive. Consequently, the vehicle exhibits understeer behavior. If the slope is infinite, then it indicates that the understeer coefficient K_{us} is zero and that the vehicle is neutral steer. On the other hand, if the slope is positive, then it implies that the understeer coefficient K_{us} is negative and that the vehicle is oversteer.

The value of the parameter $(\Omega_z L/V - \delta_f)$ is sensitive to the errors in the measurements of Ω_z, V, and δ_f. A small error in the values of Ω_z, V, and δ_f may result in a significant error in the value of the parameter $(\Omega_z L/V - \delta_f)$. For instance, if the wheelbase L of a vehicle is 2.7 m (8 ft, 10 in.), and the nominal values of Ω_z, V, and δ_f are 0.1389 rad/s (7.96 deg/s), 50 km/h (31 mph), and 0.0427 rad (2.45 deg), respectively, then an error of $\pm 1\%$ in these values will result in an error of approximately $\pm 6\%$ in the value of $(\Omega_z L/V - \delta_f)$. An error of $\pm 5\%$ in these values will result in an error in the value of $(\Omega_z L/V - \delta_f)$ ranging from -29.9% to $+31.7\%$.

Example 5.1 A passenger car has a weight of 20.105 kN (4520 lb) and a wheelbase of 2.8 m (9 ft, 2 in.). The weight distribution on the front axle is 53.5%, and that on the rear axle is 46.5% under static conditions.

(a) If the cornering stiffness of each of the front tires is 38.92 kN/rad (8750 lb/rad) and that of the rear tires is 38.25 kN/rad (8600 lb/rad), determine the steady-state handling behavior of the vehicle.
(b) If the front tires are replaced by a pair of tires, each of which has a cornering stiffness of 47.82 kN/rad (10 750 lb/rad), and the rear tires remain unchanged, determine the steady-state handling behavior of the vehicle under these circumstances.

Solution

(a) The understeer coefficient of the vehicle is

$$K_{us} = \frac{W_f}{C_{\alpha f}} - \frac{W_r}{C_{\alpha r}} = \frac{20,105 \times 0.535}{2 \times 38,920} - \frac{20,105 \times 0.465}{2 \times 38,250}$$
$$= 0.016 \text{ rad}(0.92°).$$

The vehicle is understeer, and the characteristic speed is

$$V_{\text{char}} = \sqrt{\frac{gL}{K_{us}}} = 41.5 \text{ m/s} = 149 \text{ km/h} \ (93 \text{ mph})$$

(b) When a pair of tires with higher cornering stiffness are installed in the front axle, the understeer coefficient of the vehicle is

$$K_{us} = \frac{20,105 \times 0.535}{2 \times 47,820} - \frac{20,105 \times 0.465}{2 \times 38,250} = -0.0097 \text{ rad}(-0.56°)$$

The vehicle is oversteer, and the critical speed is

$$V_{\text{crit}} = \sqrt{\frac{gL}{-K_{us}}} = 53.1 \text{ m/s} = 191 \text{ km/h} \ (119 \text{ mph})$$

5.3 Steady-State Response to Steering Input

A vehicle may be regarded as a control system upon which various inputs are imposed. During a turning maneuver, the steer angle induced by the driver can be considered as an input to the system, and the motion variables of the vehicle, such as yaw velocity, lateral acceleration, and curvature, may be regarded as outputs. The ratio of the yaw velocity, lateral acceleration, or curvature to the steering input can then be used for comparing the response characteristics of different vehicles (Bundorf 1968).

5.3.1 Yaw Velocity Response

Yaw velocity gain is an often-used parameter for comparing the steering response of road vehicles. It is defined as the ratio of the steady-state yaw velocity to the steer angle. Yaw velocity Ω_z of the vehicle under steady-state conditions is the ratio of the forward speed V to the turning radius R. From Eq. (5.10), the yaw velocity gain G_{yaw} is given by

$$G_{yaw} = \frac{\Omega_z}{\delta_f} = \frac{V}{L + K_{us}V^2/g} \tag{5.15}$$

Equation (5.15) gives the yaw velocity gain with respect to the steer angle of the front wheel. If the yaw velocity gain with respect to the steering wheel angle is desired, the value obtained from Eq. (5.15) should be divided by the steering gear ratio.

For a neutral steer vehicle, the understeer coefficient K_{us} is zero; the yaw velocity gain increases linearly with an increase of forward speed, as shown in Figure 5.12. For an understeer vehicle, the understeer coefficient K_{us} is positive. The yaw velocity gain first increases with an increase of forward speed, and reaches a maximum at a particular speed, as shown in Figure 5.12. It can be proved that the maximum yaw velocity gain occurs at the characteristic speed V_{char} mentioned previously.

For an oversteer vehicle, the understeer coefficient K_{us} is negative; the yaw velocity gain increases with the forward speed at an increasing rate, as shown in Figure 5.12. Since K_{us} is negative, at a

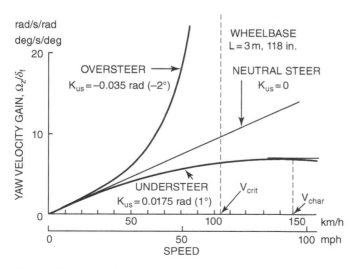

Figure 5.12 Yaw velocity gain characteristics of neutral steer, understeer, and oversteer vehicles.

particular speed, the denominator of Eq. (5.15) is zero, and the yaw velocity gain approaches infinity. This speed is the critical speed V_{crit} of an oversteer vehicle discussed previously.

The results of the above analysis indicate that from the point of view of handling response to steering input, an oversteer vehicle is more sensitive than a neutral steer one, and, in turn, a vehicle with neutral steer characteristics is more responsive than an understeer one. Since the yaw velocity of a vehicle is an easily measured parameter, the yaw velocity gain–speed characteristics can be obtained from tests. The handling behavior of a vehicle can then be evaluated from the yaw velocity gain characteristics. For instance, if the yaw velocity gain of a vehicle is found to be greater than the forward speed divided by the wheelbase (i.e., neutral steer response), the vehicle is oversteer, and if it is less, it is understeer.

5.3.2 Lateral Acceleration Response

Lateral acceleration gain, defined as the ratio of the steady-state lateral acceleration to the steer angle, is another commonly used parameter for evaluating the steering response of a vehicle. By rearranging Eq. (5.10), the lateral acceleration gain G_{acc} is given by

$$G_{acc} = \frac{V^2/gR}{\delta_f} = \frac{a_y/g}{\delta_f} = \frac{V^2}{gL + K_{us}V^2} \tag{5.16}$$

where a_y is the lateral acceleration.

Equation (5.16) gives the lateral acceleration gain with respect to the steer angle of the front wheel. If the acceleration gain with respect to the steering wheel angle is desired, the value obtained from Eq. (5.16) should be divided by the steering gear ratio.

For a neutral steer vehicle, the value of the understeer coefficient K_{us} is zero; the lateral acceleration gain is proportional to the square of forward speed, as shown in Figure 5.13(a). For an understeer vehicle, the value of the understeer coefficient K_{us} is positive; the lateral acceleration gain increases with speed, as shown in Figure 5.13(a). At very high speeds, the first term in the denominator of Eq. (5.16) is much smaller than the second term, and the lateral acceleration gain approaches a value of $1/K_{us}$ asymptotically.

For an oversteer vehicle, the value of the understeer coefficient K_{us} is negative. The lateral acceleration gain increases with an increase of forward speed at an increasing rate, as the denominator of Eq. (5.16) decreases with an increase of speed. At a particular speed, the denominator of Eq. (5.16) becomes zero, and the lateral acceleration gain approaches infinity, as shown in Figure 5.13(a). It can be shown that this speed is the critical speed of an oversteer vehicle.

5.3.3 Curvature Response

The ratio of the steady-state curvature $1/R$ to the steer angle is another parameter commonly used for evaluating the response characteristics of a vehicle. From Eq. (5.10), this parameter is expressed by

$$\frac{1/R}{\delta_f} = \frac{1}{L + K_{us}V^2/g} \tag{5.17}$$

Equation (5.17) gives the curvature response with respect to the steer angle of the front wheel. If the curvature response with respect to the steering wheel angle is desired, the value obtained from Eq. (5.17) should be divided by the steering gear ratio.

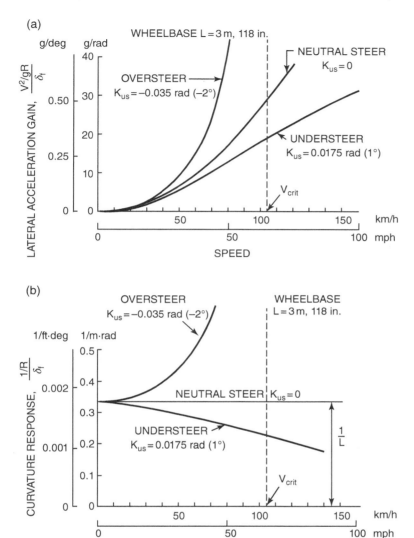

Figure 5.13 (a) Lateral acceleration gain characteristics of neutral steer, understeer, and oversteer vehicles; (b) Curvature responses of neutral steer, understeer, and oversteer vehicles.

For a neutral steer vehicle, the understeer coefficient K_{us} is zero; the curvature response is independent of forward speed, as shown in Figure 5.13(b). For an understeer vehicle, the understeer coefficient K_{us} is positive; the curvature response decreases as the forward speed increases, as shown in Figure 5.13(b).

For an oversteer vehicle, the understeer coefficient K_{us} is negative; the curvature response increases with the forward speed. At a particular speed, the curvature response approaches infinity, as shown in Figure 5.13(b). This means that the turning radius approaches zero and the vehicle spins out of control. This speed is, in fact, the critical speed V_{crit} of an oversteer vehicle discussed previously.

The results of the above analysis illustrate that from the steering response point of view, the oversteer vehicle has the most sensitive handling characteristics, while the understeer vehicle is the least responsive.

Example 5.2 A vehicle has a weight of 20.105 kN (4520 lb) and a wheelbase of 3.2 m (10.5 ft). The ratio of the distance between the center of gravity of the vehicle and the front axle to the wheelbase is 0.465. The cornering stiffness of each of the front tires is 38.92 kN/rad (8750 lb/rad) and that of the rear tires is 38.25 kN/rad (8600 lb/rad). The average steering gear ratio is 25. Determine the yaw velocity gain and the lateral acceleration gain of the vehicle with respect to the steering wheel angle.

Solution

The understeer coefficient of the vehicle is

$$K_{us} = \frac{W_f}{C_{af}} - \frac{W_r}{C_{ar}} = \frac{20{,}105 \times 0.535}{2 \times 38{,}920} - \frac{20{,}105 \times 0.465}{2 \times 38{,}250}$$

$$= 0.016 \text{ rad } (0.92°)$$

From Eq. (5.15), the yaw velocity gain with respect to the steering wheel angle is

$$G_{\text{yaw}} = \frac{\Omega_z}{\delta_f \xi_S} = \frac{V}{(L + K_{us}V^2/g)\xi_S}$$

where ξ_S is the steering gear ratio. The yaw velocity gain of the vehicle as a function of forward speed is shown in Figure 5.14.

From Eq. (5.16), the lateral acceleration gain with respect to the steering wheel angle is

$$G_{\text{acc}} = \frac{a_y/g}{\delta_f \xi_S} = \frac{V^2}{(gL + K_{us}V^2)\xi_S}$$

The lateral acceleration gain of the vehicle as a function of forward speed is also shown in Figure 5.14.

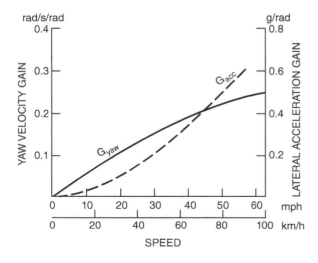

Figure 5.14 Yaw velocity gain and lateral acceleration gain characteristics of a passenger car with vehicle and tire parameters given in Example 5.2.

5.4   Testing of Handling Characteristics

To measure the handling behavior of a road vehicle under steady-state conditions, various types of test can be conducted on a skid pad, which in essence is a large, flat, paved area. Three types of test can be distinguished: the constant radius test, the constant forward speed test, and the constant steer angle test. During the tests, the steer angle, forward speed, and yaw velocity (or lateral acceleration) of the vehicle are usually measured. Yaw velocity can be measured by a rate-gyro or determined by the lateral acceleration divided by vehicle forward speed. Lateral acceleration can be measured by an accelerometer or determined by the yaw velocity multiplied by vehicle forward speed. Based on the relationship between the steer angle and the lateral acceleration or yaw velocity obtained from tests, the handling characteristics of the vehicle can be evaluated.

5.4.1   Constant Radius Test

In this test, the vehicle is driven along a curve with a constant radius at various speeds. The steer angle δ_f of the front tire or the angle of the steering wheel required to maintain the vehicle on course at various forward speeds together with the corresponding lateral acceleration are measured. The steady-state lateral acceleration can also be deduced from the vehicle forward speed and the known turning radius. The results can be plotted as shown in Figure 5.15 (Ellis 1969). The handling behavior of the vehicle can then be determined from the slope of the steer angle–lateral acceleration curve. From Eq. (5.10), for a constant turning radius, the slope of the curve is given by

$$\frac{d\delta_f}{d(a_y/g)} = K_{us} \tag{5.18}$$

This indicates that the slope of the curve represents the value of the understeer coefficient.

If the steer angle required to maintain the vehicle on a constant radius turn is the same for all forward speeds (i.e., the slope of the steer angle–lateral acceleration curve is zero), as shown in Figure 5.15, the vehicle is neutral steer. The vehicle is considered understeer when the slope of the steer angle–lateral acceleration curve is positive, which indicates the value of the understeer coefficient K_{us} is greater than zero, as shown in Figure 5.15. The vehicle is considered oversteer

Figure 5.15   Assessment of handling characteristics by constant radius test. *Source:* Ellis (1969). Reproduced with permission of Business Books.

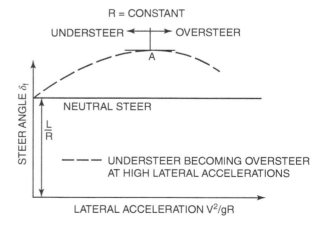

when the slope of the curve is negative, which indicates the value of the understeer coefficient K_{us} is less than zero, as illustrated in Figure 5.15.

For a practical vehicle, owing to the nonlinear behavior of tires and suspensions, load transfer, and the effects of tractive (or braking) effort, the value of the understeer coefficient K_{us} varies with operating conditions. It is possible for a vehicle to have understeer characteristics at low lateral accelerations and oversteer characteristics at high lateral accelerations, as shown in Figure 5.15.

5.4.2 Constant Speed Test

In this test, the vehicle is driven at a constant forward speed at various turning radii. The steer angle and the lateral acceleration are measured. The results can be plotted as shown in Figure 5.16 (Ellis 1969). The handling behavior of the vehicle can then be determined from the slope of the steer angle–lateral acceleration curve. From Eq. (5.10), for a constant speed turn, the slope of the curve is given by

$$\frac{d\delta_f}{d(a_y/g)} = \frac{gL}{V^2} + K_{us} \tag{5.19}$$

If the vehicle is neutral steer, the value of the understeer coefficient K_{us} will be zero and the slope of the steer angle–lateral acceleration line will be a constant of gL/V^2, as shown in Figure 5.16 (Ellis 1969).

The vehicle is considered understeer when the slope of the steer angle–lateral acceleration curve is greater than that for the neutral steer response at a given forward speed (i.e., gL/V^2), which indicates that the value of the understeer coefficient K_{us} is positive, as shown in Figure 5.16. The vehicle is considered oversteer when the slope of the curve is less than that for the neutral steer response at a given forward speed (i.e., gL/V^2), which indicates that the value of the understeer coefficient K_{us} is negative, as shown in Figure 5.16.

When the slope of the curve is zero

$$\frac{gL}{V^2} + K_{us} = 0$$

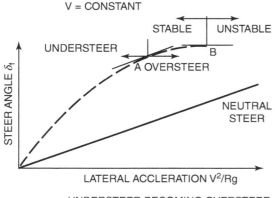

Figure 5.16 Assessment of handling characteristics by constant speed test. *Source:* Ellis (1969). Reproduced with permission of Business Books.

and

$$V^2 = \frac{gL}{(-K_{us})} = V^2_{\text{crit}}$$

this indicates that the oversteer vehicle is operating at the critical speed, and that the vehicle is at the onset of directional instability.

If, during the tests, the steer angle and yaw velocity are measured, then the slope of the steer angle–yaw velocity curve can also be used to evaluate the steady-state handling behavior of the vehicle in a similar way.

5.4.3   Constant Steer Angle Test

In this test, the vehicle is driven with a fixed steering wheel angle at various forward speeds. The lateral accelerations at various speeds are measured. From the test results, the curvature $1/R$, which can be calculated from the measured lateral acceleration and forward speed by $1/R = a_y/V^2$, is plotted against lateral acceleration, as shown in Figure 5.17. The handling behavior can then be determined by the slope of the curvature–lateral acceleration curve. From Eq. (5.10), for a constant steering wheel angle, the slope of the curve is given by

$$\frac{d(1/R)}{d(a_y/g)} = -\frac{K_{us}}{L} \tag{5.20}$$

If the vehicle is neutral steer, the value of the understeer coefficient K_{us} will be zero, and the slope of the curvature–lateral acceleration curve is zero. The characteristics of a neutral steer vehicle are therefore represented by a horizontal line, as shown in Figure 5.17.

The vehicle is considered understeer when the slope of the curvature–lateral acceleration curve is negative, which indicates that the value of the understeer coefficient K_{us} is positive, as shown in Figure 5.17. The vehicle is considered oversteer when the slope of the curvature–lateral acceleration curve is positive, which indicates that the value of the understeer coefficient K_{us} is negative.

In general, the constant radius test is the simplest and requires little instrumentation. The steer angle of the front tire (or the steering wheel angle) and forward speed are the only essential parameters to be measured during the test, as the steady-state lateral acceleration can be deduced from vehicle forward speed and the given turning radius. The constant speed test is more representative

Figure 5.17   Assessment of handling characteristics by fixed steer angle test.

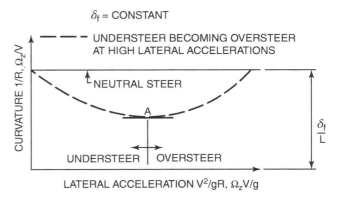

of the actual road behavior of a vehicle than the constant radius test, as the driver usually maintains a constant speed in a turn and turns the steering wheel by the required amount to negotiate the curve. The constant steer angle test, on the other hand, is easy to execute. Both the constant speed and constant steer angle tests would require, however, the measurement of the lateral acceleration or yaw velocity.

5.5 Transient Response Characteristics

Between the application of steering input and the attainment of steady-state motion, the vehicle is in a transient state. The behavior of the vehicle in this period is usually referred to as "transient response characteristics." The overall handling quality of a vehicle depends, to a great extent, on its transient behavior. The optimal transient response of a vehicle is that which has the fastest response with a minimum of oscillation in the process of approaching the steady-state motion.

In analyzing the transient response, the inertia properties of the vehicle must be taken into consideration. During a turning maneuver, the vehicle is in translation as well as in rotation. To describe its motion, it is convenient to use a set of axes fixed to and moving with the vehicle body because, with respect to these axes, the mass moments of inertia of the vehicle are constant, whereas with respect to axes fixed to earth, the mass moments of inertia vary as the vehicle changes its position.

To formulate the equations of transient motion for a vehicle during a turning maneuver, it is necessary to express the absolute acceleration of the center of gravity of the vehicle (i.e., the acceleration with respect to axes fixed to earth) using the reference frame attached to the vehicle body (Steeds 1960).

Let *ox* and *oy* be the longitudinal and lateral axes fixed to the vehicle body with the origin at the center of gravity, and let V_x and V_y be the components of the velocity V of the center of gravity along the axes *ox* and *oy*, respectively, at time *t*, as shown in Figure 5.18. As the vehicle is in both translation and rotation during a turn, at time $t + \Delta t$, the direction and magnitude of the velocity of the center of gravity as well as the orientation of the longitudinal and lateral axes of the vehicle change, as shown in Figure 5.18. The change of the velocity component parallel to the *ox* axis is

$$(V_x + \Delta V_x) \cos \Delta\theta - V_x - (V_y + \Delta V_y) \sin \Delta\theta$$
$$= V_x \cos \Delta\theta + \Delta V_x \cos \Delta\theta - V_x - V_y \sin \Delta\theta - \Delta V_y \sin \Delta\theta \qquad (5.21)$$

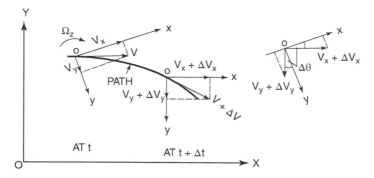

Figure 5.18 Analysis of plane motions of a vehicle using axes fixed to vehicle body.

Consider that $\Delta\theta$ is small; the above expression becomes

$$\Delta V_x - V_y\Delta\theta \qquad (5.22)$$

The component along the longitudinal axis of the absolute acceleration of the center of gravity of the vehicle can be obtained by dividing the above expression by Δt. In the limit, this gives

$$a_x = \frac{dV_x}{dt} - V_y\frac{d\theta}{dt} = \dot{V}_x - V_y\Omega_z \qquad (5.23)$$

The component dV_x/dt (or $\dot{V}_x$) is due to the change in magnitude of the velocity component V_x and is directed along the ox axis, and the component $V_y d\theta/dt$ (or $V_y\Omega_z$) is due to the rotation of the velocity component V_y. Following a similar approach, the component along the lateral axis of the absolute acceleration of the center of gravity of the vehicle a_y is

$$a_y = \frac{dV_y}{dt} + V_x\frac{d\theta}{dt} = \dot{V}_y + V_x\Omega_z \qquad (5.24)$$

The acceleration components a_x and a_y of the center of gravity of the vehicle may be derived in an alternate way. When the vehicle is moving along a curved path, the absolute acceleration a of its center of gravity may be expressed in terms of a tangential component a_t and a normal component a_n, as shown in Figure 5.19(a). The tangential component a_t is in the same direction as that of the resultant velocity V of the center of gravity and is at an angle β with the longitudinal axis ox of the vehicle, as shown in Figure 5.19(a). β is usually referred to as the vehicle sideslip angle. a_t can be resolved into two components, dV_x/dt (or $\dot{V}_x$), and dV_y/dt (or $\dot{V}_y$), directed along ox and oy axes, respectively. The normal component a_n is directed toward the turn center and its magnitude is equal to V^2/R, where R is the turning radius. It can be resolved into two components, $-(V^2/R)\sin\beta$

(a)

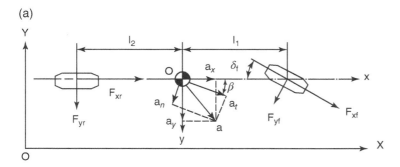

(b)

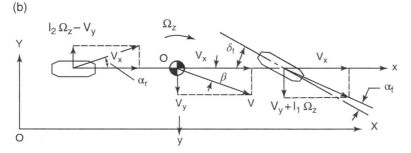

Figure 5.19 Simplified vehicle model for analysis of transient motions.

and $(V^2/R)\cos\beta$, directed along ox and oy axes, respectively. Note that $-(V^2/R)\sin\beta = -V\Omega_z\sin\beta = -V_y\Omega_z$ and that $(V^2/R)\cos\beta = V\Omega_z\cos\beta = V_x\Omega_z$, as shown in Figure 5.19(a). Combining the corresponding components of a_t and a_n along the ox and oy axes, one obtains the same expressions for a_x and a_y as those given by Eqs. (5.23) and (5.24), respectively.

Referring to Figure 5.19(a), for a vehicle having plane motion, the equations of motion with respect to the axes fixed to the vehicle body are given by

$$m\left(\dot{V}_x - V_y\Omega_z\right) = F_{xf}\cos\delta_f + F_{xr} - F_{yf}\sin\delta_f \tag{5.25}$$

$$m\left(\dot{V}_y + V_x\Omega_z\right) = F_{yr} + F_{yf}\cos\delta_f + F_{xf}\sin\delta_f \tag{5.26}$$

$$I_z\dot{\Omega}_z = l_1 F_{yf}\cos\delta_f - l_2 F_{yr} + l_1 F_{xf}\sin\delta_f \tag{5.27}$$

where I_z is the mass moment of inertia of the vehicle about the z axis (see Figure 5.1).

In the above equations, it is assumed that the vehicle body is symmetric about the longitudinal plane (i.e., the xoz plane in Figure 5.1), and that roll motion of the vehicle body is neglected.

If the vehicle is not accelerating or decelerating along the ox axis, Eq. (5.25) may be neglected, and the lateral and yaw motions of the vehicle are governed by Eqs. (5.26) and (5.27).

The slip angles α_f and α_r can be defined in terms of the vehicle motion variables Ω_z and V_y. Referring to Figure 5.19(b), and using the usual small angle assumptions,

$$\alpha_f = \delta_f - \frac{l_1\Omega_z + V_y}{V_x} \tag{5.28}$$

$$\alpha_r = \frac{l_2\Omega_z - V_y}{V_x} \tag{5.29}$$

The lateral forces acting on the front and rear tires are a function of the corresponding slip angle and cornering stiffness, and are expressed by

$$F_{yf} = 2C_{af}\alpha_f \tag{5.30}$$

$$F_{yr} = 2C_{ar}\alpha_r \tag{5.31}$$

Combining Eqs. (5.26)–(5.31), and assuming that the steer angle is small and F_{xf} is zero, the equations of lateral and yaw motions of a vehicle with steer angle as the only input variable become

$$m\dot{V}_y + \left(mV_x + \frac{2l_1 C_{af} - 2l_2 C_{ar}}{V_x}\right)\Omega_z$$
$$+ \left(\frac{2C_{af} + 2C_{ar}}{V_x}\right)V_y = 2C_{af}\delta_f(t) \tag{5.32}$$

$$I_z\dot{\Omega}_z + \left(\frac{2l_1^2 C_{af} + 2l_2^2 C_{ar}}{V_x}\right)\Omega_z + \left(\frac{2l_1 C_{af} - 2l_2 C_{ar}}{V_x}\right)V_y$$
$$= 2l_1 C_{af}\delta_f(t) \tag{5.33}$$

In the above equations, $\delta_f(t)$ represents the steer angle of the front wheel as a function of time. If, in addition to the steer angle, external forces or moments, such as aerodynamic forces and moments, are acting on the vehicle, they should be added to the right-hand side of Eqs. (5.32) and (5.33) as input variables.

When the input variables, such as the steer angle and external disturbing forces, and the initial conditions are known, the response of the vehicle, expressed in terms of yaw velocity Ω_z and lateral velocity V_y as functions of time, can be determined by solving the differential equations. As an

Figure 5.20 Yaw velocity response of a station wagon to a step input of steer angle of 0.01 rad at 97 km/h (60 mph). Reprinted with permission from SAE paper 716B © 1963 Society of Automotive Engineers, Inc.

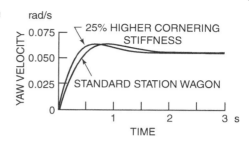

Figure 5.21 Yaw velocity response of a station wagon to a step input of lateral force of 890 N (200 lb) at 97 km/h (60 mph). Reprinted with permission from SAE paper 716B © 1963 Society of Automotive Engineers, Inc.

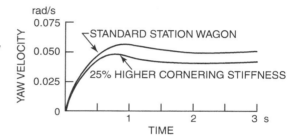

example, Figure 5.20 shows the yaw velocity responses to a step input of steer angle of 0.01 rad (0.57°) for a station wagon with two different types of tire traveling at 97 km/h (60 mph) (Bundorf et al. 1963). Figure 5.21 shows the yaw velocity responses to a step input of aerodynamic side force of 890 N (200 lb) for the same vehicle.

5.6 Directional Stability

5.6.1 Criteria for Directional Stability

The directional stability of a vehicle refers to its ability to stabilize its direction of motion against disturbances. A vehicle is considered directionally stable if, following a disturbance, it returns to a steady-state regime within a finite time. A directionally unstable vehicle diverges more and more from the original path, even after the disturbance is removed. The disturbance may arise from crosswind, momentary forces acting on the tires from the road, slight movement of the steering wheel, or a variety of other causes.

With a small perturbation about an equilibrium position, a vehicle may be regarded as a linear dynamic system. The equations of lateral and yaw motions are a set of linear differential equations with constant coefficients, as shown in Eqs. (5.32) and (5.33). Following a disturbance, the lateral and yaw velocities, V_y and Ω_z, will vary with time exponentially, $V_y = A_1 e^{\psi t}$ and $\Omega_z = A_2 e^{\psi t}$. The stability of the vehicle is determined by the value of ψ. If ψ is a real number and positive, the values of lateral and yaw velocities will increase exponentially with time, and the vehicle will be directionally unstable. A real and negative value of ψ indicates that motions of the vehicle converge to a steady state in a finite time, and that the vehicle is directionally stable. If ψ is a complex number with a positive real part, the motions will be oscillatory with increasing amplitudes, and thus the vehicle will be directionally unstable. A complex value of ψ with a negative real part indicates that the motions are oscillatory with decreasing amplitudes, and thus the vehicle is directionally stable.

To evaluate the directional stability of a vehicle, it is, therefore, necessary to determine the values of ψ. Since only the motions of the vehicle following a disturbance are of interest in the evaluation of stability, the steering input and the like are taken to be zero. This is equivalent to the examination of the free vibrations of the vehicle in the lateral direction and in yaw following the initial disturbance. To obtain the values of ψ, the following solutions to the differential equations for lateral and yaw motions (i.e., Eqs. (5.32) and (5.33)) are assumed:

$$V_y = A_1 e^{\psi t} \tag{5.34}$$

$$\Omega_z = A_2 e^{\psi t} \tag{5.35}$$

Then

$$\dot{V}_y = A_1 \psi e^{\psi t} \tag{5.36}$$

$$\dot{\Omega}_z = A_2 \psi e^{\psi t} \tag{5.37}$$

On substituting these values into Eqs. (5.32) and (5.33), and setting the right-hand sides of the equations to zero, the equations become

$$mA_1\psi + \left(\frac{2C_{af} + 2C_{ar}}{V_x}\right)A_1$$
$$+ \left(\frac{mV_x^2 + 2l_1C_{af} - 2l_2C_{ar}}{V_x}\right)A_2 = 0 \tag{5.38}$$

$$I_zA_2\psi + \left(\frac{2l_1C_{af} - 2l_2C_{ar}}{V_x}\right)A_1$$
$$+ \left(\frac{2l_1^2C_{af} + 2l_2^2C_{ar}}{V_x}\right)A_2 = 0 \tag{5.39}$$

The above equations can be rewritten as

$$mA_1\psi + a_1A_1 + a_2A_2 = 0 \tag{5.40}$$

$$I_zA_2\psi + a_3A_1 + a_4A_2 = 0 \tag{5.41}$$

where

$$a_1 = \frac{2C_{af} + 2C_{ar}}{V_x}$$

$$a_2 = \frac{mV_x^2 + 2l_1C_{af} - 2l_2C_{ar}}{V_x}$$

$$a_3 = \frac{2l_1C_{af} - 2l_2C_{ar}}{V_x}$$

$$a_4 = \frac{2l_1^2C_{af} + 2l_2^2C_{ar}}{V_x}$$

Equations (5.40) and (5.41) are known as the amplitude equations, which are linear, homogeneous, algebraic equations. To obtain a nontrivial solution for ψ, the determinant of the amplitudes must be equal to zero. Thus,

$$\begin{vmatrix} m\psi + a_1 & a_2 \\ a_3 & I_z\psi + a_4 \end{vmatrix} = 0 \tag{5.42}$$

Expanding the determinant yields the characteristic equation

$$m\,I_z\psi^2 + (I_z a_1 + ma_4)\psi + (a_1 a_4 - a_2 a_3) = 0 \tag{5.43}$$

If $(I_z a_1 + ma_4)$ and $(a_1 a_4 - a_2 a_3)$ are both positive, then ψ must be either a negative real number or a complex number having a negative real part. The terms $I_z a_1$ and ma_4 are clearly always positive; hence, it follows that the vehicle is directionally stable if $a_1 a_4 - a_2 a_3$ is positive. It can be shown that the condition for $a_1 a_4 - a_2 a_3 > 0$ is

$$L + \frac{V_x^2}{g}\left(\frac{W_f}{C_{\alpha f}} - \frac{W_r}{C_{\alpha r}}\right) > 0$$

or

$$L + \frac{V_x^2}{g}K_{us} > 0 \tag{5.44}$$

where K_{us} is the understeer coefficient defined previously. This indicates that the examination of the directional stability of a vehicle is now reduced to determining the conditions under which Eq. (5.44) is satisfied. When the understeer coefficient K_{us} is positive, Eq. (5.44) is always satisfied. This implies that when a vehicle is understeer, it is always directionally stable. When the understeer coefficient K_{us} is negative, which indicates that the vehicle is oversteer, the vehicle is directionally stable only if the speed of the vehicle is below a specific value:

$$V_x < \sqrt{\frac{gL}{-K_{us}}} \tag{5.45}$$

This specific speed is, in fact, the critical speed V_{crit} of an oversteer vehicle discussed previously. This indicates that an oversteer vehicle will be directionally stable only if it operates at a speed lower than the critical speed.

The analysis of vehicle handling presented in this chapter is based on a simplified vehicle model. In reality, the lateral and yaw motions are coupled with the motions in fore and aft (translation along the x axis shown in Figure 5.1), roll (rotation about the x axis), pitch (rotation about the y axis), and bounce (translation along the z axis). The coupling is through mechanisms such as changes in cornering properties of tires with the application of tractive or braking effort, longitudinal and lateral load transfer during a turning maneuver, and changes in the steering characteristics of the vehicle due to motions of the vehicle body relative to the unsprung parts. More complete analyses of the handling behavior of road vehicles, in which the interactions of lateral and yaw motions with those in the other directions are taken into consideration, have been made (Segel 1957; Whitcomb and Milliken 1957; Radt and Pacejka 1963; Ellis 1969; Milliken and Milliken 1995; Dixon 1996). In addition, a variety of multibody dynamics software packages is now commercially available for simulating handling behavior of passenger vehicles and light-duty trucks, such as CarSim by Mechanical Simulation Corporation, Ann Arbor, MI, USA.

It should also be mentioned that in reality the handling behavior of a vehicle involves the continuous interaction of the driver with the vehicle. To perform a comprehensive study of vehicle handling, the characteristics of the human driver should therefore be included. This topic is, however, beyond the scope of the present text.

5.6.2 Vehicle Stability Control

In a turning maneuver where tire forces are approaching or at the limits of road adhesion, the vehicle may deviate significantly from the driver's intended direction of motion and path. This may occur, for instance, when a driver attempts a sudden maneuver to avoid a crash (or an obstacle) or due to misjudgment of the severity of a curve. Under these circumstances, the driver may lose control of the vehicle. The loss of control may be a result of the vehicle changing its heading too quickly (spinning out) or not quickly enough (plowing out).

While the loss of directional stability or control of the vehicle may also be caused by locking of the wheel during braking or spinning of the driven wheel during acceleration, as described in Section 3.11.1, this can be prevented by using the antilock brake system or the traction control system, respectively, as discussed in Sections 3.11.4 and 3.11.5. However, these systems are incapable of actively controlling the directional behavior of the vehicle, such as its yaw velocity (yaw rate) and/or vehicle sideslip angle discussed previously.

With the increasing emphasis on road safety, systems designed for actively controlling the directional behavior of road vehicles have emerged. They are intended to enhance vehicle directional stability and control and tracking performance in all operating conditions, including severe turning maneuvers during acceleration, coasting, and deceleration. This type of active control system is hereinafter referred to as the vehicle stability control system, although a variety of trade names have been used, such as Active Yaw Control (AYC), AdvanceTrac, Direct Yaw Moment Control (DYMC), Dynamic Stability Control (DSC), Dynamic Stability and Traction Control (DSTC), Electronic Stability Control (ESC), Electronic Stability Program (ESP), Porsche Stability Management (PSM), Vehicle Dynamic Control (VDC), Vehicle Skid Control (VSC), Vehicle Stability Assist (VSA), Vehicle Stability Control (VSC), Vehicle Stability Enhancement (VSE), and StabiliTrak (Toyota's Vehicle Stability Control System) (Toyota Motor Corporation 1995; van Zanten et al. 1995, 1996; Yasui et al. 1996; Koibuchi et al. 1996; Abe 1999; National Highway Traffic Safety Administration 2007). The vehicle stability control system is usually integrated with the antilock brake system and traction control system.

While the design features, level of refinements, and control algorithms vary from one make to another, the basic operating principles of various types of vehicle stability control system are quite similar. In general, the intended (nominal or desired) course of the vehicle is first established from the steering wheel angle, wheel rotating speeds, accelerator pedal position (representing the engine torque delivered to the driven wheels), brake pressure, etc. The actual course of the vehicle is deduced from measured yaw rate, lateral acceleration, etc. using onboard sensors. The vehicle stability control system monitors the directional behavior of the vehicle continually and compares the intended course with the actual course of the vehicle every few milliseconds. If the difference between them is more than a prescribed level, the vehicle stability control system will intervene and regulate the brake pressures on specific tires. The aim is to generate a restoring yaw moment to minimize the deviation of the course of the vehicle from the intended one. If necessary, the system will also reduce the engine torque transmitted to the driven wheels, to reduce vehicle speed and to help restore the vehicle to the intended course. A block diagram illustrating the basic operating principles of the vehicle stability control system is shown in Figure 5.22.

In most current vehicle stability control systems, vehicle yaw rate Ω_z and sideslip angle β, shown in Figure 5.19, are used as the basic parameters for identifying the directional behavior of the vehicle. Consequently, they are the parameters that the vehicle stability control system is designed to control. Yaw rate control will enable the vehicle to maintain the desired rate and direction of rotation about its vertical axis. However, yaw rate control alone is inadequate for keeping the vehicle

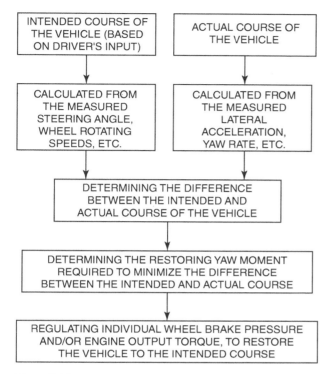

Figure 5.22 Operating principles of a vehicle stability control system.

moving along the intended path. For instance, on a slippery road, exercising yaw rate control together with steering correction can only maintain the vehicle with the desired yaw rate and orientation, but the vehicle sideslip angle may be significantly different from the intended one. As a result, the vehicle may deviate considerably from its intended path, as shown in Figure 5.23 (van Zanten et al. 1995). This indicates that controlling both the yaw rate and vehicle sideslip angle is required.

After determining the intended (nominal or desired) and actual yaw rates and vehicle sideslip angles, they are compared. If the differences between them are found to be higher than prescribed values, the control unit of the vehicle stability control system will send command signals to the actuators to modulate the brake pressures on select tires, to generate a restoring yaw moment to keep the vehicle on the intended course. Under certain circumstances, the vehicle stability control system will also reduce the engine torque to help restore the directional behavior of the vehicle, as noted previously. For instance, if a rear-wheel drive vehicle on a left-hand turn is at the verge of losing directional stability due to the driven rear tires sliding outward, then for a vehicle not equipped with the vehicle stability control system, the yaw rate and vehicle sideslip angle will become excessive and the vehicle will be "spinning out," as shown on the left of Figure 5.24(a) (Toyota Motor Corporation 1995). However, a vehicle equipped with the vehicle stability control system immediately detects that the yaw rate and vehicle sideslip angle are changing faster than appropriate for the driver's intended heading and path. It momentarily applies brake to the right front tire to generate a yaw moment to restore the vehicle back to the intended course, as shown on the right of Figure 5.24(a). In some cases, the vehicle stability control system also reduces the torque

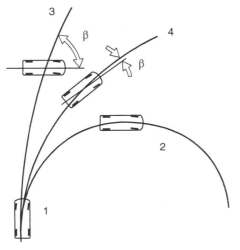

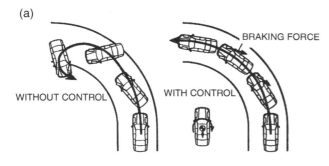

Figure 5.23 Comparisons of the handling behavior of a vehicle with yaw rate control to that with both yaw rate and sideslip angle control. Reprinted with permission from SAE paper No. 950759, © 1995 Society of Automotive Engineers, Inc.

1 STEP INPUT AT STEERING WHEEL
2 ON HIGH FRICTION ROAD
3 ON SLIPPERY ROAD WITH STEERING
  CORRECTION AND YAW RATE CONTROL
4 ON SLIPPERY ROAD WITH BOTH YAW
  RATE AND SIDE SLIP ANGLE CONTROL

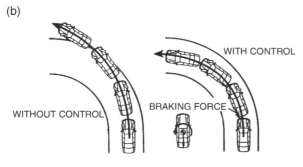

Figure 5.24 Comparisons of the handling behavior of a car with and without vehicle stability control system on the verge of (a) lateral sliding of rear tires and (b) lateral sliding of front tires. Reproduced by permission of Toyota Motor Corporation from "Toyota Vehicle Stability Control System", *Automotive Engineering*, August 1995.

transmitted to the driven tire, by means of adjusting the throttle valve, spark retardation of the engine, or by shutting off fuel supply to some cylinders. This results in the reduction of the tractive effort on the driven tires and vehicle speed, which leads to the decrease in the side force acting on the tire due to centrifugal effects. The reduction of both the tractive effort and side force on the driven tires will decrease the propensity of the tires to slide laterally, as discussed in Section 1.4. All of these help restore the directional behavior of the vehicle. The intervention by the vehicle stability control system happens quickly and smoothly and thus in many cases will go undetected by the driver.

On the other hand, if a front-wheel drive vehicle on a left-hand turn is at the verge of losing directional control due to the driven front tires sliding outward, a vehicle not equipped with the vehicle stability control system will be "plowing out," as shown on the left of Figure 5.24(b) (Toyota Motor Corporation 1995). A vehicle equipped with the vehicle stability control system immediately detects that the yaw rate and vehicle sideslip angle are changing more slowly than appropriate for the driver's intended heading and path. It momentarily applies brake to the left rear tire to generate a yaw moment to restore the vehicle back to the intended course, as shown on the right of Figure 5.24(b) Like the situation described above, under certain circumstances, the vehicle stability control system will reduce the torque transmitted to the driven front tires, to decrease the propensity of the driven front tires to slide laterally. This helps restore the directional control of the vehicle. Note that the response of the vehicle to brake pressure modulation is much faster than that to engine intervention.

In many vehicle stability control systems currently in use, the steering wheel angle, accelerator pedal position, brake pressure, wheel rotating speeds, yaw rate, and lateral acceleration of the vehicle are continuously monitored. The intended (nominal or desired) yaw rate and vehicle sideslip angle are derived from the steering wheel angle, wheel rotating speeds, and lateral acceleration. The actual value of the yaw rate is measured by an onboard sensor, while the value of vehicle sideslip angle is derived from the measured steering wheel angle, yaw rate, lateral acceleration, and estimated longitudinal speed of the vehicle (van Zanten et al. 1995). The control algorithms of commercial vehicle stability control systems are proprietary, and their details are seldom revealed.

To illustrate the approach to the development of control algorithms for vehicle stability control systems, an example is presented below. The algorithm described in the example is proposed for use in a vehicle stability control system for front-wheel drive vehicles without using an onboard yaw rate sensor. To estimate vehicle yaw rate and sideslip angle, the algorithm uses the steering wheel angle, wheel rotating speeds, and lateral acceleration as inputs (Hac and Simpson 2000).

The algorithm first generates two initial estimates of vehicle yaw rate: one is derived from the relative forward speed between the two nondriven rear wheels, based on the measured rotating speeds of the wheels, and the other is based on the measured vehicle lateral acceleration.

Assuming that the nondriven rear tires are free rolling, the vehicle yaw rate Ω_z can be estimated by (referring to Figure 5.2)

$$\Omega_z = (V_{ro} - V_{ri})/B \tag{5.46}$$

where V_{ro} and V_{ri} are the circumferential speed of the outside rear tire and that of the inside rear tire, respectively, and B is the tread (i.e., the transverse distance between the centers of the two rear tires). V_{ro} and V_{ri} can be determined from the rolling radius and measured rotating speed of the outside rear tire and those of the inside rear tire, respectively.

During cornering there is a lateral load transfer from the inside tire to the outside tire due to the roll moment caused by the centrifugal force acting at the center of the vehicle sprung mass. This will change the rolling radius of the outside rear tire and that of the inside rear tire. A correction factor is, therefore, introduced into the calculation of the rolling radii of the outside and inside rear tires, considering the measured lateral acceleration, the vehicle sprung mass, the height of its mass center above ground, the tire radial stiffness, and the ratio of the roll stiffness of the rear suspension to the combined roll stiffness of the front and rear suspensions.

The approximate value of vehicle yaw rate Ω_z may also be estimated from the measured lateral acceleration using the following relationship:

$$\Omega_z = a_y/V_x \tag{5.47}$$

where a_y is the lateral acceleration and V_x is the longitudinal velocity of the vehicle, as shown in Figure 5.19.

Equation (5.47) is a reasonable approximation of the yaw rate only in essentially steady-state maneuvers, with constant or slowly varying steering wheel angle and vehicle speed. A more complete relationship between the yaw rate and lateral acceleration is given by Eq. (5.24), which includes an unknown term dV_y/dt. Therefore, the estimate of yaw rate using Eq. (5.47) generally deteriorates during fast transient maneuvers.

The accuracy of the initial estimate of yaw rate using either Eq. (5.46) or (5.47) depends on the extent to which the underlying assumptions are satisfied under operating conditions. If in operations any of the assumptions is violated, the confidence level in the corresponding estimate is reduced. For instance, if the wheels are being braked, causing significant skid, and are far from free rolling, then the confidence level in estimating yaw rate using Eq. (5.46) will be reduced. The confidence level for the estimate based on rear wheel rotating speeds is selected from several predefined numerical values ranging from low to high. The confidence level in the estimate obtained from vehicle lateral acceleration is high when the vehicle is in an approximately steady-state turning maneuver, and is low when it is in a fast transient turning maneuver.

The preliminary estimate of yaw rate is calculated as a weighted average of estimates based on measured wheel rotating speeds and vehicle lateral acceleration, with the weightings proportional to the corresponding confidence levels. The preliminary estimate of vehicle yaw rate so determined is then fed into an observer, which provides the final estimate of vehicle yaw rate and an estimate of vehicle lateral velocity. The observer is a simplified model of vehicle dynamics in the yaw plane, as described by Eqs. (5.26) and (5.27) (neglecting Eq. (5.25)). The lateral forces acting on the front and rear axles F_{yf} and F_{yr} and the vehicle longitudinal velocity V_x are involved in the model. When the vehicle is operating at or close to the limit of tire adhesion, which is of prime interest in the design of vehicle stability control systems, the lateral forces F_{yf} and F_{yr} may be estimated by the coefficient of road adhesion μ and the normal loads on the front and rear axles, respectively. The coefficient of road adhesion μ in the lateral direction can be estimated from the measured lateral acceleration. The longitudinal velocity of the vehicle V_x may be estimated from the measured wheel rotating speeds and steering wheel angle. The observer uses the measured steering wheel angle, the measured lateral acceleration, and the preliminary estimate of vehicle yaw rate as feedback signals. The feedback terms provide correction when the estimates deviate from actual values, thus preventing the estimates from diverging with time, because of the mismatch between the model and the vehicle and because of external disturbances (Hac and Simpson 2000).

After the observer equations (i.e., Eqs. (5.26) and (5.27)) are solved, the final estimate of vehicle yaw rate Ω_z and the estimate of vehicle lateral velocity V_y are obtained. The vehicle sideslip angle β is then estimated from

$$\beta = \arctan\left(V_y/V_x\right) \tag{5.48}$$

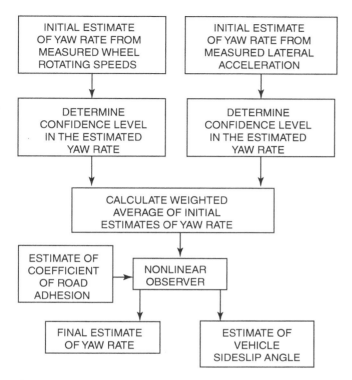

Figure 5.25 Flowchart for the algorithm for estimating vehicle yaw rate and sideslip angle used in a vehicle stability control system.

The algorithm outlined above for estimating the vehicle yaw rate and sideslip angle is illustrated by the flowchart shown in Figure 5.25. The algorithm is implemented and tested on various surfaces, including snow and ice. The results indicate that even in extreme maneuvers, it gives good measures of deviations of yaw rate and vehicle sideslip angle from the desired values, which can be used for vehicle stability control. Initial tests also indicate good robustness with respect to parameter variations, road roughness, bank angle of the road, and difference in tire inflation pressure. However, further sensitivity studies, including sensor errors, are recommended to fully establish the robustness of the algorithm (Hac and Simpson 2000).

In addition to the basic functions of the vehicle stability control system to generate a restoring yaw moment and to reduce the driving torque transmitted to the driven wheels for restoring the vehicle to its intended course, many systems have additional features. These include performing high-deceleration, automatic braking on all four wheels while maintaining uneven side-to-side braking for generating a restoring yaw moment. Under certain circumstances, this may quicken the process of restoring vehicle directional behavior. Some systems used on vehicles with a high center of gravity, such as sport utility vehicles, are programmed for an additional function known as roll stability control. It measures the roll angle of the vehicle using an additional roll-rate sensor to determine whether the vehicle is in danger of rollover. If it detects the risk of rollover, it will intervene by reducing the lateral acceleration causing the roll motion by applying hard braking to the outside front wheel or to both front wheels. In either case, the braking force generated must be large enough to cause high longitudinal skid for the front wheel(s). This greatly reduces the cornering force available to the front wheel(s) and straightens the path of the vehicle, hence increasing the effective turning radius. Furthermore, by hard braking the vehicle forward speed decreases. The combined effect of increasing the effective turning radius and reducing the forward speed will dramatically decrease the lateral acceleration and the propensity of the vehicle to roll over. The intervention of the roll stability control system may cause the deterioration of the path-following capability of the vehicle, but this is thought to carry a lower risk to the driver than rollover (National Highway Traffic Safety Administration 2007).

Studies carried out in Europe, Japan, and the United States show that the vehicle stability control system is highly effective in preventing crashes and improving road safety (National Highway Traffic Safety Administration 2007). In Germany, it has been estimated that the system would prevent 80% of skidding crashes and 35% of all vehicle fatalities. In Sweden, it would prevent 16.7% of all injury crashes excluding rear-end crashes and 21.6% of serious and fatal crashes. In Japan, it would prevent 35% of single-vehicle crashes, 50% of fatal single-vehicle crashes, 30% of head-on crashes, and 40% of fatal head-on crashes. In the United States, it is estimated that the vehicle stability control system is approximately 30% effective in preventing fatal single-vehicle crashes for passenger cars and 63% for sport utility vehicles, and that for all single-vehicle crashes, the corresponding effectiveness rates are 35% for passenger cars and 67% for sport utility vehicles.

In view of the effectiveness of the vehicle stability control system, in the United States the Federal Motor Vehicle Safety Standard (FMVSS) 126 was introduced and came into effect in 2012. It stipulates that passenger cars, multi-purpose passenger vehicles, trucks, and buses, with a gross vehicle weight rating (GVWR) of 4536 kg (10,000 lb) or less, must be equipped with electronic stability control.

5.7 Driving Automation

To improve road vehicle safety and to assist the driver in operating a motor vehicle, driver assistance systems, often referred to as intelligent vehicle systems, have been evolving over the past few decades. The cruise control was introduced in the 1960s, the antilock braking system in the 1970s, the traction control system in the 1980s, the vehicle stability control (electronic stability control) and adaptive cruise control in the 1990s, and the lane-keeping assist, automatic emergency braking, automated parking, etc. in the 2000s (Fancher and MacAdam 2014; Peng 2014). The evolution of driver assistance systems has led to the development of the automated driving system for both passenger and commercial vehicles in recent years. It has attracted intensive interest worldwide because of its potentially significant benefits to road safety and driving comfort. Prototypes of various types of automated driving system have been demonstrated by motor vehicle manufacturers, research institutions, and technology companies. This section provides a brief introduction to the potential benefits, classification, and basic enabling technologies for driving automation, and cooperative driving automation.

The prime benefits of the driving automation may be summarized as follows:

- Increasing road traffic safety. According to the U.S. Department of Transportation data, an estimated 37,133 people lost their lives as a result of motor vehicle crashes in the United States in 2017. Of these crashes, 94% involved driver-related factors, such as impaired driving, distraction, and speeding or illegal maneuvers (U.S. Department of Transportation 2018). The implementation of driving automation would significantly reduce motor vehicle accidents.

- Improving traffic flow and reducing congestion. This will not only reduce travel time but will also lead to improved fuel economy and reduction of greenhouse gas emissions from internal combustion engines which will mitigate climate change.
- Providing enhanced mobility for the elderly, disabled, and the like.
- Relieving drivers from routine driving and navigation chores.
- Facilitating development of sharing economy in transportation services.

5.7.1 Classification of Levels of Driving Automation

There have been a variety of definitions for various levels of driving automation proposed by governmental agencies and professional organizations, such as the National Highway Traffic Safety Administration (NHTSA) of the U.S. Department of Transportation, German Federal Highway Research Institute, and the Society of Automotive Engineers (SAE) International. For some time, there has been a need for standardization of the definitions of driving automation to provide clarity and consistency. The SAE International classification of levels of driving automation was adopted by NHTSA (National Highway Traffic Safety Administration 2016) and has now been widely accepted by the automotive industry.

A summary of SAE International's levels of driving automation for on-road vehicles, issued in January 2014, is given in Table 5.1.

The key definitions of the terms used in Table 5.1 are given below:

- **Dynamic driving task** includes the operational (steering, braking, accelerating, monitoring the vehicle and roadway) and tactical (responding to events, determining when to change lanes, turn, use signals, etc.) aspects of the driving task, but not the strategic (determining destinations and waypoints) aspect of the driving task.
- **Driving mode** is a type of driving scenario with characteristic *dynamic driving task* requirements (e.g., expressway merging, high-speed cruising, low-speed traffic jam, close-campus operations, etc.).
- **Request to intervene** is notification by the *automated driving system* to a *human driver* that s/he should promptly begin or resume performance of the *dynamic driving task*.
- **System** refers to the driver assistance system, combination of driver assistance systems, or *automated driving system*. Excluded are *warning and momentary intervention systems*, which do not automate any part of the *dynamic driving task* on a sustained basis and therefore do not change the *human driver's* role in performing the *dynamic driving task*.

Table 5.1 defines six levels of driving automation which span from no automation (Level 0) to full automation (Level 5). The six levels are grouped into two categories based on how the driving environment is monitored:

1) **Human driver monitoring the driving environment**

   - At Level 0, the *human driver* does all aspects of the *dynamic driving task*.
   - At Level 1, a driver assistance system on the vehicle can assist the *human driver* to perform either steering or acceleration/deceleration and with the expectation that the *human driver* performs all remaining aspects of the *dynamic driving task*.
   - At Level 2, a driver assistance system on the vehicle can perform both steering or acceleration/deceleration and with the expectation that the *human driver* performs all remaining aspects of the *dynamic driving task*

Table 5.1 Summary of SAE International's levels of driving automation for on-road vehicles (issued January 2014).

SAE Level	Name	Narrative definition	Execution steering and acceleration/ deceleration	Monitoring of driving environment	Fallback performance of dynamic driving task	System capability (driving modes)
Human driver monitors the driving environment						
0	**No Automation**	The full-time performance by the *Human driver* of all aspects of the *dynamic driving task*, even when enhanced by warning or intervention systems	Human driver	Human driver	Human driver	n/a
1	**Driver Assistance**	The *driving mode* – specific execution by a driver assistance system of either steering or acceleration/deceleration using information about the driving environment and with the expectation that the *human driver* performs all remaining aspects of the *dynamic driving task*	Human driver and system	Human driver	Human driver	Some driving modes
2	**Partial Automation**	The *driving mode* – specific execution by one or more driver assistance systems of both steering and acceleration/deceleration using information about the driving environment and with the expectation that the *human driver* performs all remaining aspects of the *dynamic driving task*	**System**	Human driver	Human driver	Some driving modes
Automated driving system ("system") monitors the driving environment						
3	**Conditional Automation**	The *driving mode* – specific performance by an *automated driving system* of all aspects of the dynamic driving task with the expectation that the *human driver* will respond appropriately to a *request to intervene*	System	**System**	Human driver	Some driving modes
4	**High Automation**	The *driving mode* – specific performance by an automated driving system of all aspects of the *dynamic driving task*, even if a *human driver* does not respond appropriately to a *request to intervene*	System	System	**System**	Some driving modes
5	**Full Automation**	The full-time performance by an *automated driving system* of all aspects of the *dynamic driving task*, under all roadway and environmental conditions that can be managed by a *human driver*	System	System	System	**All driving modes**

2) **Automated driving system monitoring the driving environment**

- At Level 3, an *automated driving system* performs all aspects of the *dynamic driving task* with the expectation that the *human driver* will respond appropriately to a *request to intervene*. A key distinction is between Level 2, where the *human driver* performs part of the *dynamic driving task*, and Level 3, where the *automated driving system* performs the entire *dynamic driving task*.
- At Level 4, an *automated driving system* performs all aspects of the *dynamic driving task*, even if a *human driver* does not respond appropriately to a *request to intervene*.
- At Level 5, an *automated driving system* performs, on a full-time basis, all aspects of the *dynamic driving task* under all roadway and environmental conditions that can be managed by a *human driver*.

To date, a variety of terms (e.g., automated, autonomous, driverless, highly automated, self-driving) have been used by governmental agencies, professional organizations, and observers to describe various levels or forms of driving automation. While no terminology is correct or incorrect, in the context of the SAE International classification that the strategic aspect of the driving task for determining destinations and waypoints is excluded from the definition of *dynamic driving task*, the term "automated vehicles" appears more appropriate than the term "autonomous vehicles." This is because the term "autonomous vehicles" may imply that these vehicles can determine the destinations and waypoints on their own.

As noted previously, the SAE classification of levels of driving automation presented in Table 5.1 was originally issued in January 2014 but has been successively updated in September 2016, June 2018, and April 2021 (Society of Automotive Engineers 2014, 2016, 2018, 2021). While in the successive editions of the recommended practice of SAE J3016 new terms and definitions have been added, and clarifications have been made to improve the utility of the documents, the classification of driving automation with six levels, from Level 0 to Level 5, remains essentially the same as that in SAE J3016 JAN2014.

5.7.2 Automated Driving Systems and Cooperative Driving Automation

1) **Automated driving systems**

The basic tasks of an automated driving system include the monitoring of the driving environment (i.e., the surroundings in all directions around the vehicle) and the tracking of the position of the vehicle. To accomplish these tasks, the following enabling technologies (sensors) may be considered:

- Lidar (light detection and ranging) may be employed to measure distances from objects (vehicles, human beings, obstacles, etc.) and relative speeds between the vehicle and moving objects using ultraviolet, infrared, or visible light (including laser). It is to provide continuous 360° visibility around the vehicle.
- As an alternative to Lidar, camera vision can be used to monitor the driving environment around the vehicle, because of its lower costs. Mono or stereo camera may also be used to support the detection of objects and hazards around the vehicle (e.g., vehicles or human beings).
- Radar (radio detection and ranging) uses electromagnetic waves in the radio or microwaves domain to detect objects and their speeds. They are usually installed in the front and rear of the vehicle to measure distances from moving vehicles in the front and at the rear, and their

relative speeds. This will provide the needed information for activating automatic emergency braking or for the adaptive cruise control, for example.

- An infrared camera is used to provide night vision for detecting human beings, animals, and other objects around the vehicle.
- Ultrasound sensors are used to measure distances from objects in close range to provide information for automated parking, for instance.
- GPS (global positioning system) is used to track the position of the vehicle.

The data collected from the devices (sensors) noted above are processed by software of the control unit of the automated driving system. While design features vary, most systems create and maintain an internal map of the driving environment around the vehicle.

An automated driving system performs the operational functions of route planning and of providing instructions to actuators, which control vehicle acceleration, braking, and steering. With recorded traffic signs, coded traffic rules, obstacle avoidance algorithms, object discrimination methods (e.g., methods for distinguishing solid objects from inflated plastic bags, for differentiating between a bicycle and a motorcycle, etc.), predictive models, etc., the system will enable the vehicle to recognize traffic signs, to follow traffic rules, to navigate around obstacles, to avoid collisions, and to operate the vehicle safely.

Like the automated driving systems for road vehicles described above, research and development of similar systems for off-road vehicles (such as agricultural vehicles and military vehicles) are also being actively pursued (Mousazadeh 2013; Roshanianfard et al. 2020; U.S. National Research Council 2002).

2) Cooperative driving automation

To integrate vehicles equipped with automated driving systems into an intelligent road transportation system, cooperative driving automation technologies are being developed. These technologies enable the sharing of driving information among individual vehicles equipped with automated driving systems, infrastructure, and other road users (including pedestrians, pedal-cyclists, and others carrying personal mobile communication devices). In other words, the technologies through wireless networks provide connectivity to all traffic participants in the proximity of the host vehicle. The intention is to provide solutions to the challenges facing road transportation, such as reducing congestion, collisions, pollution, and energy consumption. Thus, cooperative driving automation will enhance road user safety, traffic flow, and operational efficiency.

The cooperative driving automation system provides machine-to-machine communication for traffic participants, that primarily includes (Aramrattana et al. 2015):

- Vehicle-to-vehicle communication through dedicated short-range communication systems.
- Vehicle-to-infrastructure communication through either dedicated short-range communication systems or cellular modem for long range communication.

According to the Surface Vehicle Information Report SAE J3216 (Society of Automotive Engineers 2020), issued in May 2020, four classes, from Class A to D, for the cooperative driving automation are identified:

- Class A: Status-sharing – To inform the other traffic participants in the proximity, with respect to the location of the host vehicle and its observations of the surrounding environment, such as pedestrians close to or on a crosswalk.
- Class B: Intent-sharing – To inform the other traffic participants in the proximity, with respect to the intention of the host vehicle, such as to make a turn or to allow a lane change.

- Class C: Agreement-seeking – To seek agreement of the other traffic participants in the proximity, with respect to the intention of the host vehicle to perform certain operations, such as to enable coordinated intersection arrival and departure, coordinated merge, or joining platoon.
- Class D: Prescriptive – To inform the other traffic participants in the proximity, with respect to the host vehicle to accept and adhere to a prescriptive communication from traffic authorities or fleet operators.

Considerable efforts have been and are being made in the development of the cooperative automated driving system technologies (De La Fortelle et al. 2014). Governmental agencies have also developed plans for encouraging stakeholder collaboration to accelerate the development, testing, and deployment of the technologies (Federal Highway Administration 2021).

5.8 Steady-State Handling Characteristics of a Tractor–Semitrailer

An approach like that for the analysis of a two-axle vehicle described in Section 5.2 can be followed to evaluate the steady-state handling characteristics of a tractor–semitrailer with three axles, as shown in Figure 5.26. For the tractor, the equation governing its steady-state handling behavior is similar to Eq. (5.10) and is expressed by

$$
\begin{aligned}
\delta_f &= \frac{L_t}{R} + \left(\frac{W_f}{C_{af}} - \frac{W_r}{C_{ar}}\right)\frac{V^2}{gR} \\
&= \frac{L_t}{R} + K_{us.t}\frac{V^2}{gR}
\end{aligned}
\tag{5.49}
$$

where L_t is the wheelbase of the tractor and $K_{us.t}$ is the understeer coefficient of the tractor.

For most of the tractor–semitrailers, the fifth wheel is located slightly ahead of the center of the tractor rear axle. In the following simplified analysis, the fifth wheel, however, is assumed to be

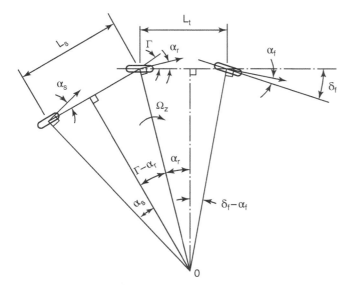

Figure 5.26 Simplified steady-state handling model for a tractor–semitrailer.

located above the center of the tractor rear axle. With this assumption, the tractor rear tire may be considered as the "steered tire" for the semitrailer, and the articulation angle Γ between the tractor and the semitrailer may be expressed by

$$\begin{aligned}\Gamma &= \frac{L_s}{R} + \left(\frac{W_r}{C_{ar}} - \frac{W_s}{C_{as}}\right)\frac{V^2}{gR}\\ &= \frac{L_s}{R} + K_{us.s}\frac{V^2}{gR}\end{aligned}$$

(5.50)

where L_s is the wheelbase of the semitrailer; W_s and C_{as} are the load and cornering stiffness of each of the tires on the semitrailer axle, respectively; and $K_{us.s}$ is the understeer coefficient for the semitrailer.

The ratio of the articulation angle Γ to the steer angle of the tractor front tire δ_f is usually referred to as the articulation angle gain, and is given by

$$\frac{\Gamma}{\delta_f} = \frac{L_s/R + K_{us.s}(V^2/gR)}{L_t/R + K_{us.t}(V^2/gR)}$$

(5.51)

An examination of the above equation reveals that five different types of steady-state handling behavior of a tractor–semitrailer are possible (Ervin and Mallikarjunarao 1982).

1) **Both the tractor and semitrailer understeer**

In this case, both $K_{us.t}$ and $K_{us.s}$ are positive, and the articulation angle gain is finite and positive for all values of forward speed, as shown in Figure 5.27. Consequently, the tractor–semitrailer is directionally stable.

2) **Both the tractor and semitrailer oversteer**

In this case, $K_{us.t}$ is positive, whereas $K_{us.s}$ is negative, and the articulation angle gain remains finite for all values of forward speed, as shown in Figure 5.28. However, when the forward speed V is greater than V_{ct} given below, the articulation angle gain changes from positive to negative:

$$V_{ct} = \sqrt{\frac{gL_s}{-K_{us.s}}}$$

(5.52)

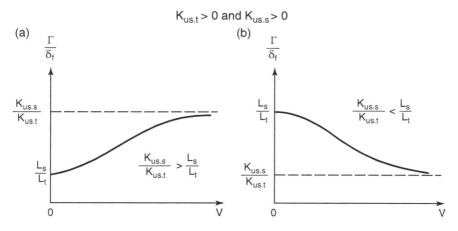

Figure 5.27 Steady-state handling characteristics of a tractor–semitrailer with $K_{us.t} > 0$ and $K_{us.s} > 0$.

Figure 5.28 Steady-state handling characteristics of a tractor–semitrailer with $K_{us.t} > 0$ and $K_{us.s} < 0$.

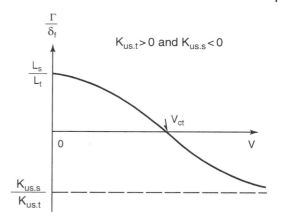

Figure 5.29 Steady-state handling characteristics of a tractor–semitrailer with $K_{us.t} < 0$, and $K_{us.s} > 0$.

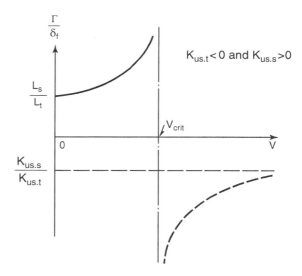

This indicates that when the forward speed V approaches V_{ct}, the articulation angle Γ approaches zero, and when $V > V_{ct}$, the orientation of the semitrailer with respect to the tractor will be opposite that shown in Figure 5.26.

3) The tractor oversteer, while the semitrailer understeer

In this case, $K_{us.t}$ is negative and $K_{us.s}$ is positive. When the forward speed approaches the critical speed V_{crit} given below, the denominator in Eq. (5.51) approaches zero and the articulation angle gain approaches infinity, as shown in Figure 5.29:

$$V_{\text{crit}} = \sqrt{\frac{gL_t}{-K_{us.t}}} \tag{5.53}$$

This indicates that when the forward speed V approaches V_{crit}, the tractor longitudinal axis becomes increasingly oriented toward the center of the turn, resulting in jackknifing.

4) **Both the tractor and semitrailer oversteer, and the ratio of the understeer coefficient of the semitrailer to that of the tractor is less than the ratio of the semitrailer wheelbase to the tractor wheelbase**

In this case, $K_{us.s} < 0$, $K_{us.t} < 0$, and $(K_{us.s}/K_{us.t}) < (L_s/L_t)$, the variation of the articulation angle gain with forward speed is shown in Figure 5.30. Like Case 3), when the forward speed approaches the critical value defined by Eq. (5.53), jackknifing will occur.

5) **Both the tractor and semitrailer oversteer, and the ratio of the understeer coefficient of the semitrailer to that of the tractor is greater than the ratio of the semitrailer wheelbase to the tractor wheelbase**

In this case, $K_{us.s} < 0$, $K_{us.t} < 0$, and $(K_{us.s}/K_{us.t}) > (L_s/L_t)$, the variation of the articulation angle gain with forward speed is shown in Figure 5.31. The articulation angle gain decreases with increasing forward speed. When the forward speed V approaches V_{ct} defined by Eq. (5.52), the gain approaches zero. With a further increase in the forward speed, the gain becomes negative and approaches minus infinity as the forward speed approaches V_{crit}, defined by Eq. (5.53). In this case, the semitrailer longitudinal axis becomes increasingly oriented toward the center of the turn, resulting in trailer swing.

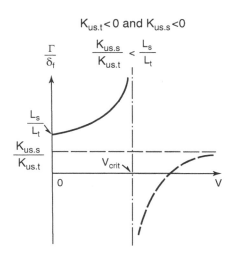

Figure 5.30 Steady-state handling characteristics of a tractor–semitrailer with $K_{us.t} < 0$, $K_{us.s} < 0$, and $K_{us.s}/K_{us.t} < L_s/L_t$.

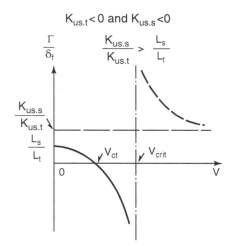

Figure 5.31 Steady-state handling characteristics of a tractor–semitrailer with $K_{us.t} < 0$, $K_{us.s} < 0$, and $K_{us.s}/K_{us.t} > L_s/L_t$.

The results of the above analysis indicate that for any form of directional instability (jackknifing or trailer swing) to occur, the tractor must be oversteer. Jackknifing can occur when the semitrailer is either understeer or oversteer. However, for trailer swing to occur, in addition to the condition that the semitrailer must be oversteer, it is required that the ratio of the understeer coefficient of the semitrailer to that of the tractor be greater than the ratio of the semitrailer wheelbase to the tractor wheelbase. Further analysis of the steering response of articulated vehicles, including truck-trailers, may be found in Ellis (1969, 1994) and El-Gindy and Wong (1985a, b).

5.9 Simulation Models for the Directional Behavior of Articulated Road Vehicles

With the increasing use in road transport of heavy articulated vehicles or road trains, which consist of a tractor unit and one or more semitrailers or full trailers, concerns for their safety in operation have been growing. This has stimulated intensive theoretical and experimental studies of the directional control and stability of this type of vehicle. A number of computer simulation models have been developed. A brief description of the basic features of some of the models is given below (Leucht 1970; Moncarz et al. 1975; Ervin et al. 1979; Fancher et al. 1979; Winkler et al. 1981; Wong and El-Gindy 1985, 1986; El-Gindy and Wong 1987).

5.9.1 The Linear Yaw Plane Model

This model is a linear mathematical model for studying the directional behavior of multiple articulated vehicles. It was developed at the University of Michigan Transportation Research Institute (UMTRI), originally for the purpose of analyzing the directional behavior of double-bottom tankers.

In developing the equations of motion for the model, the roll dynamics of the vehicle are neglected. Furthermore, the vehicle is assumed to travel at a constant forward speed. The degrees of freedom considered in the model are limited to the lateral and yaw motions of the tractor and articulation in the horizontal plane of the other sprung masses of a multiple articulated vehicle.

The following are the major assumptions made in deriving the equations of motion:

- The cornering (lateral) force and aligning moment (torque) generated at the tire–road interface are assumed to be linear functions of the slip angle of the tire.
- Articulation angles made by the various units of the vehicle train are small.
- The motion of the vehicle takes place on a horizontal surface.
- No significant tire forces are present in the longitudinal direction (either tractive or braking).
- Pitch and roll motions of the sprung masses are small, and hence neglected.
- All joints are frictionless, and articulation takes place about the vertical axis.
- Each unit of the articulated vehicle is assumed to be a rigid body, and the unsprung masses are assumed to be rigidly attached to their respective sprung masses.

5.9.2 TBS Model

TBS is a simplified nonlinear mathematical model, originally formulated by Leucht (1970). An interactive computer program based on Leucht's model was developed by Moncarz et al. (1975).

In developing the equations of motion for the model, the basic assumptions made are like those for the linear yaw plane model. However, the following improvements have been introduced:

- A nonlinear tire model is used to represent the cornering force–slip angle relationship of a tire.
- The dynamic load transfers (both longitudinal and lateral) have been taken into account in determining the normal load on each tire.

5.9.3 Yaw/Roll Model

The yaw/roll model was developed at UMTRI for the purpose of predicting the directional and roll responses of articulated vehicles in turning maneuvers which approach the rollover condition.

In the model, the forward speed of the lead unit is assumed to remain constant during the maneuver. Each sprung mass is treated as a rigid body with up to five degrees of freedom (dependent on the constraints at the hitch): lateral, vertical, yaw, roll, and pitch. The axles are treated as beam axles, which are free to roll and bounce with respect to the sprung mass to which they are attached.

The basic assumptions for this model are as follows:

- The relative roll motion between the unsprung and sprung masses takes place about roll centers, which are located at fixed distances beneath the sprung masses.
- Nonlinearities in the force–displacement relationship of a suspension, such as suspension lash, are taken into account.
- The cornering force and aligning moment produced by a given tire is a nonlinear function of the slip angle and vertical load. The influence of wheel camber on lateral force generation is neglected.
- The model permits the analysis of articulated vehicles that are equipped with any of the four coupling mechanisms, namely, conventional fifth wheel, inverted fifth wheel, pintlehook, and kingpin.
- Both closed-loop (defined path input) and open-loop (defined steer angle input) modes of steering input can be accommodated, and the effects of the steering system compliance are taken into account.

The running time of this model is about five times that of the linear yaw plane model, and it requires a large amount of input data.

5.9.4 The Phase 4 Model

This model was originally developed at UMTRI in 1980 for simulating the braking and steering dynamics of trucks, tractor–semitrailers, doubles, and triples. It is a comprehensive computer model for simulating the braking and steering response of commercial vehicles.

The Phase 4 model is a time-domain mathematical simulation of a truck/tractor, a semitrailer, and up to two full trailers. The motions of the vehicles are represented by differential equations derived from Newtonian mechanics that are solved for successive time increments by digital integration.

The mathematical model incorporates up to 71 degrees of freedom. The number of degrees of freedom is dependent on vehicle configuration and is derived from the following:

- Six degrees of freedom (three translational and three rotational) for the truck/tractor sprung mass.

- Three rotational degrees of freedom for the semitrailer (the other three translational degrees of freedom of the semitrailer are effectively eliminated by dynamic constraints at the hitch).
- Five degrees of freedom for each of the two full trailers allowed.
- Two degrees of freedom (bounce and roll) for each of the 13 axles allowed.
- A wheel rotational degree of freedom for each of the 26 wheels allowed.

For the simulation of lateral dynamic behavior, the model incorporates realistic representations of the truck tire cornering force characteristics and vehicle suspension properties of significance to cornering behavior. The program can be operated open-loop or closed-loop, and on roads of specified grade or cross-slope.

5.9.5 Summary

The models presented above vary greatly in capability, in complexity, in the number of degrees of freedom considered, and in the amount of input data required. For instance, the Phase 4 model incorporates up to 71 degrees of freedom and requires up to approximately 2300 lines of input data, dependent on vehicle configuration. On the other hand, the linear yaw plane model only includes the lateral and yaw motions of the tractor and articulation in the horizontal plane of the other sprung masses of the articulated vehicle, and only requires up to 35 lines of input data.

The capabilities and limitations of the simulation models described above have been examined (Wong and El-Gindy 1986; El-Gindy and Wong 1987). The steady-state steering response and the lateral dynamic behavior in a lane-change maneuver of a representative five-axle tractor–semitrailer have been predicted using the four models and compared with available experimental data. Figure 5.32 shows the lateral acceleration responses to steering input of the tractor–semitrailer in steady-state turns at a forward speed of 69 km/h (43 mph) on a dry, smooth asphalt surface predicted using the four models. The measured values are also shown in the figure. Based on the predicted lateral accelerations and yaw rates of the tractor, a handling diagram for the vehicle is drawn, as shown in Figure 5.33. For comparison, the measured data are also shown. The square symbol in

Figure 5.32 Comparisons of steady-state lateral acceleration response to steering input of a tractor–semitrailer predicted by various models.

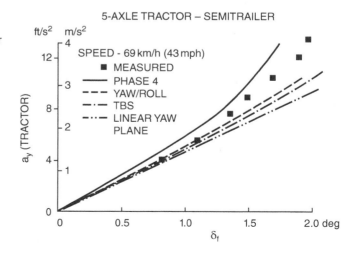

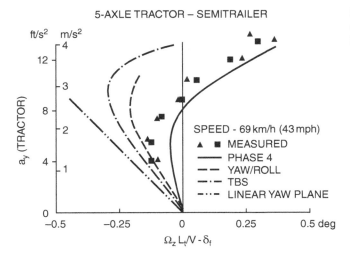

5-AXLE TRACTOR – SEMITRAILER

SPEED - 69 km/h (43 mph)
▲ ■ MEASURED
— PHASE 4
-- YAW/ROLL
-·- TBS
-··- LINEAR YAW PLANE

Figure 5.33 Comparisons of handling characteristics of a tractor–semitrailers predicted by various models.

the figure represents the value calculated from the measured yaw rate of the tractor at a forward speed of 69 km/h (43 mph), whereas the triangular symbol represents the value taken from Ervin et al. (1979).

It can be seen from Figure 5.32 that the lateral accelerations of the tractor predicted using the linear yaw plane model, the TBS model, and the yaw/roll model are reasonably close within the range up to 2° of front wheel steer angle (equivalent to a lateral acceleration of approximately 0.3g). However, there is a significant difference between the lateral accelerations predicted using the Phase 4 model and those predicted using the other three models for front wheel steer angles greater than 1.5° (equivalent to a lateral acceleration of approximately 0.2g). At an average front wheel steer angle $\delta_f = 1.0°$, the differences between the measured lateral acceleration and the predicted ones using the Phase 4 model, the yaw/roll model, the TBS model, and the linear yaw plane model are 17.5, 1.2, 2.3, and 7.7%, respectively. At an average front wheel steer angle of $\delta_f = 1.5°$, the corresponding differences are 10.3, 13.1, 13.8, and 22.3%, respectively. This indicates that the Phase 4 model gives a better prediction of lateral acceleration than the other three models when the lateral acceleration is greater than about 0.2g.

Based on the data shown in Figure 5.32, it appears that for lateral accelerations below 0.2g, the four computer simulation models give similar predictions, and the predicted values agree reasonably well with the measured ones. For lateral accelerations greater than 0.2g, the Phase 4 model gives a better prediction of the trend of yaw divergence than the other three models in comparison with the measured values. However, the Phase 4 model overestimates the response, while the yaw/roll model and the TBS model underestimate it. For lateral accelerations higher than 0.2g, the linear yaw plane model gives the highest error of prediction among the four models. This is primarily because in the linear yaw plane model, a linear tire model is adopted, and the cornering stiffness of the tire obtained at zero slip angle is used in the predictions. Furthermore, the load transfer and its effects on tire characteristics have been entirely neglected.

It should also be mentioned that the lateral accelerations that cause an inside tire to lift off the ground, predicted using the Phase 4 model, the yaw/roll, and the TBS model, are considerably lower than the measured one presented in the report by Ervin et al. (1979).

Figure 5.33 illustrates the steady-state handling characteristics of the vehicle as predicted by the four computer simulation models. The lateral acceleration at which the vehicle changes from understeer to oversteer, referred to as the "transition acceleration," predicted using the Phase 4 model is just under 0.2g. The transition accelerations predicted using the yaw/roll model and the TBS model are approximately 0.25g and 0.3g, respectively, while the measured one is approximately 0.2g. Below the transition acceleration, the Phase 4 model underestimates the understeer level (or understeer coefficient) as compared with the measured data, whereas the TBS model and the linear yaw plane model overestimate the understeer level to varying degrees. Since a linear tire model is used, the linear yaw plane model is unable to predict any variation of the handling behavior of the vehicle with lateral acceleration, and the predicted understeer level remains a constant.

In general, among the four models studied, the Phase 4 model gives the best overall prediction of the variation of handling behavior with lateral acceleration for the five-axle tractor–semitrailer examined, although there is still a noticeable difference between the predicted and measured values, as shown in Figure 5.33.

The lateral acceleration, yaw rate, and articulation angle of the tractor–semitrailer in a lane-change maneuver were predicted using the Phase 4 model, the yaw/roll model, the TBS model, and the linear yaw plane model, and are shown in Figures 5.34–5.38. The simulated results were compared with the measured data presented in the report by Fancher et al. (1979).

The responses of the tractor and semitrailer predicted by the four models generally follow the same trend as that measured. However, there are differences between the predicted peak values and the measured ones. For instance, the differences between the measured peak value of tractor lateral acceleration and those predicted using the Phase 4 model, the yaw/roll model, the TBS model, and the liner yaw plane model are approximately 16, 10, 33, and 46%, respectively, as shown in Figure 5.34. The agreement between the measured peak value of semitrailer lateral acceleration and those predicted is better than that for the tractor lateral acceleration. The differences between the measured peak value of semitrailer lateral acceleration and those predicted using the Phase 4 model, the yaw/roll model, the TBS model, and the linear yaw plane model are approximately 20, 8, 12, and 20%, respectively, as shown in Figure 5.35. The agreement between the measured tractor yaw rate response and those predicted using the four models appears to be reasonable. The differences between the measured peak value of tractor yaw rate and those predicted using the Phase 4

Figure 5.34 Variations of tractor lateral acceleration with time in a lane-change maneuver predicted by various models.

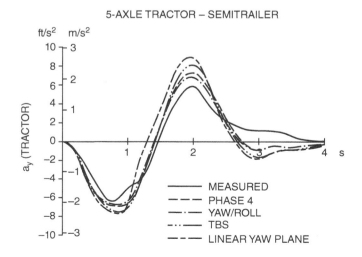

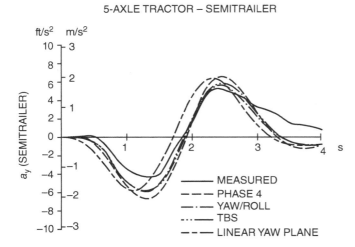

5-AXLE TRACTOR – SEMITRAILER

Figure 5.35 Variations of semitrailer lateral acceleration with time in a lane-change maneuver predicted by various models.

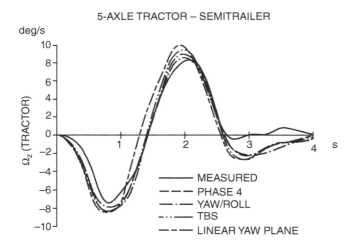

5-AXLE TRACTOR – SEMITRAILER

Figure 5.36 Variations of tractor yaw rate with time in a lane-change maneuver predicted by various models.

model, the yaw/roll model, the TBS model, and the linear yaw plane model are approximately 9.8, 4.9, 15.8, and 21.9%, respectively, as shown in Figure 5.36. The measured semitrailer yaw rate response and those predicted using the four models again show reasonable agreement. The differences between the measured peak value of the semitrailer yaw rate and those predicted using the Phase 4 model, the yaw/roll model, the TBS model, and the linear yaw plane model are approximately 13.3, 0, 3.3, and 13.3%, respectively, as shown in Figure 5.37. The articulation angle responses predicted using the four models are reasonably close. The differences between the peak values of articulation angle predicted using the four models are within 10%. However, there is a noticeable difference between the measured and predicted peak values of articulation angle. For

Figure 5.37 Variations of semitrailer yaw rate with time in a lane-change maneuver predicted by various models.

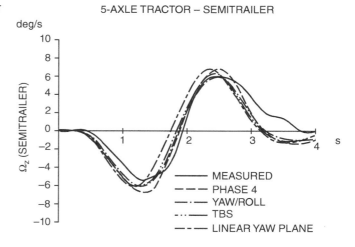

Figure 5.38 Variations of articulation angle with time in a lane-change maneuver predicted by various models.

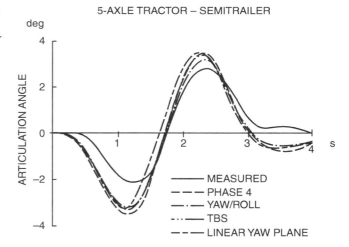

instance, the difference between the measured and the predicted peak value of articulation angle using the Phase 4 model is approximately 22%, as shown in Figure 5.38.

It is noted from the figures that there is a phase shift between the measured and predicted responses, and that there is a significant difference between the measured and predicted responses during the period from three to four seconds.

In summary, it appears that in comparison with the measured data, the steady-state steering responses of a representative tractor–semitrailer predicted using the four simulation models all have varying degrees of discrepancy, and that there are no significant differences in the steady-state steering responses predicted using the four models in the lateral acceleration range up to approximately 0.25g. There are, however, significant differences in the handling characteristics predicted using the four simulation models in most cases, as shown in the handling diagram (Figure 5.33).

Since the linear yaw plane model does not consider the effects of load transfer and uses a linear tire model, it is not capable of predicting changes in handling behavior with lateral acceleration. On the other hand, the Phase 4 model, the yaw/roll model, and the TBS model consider the effects of load transfer and the nonlinear behavior of tires to varying degrees. Consequently, these three models can predict changes in handling behavior with lateral acceleration. However, the predictions made by these three models are still noticeably different from the measured data available. For a lane-change maneuver, the responses of the representative tractor–semitrailer predicted by the four simulation models generally follow the same trend as that measured. However, there are noticeable discrepancies between the measured and predicted values.

A variety of multibody dynamics software packages is now commercially available, such as TruckSim by Mechanical Simulation Corporation, Ann Arbor, MI, USA, for simulating handling behavior of commercial vehicles, including trucks, tractor–semitrailers, and tractor–trailer combinations with multiple trailers.

References

Abe, M. (1999). Vehicle dynamics and control for improving handling and active safety: From four-wheel steering to direct yaw moment control. *Proceedings of the Institution of Mechanical Engineers, Part K: Journal of Multi-body Dynamics* **213** (4).

Aramrattana, M., Larson, T., Jansson, J., and Englund, C. (2015). Dimensions of cooperative driving, ITS and automation. Paper presented at *IEEE Intelligent Vehicle Symposium*, Seoul, Korea.

Bundorf, R.T. (1968). The influence of vehicle design parameters on characteristic speed and understeer. *SAE Transactions* **76**.

Bundorf, R.T., Pollock, D.E., and Hardin, M.C. (1963). Vehicle handling response to aerodynamic inputs. Society of Automotive Engineers paper 716B.

Bunker, K.J. (1968). Theoretical and practical approaches to motor vehicle steering mechanisms. In: *Steering, Suspension and Tyres* (ed. J.G. Giles). London: Butterworths.

De La Fortelle, A., Qian, X., Diemer, S., Gregoire, J., Moutarde, F., Bonnabel, S., Marjovi, A., Martinoli, A., Llatser, I., Festag, A., Katrin, S. (2014). Network of automated vehicles: The Autonet2030 vision. In: *Proc. 21st World Congress on Intelligent Transport Systems*, Detroit, MI. Available at: https://www.scipedia.com/public/De_La_Fortelle_et_al_2014a (accessed 14 September 2021).

Dixon, J.C. (1996). *Tires, Suspension and Handling*, 2e. Society of Automotive Engineers.

El-Gindy, M. and Wong, J.Y. (1985a). Steering response of articulated vehicles in steady-state turns. Society of Automotive Engineers paper 852335.

El-Gindy, M. and Wong, J.Y. (1985b). Steady-state steering response of an articulated vehicle with a multi-axle steering dolly. Society of Automotive Engineers paper 850537.

El-Gindy, M. and Wong, J.Y. (1987). A comparison of various computer simulation models for predicting the directional responses of articulated vehicles. *Vehicle System Dynamics* **16** (5–6).

Ellis, J.R. (1969). *Vehicle Dynamics*. London: Business Books.

Ellis, J.R. (1994). *Vehicle Handling Dynamics*. London: Mechanical Engineering Publications Limited.

Ervin R.D. and Mallikarjunarao, C. (1982). A study of the yaw stability of tractor–semitrailer combinations, *Proc. 7th IAVSD Symposium on Dynamics of Vehicles on Roads and Tracks,* Amsterdam, Netherlands.

Ervin, R.D., Nisonger, R.L., Mallikarjunarao, C., and Gillespie, T.D. (1979). The yaw stability of tractor–semitrailers during cornering. Report DOT HS-805 141, PB80-116775. National Technical Information Service, US Department of Commerce.

Fancher, P. and MacAdam, C. (2014). Intelligent vehicle systems: Longitudinal control. In: *Road and Off-Road Vehicle System Dynamics Handbook* (eds. G. Mastinu and M. Ploechl). Boca Raton, FL: CRC Press.

Fancher, P.S., Mallikarjunarao, C., and Nisonger, R.L. (1979). Simulation of the directional response characteristics of tractor–semitrailer vehicles. Report UM-HSRI-79-9, PB 80-189632. National Technical Information Service, US Department of Commerce.

Federal Highway Administration (2021). CARMA Program Overview. U.S. Department of Transportation. Available at: https://highways.dot.gov/research/operations/CARMA (accessed 21 April 2021).

Fenton, J. (ed.) (1996). *Handbook of Automotive Design Analysis*. London: Butterworths.

Hac, A. and Simpson, M.D. (2000). Estimation of vehicle side slip angle and yaw rate. Society of Automotive Engineers paper 2000-01-0696.

Koibuchi, K., Yamamoto, M., Fukada, Y., and Inagaki, S. (1996). Vehicle stability control in limit cornering by active brake. Society of Automotive Engineers paper 960487.

Leucht, P.M. (1970). The directional dynamics of the commercial tractor–semitrailer vehicle during braking. Society of Automotive Engineers paper 700371.

Milliken, W.F. and Milliken, D.L. (1995). *Race Car Vehicle Dynamics*. Society of Automotive Engineers.

Moncarz, H.T., Bernard, J.E., and Fancher, P.S. (1975). A simplified, interactive simulation for predicting the braking and steering response of commercial vehicles. Report UMHSRI-PF-75-8. Highway Safety Research Institute, University of Michigan, Ann Arbor, MI.

Mousazadeh, H. (2013). A technical review on navigation systems of agricultural autonomous off-road vehicles. *Journal of Terramechanics* **50** (3): 211–232.

National Highway Traffic Safety Administration (2007). FMVSS No. 126, Electronic Stability Control Systems: Final Regulatory Impact Analysis. U.S. Department of Transportation.

National Highway Traffic Safety Administration (2016). Federal automated vehicles policy. U.S. Department of Transportation.

Peng, H. (2014). Intelligent vehicle systems: Automatic lateral vehicle control. In: *Road and Off-Road Vehicle System Dynamics Handbook* (eds. G. Mastinu and M. Ploechl). Boca Raton, FL: CRC Press.

Radt, H.S. and Pacejka, H.B. (1963). Analysis of the steady state turning behavior of an automobile. *Proc. Symposium on the Control of Vehicles During Braking and Cornering*. London: Institution of Mechanical Engineers.

Roshanianfard, A., Noguchi, N., Okamoto, H., and Ishii, K. (2020). A review of autonomous agricultural vehicles (The experience of Hokkaido University). *Journal of Terramechanics* **91**: 155–183.

Segel, L. (1957). Theoretical prediction and experimental substantiation of the response of the automobile to steering control. *Proceedings of the Institution of Mechanical Engineers, Automobile Division*, 1956–1957.

Shibahata, Y., Shimada, K., and Tomari, T. (1993). Improvement of vehicle maneuverability by direct yaw moment control. *Vehicle System Dynamics* **22** (5–6): 465–481.

Society of Automotive Engineers (2014). J3016 JAN2014 (Surface Vehicle Information Report: Taxonomy and definitions for terms related to on-road motor vehicle automated driving systems).

Society of Automotive Engineers (2016). J3016 SEP2016 (Surface Vehicle Recommended Practice: (R) Taxonomy and definitions for terms related to driving automation systems for on-road motor vehicles).

Society of Automotive Engineers (2018). J3016 JUN2018 (Surface Vehicle Recommended Practice: (R) Taxonomy and definitions for terms related to driving automation systems for on-road motor vehicles).

Society of Automotive Engineers (2020). J3216 MAY2020 (Surface Vehicle Information Report: Taxonomy and definitions for terms related to cooperative driving automation for on-road motor vehicles).

Society of Automotive Engineers (2021). J3016 APR2021 (Surface Vehicle Recommended Practice: (R) Taxonomy and definitions for terms related to driving automation systems for on-road motor vehicles).

Steeds, W. (1960). *Mechanics of Road Vehicles*. London: Iliffe & Sons.

Toyota Motor Corporation (1995). Vehicle Stability Control System: Technical brief. *Automotive Engineering*, August issue.

U.S. Department of Transportation (2018). Preparing for the future of transportation: Automated Vehicles 3.0. Washington, DC.

U.S. National Research Council (2002). Technology development for army unmanned ground vehicles. Washington, DC.

van Zanten, A.T., Erhardt, R., and Pfaff, G. (1995). VDC, the vehicle dynamics control system of Bosch. Society of Automotive Engineers paper 950759.

van Zanten, A.T., Erhardt, R., A. Lutz, A., Neuwald, W., and Bartels, H. (1996). Simulation for the development of the Bosch-VDC. Society of Automotive Engineers paper 960486.

Whitcomb, D.W. and Milliken, Jr., W.F. (1957). Design implications of a general theory of automobile stability and control. *Proceedings of the Institution of Mechanical Engineers, Automobile Division*, 1956–1957.

Winkler, C.B., Mallikarjunarao, C., and MacAdam, C.C. (1981). Analytical test plan, Part I: Description of simulation models for parameter analysis of heavy truck dynamic stability. Report of the Transportation Research Institute, University of Michigan, Ann Arbor, MI.

Wong, J.Y. and El-Gindy, M. (1985). Computer simulation of heavy vehicle dynamic behavior: User's guide to the UMTRI models. Technical Report 3. Vehicle Weights and Dimensions Study, Road and Transportation Association of Canada.

Wong, J.Y. and El-Gindy, M. (1986). A comparison of various computer simulation models for predicting the lateral dynamic behavior of articulated vehicles. Technical Report 16. Vehicle Weights and Dimensions Study, Roads and Transportation Association of Canada.

Yasui, Y., Tozu, K., Hattori, N., and Sugisawa, M. (1996). Improvement of vehicle directional stability for transient steering maneuvers using active brake control. Society of Automotive Engineers paper 960485.

Problems

5.1 A passenger car weights 20.02 kN (4500 lb) and has a wheelbase of 279.4 cm (110 in.). The center of gravity is 127 cm (50 in.) behind the front axle. If a pair of radial-ply tires, each of which has a cornering stiffness of 45.88 kN/rad (180 lb/deg), are installed in the front, and a pair of bias-ply tires, each of which has a cornering stiffness of 33.13 kN/rad (130 lb/deg), are installed in the rear, determine whether the vehicle is understeer or oversteer. What would happen to the steady-state handling characteristics of the vehicle, if the front and rear tires are interchanged? Also, determine the critical or characteristic speed of the vehicle as appropriate.

5.2 A sports car weighs 9.919 kN (2230 lb) and has a wheelbase of 2.26 m (7.4 ft). The center of gravity is 1.22 m (4 ft) behind the front axle. The cornering stiffness of each front tire is 58.62 kN/rad (230 lb/deg) and that of each rear tire is 71.36 kN/rad (280 lb/deg). The steering gear ratio is 20:1. Determine the steady-state yaw velocity gain and lateral acceleration gain of the vehicle in the forward speed range 10–160 km/h (6.2–99.4 mph).

5.3 The sports car described in Problem 5.2 has a mass moment of inertia about a vertical axis passing through its center of gravity of 570 kg · m^2 (420 slug · ft^2). If the car is given a step input of steering wheel angle of 30° at a speed of 80.5 km/h (50 mph), determine the rise time for the yaw velocity response.

5.4 The front and rear tires of the passenger car described in Problem 5.1 are replaced by radial-ply tires of the same type, of which the relationship between cornering coefficient and inflation pressure is shown in Figure 1.29. Determine the steady-state handling behavior of the vehicle when the inflation pressure on the front tires is 276 kPa (40 psi) and that on the rear tires is 220 kPa (32 psi).

6

Steering of Tracked Vehicles

Steering of tracked vehicles is different from that of wheeled vehicles discussed in the previous chapter. Tracked vehicle steering can be accomplished by one of the following methods:

1) Skid-steering

In skid-steering, the thrust of one track is increased and that of the other is reduced, so as to create a turning moment to overcome the moment of turning resistance due to skidding of the tracks on the ground and the rotational inertia of the vehicle in yaw, as shown in Figure 6.1. Since the moment of turning resistance is usually considerable, significantly higher thrust on the outside track than that in straight-line motion and braking of the inside track are often required in making a turn. Over weak terrain, the required thrust on the outside track and the braking force on the inside track may exceed the maximum shear forces that can be developed on the track–terrain interface. As a result, it would lead to immobilization of the vehicle in executing a tight turn over soft ground.

2) Articulated steering

For tracked vehicles with two or more separate units (chassis), steering may be accomplished by rotating one unit against the other using a steering joint to make the vehicle follow a prescribed, curved path, as shown in Figure 6.2 (Nuttall 1964). In articulated steering, turning is initiated by activating the steering joint between the two units, and no adjustment of the thrusts of the outside and inside tracks of the vehicle units is required. Thus, articulated steering can provide tracked vehicles with better mobility than skid-steering for turning maneuvers, particularly over soft ground.

3) Curved track steering

Another method for directional control of tracked vehicles is that of curved track steering. To initiate a turn, the laterally flexible track is laid down on the ground in a curve, as shown in Figure 6.3 (Little 1964). This can be achieved using various kinds of mechanical arrangement, one of which is illustrated in Figure 6.3. In this arrangement, each roadwheel of the track can rotate around an axis inclined at a suitable angle from the vertical in a longitudinal plane, so that rotation around the axes displaces the lower part of the wheel to form a curved track. Steering movement of the roadwheels may be activated by a conventional steering wheel through racks and pinions and individual push rods (Little 1964). The main advantage of this steering method is that less power is required in

Theory of Ground Vehicles, Fifth Edition. J.Y. Wong.
© 2022 John Wiley & Sons, Inc. Published 2022 by John Wiley & Sons, Inc.
Companion website: www.wiley.com/go/wong/TGV5e

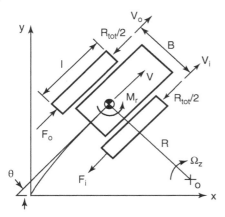

Figure 6.1 Principles of skid-steering. *Source: Theory of Land Locomotion* by M.G. Bekker, copyright © by the University of Michigan, 1956. Reproduced with permission of the University of Michigan Press.

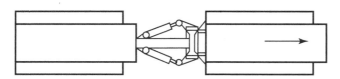

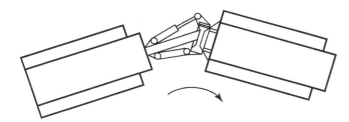

Figure 6.2 Articulated steering. *Source:* Nuttall (1964). Reproduced with permission of the International Society for Terrain–Vehicle Systems.

making a turn as compared with skid-steering. However, owing to the limitations of the lateral flexibility of the track, the minimum turning radius of the vehicle is quite large. To achieve a smaller turning radius, a supplementary steering mechanism, such as skid-steering, must be provided. Thus, not only does the complexity of design of the track system increase, but also the potential advantage of curved track steering in power saving may not be fully realized.

Among the various steering methods for tracked vehicles, skid-steering and articulated steering are commonly used. In this chapter, discussions are, therefore, focused on these two types of steering method.

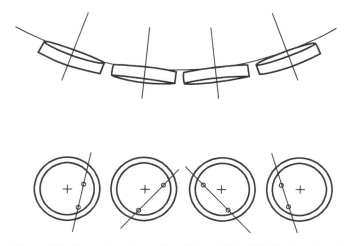

Figure 6.3 Curved track steering. *Source:* Little (1964). Reproduced with permission of the International Society for Terrain–Vehicle Systems.

6.1 Simplified Analysis of the Kinetics of Skid-Steering

The turning behavior of a tracked vehicle using skid-steering depends on the thrusts of the outside and inside tracks F_o and F_i, the resultant resisting force R_{tot}, the moment of turning resistance M_r exerted on the track by the ground, and vehicle parameters as shown in Figure 6.1. The simple case of steering at low speeds on a level ground will be examined first. The case of tracked vehicle steering at high speeds will be discussed later. At low speeds, the centrifugal force may be neglected, and the behavior of the vehicle can be described by the following two equations of motion:

$$m\frac{d^2s}{dt^2} = F_o + F_i - R_{\text{tot}} \tag{6.1}$$

$$I_z\frac{d^2\theta}{dt^2} = \frac{B}{2}(F_o - F_i) - M_r \tag{6.2}$$

where s is the displacement of the center of gravity of the vehicle; θ is the angular displacement of the vehicle; B is the tread of the vehicle (i.e., the distance between the centerlines of the two tracks); and I_z and m are the mass moment of inertia of the vehicle about the vertical axis passing through its center of gravity and the mass of the vehicle, respectively. With known initial conditions, the above two differential equations can be integrated, and the trajectory of the center of gravity and the orientation of the vehicle can be determined as discussed by Bekker (1956).

Under steady-state conditions, there are no linear and angular accelerations:

$$F_o + F_i - R_{\text{tot}} = 0 \tag{6.3}$$

$$\frac{B}{2}(F_o - F_i) - M_r = 0 \tag{6.4}$$

The thrusts of the outside and inside tracks required to execute a steady-state turn are therefore expressed by

$$F_o = \frac{R_{\text{tot}}}{2} + \frac{M_r}{B} = \frac{f_r W}{2} + \frac{M_r}{B} \tag{6.5}$$

$$F_i = \frac{R_{\text{tot}}}{2} - \frac{M_r}{B} = \frac{f_r W}{2} - \frac{M_r}{B} \tag{6.6}$$

where f_r is the coefficient of motion resistance of the vehicle in the longitudinal direction and W is the vehicle weight.

To determine the values of the thrusts F_o and F_i, the moment of turning resistance M_r must be known. This can be determined experimentally or analytically. If the normal pressure is uniformly distributed along the track, the lateral resistance per unit length of the track R_l can be expressed by

$$R_l = \frac{\mu_t W}{2l} \tag{6.7}$$

where μ_t is the coefficient of lateral resistance and l is the contact length of each of the two tracks.

The value of μ_t depends not only on the terrain, but also on the design of the track. Over soft terrain, the vehicle sinks into the ground, and the tracks together with grousers will be sliding on the surface, as well as displacing the soil laterally during turning maneuvers, commonly referred to as the bulldozing effects. The lateral forces acting on the tracks and the grousers due to displacing the soil laterally form part of the lateral resistance. It has been shown that under certain circumstances, the lateral resistance of a track may also depend on the skid of the track in the lateral direction and the turning radius (Kar 1987). Table 6.1 shows the average values of μ_t for steel and rubber tracks over various types of ground (Hayashi 1975).

If the coefficient of lateral resistance μ_t is a constant, the resultant moment of the lateral resistance about the centers of the two tracks M_r (i.e., moment of turning resistance) can be expressed by (Figure 6.4)

$$M_r = 4 \frac{W \mu_t}{2l} \int_0^{l/2} x \, dx = \frac{\mu_t W l}{4} \tag{6.8}$$

Accordingly, Eqs. (6.5) and (6.6) can be rewritten in the following form:

$$F_o = \frac{f_r W}{2} + \frac{\mu_t W l}{4B} \tag{6.9}$$

$$F_i = \frac{f_r W}{2} - \frac{\mu_t W l}{4B} \tag{6.10}$$

Table 6.1 Values of lateral resistance coefficient of tracks over various surfaces.

	Coefficient of lateral resistance μ_t		
Track material	Concrete	Hard ground (not paved)	Grass
Steel	0.50–0.51	0.55–0.58	0.87–1.11
Rubber	0.90–0.91	0.65–0.66	0.67–1.14

Source: Hayashi (1975).

Figure 6.4 Moment of turning resistance of a track with uniform pressure distribution.

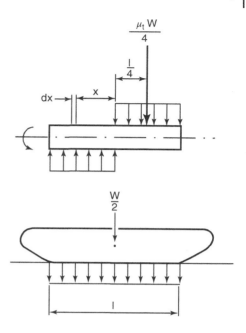

It should be emphasized that the value of M_r as calculated by Eq. (6.8) is for a vehicle with a uniform normal pressure distribution, turning at low speeds on level ground. Methods for predicting the moment of turning resistance of a track with a trapezoidal or triangular shape of normal pressure distribution or with normal loads concentrated under the roadwheels have been proposed or developed (Crosheck 1975; Kitano and Kuma 1978).

Equations (6.9) and (6.10) are of fundamental importance, and they lead to conclusions of practical significance regarding the steerability of a tracked vehicle. As discussed in Chapter 2, the maximum thrust of a track is limited by terrain properties and vehicle parameters. For the outside track,

$$F_o \leq cbl + \frac{W \tan \phi}{2} \tag{6.11}$$

where b is the track width and c and ϕ are the cohesion and angle of internal shearing resistance of the terrain, respectively.

Substituting Eq. (6.9) into Eq. (6.11),

$$\frac{f_r W}{2} + \frac{\mu_t W l}{4B} \leq cbl + \frac{W \tan \phi}{2}$$

and

$$\frac{l}{B} \leq \frac{1}{\mu_t} \left(\frac{4cA}{W} + 2 \tan \phi - 2f_r \right)$$

where A is the contact area of one track.

This indicates that to enable a tracked vehicle to steer without spinning the outside track, the ratio of track length to tread of the vehicle, l/B, must satisfy the following condition:

$$\frac{l}{B} \leq \frac{2}{\mu_t} \left(\frac{c}{p} + \tan \phi - f_r \right) \tag{6.12}$$

where p is the average normal pressure on the track, which is equal to $W/2A$.

On a sandy terrain with $c = 0$, $\phi = 30°$, $\mu_t = 0.5$, and $f_r = 0.1$, the value of l/B should be less than 1.9. In other words, if the ratio of contact length to tread of a tracked vehicle is greater than 1.9, the vehicle will not be able to steer on the terrain specified. On a clayey terrain, with $c = 3.45$ kPa (0.5 psi), $\phi = 10°$, $p = 6.9$ kPa (1 psi), $\mu_t = 0.4$, and $f_r = 0.1$, the value of l/B must be less than 2.88. These examples show the importance of the ratio of contact length to tread of a tracked vehicle to its steerability.

From Eq. (6.10), it can also be seen that if $\mu_t l/2B > f_r$, the thrust of the inside track F_i will be negative. This implies that to achieve a steady-state turn, braking of the inside track is required. For instance, with $\mu_t = 0.5$, $f_r = 0.1$, and $l/B = 1.5$, the value of $\mu_t l/2B$ will be greater than that of f_r, which indicates that a braking force has to be applied to the inside track. Since the forward thrust of the outside track of a given vehicle is limited by terrain properties as shown in Eq. (6.11), the application of a braking force to the inside track during a turn reduces the maximum resultant forward thrust of the vehicle, and thus the mobility of the vehicle over weak terrain will be adversely affected. Figure 6.5 shows the ratio of the maximum resultant forward thrust as limited by the track–terrain interaction during a turn to that in a straight-line motion as a function of the coefficient of lateral resistance μ_t for a tracked vehicle with $l/B = 1.5$ operating over a terrain with $c = 0$, $\phi = 30°$, and $f_r = 0.1$. As the value of μ_t increases from 0.2 to 0.5, the maximum resultant forward thrust available during a steady-state turn decreases from approximately 74% to 35% of that when traveling in a straight line.

On hard grounds, the resultant of the longitudinal and lateral forces acting on a track during a turn may be assumed to obey the law of Coulomb friction. The resultant shear force on the track–ground interface is limited by the coefficient of friction and the normal load on the track, and acts in the opposite direction to the relative motion of the track with respect to the ground. Based on these assumptions, the steering characteristics of tracked vehicles have been analyzed in detail by Steeds (1950). However, predictions based on Steeds's analysis deviate significantly from field observations. A general theory for skid-steering on firm ground has, therefore, been developed, which is discussed in detail in Section 6.4. The new theory offers significant improvement on the prediction of steering behavior of tracked vehicles over Steeds's method.

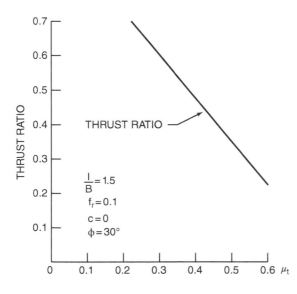

Figure 6.5 Effects of lateral resistance coefficient on the maximum thrust available during a turn.

Example 6.1   A tracked vehicle weighs 155.68 kN (35 000 lb) and has a tread of 203.2 cm (80 in.). Each of the two tracks has a contact length of 304.8 cm (120 in.) and a width of 76.2 cm (30 in.). The contact pressure is assumed to be uniform. The vehicle travels over a terrain with a cohesion $c = 3.45$ kPa (0.5 psi) and an angle of internal shearing resistance $\phi = 25°$. Over this terrain, the coefficient of motion resistance f_r is 0.15 and the average coefficient of lateral resistance μ_t is 0.5.

(a) Determine the steerability of the vehicle over the terrain specified if the skid-steering method is employed.
(b) Determine the required thrusts of the outside and inside tracks during a steady-state turn.

Solution

(a) From Eq. (6.12), the limiting value for the ratio of track length to tread is

$$\frac{l}{B} = \frac{2}{\mu_t}\left(\frac{c}{p} + \tan\phi - f_r\right) = 1.67$$

Since the ratio of track length to tread of the vehicle l/B is 1.5, which is less than the limiting value of 1.67, the vehicle is steerable over the terrain specified.

(b) The thrusts of the outside and inside tracks required during a steady-state turn can be determined using Eqs. (6.9) and (6.10):

$$F_o = \frac{f_r W}{2} + \frac{\mu_t W l}{4B} = 40.87 \text{ kN (9188 lb)}$$
$$F_i = \frac{f_r W}{2} - \frac{\mu_t W l}{4B} = -17.52 \text{ kN } (-3938 \text{ lb)}$$

The results indicate that a braking force must be applied to the inside track during the turn.

6.2 Kinematics of Skid-Steering

Figure 6.1 shows a tracked vehicle turning about a center O. If the sprocket of the outside track is rotating at an angular speed of ω_o and that of the inside track is rotating at an angular speed of ω_i and the tracks do not slip (or skid), the turning radius R and the yaw velocity of the vehicle Ω_z can be expressed by

$$R = \frac{B}{2}\frac{(r\omega_o + r\omega_i)}{(r\omega_o - r\omega_i)} = \frac{B(K_s + 1)}{2(K_s - 1)} \tag{6.13}$$

$$\Omega_z = \frac{r\omega_o + r\omega_i}{2R} = \frac{r\omega_i(K_s - 1)}{B} \tag{6.14}$$

where r is the radius of the sprocket and K_s is the angular speed ratio ω_o/ω_i.

However, during a turning maneuver, an appropriate thrust or braking force must be applied to the track, as described previously. Consequently, the track will either slip or skid, depending on whether a forward thrust or a braking force is applied. The outside track always develops a forward thrust, and therefore it slips. On the other hand, the inside track may develop a forward thrust or a braking force, depending on the magnitude of the turning resistance moment M_r and other factors as defined by Eq. (6.6). When the slip (or skid) of the track is taken into consideration, the turning radius R' and yaw velocity Ω_z' are given by.

$$
\begin{aligned}
R' &= \frac{B[r\omega_o(1-i_o) + r\omega_i(1-i_i)]}{2[r\omega_o(1-i_o) - r\omega_i(1-i_i)]} \\
&= \frac{B(K_s(1-i_o) + (1-i_i))}{2[K_s(1-i_o) - (1-i_i)]}
\end{aligned}
\tag{6.15}
$$

$$
\begin{aligned}
\Omega_z' &= \frac{r\omega_o(1-i_o) + r\omega_i(1-i_i)}{2R'} \\
&= \frac{r\omega_i[K_s(1-i_o) - (1-i_i)]}{B}
\end{aligned}
\tag{6.16}
$$

where i_o and i_i are the slip of the outside track and that of the inside track, respectively. For a given vehicle over a particular terrain, the values of i_o and i_i depend on the thrusts F_o and F_i. The relationship between thrust and slip (or skid) can be determined using the methods described in Chapter 2. When a braking force is applied to the inside track, the track skids. Equations (6.15) and (6.16) still hold; however, i_i will have a negative value.

To illustrate the effect of track slip on the steering characteristics of a tracked vehicle, the ratio of the turning radius with track slipping R' to that without slipping R is plotted against the speed ratio $K_s = \omega_o/\omega_i$ in Figure 6.6. Curve 1 shows the relationship between R'/R and K_s when the outside track slips at 20% and the inside track is disconnected from the transmission by declutching. Curve 2 shows the variation of the value of R'/R with K_s when the outside track slips and the inside track skids. This occurs when the outside track develops a forward thrust and a braking force is applied to the inside track.

It is shown that the value of R'/R is always greater than unity. Thus, the effect of track slip (or skid) is to increase the turning radius for a given speed ratio K_s.

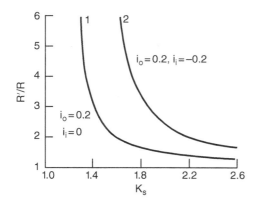

Figure 6.6 Effects of track slip on turning radius.

6.3 Skid-Steering at High Speeds

In the above analysis of the mechanics of skid-steering, low-speed operation is assumed and the effect of the centrifugal force is neglected. When a tracked vehicle is turning at moderate and higher speeds, or with a relatively small turning radius, the centrifugal force may be significant, and its effect should be taken into consideration.

Consider that a tracked vehicle is in a steady-state turn on level ground. To achieve equilibrium in the lateral direction, the resultant lateral force exerted on the track by the ground must be equal to the centrifugal force, as shown in Figure 6.7. Assume that the normal pressure distribution along the track is uniform and that the coefficient of lateral resistance μ_t is a constant; then to satisfy the equilibrium condition in the lateral direction, the center of turn must lie at a distance s_0 in front of the transverse centerline of the track–ground contact area AC, as shown in Figure 6.7. The distance s_0 can be determined by the following equation (Bekker 1956):

$$\left(\frac{l}{2} + s_0\right)\frac{\mu_t W}{l} - \left(\frac{l}{2} - s_0\right)\frac{\mu_t W}{l} = \frac{WV^2}{gR'}\cos\beta$$

$$s_0 = \frac{lV^2}{2\mu_t gR'}\cos\beta = \frac{la_y}{2\mu_t g}\cos\beta$$

(6.17)

where a_y is the lateral acceleration of the center of gravity of the vehicle.

Since the turning radius R' is usually large compared with the contact length of the track l, β would be small, and accordingly $\cos\beta$ may be assumed to be equal to 1. Equation (6.17) can be rewritten as follows:

$$s_0 = \frac{la_y}{2\mu_t g}$$

(6.18)

Figure 6.7 Forces acting on a tracked vehicle during a turn at high speeds. *Source: Theory of Land Locomotion* by M.G. Bekker, copyright © 1956 by the University of Michigan. Reproduced with permission of the University of Michigan Press.

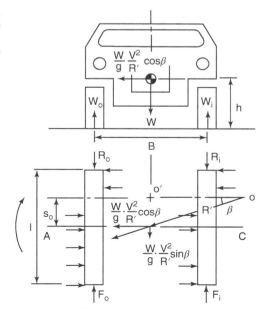

As a consequence of the shifting of the center of turn, the equivalent moment of turning resistance M_r will have two components: one is the moment of the lateral resistance exerted on the tracks by the ground about $0'$; the other is the moment of the centrifugal force about $0'$:

$$M_r = \frac{\mu_t W}{l} \left[\int_0^{l/2 + s_0} x \, dx + \int_0^{-(l/2 - s_0)} x \, dx \right] - \frac{WV^2 s_0}{gR'}$$

$$= \frac{\mu_t W}{2l} \left(\frac{l^2}{2} + 2s_0^2 \right) - \frac{WV^2 s_0}{gR'} \tag{6.19}$$

Substituting Eq. (6.18) into Eq. (6.19), the equivalent moment of turning resistance M_r becomes

$$M_r = \frac{\mu_t Wl}{4} \left(1 - \frac{V^4}{g^2 R'^2 \mu_t^2} \right)$$

$$= \frac{\mu_t Wl}{4} \left(1 - \frac{a_y^2}{g^2 \mu_t^2} \right) \tag{6.20}$$

The above equation indicates that when the centrifugal force is taken into consideration, the equivalent moment of turning resistance is reduced.

The centrifugal force also causes lateral load transfer. Thus, the longitudinal motion resistances of the outside and inside track R_o and R_i will not be identical:

$$R_o = \left(\frac{W}{2} + \frac{hWV^2}{BgR'} \right) f_r \tag{6.21}$$

$$R_i = \left(\frac{W}{2} - \frac{hWV^2}{BgR'} \right) f_r \tag{6.22}$$

where h is the height of the center of gravity of the vehicle.

Furthermore, the centrifugal force has a component along the longitudinal axis of the vehicle, $WV^2 s_0/gR'^2$. This component must be balanced by the thrusts developed by the tracks. Therefore, when the centrifugal force is considered, the thrusts required to maintain the vehicle in a steady-state turn are expressed by (Bekker 1956)

$$F_o = \left(\frac{W}{2} + \frac{h\,W\,V^2}{B\,g\,R'} \right) f_r + \frac{W\,V^2 s_0}{2g\,R'^2} + \frac{\mu_t\,W\,l}{4B} \left[1 - \left(\frac{V^2}{g\,R'\mu_t} \right)^2 \right]$$

$$= \left(\frac{W}{2} + \frac{h\,W a_y}{Bg} \right) f_r + \frac{W\,a_y s_0}{2g\,R'} + \frac{\mu_t W\,l}{4B} \left[1 - \left(\frac{a_y}{g\,\mu_t} \right)^2 \right] \tag{6.23}$$

$$F_i = \left(\frac{W}{2} - \frac{h\,W\,V^2}{B\,g\,R'} \right) f_r + \frac{W\,V^2 s_0}{2g\,R'^2} - \frac{\mu_t\,W\,l}{4B} \left[1 - \left(\frac{V^2}{gR'\mu_t} \right)^2 \right]$$

$$= \left(\frac{W}{2} - \frac{h\,W\,a_y}{Bg} \right) f_r + \frac{W\,a_y s_0}{2g\,R'} - \frac{\mu_t W\,l}{4B} \left[1 - \left(\frac{a_y}{g\mu_t} \right)^2 \right] \tag{6.24}$$

Figure 6.8 illustrates the required ratios of the thrust to vehicle weight for the outside and inside tracks, F_o/W and F_i/W, as a function of lateral acceleration a_y/g in g units, for a given vehicle on a particular terrain. As the lateral acceleration increases, the ratio of the thrust to vehicle weight for the outside track F_o/W decreases. This is because the moment of the centrifugal force about the

Figure 6.8 Thrusts on the outside and inside tracks required during a turn as a function of lateral acceleration.

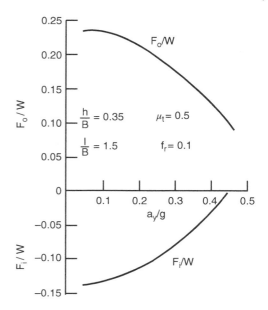

center of turn increases with an increase of lateral acceleration. Consequently, the equivalent moment of turning resistance decreases with an increase of lateral acceleration. Also note that the ratio of the thrust to vehicle weight for the inside track F_i/W is usually negative, which implies that braking of the inside track is required to maintain a steady-state turn. The magnitude of the braking force of the inside track decreases, however, as the lateral acceleration increases. This is again mainly due to the decrease of the equivalent moment of turning resistance with the increase of lateral acceleration.

Equations (6.23) and (6.24) specify the thrusts of the outside and inside tracks required under a steady-state turn for a given vehicle speed and turning radius. However, to achieve a specific turning radius and vehicle speed, certain kinematic relationships must be satisfied. These include the relationship among turning radius, vehicle speed, track slips, and sprocket speeds. To determine the required sprocket speeds for a specific turning radius and vehicle speed, the slips of the outside and inside tracks i_o and i_i should be determined. To do this, the required thrusts F_o and F_i should first be calculated using Eqs. (6.23) and (6.24). Then from the relationship between thrust and slip discussed in Chapter 2, the values of i_o and i_i can be obtained. The required angular speed ratio K_s for a given turning radius R' can be determined from Eq. (6.15):

$$K_s = \frac{(2R' + B)(1 - i_i)}{(2R' - B)(1 - i_o)} \tag{6.25}$$

The angular speeds of the sprockets ω_o and ω_i required to achieve a specific vehicle forward speed V can then be obtained from Eq. (6.16):

$$\omega_i = \frac{2V}{r[K_s(1 - i_o) + (1 - i_i)]} \quad \text{and} \quad \omega_o = K_s\omega_i \tag{6.26}$$

When the effect of the centrifugal force is taken into consideration, the analysis of the turning maneuver of a tracked vehicle becomes more involved.

6.4   A General Theory for Skid-Steering on Firm Ground

The mechanics of skid-steering of tracked vehicles on firm ground where track sinkage is negligible has been studied by Steeds, as noted previously (Steeds 1950). In his analysis, the shear stress developed on the track–ground interface is assumed to obey the Coulomb law of friction. The friction may be either isotropic or anisotropic. In the latter case, different values of coefficient of friction are assigned to the longitudinal and lateral directions of the track. The Coulomb law of friction implies that the resultant shear stress on a track element acts in a direction opposite to that of the relative motion between the track element and the ground. It also assumes that the shear stress reaches its maximum value instantly, as soon as a small relative movement between the track and the ground is initiated. Experimental evidence shows that the shear stress developed on the track–ground interface is dependent on the shear displacement, as described in Section 2.4.3. It indicates that the shear stress will reach its maximum value only after a certain shear displacement has taken place, as shown in Figure 6.9.

Field tests show that considerable discrepancy exists between the steering behavior of tracked vehicles on paved road predicted using Steeds's method and that measured. A general theory for the mechanics of skid-steering on firm ground has, therefore, been developed (Wong and Chiang 2001). It is based on the following assumptions:

- The ground is firm. Consequently, track sinkage and associated bulldozing effects of the track in the lateral direction during a turning maneuver are neglected.
- The shear stress developed at a given point on the track–ground interface during a turn is dependent on the shear displacement at that point, measured from its initial contact with the ground. For rubber belt tracks or steel link tracks with rubber pads, the shear stress is that developed between the rubber and the ground.
- The direction of the shear stress at a point on the track–ground interface is opposite to that of the sliding velocity between the track and the ground at that point.
- The component of the shear stress along the longitudinal direction of the track constitutes the tractive or braking effort, while the lateral component forms the lateral resistance of the track. The moment of the lateral resistance about the turn center of the track constitutes the moment of turning resistance.

An outline of the general theory is given below.

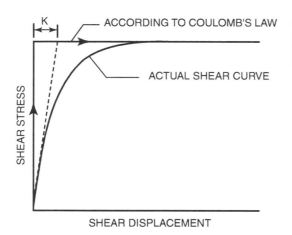

Figure 6.9 Comparisons between the shear stress and shear displacement relationship measured in the field and that based on Coulomb's law of friction.

6.4.1 Shear Displacement on the Track–Ground Interface

As noted previously, the shear stress developed on a track element is related to its shear displacement measured from its initial contact point with the ground. Consequently, it is essential to analyze the development of shear displacement of a track element during a turning maneuver.

Consider that a tracked vehicle with track width b is in a steady-state turn about O, as shown in Figure 6.10(a). Let O_1 be the origin of a frame of reference x_1y_1 fixed to and moving with the vehicle hull and located on the longitudinal centerline of the outside track at a distance s_0 from the center of gravity (CG) of the vehicle. As noted in Section 6.3, s_0 is determined from the dynamic equilibrium of the vehicle in the lateral direction during a turn. As the vehicle hull is rotating about turning center O with an angular speed (yaw velocity) Ω_z, the absolute velocity $V_{o_1y_1}$ of O_1 in the y_1 direction can be expressed by (Referring to Figure 6.10 (a) and (b))

$$V_{o1y1} = \left(R'' + \frac{B}{2} + c_x \right) \Omega_z \qquad (6.27)$$

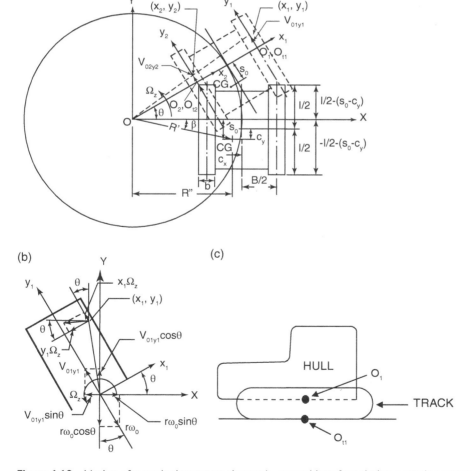

Figure 6.10 Motion of a track element on the track–ground interface during a turning maneuver.

where R'' is the lateral distance between the center of turn O and center of gravity of the vehicle and is equal to $R'\cos\beta\left(\text{or }\sqrt{R'^2 - s_0^2}\right)$, as shown in Figure 6.10(a); R' is the turning radius of the vehicle; c_x is the lateral distance between the vehicle CG and longitudinal centerline of vehicle hull (or the lateral offset of the vehicle CG with respect to the geometric center of the vehicle); and B is the tread, which is the distance between the centerlines of the outside and inside tracks.

A point O_{t1} (Figure 6.10(c)) on the outside track in contact with the ground coincident with O_1 on the plane view has a relative velocity V_{t1/O_1} with respect to O_1, which is expressed by

$$V_{t1/O_1} = r\omega_o \tag{6.28}$$

where r and ω_o are pitch radius and angular speed of the outside track sprocket, respectively.

As a result, the sliding velocity V_{t1j} of point O_{t1} on the ground along the longitudinal direction of the outside track is expressed by

$$V_{t1j} = V_{o1y1} - r\omega_o \tag{6.29}$$

Consider an arbitrary point defined by (x_1, y_1) on the outside track in contact with the ground, as shown in Figures 6.10(a) and (b). Since the track is rotating with the vehicle about the turn center O at an angular speed Ω_z, the relative velocity components of point (x_1, y_1) with respect to O_{t1} in the longitudinal and lateral directions of the track are given by $x_1\Omega_z$ and $y_1\Omega_z$, respectively.

Based on the above analysis, the sliding velocity V_{jo} (Figure 6.11) of the point defined by (x_1, y_1) on the outside track in contact with ground, with respect to the frame of reference XY fixed to the earth (hereinafter referred to as the fixed frame of reference), can be expressed by (Figures 6.10(b) and 6.11)

X component of sliding velocity V_{jo}:

$$\begin{aligned} V_{jXo} &= -V_{o1y1}\sin\theta + r\omega_o\sin\theta - x_1\Omega_z\sin\theta - y_1\Omega_z\cos\theta \\ &= -\left[\left(R'' + \frac{B}{2} + c_x + x_1\right)\Omega_z - r\omega_o\right]\sin\theta - y_1\Omega_z\cos\theta \end{aligned} \tag{6.30}$$

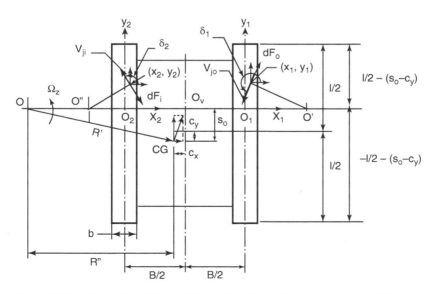

Figure 6.11 Kinematics of the outside and inside tracks during a steady-state turn.

Y component of sliding velocity V_{jo}:

$$V_{jYo} = V_{o1y1} \cos\theta - r\omega_o \cos\theta + x_1\Omega_z \cos\theta - y_1\Omega_z \sin\theta$$
$$= \left[\left(R'' + \frac{B}{2} + c_x + x_1 \right)\Omega_z - r\omega_o \right] \cos\theta - y_1\Omega_z \sin\theta \tag{6.31}$$

O' and O'' shown in Figure 6.11 are the instantaneous centers for the parts of the outside and inside tracks in contact with and sliding on the ground, as discussed by Steeds (1950).

The angle θ shown in Figure 6.10(a) and (b) is the angular displacement of the vehicle and can be determined by the integration of yaw velocity Ω_z with respect to time t that it takes for the point (x_1, y_1) to travel from the initial point of contact at the front of the track (at $y_1 = l/2 + c_y - s_0$), that is

$$\theta = \int_0^t \Omega_z dt = \Omega_z t \tag{6.32}$$

and

$$t = \int_0^t dt = \int_{y_1}^{l/2 + c_y - s_0} \frac{dy_1}{r\omega_o} = \frac{l/2 + c_y - s_0 - y_1}{r\omega_o} \tag{6.33}$$

where c_y is longitudinal distance between the CG and lateral centerline of the vehicle hull, or the longitudinal offset of the center of gravity with respect to the geometric center of the vehicle (see Figure 6.10(a)).

As a result, the shear displacement j_{Xo} at the point (x_1, y_1) on the outside track in contact with the ground along the X direction, with respect to the fixed frame of reference XY, can be determined by

$$\begin{aligned}
j_{Xo} &= \int_0^t V_{jXo}\, dt = \int_{y_1}^{l/2 + c_y - s_0} \left\{ -\left[\left(R'' + \frac{B}{2} + c_x + x_1 \right)\Omega_z - r\omega_o \right] \sin\theta \right. \\
&\quad \left. - y_1\Omega_z \cos\theta \right\} \frac{dy_1}{r\omega_o} \\
&= \left(R'' + \frac{B}{2} + c_x + x_1 \right)\left\{ \cos\left[\frac{(l/2 + c_y - s_0 - y_1)\Omega_z}{r\omega_o} \right] - 1 \right\} \\
&\quad - y_1 \sin\left[\frac{(l/2 + c_y - s_0 - y_1)\Omega_z}{r\omega_o} \right]
\end{aligned} \tag{6.34}$$

and shear displacement j_{Yo} along the Y direction is given by

$$\begin{aligned}
j_{Yo} &= \int_0^t V_{jYo}\, dt = \int_{y_1}^{l/2 + c_y - s_0} \left\{ \left[\left(R'' + \frac{B}{2} + c_x + x_1 \right)\Omega_z - r\omega_o \right] \cos\theta \right. \\
&\quad \left. - y_1\Omega_z \sin\theta \right\} \frac{dy_1}{r\omega_o} \\
&= \left(R'' + \frac{B}{2} + c_x + x_1 \right) \sin\left[\frac{(l/2 + c_y - s_0 - y_1)\Omega_z}{r\omega_o} \right] \\
&\quad - \left(\frac{l}{2} + c_y - s_0 \right) \\
&\quad + y_1 \cos\left[\frac{(l/2 + c_y - s_0 - y_1)\Omega_z}{r\omega_o} \right]
\end{aligned} \tag{6.35}$$

The resultant shear displacement j_o of the point at (x_1, y_1) on the outside track in contact with the ground is given by

$$j_o = \sqrt{j_{Xo}^2 + j_{Yo}^2} \tag{6.36}$$

Similarly, let O_2 (Figures 6.10(a) and 6.11) be the origin for a frame of reference x_2y_2 fixed to and moving with the vehicle hull and located on the longitudinal centerline of the inside track. The absolute velocity V_{o2y2} of O_2 in the y_2 direction can be expressed by

$$V_{o2y2} = \left(R'' - \frac{B}{2} + c_x \right) \Omega_z \tag{6.37}$$

Following a similar approach, the sliding velocity V_{ji} (Figure 6.11) of a point (x_2, y_2) on the inside track in contact with the ground with respect to the fixed frame of reference XY can be expressed by X component of the sliding velocity V_{ji}:

$$
\begin{aligned}
V_{jXi} &= -V_{o2y2} \sin\theta + r\omega_i \sin\theta - x_2\Omega_z \sin\theta - y_2\Omega_z \cos\theta \\
&= -\left[\left(R'' - \frac{B}{2} + c_x + x_2 \right)\Omega_z - r\omega_i \right]\sin\theta - y_2\Omega_z \cos\theta
\end{aligned} \tag{6.38}
$$

Y component of the sliding velocity V_{ji}:

$$
\begin{aligned}
V_{jYi} &= V_{o2y2} \cos\theta - r\omega_i \cos\theta + x_2\Omega_z \cos\theta - y_2\Omega_z \sin\theta \\
&= \left[\left(R'' - \frac{B}{2} + c_x + x_2 \right)\Omega_z - r\omega_i \right]\cos\theta - y_2\Omega_z \sin\theta
\end{aligned} \tag{6.39}
$$

where ω_i is the angular speed of the inside track sprocket.

The contact time t elapsed for the point defined by (x_2, y_2) will be equal to $(l/2 + c_y - s_0 - y_2)/r\omega_i$. Therefore, the shear displacement j_{Xi} on the inside track of the vehicle along the X direction with respect to the fixed frame of reference XY can be determined by

$$
\begin{aligned}
j_{Xi} &= \int_0^t V_{jXi}dt = \int_{y2}^{l/2 + c_y - s_0} \left\{ -\left[\left(R'' - \frac{B}{2} + c_x + x_2 \right)\Omega_z - r\omega_i \right]\sin\theta \right. \\
&\quad \left. - y_2\Omega_z \cos\theta \right\} \frac{dy_2}{r\omega_i} \\
&= \left(R'' - \frac{B}{2} + c_x + x_2 \right)\left\{ \cos\left[\frac{(l/2 + c_y - s_0 - y_2)\Omega_z}{r\omega_i} \right] - 1 \right\} \\
&\quad - y_2 \sin\left[\frac{(l/2 + c_y - s_0 - y_2)\Omega_z}{r\omega_i} \right]
\end{aligned} \tag{6.40}
$$

and shear displacement j_{Yi} along the Y direction is given by

$$
\begin{aligned}
j_{Yi} &= \int_0^t V_{jYi} \, dt = \int_{y2}^{l/2 + c_y - s_0} \left\{ \left[\left(R'' - \frac{B}{2} + c_x + x_2 \right)\Omega_z - r\omega_i \right]\cos\theta \right. \\
&\quad \left. - y_2\Omega_z \sin\theta \right\} \frac{dy_2}{r\omega_i} \\
&= \left(R'' - \frac{B}{2} + c_x + x_2 \right) \sin\left[\frac{(l/2 + c_y - s_0 - y_2)\Omega_z}{r\omega_i} \right] \\
&\quad - \left(\frac{l}{2} + c_y - s_0 \right) \\
&\quad + y_2 \sin\left[\frac{(l/2 + c_y - s_0 - y_2)\Omega_z}{r\omega_i} \right]
\end{aligned} \tag{6.41}
$$

The resultant shear displacement j_i of the point at (x_2, y_2) on the inside track in contact with the ground is given by

$$j_i = \sqrt{j_{Xi}^2 + j_{Yi}^2} \tag{6.42}$$

6.4.2 Kinetics in a Steady-State Turning Maneuver

As noted above, the shear stress developed at a given point on the track–ground interface is dependent on the shear displacement at that point. For different types of terrain, the relationships between shear stress and shear displacement take different forms, as discussed in Section 2.4.3. The sliding velocity between the track and the ground may also affect these relationships under certain circumstances. For instance, if the shear stress–shear displacement relationship is described by Eq. (2.58) and the effects of sliding velocity and adhesion on the track–ground interface are negligible, then it may be expressed by

$$\tau = \sigma \tan \phi \left(1 - e^{-j/K}\right) = \sigma \mu \left(1 - e^{-j/K}\right) \tag{6.43}$$

where σ is the normal pressure; μ is the coefficient of friction between the track and the ground; j is the shear displacement; and K is the shear deformation parameter.

Therefore, the shear force developed on an element dA of the track in contact with the ground can be expressed by (Figure 6.11)
On the outside track:

$$dF_o = \tau_o dA = \sigma_o \mu \left(1 - e^{-j_o/K}\right) dA \tag{6.44}$$

On the inside track:

$$dF_i = \tau_i dA = \sigma_i \mu \left(1 - e^{-j_i/K}\right) dA \tag{6.45}$$

where τ_o and τ_i are the shear stresses; σ_o and σ_i are normal pressures; and j_o and j_i are shear displacements of the elements of the outside and inside tracks, respectively.

The assumption used in Steeds's model that the shear stress (force) between the track and the ground obeys Coulomb's law of friction is a special case of Eq. (6.43). Coulomb's law assumes that full frictional stress (force) is reached as soon as a small relative motion between the track and the ground takes place. This is equivalent to the value of K in Eq. (6.43) being equal to zero. This indicates that Eq. (6.43) is a more general representation of the characteristics of the shear stress (force) between the track and the ground.

As shown in Figure 6.12, the longitudinal forces F_{yo} and F_{yi} acting on the outside and inside tracks can be expressed, respectively, by

$$
\begin{aligned}
F_{yo} &= \int dF_o \sin \left(\pi + \delta_1\right) \\
&= -\int_{-l/2 + c_y - s_0}^{l/2 + c_y - s_0} \int_{-b/2}^{b/2} \sigma_o \mu \left(1 - e^{-j_o/K}\right) \sin \delta_1 \, dx_1 \, dy_1
\end{aligned} \tag{6.46}
$$

$$
\begin{aligned}
F_{yi} &= \int dF_i \sin \left(\pi + \delta_2\right) \\
&= -\int_{-l/2 + c_y - s_0}^{l/2 + c_y - s_0} \int_{-b/2}^{b/2} \sigma_i \mu \left(1 - e^{-j_i/K}\right) \sin \delta_2 \, dx_2 \, dy_2
\end{aligned} \tag{6.47}
$$

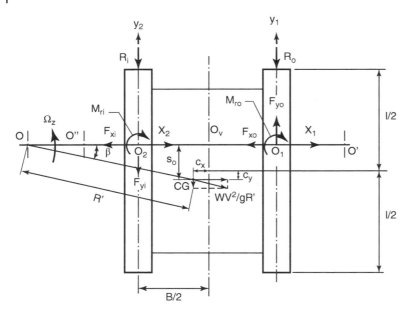

Figure 6.12 Forces and moments acting on a tracked vehicle during a steady-state turn.

where δ_1 and δ_2 shown in Figure 6.11 are the angles between the resultant sliding velocities of the points on the outside and inside tracks and the lateral directions of the tracks (i.e., x_1 and x_2 axes), respectively. Following Coulomb's law of friction, the shear force acting on the track will be in the opposite direction of the resultant sliding velocity.

The lateral forces F_{xo} and F_{xi} acting on the outside and inside tracks are given, respectively, by

$$
\begin{aligned}
F_{xo} &= \int dF_o \cos{(\pi + \delta_1)} \\
&= -\int_{-l/2 + c_y - s_0}^{l/2 + c_y - s_0} \int_{-b/2}^{b/2} \sigma_o \mu \left(1 - e^{-j_o/K}\right) \cos \delta_1 \, dx_1 \, dy_1
\end{aligned}
\tag{6.48}
$$

$$
\begin{aligned}
F_{xi} &= \int dF_i \cos{(\pi + \delta_2)} \\
&= -\int_{-l/2 + c_y - s_0}^{l/2 + c_y - s_0} \int_{-b/2}^{b/2} \sigma_i \mu \left(1 - e^{-j_i/K}\right) \cos \delta_2 \, dx_2 \, dy_2
\end{aligned}
\tag{6.49}
$$

The turning moments M_{Lo} and M_{Li} due to the longitudinal shear forces acting on the outside and inside tracks with respect to O_v (Figure 6.12) can be expressed, respectively, by

$$
M_{Lo} = -\int_{-l/2 + c_y - s_0}^{l/2 + c_y - s_0} \int_{-b/2}^{b/2} \left(\frac{B}{2} + x_1\right) \sigma_o \mu \left(1 - e^{-j_o/K}\right) \sin \delta_1 \, dx_1 \, dy_1
\tag{6.50}
$$

$$
M_{Li} = -\int_{-l/2 + c_y - s_0}^{l/2 + c_y - s_0} \int_{-b/2}^{b/2} \left(\frac{B}{2} - x_2\right) \sigma_i \mu \left(1 - e^{-j_i/K}\right) \sin \delta_2 \, dx_2 \, dy_2
\tag{6.51}
$$

The moments of turning resistance M_{ro} and M_{ri} due to the lateral shear forces acting on the outside and inside tracks with respect to O_1 and O_2 (Figure 6.12) can be expressed, respectively, by

$$M_{ro} = \int dF_o \cos(\pi + \delta_1) y_1$$

$$= -\int_{-l/2 + c_y - s_0}^{l/2 + c_y - s_0} \int_{-b/2}^{b/2} y_1 \sigma_o \mu \left(1 - e^{-j_o/K}\right) \cos \delta_1 \, dx_1 \, dy_1 \tag{6.52}$$

$$M_{ri} = \int dF_i \cos(\pi + \delta_2) y_2$$

$$= -\int_{-l/2 + c_y - s_0}^{l/2 + c_y - s_0} \int_{-b/2}^{b/2} y_2 \sigma_i \mu \left(1 - e^{-j_i/K}\right) \cos \delta_2 \, dx_2 \, dy_2 \tag{6.53}$$

To determine δ_1 and δ_2, the longitudinal sliding velocities V_{jyo} and V_{jyi} of the elements on the outside and inside tracks with respect to the moving frames of reference $x_1 y_1$ and $x_2 y_2$, respectively, are first calculated as follows (see Figure 6.11):

$$V_{jyo} = \left(R'' + \frac{B}{2} + c_x + x_1\right)\Omega_z - r\omega_o \tag{6.54}$$

$$V_{jyi} = \left(R'' - \frac{B}{2} + c_x + x_2\right)\Omega_z - r\omega_i \tag{6.55}$$

The lateral sliding velocities V_{jxo} and V_{jxi} of the track elements on the outside and inside tracks can be expressed, respectively, by

$$V_{jxo} = -y_1 \Omega_z \tag{6.56}$$

$$V_{jxi} = -y_2 \Omega_z \tag{6.57}$$

Therefore, the sine and cosine of angles δ_1 and δ_2 can be defined, respectively, by the following equations:

$$\sin \delta_1 = \frac{V_{jyo}}{\sqrt{V_{jxo}^2 + V_{jyo}^2}}$$

$$= \frac{(R'' + B/2 + c_x + x_1)\Omega_z - r\omega_o}{\sqrt{[(R'' + B/2 + c_x + x_1)\Omega_z - r\omega_o]^2 + (y_1 \Omega_z)^2}} \tag{6.58}$$

$$\sin \delta_2 = \frac{V_{jyi}}{\sqrt{V_{jxi}^2 + V_{jyi}^2}}$$

$$= \frac{(R'' - B/2 + c_x + x_2)\Omega_z - r\omega_i}{\sqrt{[(R'' - B/2 + c_x + x_2)\Omega_z - r\omega_i]^2 + (y_2 \Omega_z)^2}} \tag{6.59}$$

$$\cos \delta_1 = \frac{V_{jxo}}{\sqrt{V_{jxo}^2 + V_{jyo}^2}}$$

$$= \frac{-y_1 \Omega_z}{\sqrt{[(R'' + B/2 + c_x + x_1)\Omega_z - r\omega_o]^2 + (y_1 \Omega_z)^2}} \tag{6.60}$$

$$\cos \delta_2 = \frac{V_{jxi}}{\sqrt{V_{jxi}^2 + V_{jyi}^2}}$$

$$= \frac{-y_2 \Omega_z}{\sqrt{[(R'' - B/2 + c_x + x_2)\Omega_z - r\omega_i]^2 + (y_2 \Omega_z)^2}} \tag{6.61}$$

From the above analysis, one can derive the equilibrium equations for the tracked vehicle during a steady-state turn as follows (Figure 6.12):

$$\sum F_x = 0 \qquad F_{xo} + F_{xi} = \frac{WV^2}{gR'}\cos\beta \tag{6.62}$$

$$\sum F_y = 0 \qquad F_{yo} + F_{yi} = \frac{WV^2}{gR'}\sin\beta + (R_o + R_i) \tag{6.63}$$

$$\sum M_{ov} = 0 \qquad M_{Lo} - M_{Li} - \frac{B}{2}(R_o - R_i)$$
$$+ (s_0\cos\beta + c_x\sin\beta)\frac{WV^2}{gR'} = M_{ro} + M_{ri} \tag{6.64}$$

where R_o and R_i are external motion resistance on the outside and inside tracks, respectively.

The forces and moments are functions of the theoretical speeds $r\omega_o$, $r\omega_i$, and the offset s_0. With the other parameters known or given, such as the coefficient of friction μ, shear deformation parameter K, coefficient of motion resistance f_r, tread B, track–ground contact length l, track width b, forward speed V, turning radius R', weight W, longitudinal offset c_y and lateral offset c_x of the center of gravity with respect to geometric center of the vehicle, and CG height h, the three unknown parameters, $r\omega_o$, $r\omega_i$, and s_0, can be determined by solving the three simultaneous equations, Eqs. (6.62)–(6.64). Thus, all the forces and moments during a given steady-state turn can be completely defined.

Various normal pressure distributions under the tracks can be accommodated in the general theory. They include normal loads on the tracks concentrated under the roadwheels, normal loads supported only by the track links immediately under the roadwheels, normal loads distributed continuously over the entire track contact length, and so on.

6.4.3 Experimental Substantiation

To demonstrate the application of the general theory described above, the steering behavior of a military tracked vehicle, known as the Jaguar (Foss 1993), was simulated and the results were compared with available measured data obtained on paved road and reported in the paper by Ehlert et al. (1992).

Using the vehicle design parameters given in Foss (1993) and ground parameters $\mu = 0.9$ and $K = 0.075$ m (3 in.) deduced from Ehlert et al. (1992), the three simultaneous equations, Eqs. (6.62), (6.63), and (6.64) governing the steady-state turning behavior of the tracked vehicle are solved. The circumferential speed of the sprocket on the outside track $r\omega_o$ and that on the inside track $r\omega_i$, and the offset s_0 defining the longitudinal location of the turn center are obtained. The sprocket torques and moments of turning resistance as functions of turning radius can then be determined. Figure 6.13 shows the variations of the predicted sprocket torques for the outside and inside tracks with theoretical turning radius (derived from the circumferential speeds $r\omega_o$ and $r\omega_i$ of the outside and inside tracks, respectively) at various vehicle forward speeds with different types of normal load distribution. In the figure, the measured data for the tracked vehicle, the Jaguar on paved road, reported in the paper by Ehlert et al. (1992) are also shown. For comparison, the predictions obtained using Steeds's theory are presented as well.

Predictions obtained using the general theory bear a strong resemblance to the measured data. Both show that the magnitudes of the sprocket torques of the outside and inside tracks generally decrease with the increase of theoretical turning radius. In contrast, predictions from Steeds's theory differ greatly from the measured data. The sprocket torques predicted by Steeds's method remain essentially constant over a wide range of turning radii, particularly at low speeds.

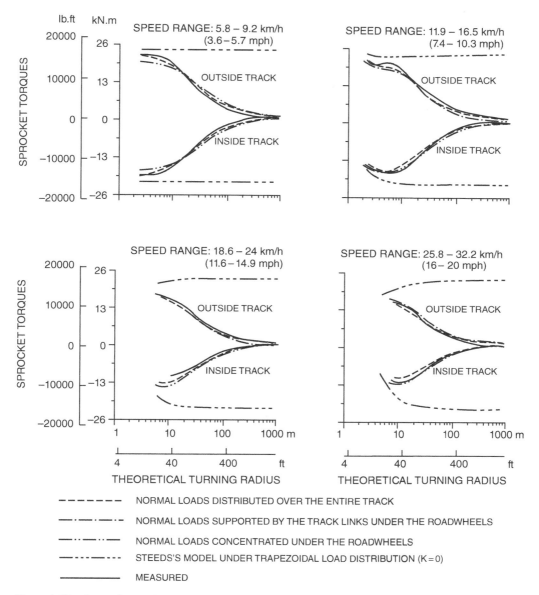

Figure 6.13 Comparisons of the measured and predicted relationships between sprocket torques and theoretical turning radii for a tracked vehicle at various forward speeds using the general theory.

The reason for the poor predictions using Steeds's theory is primarily due to the assumption that the shear stress on the track–ground interface reaches its maximum value instantly, as soon as a small relative movement between the track and the ground takes place. Figures 6.14(a) and (b) show the lateral shear stress distributions along the longitudinal centerline of the outside and inside tracks, respectively, at various turning radii. Even at the front contact point where a track element just meets the ground, the lateral stress is at its maximum. In contrast, based on the general theory, the lateral shear stress at the front contact point of the outside and inside tracks is zero, as shown in

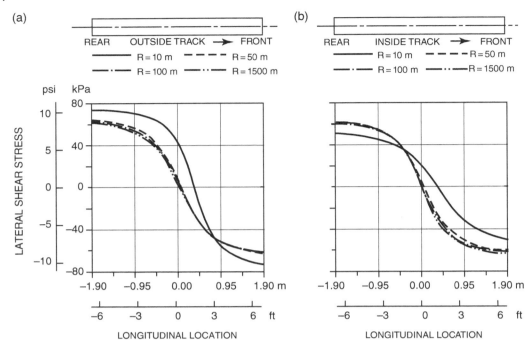

Figure 6.14 Lateral shear stress distributions along the longitudinal centerline of (a) the outside track and (b) the inside track with different turning radii according to Steeds's theory.

Figures 6.15(a) and (b), respectively. The lateral shear stress distributions on the outside and inside tracks predicted using Steeds's theory remain essentially the same when the turning radius is larger than a certain value, such as 50 m (164 ft) shown in Figure 6.14. Since the moment of the lateral shear force about the turn center of the track (i.e., O_1 or O_2 in Figure 6.12) constitutes the moment of turning resistance, this leads to essentially constant moments of turning resistance for the outside and inside tracks over a wide range of turning radii, based on Steeds's theory.

During a steady-state turn, the sprocket torques at the outside and inside tracks are to generate the required tractive and braking efforts, which form a turning moment for the tracked vehicle to overcome primarily the resultant moment of turning resistance caused by the lateral sliding of the tracks on the ground. Consequently, as shown in Figure 6.13, based on Steeds's theory, the sprocket torques for the outside and inside tracks remain essentially constant when the turning radius is larger than a certain value. In contrast, based on the general theory, the magnitude of the lateral shear stress decreases with the increase of turning radius, as shown in Figure 6.15. As a result, the magnitudes of the sprocket torque for the outside and inside tracks decrease with the increase of turning radius, as shown in Figure 6.13. This shows that the general theory provides much more realistic and accurate predictions of steering behavior of tracked vehicles than Steeds's theory.

It is also interesting to note from Figure 6.13 that the normal load distribution on the track does not have a significant effect on sprocket torques. The sprocket torques predicted under three different types of normal load distribution, namely, normal loads concentrated under the roadwheels, normal loads supported only by the track links immediately under the roadwheels, and normal loads distributed continuously over the entire track contact length with trapezoidal pressure distribution, are very close.

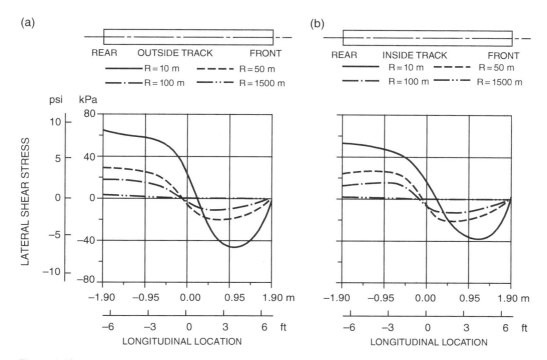

Figure 6.15 Lateral shear stress distributions along the longitudinal centerline of (a) the outside track and (b) the inside track with different turning radii according to the general theory.

6.4.4 Coefficient of Lateral Resistance

In the simplified analysis of skid-steering given in Section 6.1, a coefficient of lateral resistance μ_t (also known as the coefficient of lateral friction) is introduced in predicting the lateral force and moment of turning resistance acting on a track during a turning maneuver. Consequently, if normal pressure is assumed to be uniformly distributed along the track, the moment of turning resistance M_r due to track sliding laterally on the ground is given by Eq. (6.8).

Equation (6.8) indicates that the turning resistance moment M_r is not related to turning radius. However, experimental evidence shows that in practice M_r varies with turning radius. As an expedient, it has been suggested that μ_t in Eq. (6.8) be expressed as a function of turning radius. A number of empirical equations have been proposed to correlate μ_t with turning radius (Ehlert et al. 1992). As these equations contain a number of empirical coefficients, it is uncertain whether these empirical relations can be generally applied.

As described in Section 6.4.2, using the general theory, the moments of turning resistance for the outside and inside tracks can be predicted analytically using Eqs. (6.52) and (6.53), respectively. In predicting the moments of turning resistance using these equations, the only ground parameters required are the coefficient of friction μ and the shear deformation parameter K. This means that using the general theory, it is not necessary to introduce a coefficient of lateral resistance μ_t in predicting the moment of turning resistance. In fact, the equivalent coefficient of lateral resistance μ_t in Eq. (6.8) can be derived quantitatively by equating the sum of the moments of turning resistance M_{ro} and M_{ri} calculated from Eqs. (6.52) and (6.53) to M_r in Eq. (6.8). The variations of the equivalent coefficient of lateral resistance μ_t derived from this method with theoretical turning radius at various vehicle forward speeds for the tracked vehicle, the Jaguar, are shown in Figure 6.16.

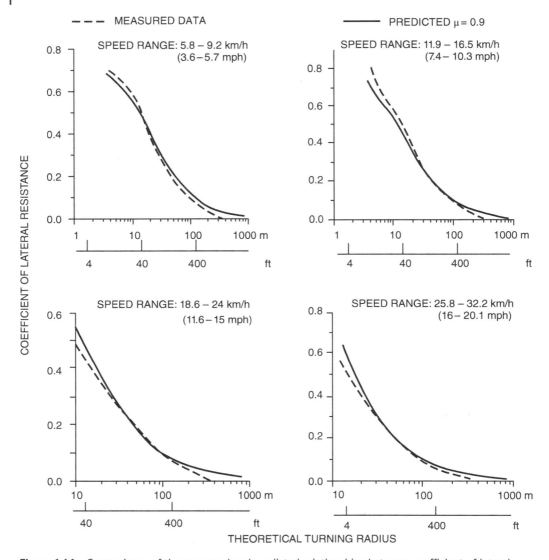

Figure 6.16 Comparisons of the measured and predicted relationships between coefficient of lateral resistance and turning radius for a tracked vehicle at various forward speeds (μ = 0.9 and K = 0.075 m).

The corresponding variations of the equivalent coefficient of lateral resistance derived from the experimental data reported in the paper by Ehlert et al. (1992) are also shown in the figure. They are obtained using Eq. (6.64), with measured data on sprocket torques and external track motion resistance and calculated centrifugal force based on vehicle forward speed and turning radius. It is shown that there is a reasonably close agreement between the measured and predicted results obtained using the general theory.

The above analysis demonstrates that if the simplified method (i.e., Eq. (6.8)) is used to predict the moment of turning resistance, the relationship between μ_t and turning radius can be derived from the general theory for a given ground condition, with known coefficient of friction μ and shear

Figure 6.17 Variations of coefficient of lateral resistance with turning radius predicted using the general theory for a tracked vehicle on grounds with different values of coefficient of friction.

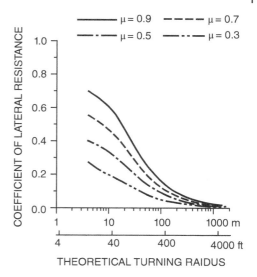

deformation parameter K. This means that by applying the general theory, the time-consuming field experiments for defining the relation between μ_t and turning radius can be eliminated.

As an example, Figure 6.17 shows the variations of the equivalent coefficient of lateral resistance μ_t with theoretical turning radius for the tracked vehicle, the Jaguar, at a speed of 7.5 km/h (4.7 mph) under different ground conditions, with the value of coefficient of friction μ varying from 0.3 to 0.9 and shear deformation parameter $K = 0.075$m (3 in). The equivalent coefficient of lateral resistance decreases with the increase of theoretical turning radius. This trend is consistent with experimental observations reported in the paper by Ehlert et al. (1992).

In summary, the general theory presented above has provided a unified approach to the study of the mechanics of skid-steering of tracked vehicles on firm ground. Following a similar approach, elucidations of the mechanics of skid-steering of tracked vehicles on deformable surfaces, based on the principles of terramechanics outlined in Chapter 2, may be accomplished.

6.5 Power Consumption of Skid-Steering

When a tracked vehicle is traveling in a straight line, the power consumption P_{st} due to the motion resistance R_{tot} is

$$P_{st} = R_{tot}V_{st} = f_r W V_{st} \tag{6.65}$$

where V_{st} is the vehicle speed in straight-line motion.

Power loss due to slip of the vehicle running gear may also be significant over unprepared terrain. In the following analysis of power consumption during a turning maneuver, the power loss due to slip is, however, neglected to simplify the analysis.

When a tracked vehicle is making a steady-state turn, power is consumed by the motion resistance R_{tot}, the moment of turning resistance M_r, and the braking torque M_b in the steering system (usually applied to the inside track). The power required during a turn P_t can be expressed by (Lvov 1960)

$$P_t = R_{tot}V + M_r\Omega_z + M_b\omega_b \tag{6.66}$$

where V is the speed of the center of gravity of the vehicle during a turn; Ω_z is the turning (rotating) speed of the vehicle about the vertical axis through the vehicle center of gravity; and ω_b is the relative angular speed of the frictional elements in the brake. When the brake is fully applied and there is no relative motion between the frictional elements, the power loss in the brake will be zero.

The ratio of the power consumption during a steady-state turn to that in straight-line motion can be expressed by

$$
\begin{aligned}
\frac{P_t}{P_{st}} &= \frac{V}{V_{st}} + \frac{M_r\Omega_z}{f_rWV_{st}} + \frac{M_b\omega_b}{f_rWV_{st}} \\
&= \frac{V}{V_{st}}\left(1 + \frac{M_r}{f_rWR} + \frac{M_b\omega_b}{f_rWV}\right)
\end{aligned}
\tag{6.67}
$$

For a given tracked vehicle on a particular terrain, the power ratio P_t/P_{st} depends on the ratios of V/V_{st}, M_r/f_rWR, and $M_b\omega_b/f_rWV$, which in turn are dependent, to a great extent, on the characteristics of the steering system used. The characteristics and the corresponding power ratio P_t/P_{st} of some typical steering systems for tracked vehicles are discussed in the next section.

6.6 Skid Steering Systems for Tracked Vehicles

Various types of steering mechanism are available for tracked vehicles using the principles of skid-steering.

6.6.1 Clutch/Brake Steering System

This system is shown schematically in Figure 6.18. To initiate a turn, the inside track is disconnected from the driveline by declutching and the brake is usually applied. The outside track is driven by the engine and generates a forward thrust. The thrust on the outside track and the braking force on the inside track form a turning moment that steers the vehicle. This steering system is very simple, but the steering brake usually absorbs considerable power during a turn. The clutch/brake steering system is therefore mainly used in low-speed tracked vehicles such as farm tractors and construction vehicles.

Under certain circumstances, the clutch/brake steering system may cause the so-called reversed steering (i.e., the controls are set to initiate an intended right-hand turn but the vehicle turns to the left or vice versa). For instance, if the vehicle is descending a slope with the throttle closed, the disengaging of the steering clutch on one side will free the associated track, while the retarding torque from the engine applies on the other track (if, at the instant of steering, the initial vehicle speed is faster than that corresponding to the engine speed with the throttle closed). If the

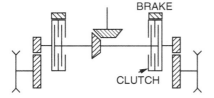

Figure 6.18 Schematic view of a clutch–brake steering system.

coefficient of lateral resistance is low, this will cause the vehicle to make a skid turn in the direction opposite that intended. Reversed steering can be eliminated by arranging the disengagement of the steering clutch and the application of the brake to overlap so that the disengaging of the clutch will be immediately followed by a braking action on the sprocket.

Consider a turning maneuver in which the inside track of the vehicle is disconnected from the driveline by declutching and the brake is fully applied. The inside track thus has zero forward speed. The vehicle will be turning about the center of the inside track, and the minimum turning radius R_{min} will be equal to $B/2$. Assume that, during the turn, the engine is running at the same speed as that prior to turning. It is obvious that with the clutch/brake steering system, the forward speed of the center of gravity of the vehicle at the minimum turning radius will be half of that prior to turning, and $V/V_{st} = 0.5$. Since the brake of the inside track is fully applied, there will be no power loss in the brake. The power ratio P_t/P_{st} for a tracked vehicle with a clutch/brake steering system at the minimum turning radius is therefore given by

$$\frac{P_t}{P_{st}} = 0.5\left(1 + \frac{M_r}{f_r WB/2}\right) \tag{6.68}$$

Assume that the normal pressure under the track is uniformly distributed and the vehicle is turning at low speeds. The moment of turning resistance M_r is given by Eq. (6.8), and Eq. (6.68) can be rewritten as follows:

$$\frac{P_t}{P_{st}} = 0.5\left(1 + \frac{\mu_t l}{2f_r B}\right) \tag{6.69}$$

For a tracked vehicle with a clutch/brake steering system, having $l/B = 1.5$ and operating over a terrain with a coefficient of lateral resistance $\mu_t = 0.5$ and a coefficient of motion resistance $f_r = 0.1$, the power consumption during a steady-state turn at the minimum turning radius will be 2.375 times that when the vehicle is traveling in a straight line. This indicates that considerably more power is required during a turning maneuver as compared to that in a straight-line motion. If the power loss due to track slip is included, the total power consumption during a turn will be even higher.

6.6.2 Controlled Differential Steering System

This type of steering system is shown schematically in Figure 6.19. Gear A is driven through a gearbox by the engine. In straight-line motion, brakes B_1 and B_2 are not applied, and gears C_1, C_2, D_1, and D_2 form an ordinary differential. For steering, the brake of the inside track, such as B_2, is

Figure 6.19 Schematic view of a controlled differential steering system.

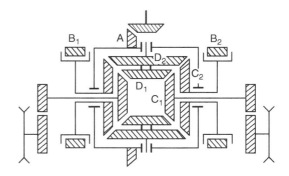

applied. This results in a reduction of the speed of the inside track and a corresponding increase of the speed of the outside track. Thus, the forward speed of the center of gravity of the vehicle during a turn will be the same as that in a straight-line motion for a given engine speed. A kinematic analysis of the controlled differential will show that the relationship between the angular speed of the sprocket of the outside track ω_o to that of the inside track ω_i can be expressed by

$$K_s = \frac{\omega_o}{\omega_i} = \frac{K_{di} + 1 - K_{di}\omega_{B2}}{K_{di} - 1 + K_{di}\omega_{B2}} \tag{6.70}$$

where K_{di} is the gear ratio of the differential and is equal to $N_{D2}N_{C1}/N_{D1}N_{C2}$, where N_{C1}, N_{C2}, N_{D1}, and N_{D2} are the number of teeth of the gears C_1, C_2, D_1, and D_2 in the differential, respectively; and ω_{B2} is the angular speed of the brake drum B_2. If brake B_2 is fully applied and the drum does not slip, Eq. (6.70) can be rewritten as follows:

$$K_s = \frac{K_{di} + 1}{K_{di} - 1} \tag{6.71}$$

When brake B_2 is fully applied, the minimum turning radius is achieved. From Eq. (6.13), the minimum turning radius $R_{\min}$ of a tracked vehicle with a controlled differential steering system is therefore expressed by

$$R_{min} = \frac{B}{2}\left(\frac{K_s + 1}{K_s - 1}\right) = \frac{BK_{di}}{2} \tag{6.72}$$

The power ratio P_t/P_{st} for a tracked vehicle with a controlled differential steering system at the minimum turning radius is given by

$$\frac{P_t}{P_{st}} = \frac{V}{V_{st}}\left(1 + \frac{M_r}{f_r W R_{min}}\right) = 1 + \frac{M_r}{f_r W B K_{di}/2} \tag{6.73}$$

where $V/V_{st} = 1$, as mentioned previously.

If the normal pressure under the track is uniformly distributed, the moment of turning resistance is expressed by Eq. (6.8), and if the vehicle is turning at low speeds, Eq. (6.73) can be rewritten as

$$\frac{P_t}{P_{st}} = 1 + \frac{\mu_t l}{2 f_r B K_{di}} \tag{6.74}$$

For a vehicle with $l/B = 1.5$ and $K_{di} = 2.0$, and operating over a terrain with $\mu_t/f_r = 5$, the power consumption during a steady-state turn at the minimum turning radius will be 2.875 times that when the vehicle is traveling in a straight line.

6.6.3 Planetary Gear Steering System

One of the simplest forms of planetary gear steering system for tracked vehicles is shown schematically in Figure 6.20. The input from the engine is through bevel gearing to shaft A, which is connected, through the planetary gear train, to the sprockets of the tracks.

In the system shown, the input to the gear train is through sun gear B and the output is through carrier C, which is connected to the sprocket. In a straight-line motion, both clutches are engaged, and the brakes are released. For steering, the clutch on the inside track is disengaged and the brake is applied to ring gear D. If the brake is fully applied to hold the ring gear fixed, the angular speed of the sprocket of the inside track is determined by

Figure 6.20 Schematic view of a planetary gear steering system.

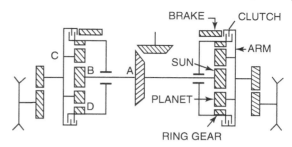

$$\omega_i = \omega_a \left(\frac{n_s}{n_s + n_r} \right) \tag{6.75}$$

where ω_a is the angular speed of shaft A; n_s and n_r are the number of teeth of the sun gear B and ring gear D, respectively.

Since the angular speed of the sprocket of the outside track ω_o is the same as ω_a, with the corresponding clutch engaged and the brake released, the speed ratio K_s can be expressed by

$$K_s = \frac{\omega_0}{\omega_i} = \frac{n_s + n_r}{n_s} \tag{6.76}$$

If the engine speed is kept constant, then the forward speed of the center of gravity of the vehicle will be less during a turn than in a straight-line motion. The forward speed of the center of gravity of the vehicle V during a turn is determined by

$$V = \frac{(\omega_o + \omega_i)r}{2} = \frac{\omega_i r(K_s + 1)}{2} \tag{6.77}$$

Also, when the brake is fully applied on the inside track, the minimum turning radius is achieved. The minimum turning radius R_{min} is expressed by

$$R_{min} = \frac{B}{2} \left(\frac{K_s + 1}{K_s - 1} \right) = \frac{B}{2} \left(\frac{2n_s + n_r}{n_r} \right) \tag{6.78}$$

When a tracked vehicle with a planetary gear steering system is turning at the minimum turning radius, the power ratio P_t/P_{st} is expressed by

$$\frac{P_t}{P_{st}} = \frac{V}{V_{st}} \left(1 + \frac{M_r}{f_r W R_{min}} \right) = \frac{(K_s + 1)}{2K_s} \left(1 + \frac{M_r}{f_r W B (K_s + 1)/(2K_s - 1)} \right) \tag{6.79}$$

If the normal pressure is uniformly distributed, the moment of turning resistance is given by Eq. (6.8), and if the vehicle is turning at low speeds, Eq. (6.79) can be rewritten as

$$\frac{P_t}{P_{st}} = \frac{1}{2K_s} \left[(K_s + 1) + \frac{\mu_t l (K_s - 1)}{2 f_r B} \right] \tag{6.80}$$

For $K_s = 2$, $l/B = 1.5$, and $\mu_t/f_r = 5$, the power consumption during a steady-state turn at the minimum turning radius will be 1.68 times that when the vehicle is in a straight-line motion.

The results of the analysis of the characteristics of various steering systems described above indicate that considerably more power is required during a turn than in a straight-line motion. To reduce the power requirement during a turn using the skid-steering principle, a number of regenerative steering systems have been developed for high-speed tracked vehicles (Steeds 1960; Ogorkiewicz 1991).

In a regenerative steering system, the power generated during a turn by the inside track from the braking force can be transferred through the system to the sprocket of the outside track. This supplies part of the power required by the sprocket of the outside track. The engine provides only the difference between the sprocket powers of the outside and inside tracks. Regenerative steering systems developed earlier are entirely mechanical. Lately, hydromechanical systems are widely used. They combine the advantages of infinitely variable drive and ease of control of hydrostatic transmissions with those of high efficiency of mechanical transmissions (Ogorkiewicz 1991).

6.7 Articulated Steering

For vehicles consisting of two or more units (chassis), steering can be accomplished by rotating one unit against the other to make the vehicle follow a prescribed curved path. This kind of steering method is referred to as "steering by articulation" or "articulated steering." There are two principal configurations of articulated steering:

1) Wagon steer

The method of wagon steer is schematically shown in Figure 6.21. This steering method is for vehicles having a common body frame (or loading platform) but with two separate chassis. Steering is accomplished by rotating both chassis about their respective vertical axes, or by rotating one of the two tracked chassis about its vertical axis. Usually, both tracked chassis have freedom in pitch to allow good ground contact over rough surfaces. The wagon steer configuration has been adopted in some heavy tracked transporters with long loading platforms. Figure 6.22 shows a tracked vehicle with wagon steer, designed for transporting long pipes and the like for the oil and gas industry.

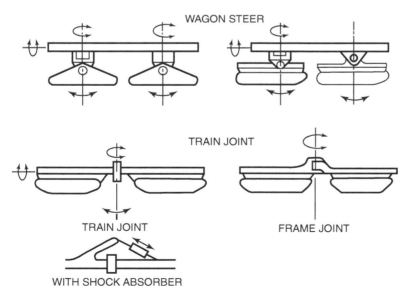

Figure 6.21 Two configurations of articulated steering: wagon steer and train joint. *Source:* Nuttall (1964). Reproduced with permission of the International Society for Terrain–Vehicle Systems.

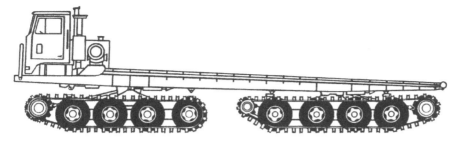

Figure 6.22 An off-road transporter having a long loading platform with two tracked chassis using wagon steer – Foremost Husky Eight. *Source:* Courtesy of Canadian Foremost Ltd.

2) Steering by train (articulation) joint

This method uses a train joint (articulation joint) to connect two (or more) separate vehicle units. Steering is accomplished by rotating one unit against the other by activating the articulation joint, as schematically shown in Figure 6.21. Usually, the design of the joint allows two adjacent units to have freedom in pitch and roll within a certain range. This method of steering has been adopted in vehicles designed for use on marginal terrain, as shown in Figure 6.23. Articulated steering has also been employed in off-road wheeled vehicles, as shown in Figure 6.24 (Dudzinski 1981; Oida 1983, 1987).

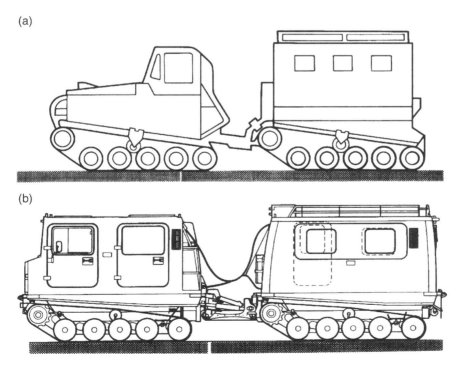

(a)

(b)

Figure 6.23 Articulated vehicles with train joint: (a) Volvo BV 202, *Source:* Courtesy of Volvo BM AB; (b) Hagglunds BV 206, *Source:* Courtesy of Hagglunds Vehicle AB.

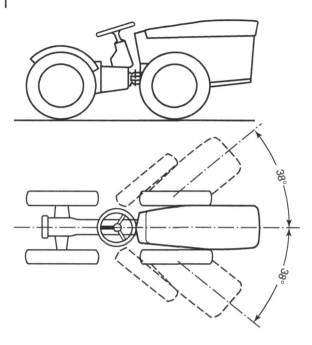

Figure 6.24 An off-road wheeled vehicle with articulated steering.

In comparison with the skid-steering method, articulated steering requires much less power to execute a turn, as there is no need for increasing the thrust on the outside track and for applying braking force to the inside track to create a turning moment to overcome the turning resistance moment caused by skidding of the tracks on the ground. Thus, articulated steering enables tracked vehicles to have better mobility than skid-steering over soft terrain, as mentioned previously. Using articulated steering, the required thrusts of the outside and inside tracks are primarily for overcoming the motion resistance of the tracks.

In addition, for skid-steering to satisfy the steerability criterion, the ratio of track contact length to tread of a vehicle must be within a certain range, as discussed in Section 6.1. For a tracked vehicle with a long loading platform using skid-steering, to meet the required ratio of track contact length to tread, the vehicle will be too wide to be practical. This is the reason why wagon steer is widely used in tracked transporters with a long loading platform, as shown in Figure 6.22. Articulated steering also makes it possible for tracked vehicles to achieve a more rational form, since a long, narrow vehicle encounters less obstacle resistance and lower motion resistance over unprepared terrain than a short, wide vehicle with the same track contact area. Field experience has also shown that the handling quality of articulated vehicles is satisfactory, even at speeds up to 72 km/h (45 mph) in some cases (Nuttall 1964).

Analytical and experimental studies of the turning behavior of the articulated tracked vehicle with two identical units connected with an articulation joint have been performed (Watanabe and Kitano 1986). Figure 6.25 shows a comparison of the variations of the predicted (represented by solid line) with the measured (represented by dots) thrust-to-weight ratio for the outside track F_o/W and that for the inside track F_i/W with the ratio of turning radius to track contact length R/l, under steady-state turning maneuvers on a hard, level ground. The measured data shown in the figure were obtained using a scale model operating on a hard surface. The drive axles of both

Figure 6.25 Variations of the thrust to weight ratios for the outside and inside tracks with turning radius to track contact length ratio for a two-unit articulated vehicle (solid line) and that for a single-unit vehicle with skid-steering (dashed lines). *Source:* Watanabe and Kitano (1986). Reproduced with permission of the International Society for Terrain–Vehicle Systems.

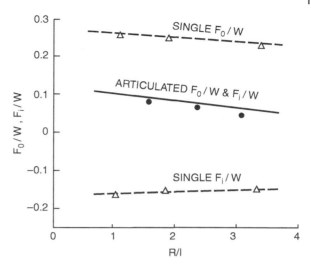

the front and rear units of the articulated vehicle used in the study have a simple differential, which results in identical thrust-to-weight ratios for the outside and inside tracks, as shown in the Figure 6.25. It indicates that the thrust-to-weight ratios of the outside and inside tracks are both positive and that the thrusts developed by the outside and inside tracks are primarily for overcoming the motion resistance of the tracks. For comparison, the figure also shows the predicted (represented by dashed lines) and the measured (represented by triangles) thrust-to-weight ratios for the outside and inside tracks of a single unit tracked vehicle with skid-steering as a function of the turning radius to track contact length ratio R/l. Both the predictions and measured data confirm that a high forward thrust-to-weight ratio of the outside track F_o/W (expressed as a positive value) and a high braking force-to-weight ratio of the inside track F_i/W (expressed as a negative value) are required to execute a steady-state turn with skid-steering. As mentioned previously, the required high forward thrust on the outside track and the high braking force on the inside track would lead to immobilization of the vehicle during a turning maneuver on weak terrain.

The minimum turning radius of an articulated tracked vehicle usually is larger than that of an equivalent vehicle with skid-steering. The initial cost of the articulated vehicle is usually higher than that of a similar vehicle with skid-steering, since the articulated vehicle has two or more separate units or chassis, which necessitates the replication of the suspension and track system (Nuttall 1964).

References

Bekker, M.G. (1956). *Theory of Land Locomotion*. Ann Arbor, MI: University of Michigan Press.

Crosheck, J.E. (1975). Skid-steering of crawlers. Society of Automotive Engineers paper 750552.

Dudzinski, P.A. (1981). Problems of turning process in articulated terrain vehicles. *Proc. 7th Int. Conf. of the International Society for Terrain–Vehicle Systems* **1**, Calgary, Canada.

Ehlert, W., Hug, B., and Schmid, I.C. (1992). Field measurements and analytical models as a basis of test stand simulation of the turning resistance of tracked vehicles. *Journal of Terramechanics* **29** (1).

Foss, C.F. (1993). *Jane's Armour and Artillery*, 14ee. Surrey, UK: Jane's Information Group.

Hayashi, I. (1975). Practical analysis of tracked vehicle steering depending on longitudinal track slippage. *Proc. 5th Int. Conf. of the International Society for Terrain–Vehicle Systems* II, Detroit-Houghton, MI.

Kar, M.K. (1987). Prediction of track forces in skid-steering of military tracked vehicles. *Journal of Terramechanics* **24** (1).

Kitano, K. and Kuma, M. (1978). An analysis of horizontal plane motion of tracked vehicles. *Journal of Terramechanics* **14** (4).

Little, L.F. (1964). The Alecto tracklayer. *Journal of Terramechanics* **1** (2).

Lvov, I.D. (1960). *Theory of Tractors (in Russian)*. Moscow: National Scientific and Technical Publishers.

Nuttall, C.J. (1964). Some notes on the steering of tracked vehicles by articulation. *Journal of Terramechanics* **1** (1).

Ogorkiewicz, R.M. (1991). *Technology of Tanks*. Surrey, UK: Jane's Information Group.

Oida, A. (1983). Turning behavior of articulated frame steering tractor. Part 1: Motion of tractor without traction. *Journal of Terramechanics* **20** (3/4).

Oida, A. (1987). Turning behavior of articulated frame steering tractor. Part 2: Motion of tractor with drawbar pull. *Journal of Terramechanics* **24** (1).

Steeds, W. (1950). Tracked vehicles, *Automobile Engineer*, April.

Steeds, W. (1960). *Mechanics of Road Vehicles*. London: Iliffe & Sons.

Watanabe, K. and Kitano, M. (1986). Study on steerability of articulated tracked vehicles. Part 1: Theoretical and experimental analysis. *Journal of Terramechanics* **23** (2).

Wong J.Y. and Chiang, C.F. (2001). A general theory for skid-steering of tracked vehicles on firm ground. *Proceedings of the Institution of Mechanical Engineers, Part D: Journal of Automobile Engineering* **215** (D3).

Problems

6.1 A tracked vehicle with skid-steering is to be designed for operation over various types of terrain ranging from desert sand with $c = 0$ and $\phi = 35°$ to heavy clay with $c = 20.685$ kPa (3 psi) and $\phi = 6°$. The average value of the coefficient of motion resistance is 0.15, and that of the coefficient of lateral resistance is 0.5. The vehicle has a uniform contact pressure of 13.79 kPa (2 psi). Select a suitable value for the ratio of contact length to tread for the vehicle, using the simplified method described in Section 6.1.

6.2 A tracked vehicle weighs 155.68 kN (35 000 lb) and has a contact length of 304.8 cm (120 in.) and a tread of 203.2 cm (80 in.). The vehicle has a uniform contact pressure and is equipped with a clutch/brake steering system. On a sandy terrain, the value of the coefficient of motion resistance is 0.15, and that of the coefficient of lateral resistance is 0.5. The angle of internal shearing resistance of the terrain ϕ is 30°.

(a) Using the simplified method described in Section 6.1, determine the thrusts of the outside and inside tracks required to execute a steady-state turn.

(b) If during the turn the sprocket of the outside track, with a radius of 0.305 m (1 ft), is rotating at 10 rad/s, and the inside track is disconnected from the driveline by declutching and the brake is applied, determine the turning radius and yaw velocity of the vehicle during the turn. The slip of the running gear during the turn may be neglected in the calculations.

6.3 Referring to Problem 6.2, using the simplified method described in Section 6.1, estimate the maximum drawbar pull that the tracked vehicle could develop during a steady-state turn. Also calculate the ratio of the maximum drawbar pull available during a steady-state turn to that in a straight-line motion under the conditions specified.

6.4 A tracked vehicle is equipped with a controlled differential steering system having a gear ratio of 3 : 1. The vehicle weighs 155.68 kN (35 000 lb) and has a tread of 203.2 cm (80 in.) and a contact length of 304.8 cm (120 in.). The contact pressure of the track is assumed to be uniform. On a particular terrain, the value of the coefficient of motion resistance is 0.15, and that of the coefficient of lateral resistance is 0.5. Determine the minimum turning radius of the vehicle. Also calculate the power required to maintain a steady-state turn at the minimum turning radius when the speed of the center of gravity of the vehicle is 10 km/h (6.2 mph).

7

Vehicle Ride Characteristics

Ride quality is concerned with the sensation or feel of the passenger in the environment of a moving vehicle. Ride comfort issues mainly arise from vibrations of the vehicle body, which may be induced by a variety of sources, including surface irregularities, aerodynamic forces, vibrations of the engine and driveline, and non-uniformities (imbalances) of the tire/wheel assembly. Usually, surface irregularities, ranging from potholes to random variations of the surface elevation profile, act as a major source that excites the vibration of the vehicle body through the tire/wheel assembly and the suspension system. Excitations by aerodynamic forces are applied directly to the vehicle body, while those due to engine and driveline vibrations are transmitted through engine/transmission mounts. Excitations resulting from mass imbalances and dimensional and stiffness variations of the tire/wheel assembly are transmitted to the vehicle body through the suspension.

The objective of the study of vehicle ride is to provide guiding principles for the control of the vibration of the vehicle body so that the passenger's sensation of discomfort does not exceed a certain level. To achieve this objective, it is essential to have a basic understanding of the human response to vibration, the vibrational behavior of the vehicle, and the characteristics of surface irregularities.

7.1 Human Response to Vibration

In general, passenger ride comfort (or discomfort) boundaries are difficult to determine precisely, because of the variations in individual sensitivity to vibration and of a lack of a generally accepted method of approach to the assessment of human response to vibration. Considerable research has been conducted by a number of investigators and professional organizations in an attempt to define ride comfort limits. A variety of methods for assessing human tolerance to vibration have been developed over the years (Van Deusen 1969; Lee and Pradko 1969; Smith et al. 1976; International Organization for Standardization 1985, 1997, 2010; Gillespie 1992; Hrovat 2014), as described below.

1) Subjective ride assessments

A traditional technique for comparing vehicle ride quality in the automotive industry in the past was to use a trained jury to rate the ride comfort, on a relative basis, of different vehicles driven over a range of road surfaces. With a large enough jury and a well-designed evaluation scheme, this method could provide a meaningful comparison of the ride quality of different vehicles. For instance, based on a study involving two cars, 78 passengers, and 18 different road sections (including road surfaces considered to be rough, medium, and smooth), a meaningful correlation was found to exist between the subjective ride ratings and simple root-mean-square (rms) acceleration measurements at either the vehicle floorboard or the passenger/seat interface. (Smith et al. 1976;

Theory of Ground Vehicles, Fifth Edition. J.Y. Wong.
© 2022 John Wiley & Sons, Inc. Published 2022 by John Wiley & Sons, Inc.
Companion website: www.wiley.com/go/wong/TGV5e

Hrovat 2014). The correlation between the subjective ride rating, designated as the ride discomfort index D_{ride}, and the corresponding rms acceleration measured for combined vertical and lateral directions may be expressed as follows (Smith et al. 1976):

$$D_{ride} = -0.43 + 40.0\, a_{vl} \tag{7.1}$$

where a_{vl} is the square root of the sum of the squares of the vertical and lateral rms accelerations in g units, and g is the acceleration due to gravity.

The lower the value of D_{ride} indicates the lower the ride discomfort rating. For instance, if the value of a_{vl} is $0.01075\,g$ (or 0.01075 in g units), then from Eq. (7.1) the value of $D_{ride} = 0$, which may be considered as an excellent ride.

On the other hand, based on the value of D_{ride} of the test vehicles used in the study, the road roughness may be assessed. Comparing the value of D_{ride} with the road sections used in the study, it was found that: $D_{ride} < 1$ corresponds approximately to a ride on a very smooth interstate highway in the U.S.; $1 < D_{ride} < 2$ corresponds to a ride on a typical state highway; and $2 < D_{ride} < 3$ corresponds to a ride on a rough secondary road (Smith et al. 1976).

2) Shake table tests

To quantitatively study human response to vibration, a large number of shake table experiments have been performed over the years. Most of this research pertains to human response to sinusoidal excitation. It is intended to identify zones of comfort (or discomfort) for humans in terms of vibration amplitude, velocity, or acceleration in each direction (such as foot-to-head, side-to-side, or back-to-chest) over a specific frequency range.

3) Ride simulator tests

In these tests, ride simulators are used to replicate the vibration of the vehicle traveling over different road surfaces. In some facilities, an actual vehicle body is mounted on hydraulic actuators, which reproduce vehicle motions in pitch, roll, and bounce (heave). Road inputs are fed into the actuators. Using the simulator, it is possible to establish human tolerance limits in terms of vibration parameters in various directions.

4) Ride measurements in vehicles

Shake table tests and ride simulator tests described above are conducted under laboratory conditions. They do not necessarily provide the same vibration environments to which the passenger is subject while driving on the road. Therefore, on-the-road ride measurements, particularly for passenger cars, have been performed. This test method attempts to correlate the response of test subjects in qualitative terms, such as "unpleasant" or "intolerable," with vibration parameters measured under actual driving conditions.

The assessment of human response to vibration is complex in that the results are influenced by the variations in individual sensitivity, and by the test methods used by different investigators. Over the years, numerous ride comfort criteria have been proposed. Figure 7.1 shows one of such criteria for vertical vibration described in the SAE *Ride and Vibration Data Manual* (Society of Automotive Engineers J6a 1965). The recommended limits shown in the figure are also referred to as the Janeway comfort criteria. It defines the acceptable amplitude of vibration as a function of frequency. As the frequency increases, the allowable amplitude decreases considerably. The Janeway comfort criteria consist of three simple relationships, each of which covers a specific frequency range, as shown in Figure 7.1. In the frequency range 1–6 Hz, the peak value of jerk, which is the product of the amplitude and the cube of the circular frequency, should not exceed 12.6 m/s^3 (496 in./s^3). For instance, at 1 Hz (2π rad/s), the recommended limit for amplitude is 12.6 m ·

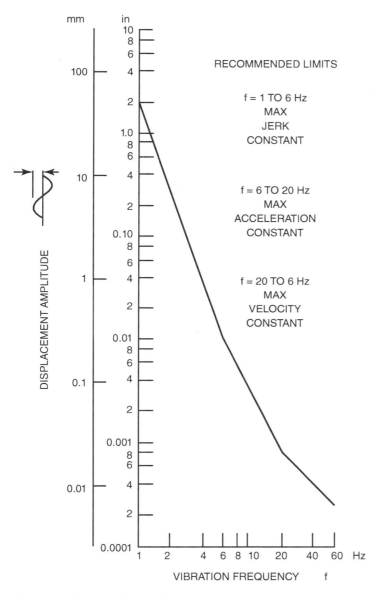

Figure 7.1 Vertical vibration limits for passenger comfort proposed by Janeway. Reprinted with permission from *Ride and Vibration Data Manual*, SAE J6a © 1965 Society of Automotive Engineers, Inc.

$s^{-3}/(2\pi s^{-1})^3 = 0.0508$ m(2 in). In the frequency range 6–20 Hz, the peak value of acceleration, which is the product of the amplitude and the square of the circular frequency, should be less than 0.33 m/s^2 (13 in. /s^2), whereas in the range 20–60 Hz, the peak value of velocity, which is the product of the amplitude and the circular frequency, should not exceed 2.7 mm/s (0.105 in/s). Janeway's comfort criteria are based on data for vertical sinusoidal vibration of a single frequency. When two or more components of different frequencies are present, there is no established basis on which to evaluate the resultant effect. It is probable, however, that the component that taken alone represents the highest sensation level will govern the sensation. Furthermore, all the data used to establish the ride comfort boundaries were obtained with test subjects standing or sitting on a hard seat.

7.1.1 International Standard ISO 2631/1:1985

The International Organization for Standardization (ISO) issues standards on mechanical vibration and shock for evaluation of human exposure to whole-body vibration. For instance, the standard ISO 2631 Part 1: General requirements (known as ISO 2631/1), issued in 1985, provides general guidance for evaluating human exposure to vibration that is of relevance to assessing vehicle ride quality. As progress is made in research in the field and additional data become available, revised editions of the standard are released from time to time.

In ISO 2631/1:1985, there are four physical factors of primary importance in determining human response to vibration: the intensity, the frequency, the direction, and the duration (exposure time) of the vibration. In practical evaluation of any vibration of which a physical description can be given in terms of those factors, three main human criteria can be distinguished. These are:

- The preservation of working efficiency ("fatigue-decreased proficiency boundary")
- The preservation of health or safety ("exposure limit")
- The preservation of comfort ("reduced comfort boundary")

To define the direction of vibration transmitted to the human body, an orthogonal coordinate system with x-, y-, and z-basicentric axes for the human body is used. The x-axis is directed from the back to chest, the y-axis from the right side to left side (side to side), and the z-axis from the foot to head. This axis system is used for seated, standing, and recumbent positions of the human body (ISO 2631/1:1985).

Figure 7.2(a) shows the fatigue-decreased proficiency boundary, as a function of frequency and exposure time, for vibration along the z-axis (foot-to-head) for a seated or standing person. It is defined in terms of the root-mean-square (rms) value of acceleration a_z as a function of frequency for daily exposure times from one minute to eight hours. Figure 7.2(b) shows the fatigue-decreased proficiency boundary for the transverse vibration along the x-axis (back-to-chest) or the y-axis (side-to-side) for a seated or standing person. The boundary for the transverse vibration is defined in terms of rms value of acceleration a_x or a_y as a function of frequency for daily exposure times from one minute to eight hours. As the daily exposure time increases, the fatigue-decreased proficiency boundary lowers. The boundary specifies a limit beyond which exposure to vibration can be regarded as carrying a significant risk of impaired working efficiency in many kinds of task, such as in vehicle driving. The actual degree of task interference in any situation depends on many factors, including individual characteristics, as well as the nature and difficulty of the task. Nevertheless, the limits shown in Figures 7.2(a) and (b) indicate the general level of onset of such interference. The data upon which these limits are based come mainly from studies on aircraft pilots and vehicle drivers.

It should be noted that for the human body the most sensitive frequency ranges are 4 to 8 Hz for vibration along the z-axis and below 2 Hz for vibration along the x- and y-axes. Accordingly, the limits in these frequency ranges are set lowest. It is shown that human tolerance to vibration decreases in a characteristic way with increasing exposure time. A comparison of Figures 7.2(a) and (b) indicates that whereas the tolerance for transverse vibration (along the x-axis or y-axis) is lower than that for vertical vibration (along the z-axis) at very low frequencies, the converse is the case for higher frequencies above approximately 2.8 Hz.

The exposure limit for the preservation of health or safety, as a function of frequency and exposure time, is of the same general form as the fatigue-decreased proficiency boundary, but the corresponding levels are raised by a factor of 2 (6 dB higher). In other words, the maximum safe exposure is determined, for any frequency, duration, and direction, by doubling the values allowed,

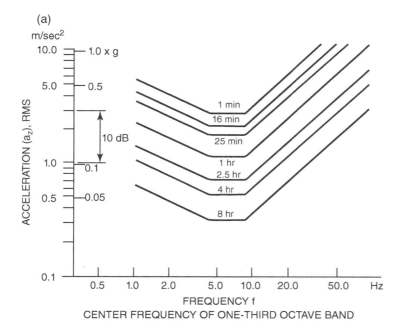

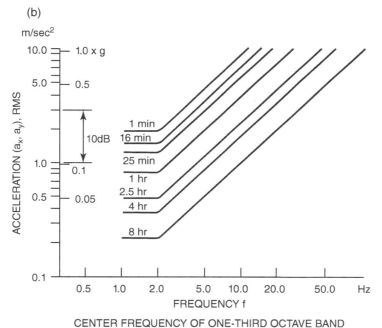

Figure 7.2 ISO 2631/1:1985 "fatigue-decreased proficiency boundary": (a) longitudinal a_z acceleration limits along the z-axis; (b) transverse a_x and a_y acceleration limits along the x-axis and y-axis, as a function of frequency and exposure time. Copied with permission of the Standards Council of Canada (SCC) on behalf of ISO. The standard can be purchased from the national member in your country or the ISO store. Copyright remains with ISO.

according to the criterion of fatigue-decreased proficiency (see Figures 7.2(a) and (b)). Exceeding the exposure limit is not recommended without special justification and precautions, even if no task is to be performed by the exposed individual.

The reduced comfort boundary, which is derived from various studies conducted for transport industries, is assumed to lie at approximately one-third of the corresponding levels of the fatigue-decreased proficiency boundary and to follow the same time and frequency dependence. Values for the reduced comfort boundary are, accordingly, obtained from the corresponding values for the fatigue-decreased proficiency boundary by a reduction of 10 dB (see Figures 7.2(a) and (b)). In transport vehicles, the reduced comfort boundary is related to difficulties of carrying out such tasks as eating, reading, and writing.

If vibrations occur in more than one direction simultaneously (i.e., "multi-axis" or "multiplanar" vibration), the corresponding limits apply separately to each vectorial component in the three axes.

It should be noted that the numerical values for limits of exposure presented above apply to vibrations transmitted from solid surfaces to the human body in the frequency range 1–80 Hz. They may be applied, within the specified frequency range, to periodic, random, and nonperiodic vibrations with a distributed frequency spectrum. ISO 2631/1 : 1985 does not cover vibrations in the frequency range below 1 Hz, associated with symptoms of motion sickness, which are of a character different from the effects of higher frequency vibrations.

7.1.2 International Standard ISO 2631-1 : 1997/Amd.1 : 2010

The ISO 2631/1:1985, presented in the previous section, was superseded by ISO 2631-1 : 1997, which incorporates new experience and research results reported in the literature. These results made it desirable to change the methods of measurement and analysis of the vibration environment and the approach to the application of the results. The fatigue-decreased proficiency boundary for preservation of working efficiency, the exposure limit for the preservation of health or safety, and the reduced comfort boundary for the preservation of comfort adopted in 2631/1:1985 are no longer used in ISO 2631-1 : 1997.

Despite these substantial changes, the majority of reports or research studies indicate that the guidance and exposure boundaries recommended in ISO 2631/1:1985 were safe and preventive of undesired effects.

The ISO 2631-1:1997 was amended in 2010. The amendment, known as Amendment 1 and designated as ISO 2631-1:1997/Amd.1:2010, includes a revised graph on health guidance caution zones, certain amended clauses, and a few new clauses. In this section, the basic features of ISO 2631-1:1997 with Amendment 1 are summarized. It defines the methods for measurement and evaluation of periodic, random, and transient whole-body vibration, and indicates the principal factors that combine to determine the degree to which a vibration exposure will be acceptable. It represents current opinion and provides guidance on the possible effects of vibration on health, comfort, perception, and motion sickness. The frequency range considered is:

- 0.5–80 Hz for health, comfort, and perception, and
- 0.1–0.5 Hz for motion sickness.

It is applicable to evaluating vibration transmitted to the human body as a whole through the supporting surfaces: the feet of a standing person, the buttocks, back and feet of a seated person, or the supporting area of a recumbent person.

In this section, the basic evaluation method using frequency-weighted rms acceleration is outlined. For vibration with crest factors below or equal to 9, the basic evaluation method is normally sufficient. The crest factor is defined as the modulus of the ratio of the maximum instantaneous peak value of the frequency-weighted acceleration signal to its rms value. The peak value shall be determined over the duration of measurement used for the integration of the rms value.

Vibration evaluation using the basic method shall always include the measurements of weighted rms acceleration, which is expressed in meters per second squared (m/s^2) for translational vibration or radians per second squares (rad/s^2) for rotational vibration. The weighted rms acceleration shall be calculated in accordance with the following equation:

$$a_w = \left[\frac{1}{T} \int_0^T a_w^2(t) dt \right]^{1/2} \tag{7.2}$$

where $a_w(t)$ is the weighted acceleration for translational vibration or rotational vibration as a function of time (time history), in meters per second squared (m/s^2) or in radians per second squared (rad/s^2), respectively; and T is the duration of the measurement, in seconds.

Frequency-weighting factors recommended and/or used for the various directions and their applications are listed in Table 7.1. As noted in the previous section, the directions of translational vibration of the human body are defined using an orthogonal coordinate system with the x-, y-, and z-basicentric axes. The numerical values of the frequency-weighting factors (W_k, W_d, and W_f) for frequencies from 0.1 to 80 Hz are shown in Table 7.2.

Table 7.2 shows that the principal frequency weighting W_k along z-axis increases with frequency from 0.1 Hz, reaches the highest value at frequency $f = 6.3$ Hz, and then decreases with the further increase of frequency. The principal weighting W_d along the x-axis or y-axis exhibits a similar trend and reaches the maximum value at frequency $f = 1$ Hz. The principal frequency weighting W_f in the vertical direction for motion sickness reaches the highest value at frequency $f = 0.16$ Hz approximately.

In case where the basic evaluation method may underestimate the effects of vibration (such as vibration with high crest factors, occasional shocks, and transient vibration), additional evaluation methods may be used. These methods, however, will not be discussed in this section. For readers interested in the additional evaluation methods, please refer to the original document of ISO 2631-1:1997/Amd.1:2010.

Table 7.1 Guide for the application of frequency-weighting for principal weightings.

Frequency weighting	Health	Comfort	Perception	Motion sickness
W_k	z-axis, seat surface	z-axis, seat surface	z-axis, seat surface	
		z-axis, standing vertical recumbent (except head)	z-axis, standing vertical recumbent (except head)	
		x-, y-, z-axes, feet (sitting)		
W_d	x-axis, seat surface	x-axis, seat surface	x-axis, seat surface	
	y-axis, seat surface	y-axis, seat surface	y-axis, seat surface	
		x-, y-axes, standing horizontal recumbent	x-, y-axes, standing horizontal recumbent	
		y-, z-axes, seat-back		
W_f				Vertical

Source: ISO 2631-1: 1997. Copied with permission of the Standards Council of Canada (SCC) on behalf of ISO. The standard can be purchased from the national member in your country or the ISO store. Copyright remains with ISO.

Table 7.2 Principal frequency weightings in one-third octaves.

Frequency f, Hz	W_k Factor × 1000	dB	W_d Factor × 1000	dB	W_f Factor × 1000	dB
0.1	31.2	−30.11	62.4	−24.09	695	−3.16
0.125	48.6	−26.26	97.3	−20.24	895	−0.96
0.16	79.0	−22.05	158	−16.01	1006	0.05
0.2	121	−18.33	243	−12.28	992	−0.07
0.25	182	−14.81	365	−8.75	854	−1.37
0.315	263	−11.60	530	−5.52	619	−4.17
0.4	352	−9.07	713	−2.94	384	−8.31
0.5	418	−7.57	853	−1.38	224	−13.00
0.63	459	−6.77	944	−0.50	116	−18.69
0.8	477	−6.43	992	−0.07	53	−25.51
1	482	−6.33	1011	0.10	23.5	−32.57
1.25	484	−6.29	1008	0.07		
1.6	494	−6.12	968	−0.28		
2	531	−5.49	890	−1.01		
2.5	631	−4.01	776	−2.20		
3.15	804	−1.90	642	3.85		
4	967	−0.29	512	−5.82		
5	1039	0.33	409	−7.76		
6.3	1054	0.46	323	−9.81		
8	1036	0.31	253	−11.93		
10	988	−0.10	202	−13.91		
12.5	902	−0.89	161	−15.87		
16	768	−2.28	125	−18.03		
20	636	−3.93	100	−19.99		
25	513	−5.80	80.0	−21.94		
31.5	405	−7.86	63.2	−23.98		
40	314	−10.05	49.4	−26.13		
50	246	−12.19	38.8	−28.22		
63	186	−14.61	29.5	−30.60		
80	132	−17.56	21.1	−33.53		

Source: ISO 2631-1:1997. Copied with permission of the Standards Council of Canada (SCC) on behalf of ISO. The standard can be purchased from the national member in your country or the ISO store. Copyright remains with ISO.
Note: Values of decibel (dB) shown in Table 7.2 are calculated as follows: $dB = 10 \log_{10} (factor)$.

Guidance for assessing the possible effects of vibration with crest factor below or equal to 9 on health, comfort, and perception, and on motion sickness, is briefly described below.

1) Health

This guide concerns the effects of periodic, random, and transient vibration on the health of persons in normal health exposed to whole-body vibration during travel, at work, and leisure activities. It applies primarily to seated persons in the frequency range 0.5–80 Hz, transmitted through the seat pan. Assessment of the effects of vibration on the health of those exposed while standing, reclining, or recumbent is usually carried out using the same evaluation method as that for seated persons.

The weighted rms acceleration shall be determined using Eq. (7.2) for each axis (x, y, and z) of translational vibration on the surface which supports the person. The frequency weightings are applied for seated persons as follows with the factors k as indicated below:

$$x\text{-axis} : \quad W_d, \ k = 1.4$$
$$y\text{-axis} : \quad W_d, \ k = 1.4$$
$$z\text{-axis} : \quad W_k, \ k = 1$$

Assessment of the exposure to vibration can be based on the calculation of daily vibration exposure $A_l(8)$, expressed as equivalent continuous acceleration over a period of 8 h. The daily vibration exposure $A_l(8)$ in meters per second squared, for each direction l (x, y, or z) is defined as:

$$A_l(8) = k_l \sqrt{\frac{1}{T_0} \sum_i a_{wli}^2 T_i} \tag{7.3}$$

where a_{wli} is the frequency-weighted rms value of the acceleration, determined over the time period T_i; $l = x, y, z$; $k_x = k_y = 1.4$ for the x- and y-directions; $k_z = 1$ for the z-direction; T_0 is the reference duration of eight hours (28,800 s).

The assessment of the effect of a vibration on health shall be made independently along each axis and with respect to the highest frequency-weighted acceleration determined in any axis on the seat pan. When vibration in two or more axes is comparable, the vibration total value (root-sum-of-squares) is sometimes used as an additional estimate of health risk.

Two health guidance caution zones are shown in Figure 7.3 and are described below:

- The first health guidance caution zone identified by 1 in Figure 7.3 is bounded by the dashed lines. The lower and upper bounds of the zone for exposures less than 10 minutes are 3 and 6 m/s², respectively, while those for exposures of 24 h (86,400 s) are 0.25 and 0.5 m/s², respectively. For defining this zone, responses are assumed to be related to energy and two different daily vibration exposures are equivalent when:

$$a_{w1} T_1^{1/2} = a_{w2} T_2^{1/2} \tag{7.4}$$

where a_{w1} and a_{w2} are the frequency-weighted rms acceleration values for the first and second exposures, respectively; and T_1 and T_2 are the corresponding durations for the first and second exposures.

Equation (7.4) indicates that in this health guidance caution zone, frequency-weighted rms acceleration decreases with the increase of the square root of exposure duration. The recommendation on this health caution zone is mainly based on exposures in the range of 4–8 h indicated in Figure 7.3. The validity of the results outside this range is uncertain.

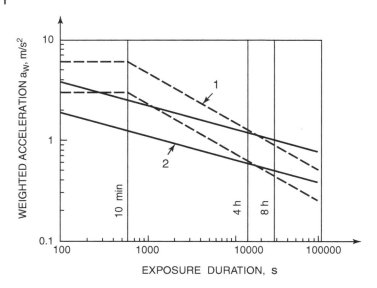

Figure 7.3 ISO 2631-1:1997/Amd.1: 2010 health guidance caution zones. Copied with permission of the Standards Council of Canada (SCC) on behalf of ISO. The standard can be purchased from the national member in your country or the ISO store. Copyright remains with ISO.

- The second health guidance caution zone identified by 2 in Figure 7.3 is bounded by the solid lines. It is based on the results of studies that two different daily vibration exposures are equivalent when:

$$a_{w1}T_1^{1/4} = a_{w2}T_2^{1/4} \tag{7.5}$$

where a_{w1} and a_{w2} are the frequency-weighted rms acceleration values for the first and second exposures, respectively; and T_1 and T_2 are the corresponding durations for the first and second exposures.

Equation (7.5) indicates that in this caution zone, the frequency-weighted rms acceleration decreases with the increase in exposure duration raised to the power of one-fourth.

As shown in Figure 7.3, the two health guidance caution zones 1 and 2 are overlapped for durations from 4–8 h, for which most occupational observations exist. The validity of the results shown in Figure 7.3 outside the range of 4–8 h is uncertain as the two zones diverge. Results for shorter durations should be treated with extreme caution (ISO 2631-1:1997/ Amd.1:2010).

2) Comfort

The guide for assessing the effects of vibration on comfort given below indicates the current consensus on the relationship between the vibration magnitude (expressed in terms of the frequency-weighted rms acceleration) and human comfort. It applies to seated, standing, and recumbent persons for periodic, random, and transient vibration. There is no conclusive evidence to support a universal time dependence of vibration effects on comfort. The frequency-weighted rms acceleration shall be determined for each axis of translational vibration (x-,y-, and z-axes) at the surface that supports the person.

For seated persons, in assessing the effects of vibration on comfort, the frequency weightings shall be applied as follows, with additional multiplying factors k as indicated:

$$x\text{-axis (supporting seat surface vibration)} : \quad W_d, \ k = 1$$
$$y\text{-axis (supporting seat surface vibration)} : \quad W_d, \ k = 1$$
$$z\text{-axis (supporting seat surface vibration)} : \quad W_k, \ k = 1$$

For standing persons, the frequency weightings shall be applied as follows, with additional multiplying factors k as indicated:

$$x\text{-axis (floor vibration)} : \quad W_d, \ k = 1$$
$$y\text{-axis (floor vibration)} : \quad W_d, \ k = 1$$
$$z\text{-axis (floor vibration)} : \quad W_k, \ k = 1$$

For recumbent persons, when measuring under the pelvis, the frequency weightings shall be applied as follows, with additional multiplying factors k as indicated:

$$\text{Horizontal axes} : \quad W_d, \ k = 1$$
$$\text{Vertical axis} : \quad W_k, \ k = 1$$

When vibrations occur in more than one direction, the total value of frequency-weighted rms acceleration a_v, determined from vibration in orthogonal coordinates, is calculated as follows:

$$a_v = \left(k_x^2 a_{wx}^2 + k_y^2 a_{wy}^2 + k_z^2 a_{wz}^2 \right)^{1/2} \tag{7.6}$$

where a_{wx}, a_{wy}, and a_{wz} are the frequency-weighted rms accelerations with respect to the orthogonal axes, x, y, and z, respectively; and k_x, k_y, and k_z are multiplying factors, the exact values of which depend on the frequency weighting identified as described above.

The use of the vibration total value a_v is recommended for evaluation of comfort. The following values give approximate indications of likely reactions to various magnitudes of overall vibration total values in public transport:

Less than 0.315 m/s^2 (approximately 1 ft/s^2):	not uncomfortable
0.315 to 0.63 m/s^2 (approximately 1 to 2 ft/s^2):	a little uncomfortable
0.5 to 1 m/s^2 (approximately 1.6 to 3.3 ft/s^2):	fairly uncomfortable
0.8 to 1.6 m/s^2 (approximately 2.6 to 5.2 ft/s^2):	uncomfortable
1.25 to 2.5 m/s^2 (approximately 4.1 to 8.2 ft/s^2):	very uncomfortable
Greater than 2 m/s^2 (approximately 6.6 ft/s^2):	extremely uncomfortable

It should be noted, however, that the reactions at various magnitudes depend on passenger expectations regarding trip duration and the type of activities passengers expect to accomplish (e.g. reading, eating, writing, etc.) and many other factors (acoustic noise, temperature, etc.).

3) Perception

The guide for assessing the perception of vibration given below applies to seated, standing, and recumbent persons for periodic and random vibration, occurring in the three translational (x, y, and z) axes. The frequency-weighted rms acceleration shall be determined for each axis (x, y, and z) on the principal surface supporting the person. The assessment of the perceptibility of the vibration shall be made with respect to the highest frequency-weighted rms acceleration determined in any axis at any point of contact at any time. Two frequency weightings, W_k for z-axis vibration and W_d for x-axis or y-axis vibration, are used for the prediction of the perceptibility of the vibration. These weightings may be applied to the following combinations of posture and vibration axes:

x-, y-, and z-axes on a supporting seat surface for sitting person	$k = 1$
x-, y-, and z-axes on a floor beneath a standing person	$k = 1$
x-, y-, and z-axes on a surface supporting a recumbent person (except head)	$k = 1$

Fifty percent of alert, fit persons can just detect a W_k frequency-weighted vibration with a peak magnitude of 0.015 m/s^2 (0.05 ft/s^2). There is a large variation between individuals in their ability to perceive vibration. When the median perception threshold is approximately 0.015 m/s^2 (0.05 ft/s^2), the interquartile range of responses may extend from about 0.01 to 0.02 m/s^2 (0.03 to 0.06 ft/s^2) peak.

The perception threshold decreases slightly with increases in vibration duration up to one second and very little with further increases in duration. Although the perception threshold does not continue to decrease with increasing duration, the sensation produced by vibration at magnitudes above threshold may continue to increase.

4) Motion sickness

Oscillatory motion at frequencies below 0.5 Hz commonly produces motion sickness. The frequency-weighted rms acceleration shall be determined for the z-axis vibration at the surface which supports the person, at frequencies between 0.1 and 0.5 Hz. The vibration shall be assessed only with respect to the overall frequency-weighted acceleration in the z-axis.

The guide for assessing the effects of vibration on motion sickness presented in ISO 2631-1 : 1997 primarily applies to motion in ships and other sea vessels, and not necessarily to ground vehicles. For this reason, it is not further discussed here. For information on evaluation of motion sickness in detail, please refer to the full document of ISO 2631-1 : 1997.

7.1.3 Absorbed Power

The concept of absorbed power, which is the product of vibration force and velocity transmitted to the human body, has also been proposed as a parameter of significance in evaluating human response to vibration (Lee and Pradko 1969). It is a measure of the rate at which vibrational energy is absorbed by a human being, and it has been used to define human tolerance to vibration for military ground vehicles negotiating rough terrain (Lee and Pradko 1969; Stikeleather et al. 1973; Murphy and Ahlvin 1975). The concept of absorbed power has been adopted by the U.S. Army AMM-75 Ground Mobility Model and subsequently by the NATO Reference Mobility Model (NRMM) for evaluating the ride quality of military ground vehicles. Presently, the tolerance limit is taken as 6 W absorbed power at the driver's position, and the ride-limiting speed is the speed at which the driver's average absorbed power over the total elapsed time reaches a sustained level of 6 W.

Having identified a specific ride comfort criterion, the designer should then select an appropriate suspension system to ensure that the level of vehicle vibration is below the specified limits when operating over a particular range of environments.

7.2 Vehicle Ride Models

To study the ride quality of ground vehicles, various ride models have been developed. For a passenger car with independent front suspensions, a seven-degrees-of-freedom model, as shown in Figure 7.4, may be used. In this model, the pitch, bounce, and roll of the vehicle body, as well as the bounce of the two front wheels and the bounce and roll (tramp) of the solid rear axle are taken into consideration. The mass of the vehicle body is usually referred to as the "sprung mass," whereas the mass of the running gear together with the associated components is referred to as the "unsprung mass." For a military tracked vehicle shown in Figure 7.5, a fifteen-degrees-of-freedom model may be used, which includes the pitch, bounce, and roll of the vehicle body and the bounce of each roadwheel.

To study the vibrational characteristics of the vehicle, equations of motion based on Newton's second law for each mass should be formulated. Natural frequencies and amplitude ratios can be determined by considering the principal modes (normal modes) of vibration (or the free

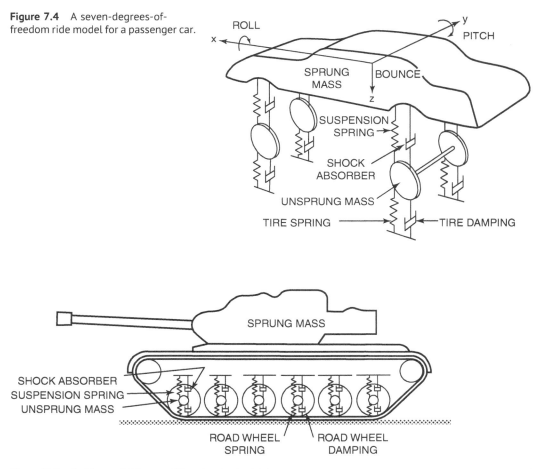

Figure 7.4 A seven-degrees-of-freedom ride model for a passenger car.

Figure 7.5 A ride model for a military tracked vehicle.

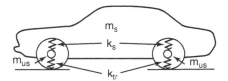

Figure 7.6 A two-degrees-of-freedom ride model for the sprung and unsprung mass.

m_s = 1814 kg, 4000 lb
m_{us} = 181 kg, 400 lb, COMBINED
k_s = 88 kN/m, 500 lb/in., COMBINED
K_{tr} = 704 kN/m, 4000 lb/in., COMBINED

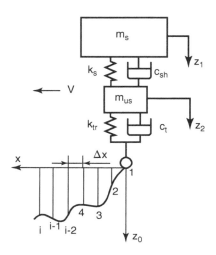

Figure 7.7 A quarter-car model.

vibration) of the system. When the excitation of the system is known, the response can, in principle, be determined by solving the equations of motion. However, as the degrees of freedom of the system increase, the analysis becomes increasingly complex and computer simulations are employed.

A vehicle represents a complex vibration system with many degrees of freedom. It is possible, however, to simplify the system by considering only some of its major motions. For instance, to obtain a qualitative insight into the functions of the suspension, particularly the effects of the sprung and unsprung mass, spring stiffness, and damping on vehicle vibrations, a linear model with two degrees of freedom, as shown in Figures 7.6 and 7.7, may be used. On the other hand, to reach a better understanding of the pitch and bounce vibration of the vehicle body, a two-degrees-of-freedom model, as shown in Figure 7.8, may be employed.

7.2.1 Two-Degrees-of-Freedom Vehicle Model for Vertical Vibrations of Sprung and Unsprung Mass

Figure 7.6 shows a two-degrees-of-freedom model for vehicle vertical vibration analysis. It includes a sprung mass representing the vehicle body and an unsprung mass representing the combination of four wheels and associated components.

Figure 7.8 A two-degrees-of-freedom ride model for bounce and pitch of the sprung mass.

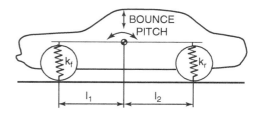

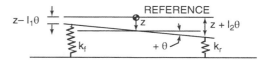

Figure 7.7 shows a two-degrees-of-freedom model representing a quarter of a passenger car. It is commonly referred to as the quarter-car model. While it is a simplified model, it can provide significant insight into the effects of suspension characteristics on vehicle vibration in the vertical direction (or heave) (Tseng and Hrovat 2015).

The quarter-car model includes the following parameters:

- sprung mass m_s – corresponding to a portion of the mass of the vehicle and load
- suspension spring rate k_s
- suspension shock absorber damping coefficient c_{sh}
- unsprung mass m_{us} – mass of the tire and associated components
- tire spring rate k_{tr}
- tire damping coefficient c_t – if its value is relatively small in comparison with c_{sh}, it may be neglected in the analysis.

The tire–ground contact can be represented by various models, including the point contact model, moving average ground profile model, and tire enveloping model (Mucka and Gagnon 2015).

The motions of the sprung and sprung mass in the vertical direction can be described by two coordinates, z_1 and z_2 (Figure 7.7), with origins at their static equilibrium positions. By applying Newton's second law to the sprung and unsprung mass separately, the equations of motion of the system can be obtained.

For vibrations excited by surface undulation, the equations of motion are as follows. For the sprung mass:

$$m_s\ddot{z}_1 + c_{sh}(\dot{z}_1 - \dot{z}_2) + k_s(z_1 - z_2) = 0 \tag{7.7}$$

For the unsprung mass:

$$m_{us}\ddot{z}_2 + c_{sh}(\dot{z}_2 - \dot{z}_1) + k_s(z_2 - z_1) + c_t\dot{z}_2 + k_{tr}z_2 = F(t) = c_t\dot{z}_0 + k_{tr}z_0 \tag{7.8}$$

where m_s is the sprung mass, m_{us} is the unsprung mass, c_{sh} is the damping coefficient of the shock absorber, c_t is the damping coefficient of the tire, k_s is the stiffness of the suspension spring, k_{tr} is the equivalent spring stiffness of the tire, and $F(t)$ is the excitation acting on the wheels and induced by surface irregularities. If z_0 is the elevation of the surface profile and $\dot{z}_0$ represents the vertical velocity of the tire at the ground contact point, which is the slope of the profile multiplied by the forward speed of the vehicle, then the excitation due to surface undulation may be expressed by $c_t\dot{z}_0 + k_{tr}z_0$, as shown in Eq. (7.8). Excitations due to aerodynamic forces and to vibrations of the engine and driveline are applied to the sprung mass, while those due to non-uniformities of the tire/wheel assembly are applied to the unsprung mass. If the excitation of the system is known, then, in principle, the resulting vibrations of the sprung and unsprung mass can be determined by solving Eqs. (7.7) and (7.8).

1) Vibration characteristics of the sprung mass and unsprung mass

The natural frequencies of the sprung and unsprung masses are of significance to their vibrational characteristics. To determine the natural frequencies of the two-degrees-of-freedom system (the quarter-car model) shown in Figure 7.7, the free vibration of the system is considered, or the principal modes of vibration are considered. The equations of motion for free vibration are obtained by setting the right-hand sides of both Eqs. (7.7) and (7.8) to zero. For an undamped system, from Eqs. (7.7) and (7.8), the equations of motion for free vibration are as follows:

$$m_s \ddot{z}_1 + k_s z_1 - k_s z_2 = 0 \tag{7.9}$$

$$m_{us} \ddot{z}_2 + k_s z_2 - k_s z_1 + k_{tr} z_2 = 0 \tag{7.10}$$

The solutions to the above differential equations can be assumed to be in the following form:

$$z_1 = Z_1 \cos \omega_n t \tag{7.11}$$

$$z_2 = Z_2 \cos \omega_n t \tag{7.12}$$

where ω_n is the undamped circular natural frequency, and Z_1 and Z_2 are the amplitudes of the sprung and unsprung mass, respectively. Substituting the assumed solutions into Eqs. 7.9 and 7.10, one obtains the following amplitude equations:

$$\left(-m_s \omega_n^2 + k_s \right) Z_1 - k_s Z_2 = 0 \tag{7.13}$$

$$-k_s Z_1 + \left(-m_{us} \omega_n^2 + k_s + k_{tr} \right) Z_2 = 0 \tag{7.14}$$

These equations are satisfied for any Z_1 and Z_2 if the following determinant is zero:

$$\begin{vmatrix} \left(-m_s \omega_n^2 + k_s \right) & -k_s \\ -k_s & \left(-m_{us} \omega_n^2 + k_s + k_{tr} \right) \end{vmatrix} = 0 \tag{7.15}$$

Expanding the determinant leads to the characteristic equation of the system:

$$\omega_n^4 (m_s m_{us}) + \omega_n^2 (-m_s k_s - m_s k_{tr} - m_{us} k_s) + k_s k_{tr} = 0 \tag{7.16}$$

The solution of the characteristic equation yields two undamped natural frequencies of the system, ω_{n1}^2 and ω_{n2}^2:

$$\omega_{n1}^2 = \frac{B_1 - \sqrt{B_1^2 - 4A_1 C_1}}{2A_1} \tag{7.17}$$

$$\omega_{n2}^2 = \frac{B_1 + \sqrt{B_1^2 - 4A_1 C_1}}{2A_1} \tag{7.18}$$

where

$$A_1 = m_s m_{us}$$
$$B_1 = m_s k_s + m_s k_{tr} + m_{us} k_s$$
$$C_1 = k_s k_{tr}$$

Although each of these leads to frequencies $\pm\omega_{n1}$ and $\pm\omega_{n2}$, the negative values are discarded as being of no physical significance. The corresponding natural frequencies in Hz (cycles/s) are expressed by

$$f_{n1} = \frac{1}{2\pi}\omega_{n1} \tag{7.19}$$

$$f_{n2} = \frac{1}{2\pi}\omega_{n2} \tag{7.20}$$

For a typical passenger car, the sprung mass m_s is an order of magnitude higher than the unsprung mass m_{us}, while the stiffness of the suspension spring k_s is an order of magnitude lower than the equivalent spring stiffness of the tire k_{tr}, as shown in Figure 7.6. In view of this, an approximate method may be used to determine the two natural frequencies of the system. The approximate values of the undamped natural frequencies in Hz of the sprung and unsprung mass, f_{n-s} and f_{n-us}, can be expressed by

$$f_{n-s} = \frac{1}{2\pi}\sqrt{\frac{k_s k_{tr}/(k_s + k_{tr})}{m_s}} \tag{7.21}$$

$$f_{n-us} = \frac{1}{2\pi}\sqrt{\frac{k_s + k_{tr}}{m_{us}}} \tag{7.22}$$

With the values of m_s, m_{us}, k_s, and k_{tr} shown in Figure 7.6, the two natural frequencies calculated using Eqs. (7.19) and (7.20) are 1.04 and 10.5 Hz, respectively, which are found to be practically identical to those obtained using Eqs. (7.21) and (7.22). The natural frequency of the unsprung mass is an order of magnitude higher than that of the sprung mass. For passenger cars, the damping ratio provided by shock absorbers is usually in the range of 0.2–0.4, and that the damping of the tire is relatively insignificant. Consequently, there is little difference between the undamped and damped natural frequencies, and undamped natural frequencies are commonly used to characterize the system.

The wide separation of the natural frequencies of the sprung and unsprung mass has a significant implication on the vibration isolation characteristics of the suspension system. For instance, if the wheel hits a bump, the impulse will set the wheel into oscillation. When the wheel passes over the bump, the unsprung mass will be in free oscillation at its own natural frequency f_{n-us}. For the sprung mass, however, the excitation will be the vibration of the unsprung mass. The ratio of the frequency of excitation to the natural frequency of the sprung mass is therefore equal to f_{n-us}/f_{n-s}. Since the value of f_{n-us} is an order of magnitude higher than that of f_{n-s}, the amplitude of oscillation of the sprung mass will be very small. As can be seen from Figure 7.9 when the ratio of the frequency of excitation to the natural frequency of the system is high, the transmissibility, which is the ratio of output to input of a vibrating system, is very low. Thus, excellent vibration isolation for the sprung mass (i.e., vehicle body) is achieved in this case.

When the vehicle travels over an undulating surface, the excitation will normally consist of a wide range of frequencies. As can be seen from Figure 7.9, high-frequency inputs can be effectively isolated through the suspension because the natural frequency of the sprung mass is low, or the ratio of frequency of excitation to the natural frequency of the sprung mass is high. Low-frequency excitations can, however, be transmitted to the vehicle body unimpeded, or even amplified, as the transmissibility is high when the frequency of excitation is close to the natural frequency of the sprung mass.

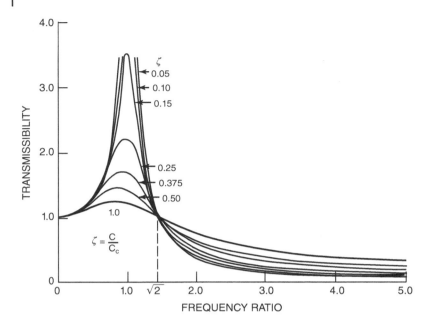

Figure 7.9 Transmissibility as a function of the ratio of frequency of excitation to natural frequency for a single-degree-of-freedom system.

If the road profile is sinusoidal, then the responses of the sprung and unsprung mass can be determined using the classical methods in vibration analysis. For the two-degrees-of-freedom system shown in Figure 7.7, with the damping of the tire neglected, the ratio of the vibration amplitude of the sprung mass Z_1 to that of the surface profile Z_0 is expressed by

$$\frac{Z_1}{Z_0} = \frac{\sqrt{A_2}}{\sqrt{B_2 + C_2}} \tag{7.23}$$

where

$$A_2 = (k_s k_{tr})^2 + (c_{sh} k_{tr} \omega)^2$$
$$B_2 = \left[(k_s - m_s \omega^2)(k_{tr} - m_{us} \omega^2) - m_s k_s \omega^2 \right]^2$$
$$C_2 = (c_{sh} \omega)^2 \left[m_s \omega^2 + m_{us} \omega^2 - k_{tr} \right]^2$$

The ratio of the vibration amplitude of the unsprung mass Z_2 to that of the surface profile Z_0 is given by

$$\frac{Z_2}{Z_0} = \frac{\sqrt{A_3}}{\sqrt{B_2 + C_2}} \tag{7.24}$$

where

$$A_3 = \left[k_{tr}(k_s - m_s \omega^2) \right]^2 + (c_{sh} k_{tr} \omega)^2.$$

In the equations above, ω is the circular frequency of excitation, which is equal to $2\pi V/l_w$, where V is the vehicle forward speed and l_w is the wavelength of the road profile.

If the damping of the shock absorber is neglected (i.e., $c_{sh} = 0$), then the expressions for the responses of the sprung and unsprung mass to the excitation of sinusoidal road profile are simplified, and the following relations are obtained:

$$\frac{Z_1}{Z_0} = \frac{k_s k_{tr}}{(k_s - m_s\omega^2)(k_{tr} - m_{us}\omega^2) - m_s k_s \omega^2}$$
$$= \frac{k_s k_{tr}}{m_s m_{us}(\omega_{n1}^2 - \omega^2)(\omega_{n2}^2 - \omega^2)} \tag{7.25}$$

$$\frac{Z_2}{Z_0} = \frac{k_{tr}(k_s - m_s\omega^2)}{m_s m_{us}(\omega_{n1}^2 - \omega^2)(\omega_{n2}^2 - \omega^2)} \tag{7.26}$$

where ω_{n1} and ω_{n2} are the undamped circular natural frequencies of the system. When the frequency of excitation ω coincides with one of the natural frequencies, resonance results. The resonance of the unsprung mass (tire/wheel assembly) is usually referred to as "wheel hop" resonance.

2) Basic functional requirements for a suspension system

Basic functional requirements for an automotive suspension are:

- to provide good vibration isolation from surface irregularities
- to avoid excessive suspension travel or related hard impact on the bump stopper
- to maintain good road holding capability

Using the quarter-car model shown in Figure 7.7, the effects of suspension characteristics on vibration isolation, suspension travel, and road holding are discussed below.

A) Vibration isolation

This can be evaluated by the response of the sprung mass (output) to the excitation from the ground (input). Usually, the transmissibility (or transfer function) can be used as a basis for assessing the vibration isolation characteristics of a linear suspension system.

Figure 7.10 shows the effect of the ratio of the unsprung mass to the sprung mass m_{us}/m_s on the transmissibility of a two-degrees-of-freedom system with $m_s = 454.5$ kg (1000 lb), $k_{tr} = 176$ kN/m (1000 lb/in.), $k_{tr}/k_s = 8$, and damping ratio $\zeta = 0.3$. In the range of frequency of excitation (hereinafter referred to as the "frequency range") below the natural frequency of the sprung mass (around 1 Hz), the mass of the unsprung parts has very little effect on the vibration of the sprung mass. When the frequency of excitation is close to but lower than the natural frequency of the unsprung mass (around 10 Hz), the lighter the unsprung mass, the lower the transmissibility will be, which implies that the vibration of the sprung mass is lower with a lighter unsprung mass. However, in the frequency range above the natural frequency of the unsprung mass (around 10 Hz), a lighter unsprung mass will lead to a slightly higher transmissibility.

Based on the results presented above, it can be said that while the unsprung mass has little influence on the vibration of the sprung mass in the low-frequency range, a lighter unsprung mass does provide better vibration isolation in the mid-frequency range. There is a slight penalty, however, in the frequency range higher than the natural frequency of the unsprung mass.

Figure 7.11 shows the effect of the ratio of the equivalent tire stiffness k_{tr} to the suspension spring stiffness k_s on the transmissibility of the system. For a given tire stiffness, a higher ratio of k_{tr}/k_s indicates a lower suspension spring stiffness. In the frequency range below the natural frequency

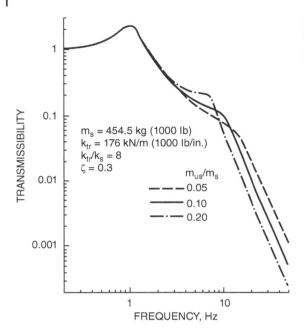

Figure 7.10 Transmissibility as a function of frequency of excitation for various ratios of unsprung mass to sprung mass of a quarter-car model.

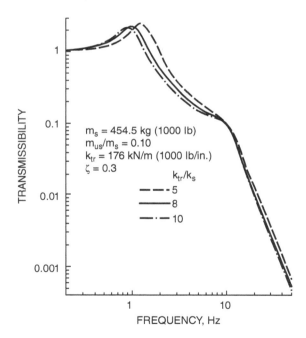

Figure 7.11 Transmissibility as a function of frequency of excitation for various ratios of tire stiffness to suspension spring stiffness of a quarter-car model.

of the sprung mass, the lower the ratio of k_{tr}/k_s, the lower the transmissibility will be. In the frequency range between the natural frequency of the sprung mass and that of the unsprung mass, a softer suspension spring (or higher k_{tr}/k_s ratio) provides better vibration isolation. In the frequency range above the natural frequency of the unsprung mass, the suspension spring stiffness has a

Figure 7.12 Transmissibility as a function of frequency of excitation for various damping ratios of a quarter-car model.

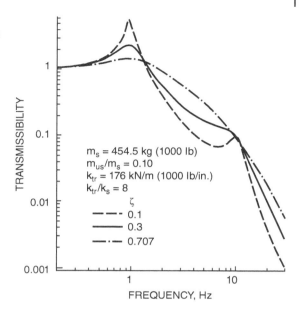

m_s = 454.5 kg (1000 lb)
m_{us}/m_s = 0.10
k_{tr} = 176 kN/m (1000 lb/in.)
k_{tr}/k_s = 8

ζ
--- $\cdot$ 0.1
——— 0.3
—$\cdot$— 0.707

relatively insignificant effect on the vibration of the sprung mass, and the transmissibility is independent of the ratio k_{tr}/k_s. The results presented above indicate that a softer suspension spring provides better vibration isolation in the mid- to high-frequency range, although there is some penalty in the frequency range below the natural frequency of the sprung mass.

Figure 7.12 shows the effect of the damping ratio of the suspension shock absorber ζ on the transmissibility of the system. In the frequency range close to the natural frequency of the sprung mass, the higher the damping ratio, the lower the transmissibility will be. In the frequency range between the natural frequency of the sprung mass and that of the unsprung mass, the lower the damping ratio, the lower the transmissibility will be. At a frequency close to the natural frequency of the unsprung mass, the damping ratio has little effect on the response of the sprung mass. However, in the frequency range above the natural frequency of the unsprung mass, the lower the damping ratio, the lower the transmissibility will be.

Based on the results described above, to provide good vibration isolation in the frequency range close to the natural frequency of the sprung mass, a high damping ratio is required. However, in the mid- to high-frequency range, a lower damping ratio is preferred.

B) Suspension travel

This is measured by the deflection of the suspension spring or by the relative displacement between the sprung and unsprung mass $(z_2 - z_1)$. It defines the space required to accommodate the suspension spring movement between bump and rebound stops, commonly known as the "rattle space."

Figure 7.13 shows the effect of the ratio of the unsprung mass to the sprung mass m_{us}/m_s on the suspension travel ratio of the system, which is defined as the ratio of the maximum relative displacement between the sprung and unsprung mass $(z_2 - z_1)_{max}$ to the amplitude of the sinusoidal road profile Z_0. For a given amplitude of the surface profile Z_0, in the frequency range below the natural frequency of the sprung mass, the mass ratio m_{us}/m_s has little effect on suspension travel. In

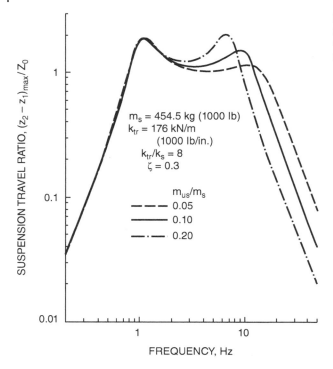

Figure 7.13 Suspension travel ratio as a function of frequency of excitation for various ratios of unsprung mass to sprung mass of a quarter-car model.

the frequency range between the natural frequency of the sprung mass and that of the unsprung mass, the increase in the mass ratio causes an increase in suspension travel. However, in the frequency range above the natural frequency of the unsprung mass, the higher the mass ratio, the lower the suspension travel will be.

Based on the results shown above, it can be said that while the unsprung mass has little effect on suspension travel in the low-frequency range, a lighter unsprung mass does reduce suspension travel in the mid-frequency range. There is some penalty, however, in the frequency range above the natural frequency of the unsprung mass.

Figure 7.14 shows the effect of the ratio of the equivalent tire stiffness k_{tr} to the suspension spring stiffness k_s on the suspension travel ratio. In the frequency range below the natural frequency of the sprung mass, a softer suspension spring leads to higher suspension travel. In the frequency range above the natural frequency of the unsprung mass, the suspension spring stiffness has little effect on suspension travel. In the mid-frequency range between the natural frequency of the sprung mass and that of the unsprung mass, the suspension travel is initially lower with a softer suspension spring, and then higher at a frequency approaching the natural frequency of the unsprung mass. The frequency at which this change-over takes place is called the "crossover" frequency and is approximately 3 Hz for the system examined.

Based on the results shown above, it can be said that in the low-frequency range, a softer suspension spring often leads to higher suspension travel. In the high-frequency range, the suspension spring stiffness has little effect on suspension travel. In the mid-frequency range from the natural frequency of the sprung mass to the crossover frequency (i.e., from 1 to 3 Hz for the system

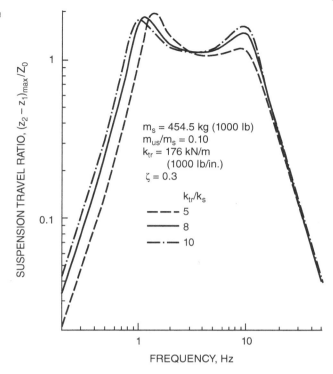

Figure 7.14 Suspension travel ratio as a function of frequency of excitation for various ratios of tire stiffness to suspension spring stiffness of a quarter-car model.

examined), a softer suspension spring leads to lower suspension travel. In the frequency range from the crossover frequency to the natural frequency of the unsprung mass, a softer suspension spring, however, leads to higher suspension travel.

Figure 7.15 shows the effect of the damping ratio ζ on the suspension travel ratio. Over the entire frequency range from below the natural frequency of the sprung mass to above the natural frequency of the unsprung mass, the higher the damping ratio, the lower the suspension travel will be. This indicates that to reduce the suspension travel, a higher damping ratio is required.

C) Road holding

When the vehicle system vibrates, the normal force acting between the tire and the road fluctuates. Since the cornering force, tractive effort, and braking effort developed by the tire are related to the normal load on the tire, the vibration of the tire affects the road holding capability and influences the handling and performance of the vehicle. The normal force between the tire and the road during vibration can be represented by the dynamic tire deflection or by the displacement of the unsprung mass relative to the road surface.

Figure 7.16 shows the effect of the ratio of the unsprung mass to the sprung mass m_{us}/m_s on the dynamic tire deflection ratio, which is the ratio of the maximum relative displacement between the unsprung mass and the road surface $(z_0 - z_2)_{max}$ to the amplitude of the sinusoidal road profile Z_0. In the frequency range below the natural frequency of the sprung mass, the mass ratio m_{us}/m_s has little effect on the dynamic tire deflection (or road holding). In the mid-frequency range between

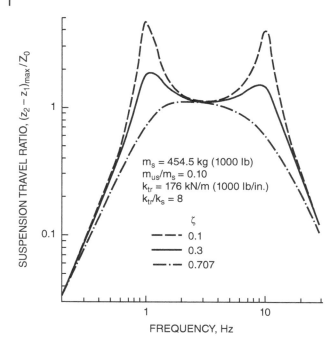

Figure 7.15 Suspension travel ratio as a function of frequency of excitation for various damping ratios of a quarter-car model.

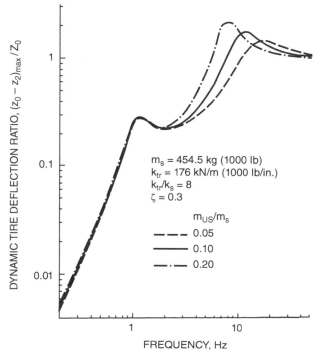

Figure 7.16 Dynamic tire deflection ratio as a function of frequency of excitation for various ratios of unsprung mass to sprung mass of a quarter-car model.

the natural frequency of the sprung mass and that of the unsprung mass, a lighter unsprung mass leads to lower dynamic tire deflection. In the frequency range above the natural frequency of the unsprung mass, the unsprung mass has a relatively insignificant effect on the dynamic tire deflection. Note that if during vibration, the relative displacement between the unsprung mass and the road surface is such that it allows the static tire deflection (i.e., the deflection of the tire under static load) to fully recover, the normal force between the tire and the road will be reduced to zero or the tire is at the verge of bouncing off the ground. This is an undesirable situation, as the tire is losing contact with the ground and the road holding capability of the vehicle is adversely affected. For the system shown in Figure 7.16, with a sprung mass of 454.5 kg (1000 lb) and a mass ratio $m_{us}/m_s = 0.2$, the static tire deflection is approximately 3 cm (5.35 kN/176 kN/m) or 1.2 in. If the vehicle travels over a sinusoidal road profile at an appropriate speed that generates a frequency of excitation of 8 Hz (i.e., $f = V/l_w$, where V is the vehicle speed in m/s (or ft/s) and l_w is the wavelength of the road profile in m (or ft)), then from Figure 7.16, the ratio of the maximum dynamic tire deflection to the amplitude of the road profile is approximately 2. It indicates that if the amplitude of the road profile is 1.5 cm (0.6 in.), the maximum dynamic tire deflection will be 3 cm (1.2 in.). Since the static tire deflection is 3 cm (1.2 in.), this implies that under these circumstances, the tire will lose contact with the ground during part of the vibration cycle.

Figure 7.17 shows the effect of the ratio of the equivalent tire stiffness k_{tr} to the suspension spring stiffness k_s on the dynamic tire deflection ratio. In the low- and high-frequency ranges, the suspension spring stiffness has a relatively insignificant influence on the dynamic tire deflection. In the mid-frequency range between the natural frequency of the sprung mass and the crossover frequency (i.e., from 1 to 6 Hz for the system shown), a softer suspension spring leads to lower dynamic

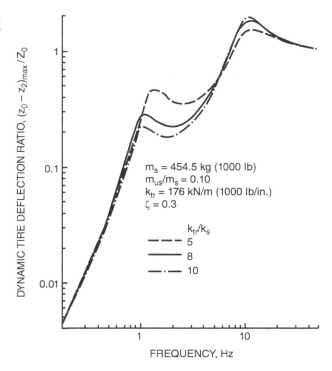

Figure 7.17 Dynamic tire deflection ratio as a function of frequency of excitation for various ratios of tire stiffness to suspension spring stiffness of a quarter-car model.

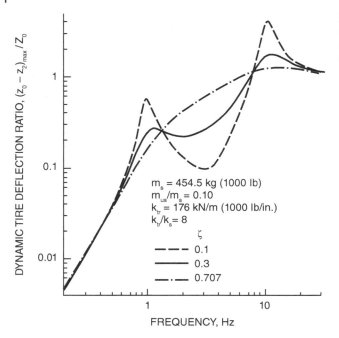

Figure 7.18 Dynamic tire deflection ratio as a function of frequency of excitation for various damping ratios of a quarter-car model.

tire deflection. However, in the frequency range close to the natural frequency of the unsprung mass, a stiffer suspension spring leads to lower dynamic tire deflection, and hence better road holding capability.

Based on the results described above, a softer suspension spring will generally provide better overall vibration isolation. However, to achieve better road holding capability at a frequency of excitation close to the natural frequency of the unsprung mass, a stiffer suspension spring is preferred. This is the reason why the suspension spring for performance cars is usually stiffer than that for ordinary passenger cars. Consequently, the natural frequency of the sprung mass for performance cars (up to 2 or 2.5 Hz) is higher than that for ordinary passenger cars (usually in the range from 1 to 1.5 Hz).

Figure 7.18 shows the effect of the damping ratio on the dynamic tire deflection ratio. In the frequency range below the natural frequency of the sprung mass, to maintain good road holding capability higher damping is required. In the mid-frequency range from just above the natural frequency of the sprung mass to just below the natural frequency of the unsprung mass, lower damping ratio is preferred. For frequency of excitation beyond that range higher damping ratio is required.

7.2.2 Numerical Methods for Determining the Response of a Quarter-Car Model to Irregular Surface Profile Excitation

In practice, road profile is usually irregular and is seldom sinusoidal. To determine the response of the quarter-car model shown in Figure 7.7 to the excitation of irregular surface profiles, numerical methods are used.

As noted previously, in Figure 7.7, z_0 describes the elevation of the surface profile, and $\dot{z}_0$ represents the vertical velocity of the tire at the ground contact point, and is expressed by

$$\dot{z}_0 = V\frac{dz_0}{dx} \tag{7.27}$$

where V is the forward speed of the vehicle and dz_0/dx is the slope of the surface profile.

When the vehicle is traveling at a constant speed, both z_0 and $\dot{z}_0$ can be considered as functions of time and are known for a given road profile. The responses of the sprung and unsprung mass, z_1, $\dot{z}_1$, $\ddot{z}_1$, z_2, $\dot{z}_2$, and $\ddot{z}_2$, at different locations (or stations, as shown in Figure 7.7) can be obtained using the following numerical procedure based on the Taylor series (Vierck 1979).

If, at the initial point (station 1 shown in Figure 7.7), $(z_0)_1$ and $(\dot{z}_0)_1$ are zero, then $(z_1)_1 = (\dot{z}_1)_1 = (\ddot{z}_1)_1 = (z_2)_1 = (\dot{z}_2)_1 = (\ddot{z}_2)_1 = 0$ (the subscript outside the parentheses indicates the station number).

At station 2,

$$(z_1)_2 = (\dot{z}_1)_2\Delta t/3 = (\ddot{z}_1)_2(\Delta t)^2/6 \tag{7.28}$$

$$(\dot{z}_1)_2 = (\ddot{z}_1)_2\Delta t/2 \tag{7.29}$$

$$(z_2)_2 = (\dot{z}_2)_2\Delta t/3 = (\ddot{z}_2)_2(\Delta t)^2/6 \tag{7.30}$$

$$(\dot{z}_2)_2 = (\ddot{z}_2)_2\Delta t/2 \tag{7.31}$$

From Eqs. (7.7) and (7.8) and neglecting the damping of the tire c_t,

$$m_s(\ddot{z}_1)_2 = c_{sh}\left[(\dot{z}_2)_2 - (\dot{z}_1)_2\right] + k_s\left[(z_2)_2 - (z_1)_2\right] \tag{7.32}$$

$$m_{us}(\ddot{z}_2)_2 = c_{sh}\left[(\dot{z}_1)_2 - (\dot{z}_2)_2\right] + k_s\left[(z_1)_2 - (z_2)_2\right] + k_{tr}\left[(z_0)_2 - (z_2)_2\right] \tag{7.33}$$

Substituting Eqs. (7.28), (7.29), (7.30), and (7.31) into the two equations above and solving them simultaneously, one obtains

$$(\ddot{z}_1)_2 = \frac{k_{tr}(z_0)_2 A_4}{B_4 C_4 - A_4^2}$$

$$(\ddot{z}_2)_2 = \frac{(\ddot{z}_1)_2 B_4}{A_4}$$

where

$$A_4 = c_{sh}\Delta t/2 + k_s(\Delta t)^2/6$$

$$B_4 = m_s + c_{sh}\Delta t/2 + k_s(\Delta t)^2/6$$

$$C_4 = m_{us} + c_{sh}\Delta t/2 + (k_s + k_{tr})(\Delta t)^2/6$$

The set of equations given above enables the values of the parameters at station 2, $(\ddot{z}_1)_2$, $(\dot{z}_1)_2$, $(z_1)_2$, $(\ddot{z}_2)_2$, $(\dot{z}_2)_2$, and $(z_2)_2$, to be determined for a given elevation of road profile at station 2, $(z_0)_2$. The time increment Δt in the above equations is taken as the increment in horizontal distance Δx (Figure 7.7) divided by vehicle speed V, that is, $\Delta t = \Delta x/V$. The selection of the value of Δt depends on the accuracy required. In general, Δt should be less than 5% of the period of the free vibration of the unsprung mass τ_{us}, where $\tau_{us} = 1/f_{n-us}$ and f_{n-us} is the undamped natural frequency of the unsprung mass.

At subsequent stations i (≥ 3),

$$(z_1)_i = (\ddot{z}_1)_{i-1}(\Delta t)^2 + 2(z_1)_{i-1} - (z_1)_{i-2} \tag{7.34}$$

$$(\dot{z}_1)_i = \left[3(z_1)_i - 4(z_1)_{i-1} + (z_1)_{i-2}\right]/2\Delta t \tag{7.35}$$

$$(\ddot{z}_1)_i = \left\{c_{sh}\left[(\dot{z}_2)_i - (\dot{z}_1)_i\right] + k_s\left[(z_2)_i - (z_1)_i\right]\right\}/m_s \tag{7.36}$$

$$(z_2)_i = (\ddot{z}_2)_{i-1}(\Delta t)^2 + 2(z_2)_{i-1} - (z_2)_{i-2} \tag{7.37}$$

$$(\dot{z}_2)_i = \left[3(z_2)_i - 4(z_2)_{i-1} + (z_2)_{i-2}\right]/2\Delta t \tag{7.38}$$

$$(\ddot{z}_2)_i = \left\{k_{tr}\left[(z_0)_i - (z_2)_i\right] - c_{sh}\left[(\dot{z}_2)_i - (\dot{z}_1)_i\right] - k_s\left[(z_2)_i - (z_1)_i\right]\right\}/m_{us} \tag{7.39}$$

As an example, Figure 7.19(a) shows the acceleration response of the sprung mass of a quarter-car model traveling over an irregular road surface shown in Figure 7.19(b) at a speed $V = 80$ km/h (50

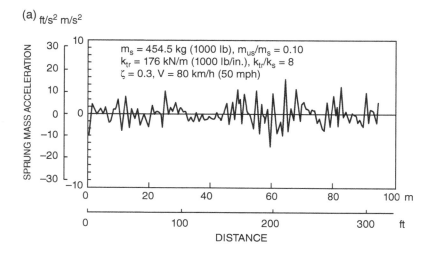

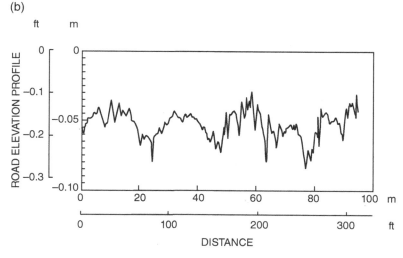

Figure 7.19 Simulations of the vibration of a quarter-car model over irregular road profiles: (a) sprung mass acceleration; (b) road elevation profile.

mph), obtained using the numerical procedure described above. The parameters of the quarter-car model used in the simulation are $m_s = 454.5$ kg (1000 lb), $m_{us}/m_s = 0.10$, $k_{tr} = 176$ kN/m (1000 lb/in.), $k_{tr}/k_s = 8$, and $\zeta = 0.3$.

7.2.3 Two-Degrees-of-Freedom Vehicle Model for Pitch and Bounce

Because of the wide separation of the natural frequencies of the sprung and unsprung mass, the up-and-down linear motion (bounce) and the angular motion (pitch) of the vehicle body and the motion of the wheels may be considered to exist almost independently. The bounce and pitch of the vehicle body can therefore be studied using the model shown in Figure 7.8. In this model, damping is neglected.

By applying Newton's second law and using the static equilibrium position as the origin for both the linear displacement of the center of gravity z and angular displacement of the vehicle body θ, the equations of motion for the system can be formulated.

For free vibration, the equation of motion for bounce is

$$m_s\ddot{z} + k_f(z - l_1\theta) + k_r(z + l_2\theta) = 0 \tag{7.40}$$

and the equation of motion for pitch is

$$I_y\ddot{\theta}\left(\text{or } m_s r_y^2\ddot{\theta}\right) - k_f l_1(z - l_1\theta) + k_r l_2(z + l_2\theta) = 0 \tag{7.41}$$

where k_f is the front spring stiffness, k_r is the rear spring stiffness, and I_y and r_y are the mass moment of inertia and radius of gyration of the vehicle body about the y axis (Figure 7.4), respectively.

By letting

$$D_1 = \frac{1}{m_s}(k_f + k_r)$$

$$D_2 = \frac{1}{m_s}(k_r l_2 - k_f l_1)$$

$$D_3 = \frac{1}{I_y}(k_f l_1^2 + k_r l_2^2) = \frac{1}{m_s r_y^2}(k_f l_1^2 + k_r l_2^2)$$

Equations 7.40 and 7.41 can be rewritten as

$$\ddot{z} + D_1 z + D_2\theta = 0 \tag{7.42}$$

$$\ddot{\theta} + D_3\theta + \frac{D_2}{r_y^2}z = 0 \tag{7.43}$$

It is evident that D_2 is the coupling coefficient for the bounce and pitch motions, and that these motions uncouple when $k_f l_1 = k_r l_2$. With $k_f l_1 = k_r l_2$, a force applied to the center of gravity induces only bounce motion, while a moment applied to the body produces only pitch motion. In this case, the natural frequencies for the uncoupled bounce and pitch motions are

$$\omega_{nz} = \sqrt{D_1} \tag{7.44}$$

$$\omega_{n\theta} = \sqrt{D_3} \tag{7.45}$$

It is found that this would result in a poor ride.

In general, the pitch and bounce motions are coupled and an impulse at the front or rear wheel excites both motions. To obtain the natural frequencies for the coupled bounce and pitch motions, the free vibration of the system is considered (or the principal modes of vibration are considered). The solutions to the equations of motion (i.e., Eqs. (7.40) and (7.41)) can be expressed in the form of

$$z = Z \cos \omega_n t \tag{7.46}$$

$$\theta = \Theta \cos \omega_n t \tag{7.47}$$

where ω_n is the undamped circular natural frequency, and Z and Θ are the amplitudes of bounce and pitch, respectively.

Substituting the above equations into Eqs. (7.42) and (7.43), one obtains the following amplitude equations:

$$\left(D_1 - \omega_n^2\right)Z + D_2\Theta = 0 \tag{7.48}$$

$$\left(\frac{D_2}{r_y^2}\right)Z + \left(D_3 - \omega_n^2\right)\Theta = 0 \tag{7.49}$$

Following an approach similar to that described in Section 7.2.1, one obtains the characteristic equation for the system:

$$\omega_n^4 - (D_1 + D_3)\omega_n^2 + \left(D_1 D_3 - \frac{D_2^2}{r_y^2}\right) = 0 \tag{7.50}$$

From Eq. (7.50), two undamped natural frequencies ω_{n1} and ω_{n2} can be obtained:

$$\omega_{n1}^2 = \frac{1}{2}(D_1 + D_3) - \sqrt{\frac{1}{4}(D_1 - D_3)^2 + \frac{D_2^2}{r_y^2}} \tag{7.51}$$

$$\omega_{n2}^2 = \frac{1}{2}(D_1 + D_3) + \sqrt{\frac{1}{4}(D_1 - D_3)^2 + \frac{D_2^2}{r_y^2}} \tag{7.52}$$

These frequencies for coupled motions, ω_{n1} and ω_{n2}, always lie outside the frequencies for uncoupled motions, ω_{nz} and $\omega_{n\theta}$.

From Eqs. (7.48) and (7.49), the amplitude ratios of the bounce and pitch oscillations for the two natural frequencies ω_{n1} and ω_{n2} can be determined.

For ω_{n1},

$$\left.\frac{Z}{\Theta}\right|_{\omega_{n1}} = \frac{D_2}{\omega_{n1}^2 - D_1} \tag{7.53}$$

and for ω_{n2},

$$\left.\frac{Z}{\Theta}\right|_{\omega_{n2}} = \frac{D_2}{\omega_{n2}^2 - D_1} \tag{7.54}$$

It can be shown that the two amplitude ratios will have opposite signs.

To further illustrate the characteristics of the bounce and pitch modes of oscillation, the concept of an oscillation center is introduced. The location of the oscillation center is denoted by l_0 measured from the center of gravity, and it can be determined from the amplitude ratios. Thus, one center is associated with ω_{n1}, and the other with ω_{n2}.

For ω_{n1},

$$l_{01} = \frac{D_2}{\omega_{n1}^2 - D_1} \tag{7.55}$$

and for ω_{n2},

$$l_{02} = \frac{D_2}{\omega_{n2}^2 - D_1} \tag{7.56}$$

When the value of the amplitude ratio is negative, the oscillation center will be located to the right of the center of gravity of the vehicle body, in accordance with the sign conventions for z and θ shown in Figure 7.20. On the other hand, when the value of the amplitude ratio is positive, the oscillation center will be located to the left of the center of gravity. In general, a road input at the front or rear wheel will cause a moment about each oscillation center, and therefore will excite both bounce and pitch oscillations. In other words, the body motion will be the sum of the oscillations about the two centers.

Usually, the oscillation center that lies outside the wheelbase is called the bounce center, and the associated natural frequency is called the bounce frequency. On the other hand, the oscillation center that lies inside the wheelbase is called the pitch center, and the associated natural frequency is called the pitch frequency.

Example 7.1   Determine the pitch and bounce frequencies and the locations of oscillation centers of an automobile with the following data:

- Sprung mass $m_s = 2120$ kg (weight 4676 lb)
- Radius of gyration $r_y = 1.33$ m (4.36 ft)
- Distance between the front axle and center of gravity $l_1 = 1.267$ m (4.16 ft)
- Distance between the rear axle and center of gravity $l_2 = 1.548$ (5.08 ft)
- Front spring stiffness $k_f = 35$ kN/m (2398 lb/ft)
- Rear spring stiffness $k_r = 38$ kN/m (2604 lb/ft)

Figure 7.20   Oscillation centers for bounce and pitch of vehicle sprung mass.

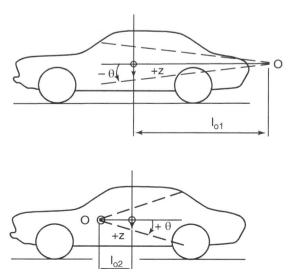

Solution

The constants D_1, D_2, and D_3 are first calculated as follows:

$$D_1 = \frac{k_f + k_r}{m_s} = \frac{35{,}000 + 38{,}000}{2120} = 34.43 \text{ s}^{-2}$$

$$D_2 = \frac{k_r l_2 - k_f l_1}{m_s} = \frac{38{,}000 \times 1.548 - 35{,}000 \times 1.267}{2120} = 6.83 \text{ m·s}^{-2}$$

$$D_3 = \frac{k_f l_1^2 + k_r l_2^2}{m_s r_y^2} = \frac{35{,}000 \times 1.267^2 + 38{,}000 \times 1.548^2}{2120 \times 1.33^2} = 39.26 \text{ s}^{-2}$$

$$\left(\frac{D_2}{r_y}\right)^2 = 26.37 \text{ s}^{-4}$$

$$D_3 + D_1 = 73.69 \text{ s}^{-2}$$

$$D_3 - D_1 = 4.83 \text{ s}^{-2}$$

$$\omega_{n1}^2 = \frac{1}{2}(D_1 + D_3) - \sqrt{\frac{1}{4}(D_1 - D_3)^2 + \left(\frac{D_2}{r_y}\right)^2}$$

$$= 36.85 - \sqrt{5.83 + 26.37} = 31.17 \text{ s}^{-2}$$

$$\omega_{n1} = 5.58 \text{ s}^{-1} \quad \text{or} \quad f_{n1} = 0.89 \text{ Hz}$$

$$\omega_{n2}^2 = \frac{1}{2}(D_1 + D_3) + \sqrt{\frac{1}{4}(D_1 - D_3)^2 + \left(\frac{D_2}{r_y}\right)^2}$$

$$= 36.85 + \sqrt{5.83 + 26.37} = 42.52 \text{ s}^{-2}$$

$$\omega_{n2} = 6.52 \text{ s}^{-1} \quad \text{or} \quad f_{n2} = 1.04 \text{ Hz}$$

The locations of the oscillation centers can be determined using Eqs. (7.55) and (7.56). For ω_{n1},

$$l_{01} = \left.\frac{Z}{\Theta}\right|_{\omega_{n1}} = \frac{D_2}{\omega_{n1}^2 - D_1} = \frac{6.83}{31.17 - 34.43} = -2.09 \text{ m} \quad (82 \text{ in.})$$

and for ω_{n2},

$$l_{02} = \left.\frac{Z}{\Theta}\right|_{\omega_{n2}} = \frac{D_2}{\omega_{n2}^2 - D_1} = \frac{6.83}{42.52 - 34.43} = +0.84 \text{ m} \quad (33 \text{ in.})$$

This indicates that one oscillation center is situated at a distance of 2.09 m (82 in.) to the right of the center of gravity, and the other is located at a distance of 0.84 m (33 in.) to the left of the center of gravity, as shown in Figure 7.20.

For most passenger cars, the natural frequency for bounce is in the range of 1.0–1.5 Hz, and the natural frequency for pitch is slightly higher than that for bounce. For cars with coupled front–rear suspension systems, the natural frequency for pitch may be lower than that for bounce. In roll, the natural frequency is usually higher than those for bounce and pitch primarily because of the effect of antiroll bars. The natural frequency for roll usually varies in the range of 1.5–2.0 Hz for cars.

The locations of the oscillation centers have practical significance to ride behavior. One case of interest is that when the motions of bounce and pitch are uncoupled (i.e., $k_f l_1 = k_r l_2$). In this case, one oscillation center will be at the center of gravity, and the other will be at an infinite distance from the center of gravity. The other case of interest is when $r_y^2 = l_1 l_2$. In this case, one oscillation

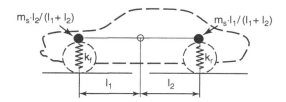

Figure 7.21 Equivalent system having two concentrated masses for the vehicle body.

center will be located at the point of attachment of the front spring to the vehicle body (or its equivalent), and the other at the point of attachment of the rear spring to the body. This can be verified by setting $l_{01} = l_2$ and $l_{02} = l_1$ in Eqs. (7.55) and (7.56), respectively. Under these circumstances, the two-degrees-of-freedom model for pitch and bounce shown in Figure 7.8 can be represented by a dynamically equivalent system with two concentrated masses at the front and rear spring attachment points (or their equivalents), as shown in Figure 7.21. The equivalent concentrated mass at the front will be $m_s l_2/(l_1 + l_2)$ and that at the rear will be $m_s l_1/(l_1 + l_2)$. The equivalent system is, in fact, two single-degree-of-freedom systems with natural frequency $\omega_{nf} = \sqrt{k_f(l_1 + l_2)/m_s l_2}$ for the front and natural frequency $\omega_{nr} = \sqrt{k_r(l_1 + l_2)/m_s l_1}$ for the rear. Thus, there is no interaction between the front and rear suspensions, and input at one end (front or rear) causes no motion of the other. This is a desirable condition for a good ride. For practical vehicles, however, this condition often cannot be satisfied. Currently, the ratios of $r_y^2/l_1 l_2$ vary from approximately 0.8 for sports cars through 0.9–1.0 for conventional passenger cars to 1.2 and above for some front-wheel-drive cars.

In considering the natural frequencies for the front and rear ends, it should be noted that excitation from the road to a moving vehicle will affect the front wheels first and the rear wheels later. Consequently, there is a time lag between the excitation at the front and that at the rear. This results in a pitching motion of the vehicle body. To minimize this pitching motion, the equivalent spring rate and the natural frequency of the front end should be slightly less than those of the rear end. In other words, the period for the front end ($2\pi/\omega_{nf}$) should be greater than that for the rear end ($2\pi/\omega_{nr}$). This ensures that both ends of the vehicle will move in phase (i.e., the vehicle body is merely bouncing) within a short time after the front end is excited. From the point of view of passenger ride comfort, pitching is more annoying than bouncing. The desirable ratio of the natural frequency of the front end to that of the rear end depends on the wheelbase of the vehicle, the average driving speed, and the wavelengths of the road profile.

As noted previously, a variety of multibody dynamics software packages, such as CarSim and TruckSim (Mechanical Simulation Corporation, Ann Arbor, MI, USA), etc. have become commercially available. They can be used to simulate vibrations of cars and commercial vehicles in detail.

7.3 Introduction to Random Vibration

7.3.1 Surface Elevation Profile as a Random Function

1) Basic concept of random functions

In early studies of vehicle ride characteristics, excitations from the ground in the forms of sine waves, step functions, or triangular waves are used. While these ground profiles could provide a basis for comparing the ride quality of vehicles with various suspension designs on a relative basis, they do not provide a realistic road input for the study of the actual ride behavior of vehicles, since surface profiles are rarely of simple forms. Later, it is recognized that ground profiles should be

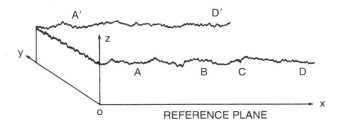

Figure 7.22 Surface elevation profile as a random function.

more realistically described as random functions (Macaulay 1963). A characteristic of a random function is that its instantaneous value cannot be predicted in a deterministic manner. For instance, the vertical road profile displacement z above a reference plane at any point, such as A, is not predictable as a function of the distance between the point in question and the origin along the x-axis shown in Figure 7.22. However, certain properties of random functions can be described statistically. For instance, the mean-square value of the vertical road profile displacement as a function of wavelength can be used to characterize the road profile as a random function.

There are certain features of random functions that are of practical interest. Referring to the surface profile shown in Figure 7.22, if the statistical properties of the portion of the road between A and B are the same as those of any other portion, such as CD, then the random function representing the surface profile is said to be stationary. This means that under these circumstances, the statistical properties of the surface profile derived from a portion of the road can be used to define the properties of the entire section of the road surface. If the statistical properties of the surface profile on one plane, such as AD, are the same as those on any parallel plane, such as $A'D'$, then the random function representing the surface profile is said to be ergodic. Thus, if the random function is both stationary and ergodic, the analysis of surface profile will be simplified to a great extent.

The availability of high-speed profilers (profilometer) has facilitated the collection of road profile data for evaluation of vehicle ride quality and vehicle suspension designs, as well as for road management and maintenance purposes (Andren 2006).t

2) Power spectral density of the vertical road profile displacement

Road profile data are commonly processed using the fast Fourier transform (FFT) technique, while other methods can also be used (Bendat and Piersol 2010; International Organization for Standardization 2016). FFT converts the road profile data to the power spectral density of the vertical road profile displacement (usually referred to as displacement power spectral density) in the spatial frequency domain. The spatial frequency Ω is the inverse of wavelength l (e.g. $\Omega = 1/l$ in cycles/m or cycles/ft). The angular spatial frequency, which is equal to $2\pi/l$ expressed in rad/m or rad/ft, is also used. In the following analysis, however, the spatial frequency Ω is primarily used. The power spectral density of the vertical road profile displacement $S_s(\Omega)$ is the mean-square value of vertical displacement $\overline{z}^2$ per unit spatial frequency bandwidth. The mean-square value of vertical displacement $\overline{z}^2_{\Omega_1 \to \Omega_2}$ in the frequency band from spatial frequency Ω_1 to Ω_2 can be obtained by integrating the power spectral density $S_s(\Omega)$ over the same bandwidth as follows:

$$\overline{z}^2_{\Omega_1 \to \Omega_2} = \int_{\Omega_1}^{\Omega_2} S_g(\Omega) d\Omega \qquad (7.57)$$

Figure 7.23 shows the variations of power spectral density of vertical road profile displacement with spatial frequency for various types of runways and highways (Van Deusen 1968). Figure 7.24 shows the variations of displacement power spectral density with spatial frequency for two types of unprepared terrain in the undisturbed state (Wong 1972). The displacement power spectral density is expressed in m^3 (ft^3) or in m^2/cycles/m (ft^2/cycles/ft).

The relationships between the displacement power spectral density $S_g(\Omega)$ and the spatial frequency Ω for the ground profiles shown in Figures 7.23 and 7.24 may be approximated by

$$S_g(\Omega) = C_{sp}\Omega^{-N} \tag{7.58}$$

where C_{sp} and N are constants.

Fitting Eq. (7.58) to the relationships between $S_g(\Omega)$ and Ω shown in Figures 7.23 and 7.24 yields the values of C_{sp} and N given in Table 7.3. N is a dimensionless constant, while the dimension of C_{sp} varies with the value of N.

It should be mentioned that in ground profile measurements, the spatial frequency need not in general be measured lower than 0.01 cycles/m (0.003 cycles/ft) for road vehicles and 0.05 cycles/m (0.015 cycles/ft) for off-road vehicles. The enveloping effect of the pneumatic tire acts as a low-pass filter for road input to the vehicle. For general road profile measurements, this results in a recommended upper limit of 10 cycles/m (3.05 cycles/ft) (ISO 8608:2016).

It should also be noted that on deformable terrain, like the multi-pass effect on terrain properties discussed in Chapter 2, the passage of the front tire can modify the terrain profile such that the profile for the following tire is usually not the same as that for the preceding tire. The issue of the extent that the tire or track modifies terrain profile has not been extensively investigated.

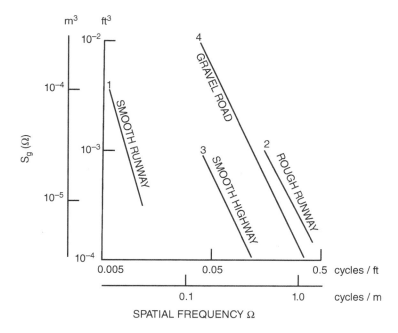

Figure 7.23 Variations of displacement power spectral density with spatial frequency for various types of road and runway surface. Reprinted with permission from SAE paper 670021 © 1968 Society of Automotive Engineers Inc.

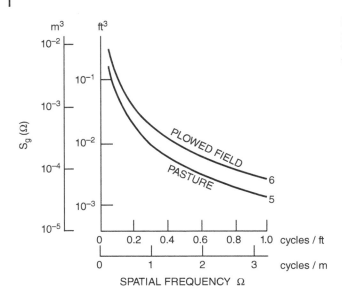

Figure 7.24 Variations of displacement power spectral density with spatial frequency for two types of unprepared terrain in the undisturbed state.

Table 7.3 Values of C_{sp} and N for power spectral density functions for various surfaces.

No.	Description	N	C_{sp}	C'_{sp}
1	Smooth runway	3.8	4.3×10^{-11}	1.6×10^{-11}
2	Rough runway	2.1	8.1×10^{-6}	2.3×10^{-5}
3	Smooth highway	2.1	4.8×10^{-7}	1.2×10^{-6}
4	Highway with gravel	2.1	4.4×10^{-6}	1.1×10^{-5}
5	Pasture	1.6	3.0×10^{-4}	1.6×10^{-3}
6	Plowed field	1.6	6.5×10^{-4}	3.4×10^{-3}

Source: Van Deusen (1969) and Wong (1972).
Note: C_{sp} is the value used for computing $S_g(\Omega)$ in m^3 or m^2/cycles/m. C'_{sp} is the value used for computing $S_g(\Omega)$ in ft^3 or in ft^2/cycles/ft. The numbers in the table refer to the curves shown in Figures 7.23 and 7.24.

In using the relationships between displacement power spectral density and spatial frequency for simulations of vehicle ride, the influence of the tire–road contact model on vehicle vibration response should be considered. Various tire–road contact models, including the point contact model, moving averaging road profile model, and tire-enveloping model, have been studied. It has been found that tire–road models have only marginal influence on predicting ride comfort. It is shown, however, that tire–road contact models would have a marked impact on predicting dynamic tire forces (Mucka and Gagnon 2015).

3) Classification of road surface profiles (roughness)

A) ISO 8608:2016

According to this ISO standard, road surface profiles (roughness) are classified into eight categories from A to H. This classification is based on the relationships between the power spectral density of

Figure 7.25 ISO classification of roads. *Source:* ISO 8608 : 2016. Copied with permission of the Standards Council of Canada (SCC) on behalf of ISO. The standard can be purchased from the national ISO members in your country or the ISO store. Copyright remains with ISO.

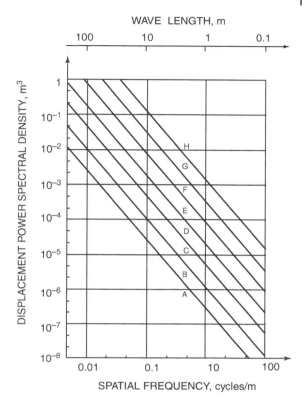

the vertical road profile displacement (displacement power spectral density) and the spatial frequency (the inverse of wavelength of road profile), as shown in Figure 7.25. For instance, for class A, the relationships between displacement power spectral density and spatial frequency are in the zone below line A shown in the figure. For class B, the relationships between the displacement power spectral density and spatial frequency are within the zone between the upper limit (boundary) defined by line B and the lower limit defined by line A shown in the figure. For class H, the relationships between the displacement power spectral density and spatial frequency are in the zone above line G.

The lower limit, upper limit, and geometric mean of the displacement power spectral density at spatial frequency of 0.1 cycles/m for road class A to class H are given in Table 7.4. The geometric mean of the lower and upper limits is the length of one side of the square whose area is equal to the area of a rectangle with sides of the lower and upper limits. In other words, the geometric mean is the square root of the product of lower and upper limits. In Table 7.4, for road class B, the lower limit of the displacement power spectral density at spatial frequency of 0.1 cycles/m is 32×10^{-6} m^3 and the upper limit is 128×10^{-6} m^3. Accordingly, the geometric mean is 64×10^{-6} m^3 ($\sqrt[2]{32 \times 128} \times 10^{-6}$ m^3), as shown in Table 7.4. For road class A, the lower limit is open and not specified, and the upper limit of the displacement power spectral density at spatial frequency of 0.1 cycles/m is 32×10^{-6} m^3. For this case, it may be considered that the rectangle referred to above is degenerated to a line representing the upper limit of 32×10^{-6} m^3. Accordingly, the geometric mean is taken as 16×10^{-6} m^3, which is half of the upper limit, as shown in Table 7.4. For class H, the upper limit is open and not identified, and the lower limit of the displacement

Table 7.4 ISO 8608:2016 road classification.

| Road class | Degree of roughness | | |
| | Lower limit | Geometric mean | Upper limit |
	Displacement power spectral density, 10^{-6} m³, at spatial frequency of 0.1 cycles/m		
A	–	16	32
B	32	64	128
C	128	256	512
D	512	1024	2048
E	2048	4094	8192
F	8192	16,384	32,768
G	32,768	65,536	131,072
H	131,072	262,144	–

Source: ISO 8608:2016. Copied with permission of the Standards Council of Canada (SCC) on behalf of ISO. The standard can be purchased from the national ISO members in your country or the ISO store. Copyright remains with ISO.

power spectral density at spatial frequency of 0.1 cycles/m is $131{,}072 \times 10^{-6}$ m³. Following a similar approach to that for class A, the geometric mean for class H is taken as $262{,}144 \times 10^{-6}$ m³, which is twice the lower limit, as shown in Table 7.4.

It is generally considered that class A road is very smooth. The roughness for road class C is considered average, and class E is considered quite rough (very poor).

For vehicle vibration analysis, it is more convenient to express the displacement power spectral density in terms of time (temporal) frequency f in Hz, rather than in spatial frequency Ω since vehicle vibration is a function of time. The transformation of the spatial frequency Ω in cycles/m (or cycles/ft) to the time frequency f in Hz is through vehicle speed:

$$
\begin{aligned}
f(\text{Hz}) &= \Omega(\text{cycles/m})V(\text{m/s}) \\
&= \Omega(\text{cycles/ft})V(\text{ft/s})
\end{aligned}
\tag{7.59}
$$

The transformation of the displacement power spectral density expressed in terms of the spatial frequency $S_g(\Omega)$ to that in terms of time frequency $S_g(f)$ is through vehicle speed:

$$
S_g(f) = \frac{S_g(\Omega)}{V}
\tag{7.60}
$$

B) International roughness index (IRI)

The International Roughness Index (IRI) is an index computed from a longitudinal road profile measurement using a quarter-car model with specific parameters at a simulation speed of 80 km/h (50 mph) (Sayers et al. 1986; Sayers 1995; American Society for Testing and Materials 2015). It is widely used in many countries for the evaluation and management of road systems (Mucka 2017).

The general procedures for determining IRI may be summarized as follows:

- Measurements of longitudinal road profiles are commonly made using precision (or Class 1) profiling devices, including laser profilers. For Class 1 profiles, the distance between longitudinal samples shall be less than or equal to 25 mm (1 in.) and the precision for the measurement of road profile shall not exceed 0.38 mm (0.015 in.) (American Society for Testing and Materials 2018).

- The measured longitudinal road profiles are used as input to a quarter-car model for calculating the value of IRI. The quarter-car model used is like that schematically shown in Figure 7.7, with the exception that the tire damping is neglected. The parameters of the quarter-car model, usually referred to as the Golden Car Model, are given as follows: $m_{us}/m_s = 0.15$; $k_{tr}/m_s = 653$ $1/s^2$; $k_s/m_s = 63.3$ $1/s^2$; and $c_{sh}/m_s = 6.0$ $1/s$. The reference forward speed V is 80 km/h (50 mph).

- The value of the IRI for one wheel track (usually the right wheel track), which is an accumulation of the simulated relative displacement (mm or m) between the sprung mass and the unpsrung mass normalized with respect to the distance L (m or km) traveled, is calculated as follows (Sayers 1995):

$$IRI = \frac{1}{L} \int_0^{L/V} |\dot{z}_1 - \dot{z}_2| dt \qquad (7.61)$$

where $\dot{z}_1$ and $\dot{z}_2$ are the vertical velocities of the sprung mass and unsprung mass, respectively; and IRI is expressed in mm/m or m/km and is considered as a nondimensional index.

- When longitudinal road profiles are measured simultaneously for both traveled wheel tracks (right and left tracks), the mean value of the IRI from both tracks is referred to as MRI (American Society for Testing and Materials 2015). MRI is used for characterizing road roughness in many states in the U.S. (Mucka 2017).

When the value of IRI is less than 2 (mm/m or m/km), the car ride is judged to be comfortable at speed over 120 km/h (75 mph). Undulation is barely perceptible at 80 km/h (50 mph) in the range of IRI from 1.3 to 1.8. In the range of IRI from 6 to 8, the ride is comfortable up to 70–90 km/h (44–56 mph) and the road surface is usually associated with defects, having frequent moderate and uneven depressions or patches, or having occasional potholes. With IRI value of 10, it is necessary to reduce vehicle speed below 50 km/h (30 mph) and the road surface is usually associated with deep depressions, potholes, and severe disintegration (American Society for Testing and Materials 2015).

While the general procedure for determining the value of IRI (or MRI) is specified above, implementations and applications of IRI, for purposes of evaluating and managing road systems, vary from country to country or from region (province or state) to region within a country (Mucka 2017). For instance, the distance L, over which the longitudinal road profile and the value of IRI are determined in different states of the U.S., varies from 16.1 m (or 0.01 mile) through 152.4 m (500 ft) to 1610 m (1 mile). Furthermore, a generally accepted classification of roads based on IRI, for purposes of road construction and maintenance, has not yet been established. The number of road classes and the values of IRI for the lower and upper limits of each class vary from country to country (Mucka 2017).

7.3.2 Frequency Response Function

For a linear system, a linear relationship between input and output exists. This relationship, which also holds for random functions, is shown in the block diagram of Figure 7.26 for a linear vehicle system. The vehicle system, characterized by its transfer function, relates the input, such as the power spectral density of the vertical road profile displacement, to the output, such as the power spectral density of the vertical displacement of vehicle sprung mass. The transfer function or frequency response function is simply the ratio of the output to input under steady-state conditions. For instance, if the vehicle is simplified to a single-degree-of-freedom system, and both the input of

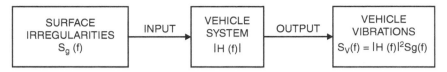

Figure 7.26 Input and output of a linear vehicle system.

road profile and the output of sprung mass are expressed in the same unit (i.e., displacement, velocity, or acceleration), the modulus of the transfer function $H(f)$ is expressed by

$$|H(f)| = \left|\sqrt{\frac{1 + (2\zeta f/f_n)^2}{\left[1 - (f/f_n)^2\right]^2 + [2\zeta f/f_n]^2}}\right| \tag{7.62}$$

where f is the frequency of excitation, f_n is the natural frequency of the vehicle sprung mass, and ζ is the damping ratio of the suspension system.

In this case, the transfer function $H(f)$ is simply the transmissibility shown in Figure 7.9. If the input is defined in terms of displacement power spectral density function and the output is defined as the acceleration power spectral density of vehicle sprung mass, then the modulus of the transfer function $H(f)$ will take the following form:

$$|H(f)| = \left|(2\pi f)^2 \sqrt{\frac{1 + (2\zeta f/f_n)^2}{\left[1 - (f/f_n)^2\right]^2 + [2\zeta f/f_n]^2}}\right| \tag{7.63}$$

The squared values of the moduli of two transfer functions representing two simplified single-degree-of-freedom vehicle models, one with a bounce natural frequency of 3.5 Hz and a damping ratio of 0.1 and the other with a bounce natural frequency of 1.0 Hz and a damping ratio of 0.5, are shown in Figure 7.27 (Wong 1972). The transfer functions shown are for predicting vehicle response having displacement as input and acceleration as output.

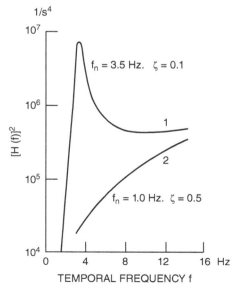

Figure 7.27 The square of the moduli of the transfer functions of two simplified vehicle models with different natural frequencies and damping ratios.

The relationship between the input power spectral density of the vertical road profile displacement $S_g(f)$ and the output power spectral density of vertical acceleration of the vehicle sprung mass $S_v(f)$ is given by

$$S_v(f) = |H(f)|^2 \, S_g(f) \tag{7.64}$$

This indicates that in this case, the output power spectral density of $S_v(f)$ is related to the input power spectral density of $S_g(f)$ through the square of the modulus of the transfer function for a linear vehicle system. In the evaluation of vehicle ride quality, the power spectral density for the acceleration of vehicle sprung mass as a function of time frequency is of prime interest.

7.3.3 Evaluation of Vehicle Vibration in Relation to Ride Comfort Criteria

After the power spectral density function of vehicle sprung mass acceleration has been obtained, further analysis is required to relate it to any ride comfort criterion that may be chosen. For instance, if the fatigue or decreased proficiency boundaries for z-axis acceleration (a_z) recommended by the International Standard ISO 2631-1 : 1985 shown in Figure 7.2(a) are selected, then the transformation of the power spectral density function of vehicle sprung mass vertical acceleration into the rms value of vertical acceleration as a function of frequency is necessary. As indicated previously by Eq. (7.57), the mean-square value of vertical displacement $\bar{z}^2_{\Omega_1 \to \Omega_2}$ in the frequency band from spatial frequency Ω_1 to Ω_2 can be obtained by integrating the power spectral density $S_s(\Omega)$ over the same bandwidth. Following this procedure, the mean-square value of vehicle sprung mass vertical acceleration can be obtained by integrating the corresponding vertical acceleration power spectral density function over a given frequency band.

In practice, a series of discrete center frequencies of one-third octave band within the range of interest is first selected. To determine the mean-square value of vehicle sprung mass vertical acceleration at a given center frequency f_c, the power spectral density function of vertical acceleration is integrated over a one-third octave band of which the upper cutoff frequency is $\sqrt[3]{2}$ times the lower. In other words, by integrating the power spectral density function of vertical acceleration over a frequency band of $0.89 - 1.12 \, f_c$, the mean-square value of vertical acceleration at a given center frequency f_c can be obtained. The rms value of vertical acceleration at each center frequency f_c can then be calculated by

$$\text{rms acceleration} = \left[\int_{0.89 \, f_c}^{1.12 f_c} S_v(f) \, df \right]^{1/2} \tag{7.65}$$

where $S_v(f)$ is the power spectral density function of vehicle sprung mass vertical acceleration.

After obtaining the rms values of vehicle sprung mass vertical acceleration of a series of center frequencies within the range of interest, one can then evaluate the vertical vibration of the vehicle against the ride comfort criterion chosen.

Figure 7.28 shows the variations of measured rms values of vertical and lateral accelerations in one-third octave band with frequency at the driver's seat of a North American passenger car traveling at a speed of 80 km/h (50 mph) over a smooth highway, as compared with the reduced comfort boundaries recommended by ISO 2631/1 : 1985 (Healy 1977). It should be noted that, as

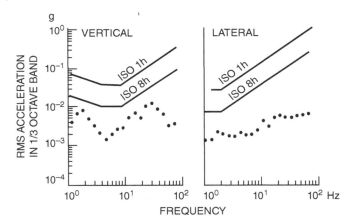

Figure 7.28 Measured vertical and lateral accelerations of a passenger car traveling at 80 km/h (50 mph) over a smooth road. *Source:* Healy (1977). Reproduced with permission of the American Society of Mechanical Engineers.

mentioned in Section 7.1.1, the values for the reduced comfort boundaries for vertical and lateral direction shown in Figure 7.28 correspond to approximately one-third of the values for the fatigue-decreased proficiency boundaries for longitudinal (vertical or z-axis) and transverse (lateral or y-axis) acceleration limits shown in Figure 7.2(a) and (b), respectively. Evaluations of vehicle ride quality following the guidance of ISO 2631-1 : 1997 for comfort have been reported (Mucka et al. 2020).

The procedure described above is for a simplified vehicle model with a single degree of freedom. A vehicle has many degrees of freedom, and between the driver and the vehicle, there is a seat suspension. In addition, more than one random input is imposed on the vehicle. In the case of a passenger car, there are four inputs, one from each of the four wheels. The interaction of the random inputs with each other becomes important in determining the output. The consideration of cross-spectral densities is essential, and the time lag of the input at the rear wheel with respect to the front wheel should also be considered. All of these would make the analysis much more complex than that described above. However, analytical techniques based on random vibration theory have been developed into a practical tool to evaluate vehicle ride quality under various operating conditions (Van Deusen 1968; Popp 2014).

7.4 Active and Semi-Active Suspensions

As discussed previously, to achieve good vibration isolation for the sprung mass over a wide range of frequency, a soft suspension spring is generally required, while to provide good road holding capability at a frequency close to the natural frequency of the unsprung mass ("wheel hop" frequency), a stiff suspension spring is preferred. To reduce the amplitude of vibration of the sprung mass at a frequency close to its natural frequency, a high damping ratio is required, while in the high-frequency range, to provide good vibration isolation for the sprung mass, a low damping ratio is preferred. On the other hand, to achieve good road holding in the high-frequency range, a high damping ratio is required. These conflicting requirements cannot be met by a conventional (passive) suspension system since the characteristics of its spring and shock absorber are fixed and cannot be modulated in accordance with vehicle operating conditions.

7.4.1 Active Suspensions

To provide a vehicle with improved ride quality, handling, and performance under various operating conditions, active suspension systems and their controls have been studied and developed (Chalasani 1986; Crolla et al. 1987; Tseng and Hrovat 2015). An active suspension for a quarter-car model is schematically illustrated in Figure 7.29. The spring and shock absorber in a conventional suspension system are replaced by a force actuator in an active system. The actuator may also be installed in parallel with a conventional suspension spring. The operating conditions of the vehicle are continuously monitored by sensors, as schematically shown in the figure. Based on the signals obtained by the sensors and the prescribed control strategy, the force in the actuator is modulated to achieve improved ride, handling, and performance. The optimum control strategy is defined as the one that minimizes the following:

- the rms value of the sprung mass acceleration
- the rms value of the suspension travel
- the rms value of the dynamic tire deflection

Usually, these quantities are multiplied by weighting factors and then combined to form an evaluation function. Various control theories have been applied to establishing the optimum control strategy to minimize the evaluation function.

An active suspension can also be used to control the height, roll, dive (forward pitching), and squat (rearward pitching) of the vehicle body. By exercising height control, the ride height of the vehicle body can be kept constant despite changes in load conditions. This ensures adequate suspension travel for negotiating bumps. To reduce aerodynamic resistance and aerodynamic lift at high speeds (see Section 3.2), the ground clearance and the angle of attack of the vehicle body can be conveniently adjusted with an active system. Over rough terrain, the ground clearance and suspension travel can be regulated to suit operating requirements. During cornering, roll control can be achieved by adjusting damping forces or by producing antiroll forces in the left and right suspensions. With an active system, it is possible to eliminate the roll of the vehicle body and the associated roll–steer (steering action induced by the roll of the vehicle body relative to the tires), thus

Figure 7.29 Schematic of an active suspension system for a quarter-car model.

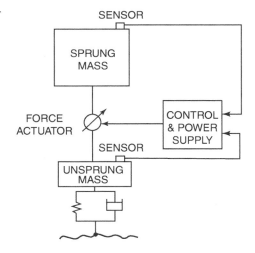

maintaining the desirable handling characteristics during cornering. During acceleration or braking, squat or dive control can be achieved by adjusting damping forces or by producing antipitch forces in the front and rear suspensions to maintain the desired attitude of the vehicle body and the required normal loads on the tires.

While an active suspension system offers opportunities for substantial improvements in ride, handling, and vehicle posture, it typically leads to higher energy consumption for operating the force actuator, increased cost, added complexity, and other operational requirements (such as robust operations and diagnostics, and fault containment and management) (Tseng and Hrovat 2015).

7.4.2 Semi-Active Suspensions

As a compromise between a passive and an active suspension, with respect to performance, complexity, energy consumption, and cost, semi-active suspensions have been developed. In a semi-active system, the conventional suspension spring is usually retained, while the damping force in the controllable shock absorber can be modulated in accordance with operating conditions. A semi-active suspension for a quarter-car model is schematically shown in Figure 7.30. The regulating of the damping force can be achieved by adjusting the orifice area in some controllable shock absorbers, thus changing the resistance to fluid flow.

Applications of electrorheological fluids and magnetorheological fluids to the development of controllable dampers have been studied (Petek 1992; Wong et al. 1993; Wu et al. 1994; Sturk et al. 1995; Mui et al. 1996; Dixon 2007). An electrorheological fluid is a mixture of a dielectric base oil and fine semiconducting particles, such as aluminum silicate. In the presence of an electrical field, the electrorheological fluid thickens, allowing for continuous control of its apparent viscosity and hence its resistance to flow. By regulating the voltage applied across the flow of the electrorheological fluid in a shock absorber, the damping force can be regulated. The process is continuous and reversible. The response is generally fast (possibly 1 ms), but is temperature sensitive, becoming rather slow at low temperatures. One of the major challenges facing the electrorheological absorber

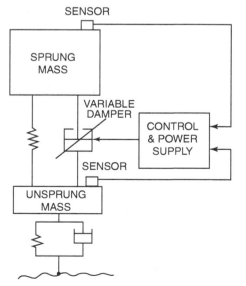

Figure 7.30 Schematic of a semiactive suspension for a quarter-car model.

Figure 7.31 A concept for an electrorheological damper.

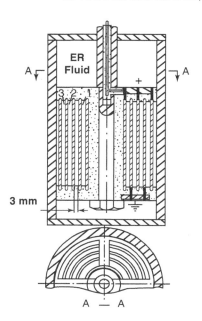

is the development of electrorheological fluids that have adequate shear strength (yield stress) and can function effectively over a sufficiently wide temperature range from −40 to +120° C. An achievable shear strength for electrorheological fluid is 4 kPa, which is low, but sufficient to be of some practical use (Dixon 2007). Figure 7.31 shows the concept of an electrorheological damper (Wu et al. 1994). Voltage is applied to the concentric cylinders to generate electrical fields. The electrorheological fluid flows through the gaps between the cylinders when the damper is in operation. Varying the voltage applied causes a change in the apparent viscosity of the fluid and hence the damping force. The electrical field strength across the gaps could be as high as a few kilovolts per millimeter of gap. Figure 7.32 shows the measured performance of an electrorheological damper on controlling the vibration of the sprung mass of a quarter-car model (Wu et al. 1994). When the electrorheological damper is activated and appropriate voltage is applied in accordance with a specific control strategy, the acceleration of the sprung mass is notably reduced in comparison with the damper acting as a passive device with no electrical field being applied.

Controllable dampers using the magnetorheological fluid as a working medium have been developed and are available for some brands of cars in production. A magnetorheological fluid is a mixture of micron-sized, in the range of 3–10 μm, magnetizable particles suspended in a carrier fluid, such as silicone oil. The magnetizable particles are basically just soft iron. Fibrous carbon may be added. A surfactant is also used to minimize the settling of the particles (Dixon 2007). The apparent viscosity of this type of fluid, and hence its resistance to flow, can be regulated by changing the magnetic field applied to the fluid. In comparison with the electrorheological fluid, the magnetorheological fluid has much higher shear strength, typically 50–100 kPa, which is at least of an order of magnitude higher than that of the electrorheological fluid. Temperature has less effect on the magnetorheological fluid than on the electrorheological fluid. Thus, it can effectively operate over a broader range of temperatures. Figure 7.33 shows the concept of a magnetorheological damper. By varying the current in the electromagnet, magnetic field strength around the orifice in the piston can be changed. The apparent viscosity of the fluid flowing through the orifice, and hence the damping force, can then be regulated. The general characteristics of a magnetorheological damper

(a)

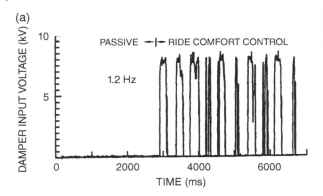

Figure 7.32 Comparisons of the measured vibration of the sprung mass of a quarter-car model before and after the activation of an electrorheological damper.

(b)

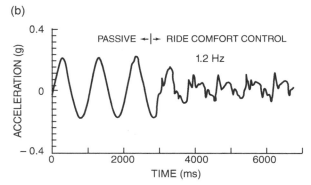

at a given magnetic field strength, expressed in terms of damping force versus relative displacement and damping force versus relative velocity between the piston and cylinder, are shown in Figure 7.34.

In comparison with a fully active system, a semi-active suspension requires much less power, and is less complex. It has also been found that when properly designed, the performance of a semi-active system may approach that of a fully active suspension under certain circumstances (Margolis 1982).

To successfully develop a semi-active suspension system, in addition to the design of the damper and the properties of the working medium used in the damper, the control strategy for modulating the damping force under various operating conditions is of great importance.

Two representative control strategies are outlined in subsections 1 and 2 below.

1) On–off control strategy

This control strategy can be described as follows: (a) if $\dot{z}_1(\dot{z}_1 - \dot{z}_2) > 0$, then the maximum (sometimes referred to as "firm") damping is required; (b) if $\dot{z}_1(\dot{z}_1 - \dot{z}_2) < 0$, then the minimum (sometimes referred to as "soft") damping is used, where $\dot{z}_1$ and $\dot{z}_2$ are the velocities of the sprung and unsprung mass, respectively (Krasnicki 1981; Margolis and Goshtasbpour 1984). This strategy indicates that if the relative velocity of the sprung mass with respect to the unsprung mass is in the same direction as that of the sprung mass absolute velocity, then a maximum damping force should be applied to reduce the acceleration of the sprung mass. On the other hand, if the two velocities are in opposite directions, the damping force should be at a minimum to minimize the acceleration of the sprung mass.

Figure 7.33 A concept of a magnetorheological damper.

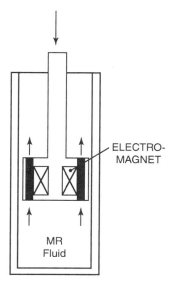

ELECTRO-
MAGNET

MR
Fluid

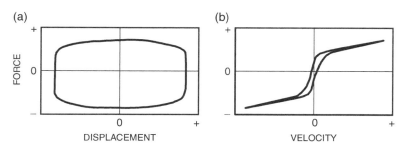

Figure 7.34 General characteristics of a magnetorheological damper: (a) damping force versus relative displacement; (b) damping force versus relative velocity between the piston and the cylinder of the damper at a given magnetic field strength.

Accurate measurement of the absolute vibration velocity of the sprung mass on a moving vehicle is difficult to achieve. Integrating the signals from an accelerometer to obtain velocity often does not yield sufficiently accurate results, particularly at low frequency.

In some suspension systems, instead of modulating the damping force continuously, the level of damping in the controllable shock absorbers is set by the driver in discrete steps in accordance with driving conditions. For instance, on relatively smooth highways and driving at high speeds, the damping may be set at a low level to provide a good ride. On the other hand, on rough roads and driving at low speeds, the damping may be set at a high level to reduce the vibration amplitude of the vehicle body.

2) Continuous control strategy

This control strategy for continuously regulating the damping force can be described as follows: (a) if $(\dot{z}_1 - \dot{z}_2)(z_1 - z_2) > 0$, then minimum damping is required; (b) if $(\dot{z}_1 - \dot{z}_2)(z_1 - z_2) < 0$, then the desired damping coefficient is $c_{sh} = k_s(z_1 - z_2)/(\dot{z}_1 - \dot{z}_2)$, where z_1 and z_2 are the displacements of the sprung and unsprung mass, respectively; $\dot{z}_1$ and $\dot{z}_2$ are velocities of the sprung and unsprung mass, respectively; k_s is the suspension spring stiffness (Alanoly and Sankar 1987; Jolly and Miller 1989). This control strategy requires only the measurements of the relative displacement and velocity between the sprung and unsprung mass, which can easily be made in practice.

This strategy indicates that if the spring force and damping force exerted to the sprung mass are in the same direction, to reduce the sprung mass acceleration, the damping force should be at a minimum. On the other hand, if the spring force and damping force are in opposite directions, then the damping force should be adjusted in such a way that it will be equal to the spring force in magnitude to produce zero acceleration for the sprung mass.

Figure 7.35 shows the ratio of the rms sprung mass acceleration with an active damper using the continuous control strategy to that using a passive damper with a fixed damping ratio $\zeta = 0.3$, as a function of vehicle speed over the road profile shown in Figure 7.19(b) based on simulation results. The damping ratio of the active damper can be continuously varied as required in the range between 0.1 and 1.0. The figure also shows the effect of time delay (from 1 to 5 ms) in modulating the damping force on the response of the sprung mass. It is shown that the semi-active suspension with the continuous control strategy offers better vibration isolation than a passive suspension system over a wide range of vehicle speeds. However, this control strategy does not necessarily offer an optimal road holding capability. As shown in Figure 7.36, the ratio of the rms dynamic tire deflection with the active damper to that with the passive damper is greater than one over a wide range of speeds, except at speeds lower than 20 km/h (12.5 mph).

Common to both active and semi-active suspension systems, the force exerted on the unsprung mass to control its motion (e.g., the normal force between the tire and the road) will act against the sprung mass. Thus, the reacting force will affect the vibration of the sprung mass. This imposes limits to what an active or semi-active suspension system can realistically achieve, in terms of providing optimal vibration isolation for the vehicle body and road holding at the same time. It indicates that in the development of the control strategy for an active or semi-active system, a proper compromise must be struck between ride comfort and road holding.

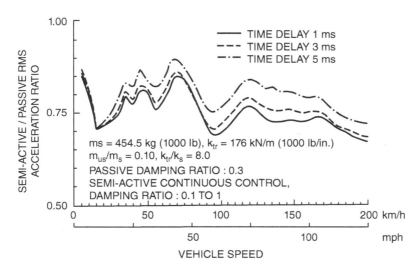

Figure 7.35 Ratios of the rms value of sprung mass acceleration with a semi-active suspension using the continuous control strategy to that with a passive suspension as a function of vehicle speed over a road with profile shown in Figure 7.19(b).

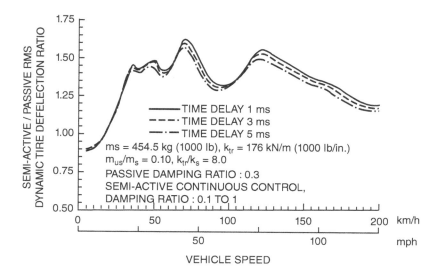

Figure 7.36 Ratios of the rms value of dynamic tire deflection with a semi-active suspension using the continuous control strategy to that with a passive suspension as a function of vehicle speed over a road with profile shown in Figure 7.19(b).

References

Alanoly, J. and Sankar, S. (1987). A new concept in semi-active vibration isolation. *Journal of Mechanisms, Transmissions and Automation in Design* **109** (2).

American Society for Testing and Materials (2015). ASTM E1926-08 (Standard practice for computing International Roughness Index of roads from longitudinal profile measurements).

American Society for Testing and Materials (2018). ASTM E950/E950M-09 (Standard test method for measuring the longitudinal profile of traveled surfaces with an accelerometer-established inertial profiling reference).

Andren, P. (2006). Power spectral density approximations of longitudinal road profiles. *International Journal of Vehicle Design* **40** (1–3).

Bendat, J.S. and Piersol, A.G. (2010). *Random Data: Analysis and Measurements Procedure*, 4e. Hoboken, NJ: Wiley.

Chalasani, R.M. (1986). Ride performance potential of active suspension systems, part I and part II. In: *Proc. ASME Symposium on Simulation and Control of Ground Vehicles and Transportation Systems* (eds. L. Segel, J.Y. Wong, E.H. Law and D. Hrovat). American Society of Mechanical Engineers publication AMD-80, DSC-2.

Crolla, D.A., Norton, D.N.L., Pitcher, R.H., and Lines, J.A. (1987). Active suspension control of an off-road vehicle. *Proceedings of the Institution of Mechanical Engineers, Part D: Journal of Automobile Engineering* **201** (D1).

Dixon, J.C. (2007). *The Shock Absorber Handbook*, 2e. Hoboken, NJ: Wiley.

Gillespie, T.D. (1992). *Fundamentals of Vehicle Dynamics*. Society of Automotive Engineers.

Healy, A.J. (1977). Digital processing of measured random vibration data for automobile ride evaluation. American Society of Mechanical Engineers publication AMD-24.

Hrovat, D. (2014). Active and semi-active suspension control. In: *Road and Off-Road Vehicle System Dynamics Handbook* (eds. G. Mastinu and M. Ploechl). Boca Raton, FL: CRC Press.

International Organization for Standardization (1985). ISO 2631/1:1985 (Evaluation of human exposure to whole-body vibration – Part 1: General requirements).

International Organization for Standardization (1997). ISO 2631-1:1997 (Mechanical vibration and shock – Evaluation of Human Exposure to Whole-body Vibration – Part 1: General requirements).

International Organization for Standardization (2010). ISO 2631-1:1997/Amd.1:2010 (Mechanical vibration and shock – Evaluation of human exposure to whole-body vibration – Part 1: General requirements, Amendment 1).

International Organization for Standardization (2016). ISO 8608:2016 (Mechanical vibration – Road surface profiles – Reporting of measured data).

Jolly, M.R. and Miller, L.R. (1989). The control of semi-active dampers using relative feedback signals. Society of Automotive Engineers paper 892483.

Krasnicki, E.J. (1981). The experimental performance of an 'on-off' active damper. *Shock and Vibration Bulletin* **51** (1).

Lee, R.A. and Pradko, F. (1969). Analytical analysis of human vibration. *SAE Transactions* **77**, Society of Automotive Engineers paper 680091.

Macaulay, M.A. (1963). Measurement of road surfaces. In: *Advances in Automobile Engineering, Part I* (ed. G.H. Tidbury). Oxford, UK: Pergamon Press.

Margolis, D.L. (1982). Semi-active heave and pitch control for ground vehicles. *Vehicle System Dynamics* **11** (1).

Margolis, D.L. and Goshtasbpour, W. (1984). The chatter of semi-active on–off suspension and its cure. *Vehicle System Dynamics* **13** (3).

Mucka, P. (2017). International Roughness Index specifications around the world. *Road Materials and Pavement Design* **18** (4).

Mucka, P. and Gagnon, L. (2015). Influence of tyre–road contact model on vehicle vibration response. *Vehicle System Dynamics* **53** (9).

Mucka, P., Stein, G.J., and Tobolka, P. (2020). Whole-body vibration and vertical road profile displacement power spectral density. *Vehicle System Dynamics* **58** (4).

Mui, G., Russell, D.L., and Wong, J.Y. (1996). Nonlinear parameter identification of an electro-rheological fluid damper. *Journal of Intelligent Material Systems and Structures* **7** (5).

Murphy N.R., Jr., and Ahlvin, R.B. (1975). Ride dynamics module for AMM-75 Ground Mobility Model. *Proc. 5th Int. Conf. of the International Society for Terrain–Vehicle Systems* IV, Detroit, MI.

Petek, N.K. (1992). Shock absorber uses electrorheological fluid. *Automotive Engineering* June: 27–30.

Popp, K. (2014). Ride comfort and road holding. In: *Road and Off-Road Vehicle System Dynamics Handbook* (eds. G. Mastinu and M. Ploechl). Boca Raton, FL: CRC.

Sayers, M.W. (1995). On the calculation of International Roughness Index from longitudinal profile. Transportation Research Record 1501.

Sayers, M.W., Gillespie, T.D., and Peterson, D.O. (1986). Guidelines for conducting and calibrating road roughness measurements. Technical paper 46. World Bank.

Smith, C.C., McGehee, D.Y., Healey, A.J. (1976). The prediction of passenger riding comfort from acceleration data. Research Report 16, Department of Transportation, Office of University Research, Washington, DC.

Society of Automotive Engineers (1965). Ride and Vibration Data Manual–SAE J6a.

Stikeleather, L.F., Hall, G.O., and Radke, A.O. (1973). A study of vehicle vibration spectra as related to seating dynamics. *SAE Transactions* **81**, Society of Automotive Engineers paper 720001.

Sturk, M., Wu, X.M., and Wong, J.Y. (1995). Development and evaluation of a high voltage supply unit for electrorheological fluid dampers. *Vehicle System Dynamics* **24** (2).

Tseng, H.E. and Hrovat, D. (2015). State of the art survey: active and semi-active suspension control. *Vehicle System Dynamics* **53** (7): 1–29.

Van Deusen, B.D. (1968). Analytical techniques for design riding quality into automotive vehicles. *SAE Transactions* **76**, Society of Automotive Engineers paper 670021.

Van Deusen, B.D. (1969). Human response to vehicle vibration. *SAE Transactions* **77**, Society of Automotive Engineers paper 680090.

Vierck, R.K. (1979). *Vibration Analysis*, 2e. New York: Harper & Row.

Wong, J.Y. (1972). Effect of vibration on the performance of off-road vehicles. *Journal of Terramechanics* **8** (4).

Wong, J.Y., Wu, X.M., Sturk, M., and Bortolotto, C. (1993). On the applications of electrorheological fluids to the development of semi-active suspension systems for ground vehicles. *Transactions of Canadian Society for Mechanical Engineering* **17** (4B).

Wu, X.M., Wong, J.Y., Sturk, M., and Russell, D.L. (1994). Simulation and experimental study of a semi-active suspension with an electrorheological damper. *International Journal of Modern Physics B* **8** (20 and 21); and. In: *Electrorheological Fluids: Mechanisms, Properties, Technology, and Applications* (eds. R. Tao and G.D. Roy). Singapore: World Scientific.

Problems

7.1 The sprung parts of a passenger car weigh 11.12 kN (2500 lb) and the unsprung parts weigh 890 N (200 lb). The combined stiffness of the suspension springs is 45.53 kN/m (260 lb/in.) and that of the tires is 525.35 kN/m (3000 lb/in.). Determine the two natural frequencies of the bounce motions of the sprung and unsprung masses. Calculate the amplitudes of the sprung and unsprung parts if the car travels at a speed of 48 km/h (30 mph) over a road of a sinewave form with a wavelength of 9.15 m (30 ft) and an amplitude of 5 cm (2 in.).

7.2 Owing to the wide separation of the natural frequency of the sprung parts from that of the unsprung parts, the bounce and pitch motions of the vehicle body and the wheel motions exist almost independently. The sprung parts of a vehicle weigh 9.79 kN (2200 lb), its center of gravity is 106.7 cm (42 in.) behind the front axle, and the wheelbase is 228.6 cm (90 in.). The combined stiffness of the springs of the front suspension is 24.52 kN/m (140 lb/in.) and that of the rear suspension is 26.27 kN/m (150 lb/in.). The radius of gyration of the sprung parts about a horizontal transverse axis through the center of gravity is 102.6 cm (40.4 in.). Calculate the natural frequencies of the pitch and bounce motions of the vehicle body. Also determine the locations of the oscillation centers.

7.3 If the vehicle described in Problem 7.2 travels over a concrete highway with expansion joints 15.24 m (50 ft) apart, calculate the speeds at which the bounce motion and pitch motion of the vehicle body are most apt to arise.

7.4 If the radius of gyration of the sprung parts of the vehicle described in Problem 7.2 can be varied, determine the conditions under which the oscillation centers of the vehicle body will be located at the points of attachment of the front and rear springs. Also calculate the natural frequencies of the sprung parts.

7.5   A tractor with a bounce natural frequency of 3.5 Hz and a damping ratio of 0.1 travels at a speed of 5 km/h (3.1 mph) over a plowed field of which the surface roughness characteristics are described in Table 7.3. Determine the root-mean-square value of vertical acceleration of the tractor at a frequency of 1 Hz. Evaluate whether the vibration of the vehicle is acceptable from a fatigue or decreased proficiency viewpoint for an eight hour duration based on the International Standard ISO 2631/1:1985.

7.6   An independent front suspension of a passenger car carries a mass (sprung mass) of 454.5 kg (or an equivalent weight of 1000 lb). The suspension spring rate is 22 kN/m (125 lb/in.). The mass of the tire/wheel assembly (unsprung mass) is 45.45 kg (or an equivalent weight of 100 lb) and the equivalent tire stiffness is 176 kN/m (1000 lb/in.). The damping ratio ζ of the suspension produced by the shock absorber is 0.3. If the car is traveling on a sinusoidal road profile with a wavelength of 5 m (16.4 ft) and an amplitude of 5 cm (2 in.), estimate the lowest vehicle speed at which the tire may lose contact with the road.

8

Introduction to Air-Cushion Vehicles

An air-cushion vehicle may be defined as a ground vehicle that is supported by a cushion of pressurized air. The cushion performs two basic functions:

- To separate the vehicle from the supporting surface, thus reducing or eliminating surface contact and the associated motion resistance
- To provide the vehicle with a suspension system

Since practical air-cushion concepts emerged in the 1950s, they have found applications in overwater as well as overland transport. Commercial passenger-carrying ferry services across stretches of water by air-cushion vehicles were initiated in the United Kingdom in the 1960s. Air-cushion vehicles with a capacity of 254 passengers and 30 cars went into service across the English Channel in 1968. A later version had a capacity of 418 passengers and 60 cars.

Owing to changing circumstances, such as the opening of the Channel Tunnel between England and France in 1994, and competition with traditional ferries and catamarans with lower fuel and maintenance costs, commercial ferry services with air-cushion vehicles across the English Channel ceased in 2000 after 32 years. While commercial services by air-cushion vehicles have declined since then, they still play an active role in defense and coast guard operations, in search and rescue missions, and in recreational activities, because of their unique capability of being able to travel across a variety of surfaces, including land, water, mud, marsh, and ice.

In this chapter, the performance of the principal types of air-cushion system is discussed. The characteristics unique to air-cushion vehicles are also examined.

8.1 Air-Cushion Systems and their Performances

There are two principal types of air-cushion system: the plenum chamber and the peripheral jet.

8.1.1 Plenum Chambers

Figure 8.1 shows the basic features of a simple plenum chamber (Elsley and Devereux 1968). Most air-cushion vehicles essentially employ a plenum chamber configuration. Pressurized air is pumped into the chamber by a fan or a compressor to form an air cushion that supports the vehicle. Under steady-state conditions, the air being pumped into the chamber is just sufficient to replace

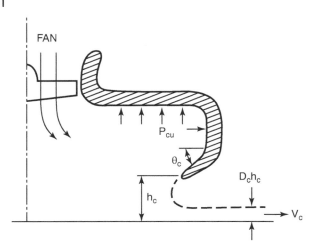

Figure 8.1 Geometry of a simple plenum chamber. Reproduced with permission from *Hovercraft Design and Construction* by G.H. Elsley and A.J. Devereux, copyright © by Elsley and Devereux (1968).

the air leaking under the peripheral gap, and the weight of the vehicle W is equal to the lift F_{cu} generated by the cushion pressure p_{cu}:

$$F_{cu} = W = p_{cu}A_c \tag{8.1}$$

where A_c is the effective cushion area.

For most current designs, the cushion pressure varies in the range $1.2 - 3.3$ kPa $(25 - 70 \text{ lb/ft}^2)$ for overwater and overland vehicles. For prototype high-speed guided ground transport vehicles, a cushion pressure of 4.2 kPa (87 lb/ft^2) has been used.

Assume that the air inside the chamber is essentially at rest. From Bernoulli's theorem, the velocity of air escaping under the peripheral gap V_c is given by

$$V_c = \sqrt{\frac{2p_{cu}}{\rho}} \tag{8.2}$$

where ρ is the mass density of air.

The total volume flow of air from the cushion Q is given by

$$Q = h_c l_{cu} D_c V_c = h_c l_{cu} D_c \sqrt{\frac{2p_{cu}}{\rho}} \tag{8.3}$$

where h_c is the clearance height; l_{cu} is the cushion perimeter; and D_c is the discharge coefficient. The discharge coefficient is primarily a function of the wall angle θ_c shown in Figure 8.1 and the length of the wall. For a long wall and non-viscous fluid, the values of D_c are as follows:

θ_c	0	45°	90°	135°	180°
D_c	0.50	0.537	0.611	0.746	1.000

In practice, because of the viscosity of the air, the values of D_c tend to be slightly less than those given above.

The power required to sustain the air cushion at the peripheral gap P_a is given by

$$P_a = p_{cu}Q$$
$$= h_c l_{cu} D_c p_{cu}^{3/2} \left(\frac{2}{\rho}\right)^{1/2} \tag{8.4}$$

Substituting Eq. (8.1) into the above equation, one obtains

$$P_a = h_c l_{cu} D_c \left(\frac{W}{A_c}\right)^{3/2} \left(\frac{2}{\rho}\right)^{1/2} \tag{8.5}$$

This equation shows that the power required to sustain the cushion in the plenum chamber varies with the clearance height and the perimeter, and that, for a given vehicle, it is proportional to the weight of the vehicle raised to the power of 3/2. In determining the power required to drive the fan, intake losses, ducting losses, diffusion losses, and fan efficiency should be taken into consideration.

Consider that an air jet with the same amount of volume flow Q and having the same air velocity V_c as those of the air cushion is directly used to generate a lift force. The lift force F_l generated by the change of momentum of the air jet is given by

$$F_l = \rho Q V_c \tag{8.6}$$

An augmentation factor K_a, which is a measure of the effectiveness of an air-cushion system as a lift generating device, can be defined as

$$K_a = \frac{F_{cu}}{F_l} = \frac{p_{cu}A_c}{\rho Q V_c} = \frac{A_c}{2h_c l_{cu} D_c} \tag{8.7}$$

Introducing the concept of hydraulic diameter D_h

$$D_h = \frac{4A_c}{l_{cu}} \tag{8.8}$$

Equation (8.7) becomes

$$K_a = \frac{D_h}{8h_c D_c} \tag{8.9}$$

This expression shows that the higher the ratio of the hydraulic diameter D_h to the clearance height h_c, the more effective the air-cushion system will be. Useful guidelines for the selection of the configuration and dimensions of an air-cushion vehicle can be drawn from this simple equation.

There are two principal forms of plenum chamber in use: one with a flexible skirt, and the other with a combination of flexible skirt and rigid sidewall, as shown in Figure 8.2. The prime reason for using the flexible skirt is to allow the vehicle to have relatively large clearance between its hard structure and the supporting surface, while at the same time keeping the clearance height under the skirt sufficiently small to enable the power required for lift to remain within reasonable limits. A combination of flexible skirt and rigid sidewall is used in marine air-cushion vehicles in which the air can only leak through the gaps in the front and rear of the vehicle. The air in the cushion is prevented from leaking along the sides by rigid sidewalls immersed in the water. Thus, the power required to sustain the cushion is reduced. The sidewalls can also contribute to the directional stability of the vehicle.

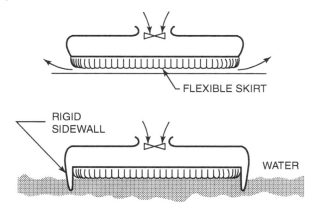

Figure 8.2 Flexible skirted plenum chamber and rigid sidewall plenum chamber.

There are many variants of the plenum chamber configuration with a flexible skirt. Figure 8.3 shows the multiple-cone skirt system used in the Bertin Terraplane BC7 (National Research Council of Canada 1969). The conical form ensures that the shape of the skirt under pressure is stable. The system can provide the vehicle with sufficient roll and pitch stability. When the vehicle rolls, the air gap of the cone on the down-going side is reduced. Consequently, the air flow from that side decreases and the cushion pressure increases. This, together with the decrease of cushion pressure in the cone on the up-going side, provides a restoring moment that tends to bring the vehicle back to its original position. This system is also less sensitive to loss of lift over ditches than the single plenum chamber configuration. The multiple-cone system shown in Figure 8.3 requires, however, more power to sustain the cushion than an equivalent single plenum chamber because the ratio of the total cushion perimeter l_{cu} to the cushion area A_c is higher than that of a comparable single plenum chamber. In other words, the equivalent hydraulic diameter of the multiple-cone skirt system is lower than that of an equivalent single plenum chamber. To reduce the volume flow, a peripheral skirt around the multiple cones may be added, as shown in Figure 8.4. This also increases the effective cushion area, although the cushion pressure between the cones and the peripheral skirt is lower than that inside the cones.

The performance of the multiple-cone system with a peripheral skirt may be evaluated analytically (Wong 1972a). Consider that the cushion pressure inside the cones is p_{cu} and that between

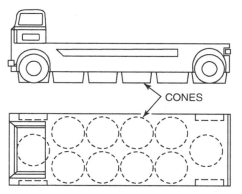

Figure 8.3 Multiple-cone skirt system used in the Bertin Terraplane BC7. *Source:* National Research Council of Canada (1969).

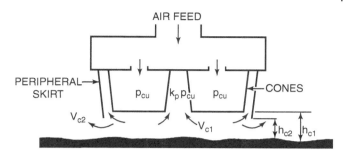

Figure 8.4 Multiple-cone system with a peripheral skirt.

the peripheral skirt and the cones is $k_p p_{cu}$, and assume that the air inside the cushion is substantially at rest; then from Bernoulli's theorem, the velocity of the air escaping under the peripheral skirt V_{c2} is given by

$$V_{c2} = \left(\frac{2k_p p_{cu}}{\rho}\right)^{1/2} \tag{8.10}$$

and the total volume flow from the peripheral skirt Q_2 is given by

$$Q_2 = h_{c2} l_{c2} D_{c2} \left(\frac{2k_p p_{cu}}{\rho}\right)^{1/2} \tag{8.11}$$

where h_{c2}, l_{c2}, and D_{c2} are the clearance height, perimeter, and discharge coefficient of the peripheral skirt, respectively.

Under steady-state conditions, the total lift force generated by the system is given by

$$F_{cu} = W = p_{cu} A_{c1} + k_p p_{cu} A_{c2} \tag{8.12}$$

where A_{c1} is the total cushion area of the cones and A_{c2} is the cushion area between the peripheral skirt and the cones. The cushion pressure p_{cu} required to support the vehicle weight is expressed by

$$p_{cu} = \frac{W}{A_{c1} + k_p A_{c2}} \tag{8.13}$$

Based on the assumptions of inviscid, incompressible flow, the total volume of air escaping under the cones Q_1 is equal to that escaping under the peripheral skirt Q_2:

$$Q_1 = n_c h_{c1} l_{c1} D_{c1} \left[\frac{2(1-k_p) p_{cu}}{\rho}\right]^{1/2} = Q_2 \tag{8.14}$$

where n_c is the number of cones, and h_{c1}, l_{c1}, and D_{c1} are the clearance height, perimeter, and discharge coefficient of the cones, respectively.

The power required to sustain the air cushion is given by

$$P_a = p_{cu} Q_1 = p_{cu} Q_2$$

$$= h_{c2} l_{c2} D_{c2} \left(\frac{W}{A_{c1} + k_p A_{c2}}\right)^{3/2} \left(\frac{2k_p}{\rho}\right)^{1/2} \tag{8.15}$$

Based on Eqs. (8.11) and (8.14), an expression for the pressure ratio k_p can be derived,

$$k_p = \frac{n_c^2 h_{c1}^2 l_{c1}^2 D_{c1}^2}{n_c^2 h_{c1}^2 l_{c1}^2 D_{c1}^2 + h_{c2}^2 l_{c2}^2 D_{c2}^2} \qquad (8.16)$$

It is found that the value of k_p calculated from the above equation is very close to that quoted in the literature. Note that the difference between the clearance heights h_{c1} and h_{c2} affects the pressure ratio k_p, and hence the characteristics of the cushion system.

The augmentation factor K_a of a multiple-cone system with a peripheral skirt is given by

$$K_a = \frac{A_{c1} + k_p A_{c2}}{2 k_p h_{c2} l_{c2} D_{c2}} \qquad (8.17)$$

It can be shown that, other conditions being equal, the augmentation factor of the multiple-cone system with a peripheral skirt would be much higher than that of an equivalent system without a peripheral skirt.

Another flexible skirt system of the plenum chamber type is the segmented skirt developed by the Hovercraft Development Ltd. (HDL), as shown in Figure 8.5(a) (Sullivan 1971; Trillo 1971). The unique feature of this type of skirt system is that the segments are unattached to one another. Consequently, when moving over a rough surface, only the segments in contact with the obstacles will deflect. When a segment is damaged or even removed, adjacent segments expand under cushion pressure and tend to fill the gap. Furthermore, the drag of the segmented skirt is found to be less than that of a continuous skirt because of its higher flexibility. The performance of the segmented skirt system may be predicted using the theory for a simple plenum chamber described previously.

Figure 8.5(b) shows the bag and finger skirt developed by the British Hovercraft Corporation (BHC) (Sullivan 1971). The fingers in this skirt system have similar characteristics to those of the segments in the segmented skirt. The cushion air is fed from the bag through holes into the fingers. A diaphragm is installed in the bag to help prevent the vertical oscillation of the skirt system.

Example 8.1 A multiple-cone system with a peripheral skirt like that shown in Figure 8.4 has the following parameters:

• Gross vehicle weight, W	48.93 kN (11,000 lb)
• Number of cones, n_c	8
• Perimeter of each cone, l_{c1}	3.6 m (11.8 ft)
• Perimeter of the peripheral skirt, l_{c2}	17.5 m (57.5 ft)
• Total cushion area of the cones, A_{c1}	8.2 m^2(88.3 ft^2)
• Cushion area between the cones and the peripheral skirt, A_{c2}	9.6 m^2(103.3 ft^2)
• Clearance heights, h_{c1} and h_{c2}	2.5 cm (1 in.)
• Discharge coefficients, D_{c1} and D_{c2}	0.60

Determine the power required to generate the lift and the augmentation factor.

(a)

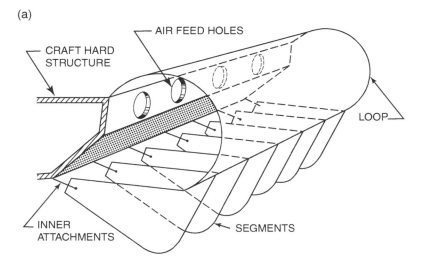

(b)

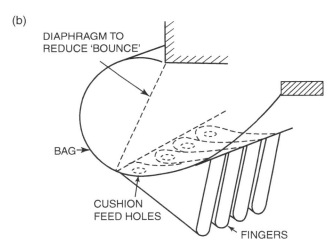

Figure 8.5 (a) Hovercraft Development Ltd. segmented skirt. *Source:* Trillo (1971). Reproduced with permission of R.L. Trillo. (b) British Hovercraft Corporation bag and finger skirt. Reprinted with permission from SAE paper 710183 © 1971 Society of Automotive Engineers, Inc.

Solution

From Eq. (8.16), for $D_{c1} = D_{c2}$ and $h_{c1} = h_{c2}$, the pressure ratio k_p is calculated as follows:

$$k_p = \frac{n_c^2 l_{c1}^2}{n_c^2 l_{c1}^2 + l_{c2}^2} = 0.73$$

From Eq. (8.13), the required cushion pressure p_{cu} is determined by

$$p_{cu} = \frac{W}{A_{c1} + k_p A_{c2}} = 3.22 \, \text{kPa} \left(67 \ \text{lb/ft}^2\right)$$

From Eq. (8.15), the power required to sustain the cushion is obtained by

$$P_a = h_{c2}l_{c2}D_{c2}\left(\frac{W}{A_{c1} + k_p A_{c2}}\right)^{3/2}\left(\frac{2k_p}{\rho}\right)^{1/2}$$

$$= 52.2\,\text{kW (70 hp)}$$

From Eq. (8.17), the augmentation factor K_a is determined as follows:

$$K_a = \frac{A_{c1} + k_p A_{c2}}{2k_p h_{c2}l_{c2}D_{c2}} = 39$$

8.1.2 Peripheral Jets

In the early days of development of the air-cushion technology, the peripheral jet system was used. This system is schematically shown in Figure 8.6. In this system, a curtain of air is produced around the periphery by ejecting air downward and inward from a nozzle. This curtain of air helps contain the cushion under the vehicle and reduces air leakage. Thus, it could offer higher operational efficiency than the simple plenum chamber.

In addition to the lift force generated by the cushion pressure, the air jet also provides a small amount of vertical lift. Under steady-state conditions, the weight of the vehicle W is balanced by the lift force F_{cu}:

$$F_{cu} = W = p_{cu}A_c + J_j l_j \sin\theta_j$$

where J_j is the momentum flux of the air jet per unit length of the nozzle, which is the product of the jet velocity and mass flow rate per unit nozzle length; l_j is the nozzle perimeter; and θ_j is the angle of the nozzle from the horizontal.

There are several theories for predicting the performance of peripheral jet systems. Among them, the so-called "exponential theory" is one of the most used. In this theory, it is assumed that from the outlet of the nozzle (point A) to the point of ground contact (point B), the jet maintains its thickness

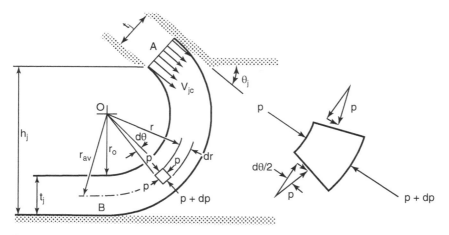

Figure 8.6 Geometry of a peripheral jet system.

as well as its circular path, and that the air is inviscid and incompressible. The total pressure p_j is assumed to be constant across the jet with a static pressure gradient within it. The distribution of static pressure p across the jet must satisfy the boundary conditions, that is, $p = 0$ on the outside and $p = p_{cu}$ on the cushion side.

Consider a small element of the jet at a distance r from the center of curvature O. The pressure difference across the element is balanced by the centrifugal force, and the equation of equilibrium for the element is given by

$$(p + dp)(r + dr)d\theta - pr \, d\theta - 2p \sin (d\theta/2)dr = \frac{\rho V_{jc}^2}{r} r \, dr \, d\theta$$

where V_{jc} is the velocity of an element of the jet.

Neglecting second-order terms and making simplifications, such as $\sin(d\theta/2) \approx d\theta/2$, one can rewrite the equation above as

$$\frac{dp}{dr} = \frac{\rho V_{jc}^2}{r} \tag{8.18}$$

Since the total pressure p_j is assumed to be constant across the jet, from Bernoulli's theorem, the following relation is obtained:

$$p_j = p + \frac{\rho V_{jc}^2}{2} \tag{8.19}$$

Substituting Eq. (8.19) into (8.18), one obtains

$$\frac{dp}{p_j - p} = \frac{2dr}{r} \tag{8.20}$$

Since the variation of r is limited, the value of r in the above equation may be considered constant and equal to the average radius of curvature of the path r_{av}. Integrating Eq. (8.20) and substituting the limits $r = r_0$, $p = 0$; $r = r_0 + t_j$, $p = p_{cu}$, one obtains the following expression relating the cushion pressure p_{cu} and the total pressure of the jet p_j:

$$\frac{p_{cu}}{p_j} = 1 - e^{-2t_j/r_{av}} \tag{8.21}$$

where t_j is the thickness of the jet or nozzle width.

Noting that $r_{av} \approx h_j/(1 + \cos \theta_j)$, one obtains

$$\frac{p_{cu}}{p_j} = 1 - e^{-2t_j(1 + \cos \theta_j)/h_j} \tag{8.22}$$

where h_j is the clearance height.

The total volume flow Q_j is given by

$$Q_j = \int_{r_0}^{r_0 + t_j} l_j V_{jc} dr$$

$$= \frac{l_j h_j}{1 + \cos \theta_j} \sqrt{\frac{2p_j}{\rho}} \left[1 - \sqrt{1 - p_{cu}/p_j} \right] \tag{8.23}$$

and the power required is

$$P_{aj} = p_j Q_j = \frac{l_j h_j (1 - e^{-x}) p_{cu}^{3/2} (2/\rho)^{1/2}}{(1 + \cos\theta_j)(1 - e^{-2x})^{3/2}} \tag{8.24}$$

where $x = t_j (1 + \cos\theta_j)/h_j$.

For given values of h_j, l_j, p_{cu}, and θ_j, the power requirement is a minimum for

$$\frac{\partial P_{aj}}{\partial x} = 0$$

which gives $x = 0.693$. The minimum power $p_{aj\,min}$ is expressed by

$$P_{aj\,min} = \frac{4 l_j h_j p_{cu}^{3/2} (2/\rho)^{1/2}}{3\sqrt{3}(1 + \cos\theta_j)} \tag{8.25}$$

Comparing the power requirement of a simple plenum chamber P_a with the minimum power of a peripheral jet $P_{aj\,min}$ having the same cushion pressure and similar dimensions ($l_j = l_c$ and $h_j = h_c$), one may obtain the following power ratio:

$$\frac{P_a}{P_{aj\,min}} = \frac{3\sqrt{3}D_c(1 + \cos\theta_j)}{4} \tag{8.26}$$

Assume that $\theta_j = 45°$ and $D_c = 0.6$, the power requirement of a simple plenum chamber will be 33% higher than that of an equivalent peripheral jet system. The augmentation factor K_{aj} for a peripheral jet system is expressed by

$$\begin{aligned} K_{aj} &= \frac{p_{cu}A_c + J_j l_j \sin\theta_j}{J_j l_j} \\ &= \frac{p_{cu}A_c}{J_j l_j} + \sin\theta_j \\ &= \frac{p_{cu}D_h}{4 J_j} + \sin\theta_j \end{aligned} \tag{8.27}$$

The moment flux per unit nozzle length J_j can be determined by

$$\begin{aligned} J_j &= \int_{r_0}^{r_0 + t_j} \rho V_{jc}^2 dr \\ &= \int_{r_0}^{r_0 + t_j} 2\left(p_j - p\right) dr \\ &= \int_0^{p_{cu}} r\, dp = r_{av} p_{cu} \end{aligned} \tag{8.28}$$

Substituting Eq. (8.28) into (8.27), one obtains

$$\begin{aligned} K_{aj} &= \frac{D_h}{4 r_{av}} + \sin\theta_j \\ &= \frac{D_h}{4 h_j}(1 + \cos\theta_j) + \sin\theta_j \end{aligned} \tag{8.29}$$

Comparing the augmentation factor of a simple plenum chamber with that of a peripheral jet system having the same hydraulic diameter and clearance height, one can obtain the following ratio:

$$\frac{K_{aj}}{K_a} = 2D_c \left(1 + \cos\theta_j + 4\sin\theta_j \frac{h_j}{D_h}\right) \tag{8.30}$$

Assume that $D_c = 0.6$, $\theta_j = 45°$ and $h_j/D_h = 0.001$, the augmentation factor of a peripheral jet system is approximately twice that of an equivalent simple plenum chamber.

While in theory the peripheral jet system appears to be superior to the plenum chamber, in practice it is not necessarily so. It has been found that using flexible nozzles for the peripheral jet system, difficulties arise in maintaining the jet width and angle, and in excessive nozzle wear. Using relatively hard nozzles, on the other hand, would induce high surface drag. Moreover, the advent of the flexible skirt enables the clearance height, and hence the power requirement for lift, to be reduced considerably, while maintaining sufficient clearance between the hard structure of the vehicle and the supporting surface. On some vehicles, the designed clearance height is only a few millimeters. This renders the power-saving aspect of the peripheral jet system rather insignificant. All of these have led to the use of an essentially plenum chamber configuration in almost all air-cushion vehicles.

8.2 Resistances of Air-Cushion Vehicles

There are drag components unique to air-cushion vehicles, which require special attention. For overland operations, in addition to aerodynamic resistance, there are momentum drag, trim drag, and skirt contact drag. For overwater operations, additional wave-making drag, wetting drag, and drag due to waves must be considered.

As mentioned previously, the introduction of flexible skirts permits a considerable reduction of clearance height, and hence power for lift. It should be pointed out, however, that the reduction of clearance height would likely increase the skirt contact drag, thus increasing the power for propulsion. Consequently, a proper balance between the reduction of lift power and the associated increase in propulsion power must be struck to achieve a minimum total power requirement.

The aerodynamic resistance of an air-cushion vehicle can be evaluated using the methods discussed in Chapter 3. Typical values for the coefficient of aerodynamic resistance C_D obtained from wind-tunnel tests range from 0.25 for SR.N2 to 0.38 for SR.N5 based on frontal area (Elsley and Devereux 1968). For a surface effect ship, a value of 0.5 for C_D has been reported (Trillo 1971).

8.2.1 Momentum Drag

To sustain the cushion, air is continuously drawn into the cushion system. When the vehicle is moving, the air is effectively accelerated to the speed of the vehicle. This generates a resisting force in the direction of the air relative to the vehicle, which is usually referred to as the momentum drag R_m. The momentum drag can be expressed by

$$R_m = \rho Q V_a \tag{8.31}$$

where V_a is the speed of the air relative to the vehicle, and Q is the volume flow of the cushion system. This momentum drag is unique to air-cushion vehicles.

Part of the power to overcome momentum drag may be recovered from utilizing the dynamic pressure of the airstream at the inlet of the fan to generate the cushion pressure. The dynamic pressure of the airstream at the intake of the fan p_d is given by

$$p_d = \frac{\rho V_a^2}{2} \tag{8.32}$$

Assume that the efficiency of the cushion system including the fan and ducting is η_{cu}, then the power that can be recovered from generating the cushion pressure is given by

$$P_r = \frac{\eta_{cu}\rho Q V_a^2}{2} \tag{8.33}$$

Since the power required to overcome the moment drag P_m is equal to $\rho Q V_a^2$, the ratio of P_r to P_m is

$$\frac{P_r}{P_m} = \frac{\eta_{cu}\rho Q V_a^2}{2\rho Q V_a^2} = \frac{\eta_{cu}}{2} \tag{8.34}$$

This indicates that if the efficiency of the cushion system η_{cu} is 100%, half of the power expended in overcoming the momentum drag can be recovered.

8.2.2 Trim Drag

If the cushion base of the vehicle is not horizontal, the lift force that is perpendicular to the cushion base will have a horizontal component. This horizontal component is given by

$$R_{tr} = p_{cu}A_c \sin \theta_t \tag{8.35}$$

where θ_t is the trim angle (i.e., the angle between the cushion base and the horizontal).

R_{tr} may be a drag or a thrust component, dependent on whether the vehicle is trimmed nose up or down.

8.2.3 Skirt Contact Drag

For overland operations, contact between the skirt and the ground may be inevitable, particularly at low clearance heights. This gives rise to a drag component commonly known as the skirt contact drag R_{sk}. The physical origin of this drag component appears to be derived from the following major sources: friction between the skirt and the ground, and the deformation of the skirt and the terrain, including vegetation, due to skirt–ground interaction (Fowler 1975). A reliable method for predicting skirt contact drag is lacking, although from experience it is known that the cushion pressure, clearance height, skirt design and material, and the strength and geometry of the terrain surface have a significant influence on the skirt contact drag. The value of the skirt contact drag is usually obtained from experiments. Table 8.1 gives the values of the coefficient of towing resistance of two air-cushion trailers and a self-propelled air-cushion vehicle over various types of terrain (Eggleton and Laframboise 1974; Liston 1973). One of the air-cushion trailers is equipped with a Bertin-type multiple-cone system having a peripheral skirt and is built by HoverJak; the other is equipped with the Hovercraft Development Ltd. segmented skirt and is built by Terracross (Eggleton and Laframboise 1974). The self-propelled air-cushion vehicle is a Bell SK-5 equipped with the type of fingered skirt developed by the British Hovercraft Corporation (Liston 1973). The coefficient of towing resistance is defined as the ratio of the towing resistance to the total vehicle weight.

For the air-cushion trailer built by HoverJak, 2% of the total vehicle weight is carried by the guided wheels, whereas for the one built by Terracross, 7% of the vehicle weight is carried by the guided wheels. The values given in Table 8.1 for the air-cushion trailers include, therefore, both the skirt contact drag and the rolling resistance of the guided wheels. The values given in Table 8.1 for the Bell SK-5 may be considered those of the coefficient of skirt contact drag since no guided wheels were used.

Table 8.1 Coefficient of towing resistance

Type of vehicle	Type of air-cushion system	Terrain	Coefficient of towing resistance	Total vehicle weight (kN)
Air-cushion trailer by HoverJak	Multiple-cone system with peripheral skirt	Concrete, dry	0.002-0.005	148.3 (33,341 lb)
		Flat rock, dry	0.014-0.018	
		Dry mud	0.011-0.016	130.8 (29,406 lb)
		Sandy road	0.023	
Air-cushion trailer by Terracross	H.D.L segmented skirt	Wet flat rock	0.018	143.5 (32,262 lb)
		Water or mud	0.015	206.9 (46,515 lb)
		Wet mud	0.019	
		Dry mud	0.022-0.037	
		Churned marsh	0.035	
Self-propelled air-cushion vehicle Bell SK-5	B.H.C fingered skirt	Rough hummocky snow, hard glazed surface	0.002	58.1 (13,062 lb)
		Rock strewn creek bed, left rough	0.012-0.022	
		Rock strewn creek bed, graded level	0.020-0.030	
		Swamp grass, tufts in water	0.006-0.034	
		Light brush on rough ground	0.075-0.25	

Sources: Eggleton and Laframboise (1974), Liston (1973).

In logged-over areas with stumps, the average values of the coefficient of towing resistance range from 0.06 to 0.24 for an air-cushion trailer equipped with a Bertin-type multiple-cone system having a peripheral skirt (Silversides et al. 1974).

Knowing the value of the coefficient of skirt contact drag C_{sk}, one can calculate the skirt contact drag R_{sk} from the following equation:

$$R_{sk} = C_{sk}W \tag{8.36}$$

where W is the total vehicle weight.

As mentioned previously, among the various factors, the skirt clearance height has a considerable influence on the skirt contact drag. To establish quantitative relationships between the clearance height of the skirt and the coefficient of skirt contact drag over various surfaces, a series of experiments were carried out (Fowler 1979). The experiments were performed using segmented skirts at a low speed of approximately 2 m/s (6.6 ft/s) over surfaces ranging from concrete through terrains covered with long grass to rough, porous ground with crushed rock. Figure 8.7 shows the variation of the coefficient of skirt contact drag with the product of the clearance height h_c and the coefficient

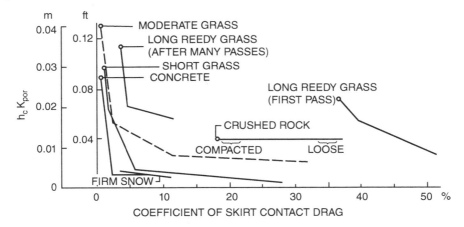

Figure 8.7 Effects of clearance and ground porosity parameter $h_c K_{por}$ on skirt contact drag coefficient over different surfaces. *Source:* Fowler (1979).

K_{por}, based on measured data. The coefficient K_{por} considers the effects of ground porosity on the volume flow of the air-cushion system. For instance, over porous ground with crushed rock, the cushion air will escape through not only the clearance between the skirt and the surface, but also the void between crushed rocks under the cushion. This indicates that the total volume flow of the air from the cushion will be higher by a factor of K_{por} than that calculated using Eqs. (8.3), (8.11), (8.14), or (8.23). Accordingly, the power required to sustain the air cushion will also be higher. The approximate values of K_{por} for various types of ground are given in Table 8.2 (Fowler 1979). It shows that over a smooth concrete, the value of K_{por} is 1, which indicates that the cushion air only escapes through the clearance between the skirt and the concrete surface. Over crushed rock, the value of K_{por} can be as high as 6, which indicates that the volume of air escaping from the cushion through the void between rocks is five times that through the clearance between the skirt and the surface.

From Figure 8.7, the coefficient of skirt contact drag increases significantly when the value of $h_c K_{por}$ falls below a threshold value. For instance, over a smooth concrete, if the value of $h_c K_{por}$ is lower than approximately 0.0035 m (0.14 in.), the value of the coefficient of skirt contact drag will increase significantly. For long reedy grass (after many passes), the threshold value of $h_c K_{por}$ is approximately 0.02 m (0.79 in.). The value of $h_c K_{por}$ determines the volume flow and power for lift, while the coefficient of skirt contact drag affects the power for propulsion. To achieve an optimum

Table 8.2 Values of coefficient K_{por} for various surfaces.

Ground	K_{por}
Smooth concrete	1
Firm snow	1.5
Short grass	6
Moderate grass	6
Long reedy grass (first pass)	6
Long reedy grass (tenth pass)	6
Crushed rock	6

Source: Fowler (1979).

operating condition for an overland air-cushion vehicle, the value of $h_c K_{por}$ must, therefore, be carefully selected so that the total power requirement (including both power for lift and for propulsion) is minimized (Wong 1973).

8.2.4 Total Overland Drag

For a vehicle wholly supported by an air cushion operating overland, the total drag consists of the aerodynamic resistance, momentum drag, trim drag, and skirt contact drag. Although wholly air-cushion-supported vehicles can function overland, they are relatively difficult to maneuver and control in constricted space and in the traverse of a slope. The longitudinal slope that this type of vehicle can negotiate is also limited. To solve these problems, surface-contacting devices such as wheels, tracks, and the like may be used. In this type of arrangement, the air cushion is used to carry a proportion of the vehicle weight, while the remaining part of the vehicle weight is carried by the surface-contacting device for directional control, positioning, and possibly for traction and braking. A vehicle that uses an air cushion together with a surface-contacting device for supporting the vehicle weight and directional control is usually referred to as a "hybrid vehicle."

For the hybrid vehicle, the resistance of the surface-contacting device must be taken into consideration in computing the total overland drag. The resistance of the wheel or the track over unprepared terrain can be predicted using the methods described in Chapter 2. It is found that, among the design parameters, the load distribution between the air cushion and the surface-contacting device has a considerable effect on the total power consumption of the hybrid vehicle (Wong 1972a). Figure 8.8 shows the variation of power consumption with the ratio of the load supported by the air cushion W_a to the total vehicle weight W for a particular hybrid vehicle equipped with tires over clay (Wong 1972a). For a given hybrid vehicle over a particular type of terrain, there is an optimum load distribution that could minimize the power consumption. Figure 8.9 shows the variation of the optimum load distribution with terrain conditions for a particular hybrid vehicle equipped with tires (Wong 1972a).

Another type of overland vehicle system employing air-cushion technology is the air-cushion trailer-towing vehicle system (Wong 1972b). It consists of two separate units: an air-cushion trailer and a towing vehicle. Figure 8.10 shows schematically an air-cushion trailer built by Terracross (Eggleton and Laframboise 1974). The towing vehicle is usually a conventional tracked or wheeled vehicle. The system offers the convenience of an ordinary tractor–trailer unit. However, since the air-cushion trailer is not self-propelled, the mobility of the system depends on that of the towing vehicle. This restricts the use of this system to areas where the conventional towing vehicle can operate effectively. The towing vehicle must develop sufficient drawbar pull to overcome the total drag acting on the air-cushion trailer, which includes the skirt contact drag, resistance of guided wheels, and trim drag. This type of system normally operates at low speeds; aerodynamic resistance and momentum drag acting on the air-cushion trailer would be insignificant and may be neglected.

As mentioned previously, one of the basic functions of the air-cushion system is to support the vehicle weight. Consequently, power is required to generate the lift of the air cushion. To compare the relative merits of an air-cushion vehicle with a conventional ground vehicle on a rational basis, the power required to generate the lift by the cushion should be considered equivalent to part of the power required to overcome the motion resistance of a conventional vehicle. The concept of the equivalent coefficient of motion resistance f_{eq} for a vehicle wholly supported by an air cushion is proposed. It is defined as

$$f_{eq} = \frac{P_a}{WV} + \frac{R_m + R_a + R_{sk}}{W} \tag{8.37}$$

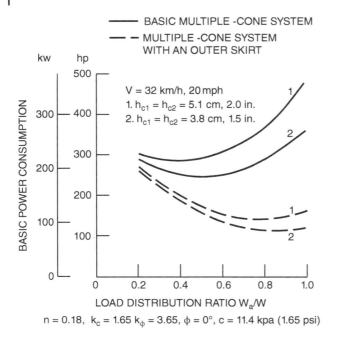

$n = 0.18$, $k_c = 1.65$ $k_\phi = 3.65$, $\phi = 0°$, $c = 11.4$ kpa (1.65 psi)

Figure 8.8 Variations of basic power consumption with load distribution ratio for a hybrid vehicle in clay; the values of k_c and k_ϕ are in U.S. customary units.

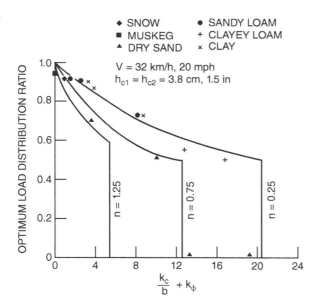

Figure 8.9 Variations of the optimum load distribution with terrain conditions for a hybrid vehicle equipped with a multiple-cone system having a peripheral skirt; the values of k_c and k_ϕ are in U.S. customary units.

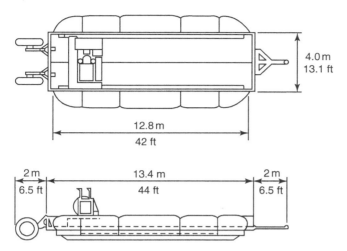

Figure 8.10 An air-cushion trailer with segmented skirt. Reproduced with permission from Eggleton and Laframboise (1974).

where P_a is the power required to sustain the air cushion, W is the total weight of the vehicle, and V is the vehicle speed.

The equivalent coefficient of motion resistance of an air-cushion vehicle depends on the operating speed. For a hybrid vehicle partly supported by an air cushion and partly supported by a surface-contacting device, such as a track or a wheel, the equivalent coefficient of motion resistance is defined as

$$f_{eq} = \frac{P_a}{WV} + \frac{R_m + R_a + R_{sk} + R_r}{W} \tag{8.38}$$

where R_r is the motion resistance of the surface-contacting device.

Figure 8.11 shows the variation of the equivalent coefficient of motion resistance with operating speed for a particular hybrid vehicle equipped with tires over clay, loose sand, and snow (Wong 1972b).

8.2.5 Wave-Making Drag

When an air-cushion vehicle travels over water, waves will be generated, as shown in Figure 8.12. The vehicle tends to align itself with the wave, and the cushion base will be inclined. Thus, the lift force produces a rearward component that is commonly known as the wave-making drag.

To better understand the mechanism that generates the wave-making drag, it is instructive to consider the nature of the interaction between the air cushion and the water. When the vehicle is on a cushion over water at zero forward speed, the water will be depressed by an amount equal to $p_{cu}/p_w g$, where g is the acceleration due to gravity and p_w is the mass density of the water, as shown in Figure 8.12(a). When the vehicle travels forward, the water under the front part of the vehicle is just coming under the action of the cushion pressure, whereas under the rear part of the vehicle, the water surface has been subjected to cushion pressure for a certain period. Consequently, the water surface will be inclined downward toward the rear. As the vehicle tends to align itself with the water surface, the cushion base will take a nose-up attitude, as shown in Figure 8.12(b). The rearward component of the lift force perpendicular to the cushion base gives rise to the wave-making drag.

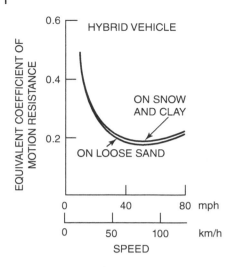

Figure 8.11 Variations of the equivalent coefficient of motion resistance with speed of a hybrid vehicle over different types of terrain.

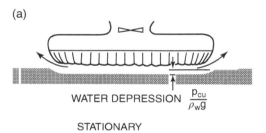

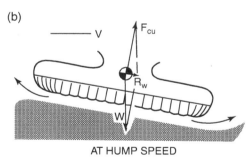

Figure 8.12 Formation of wave-making drag: (a) air-cushion vehicle stationary; (b) at hump speed.

The magnitude of this drag component increases with speed and reaches a maximum at a particular speed that is usually referred to as the "hump speed." As the vehicle speed further increases, the time during which the cushion interacts with the water surface becomes shorter. Consequently, the depression of the water becomes less, and the water surface under the vehicle begins to approach level again. The wave-making drag, therefore, decreases accordingly.

The wave-making drag may be predicted with sufficient accuracy by various methods. For a relatively long and narrow air-cushion vehicle, a two-dimensional theory for predicting the wave-making drag R_w based on Lamb's work has been proposed (Crewe and Egginton 1960; Elsley and Devereux 1968):

$$R_w = \frac{2p_{cu}^2 A_c}{l\rho_w g}\left(1 - \cos\frac{gl}{V^2}\right)$$

or

$$\frac{R_w l}{p_{cu}^2 A_c} = \frac{2}{\rho_w g}\left(1 - \cos\frac{gl}{V^2}\right) \tag{8.39}$$

where l is the length of the cushion; p_{cu} is the cushion pressure; A_c is the cushion area; V is the vehicle forward speed; and $V/\sqrt{gl}$ is the Froude number. The variation of $R_w l/p_{cu}^2 A_c$ with the Froude number is shown in Figure 8.13 (Elsley and Devereux 1968). The wave-making drag is a maximum when the Froude number is 0.56 or $\cos(gl/V^2)$ is equal to -1. This condition is commonly known as the "hump", and the associated drag is called "hump drag." Note that the wave-making drag is proportional to the square of the cushion pressure.

A more accurate method for predicting wave-making drag that takes the shape of the planform of the vehicle into account has been developed (Newman and Poole 1962). The water depth also affects the wave-making drag. Over shallow water, the wave-making drag is higher than that over deep water (Newman and Poole 1962).

8.2.6 Wetting Drag

The wetting drag is a drag component due to water spray striking the skirt and skirt–water contact. Although it is known that the clearance height, cushion pressure, vehicle size and shape, and vehicle speed have varying degrees of influence over the magnitude of the wetting drag, no satisfactory analytical method exists for the prediction of this drag component. A common practice to determine the wetting drag is to measure the total drag over calm water by model or full-scale testing,

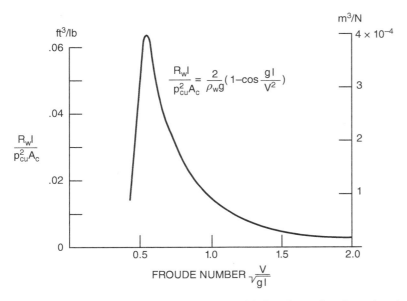

Figure 8.13 Variations of wave-making drag with Froude number. Reproduced with permission from *Hovercraft Design and Construction* by G.H. Elsley and A.J. Devereux, copyright © by Elsley and Devereux (1968).

and then to subtract those drag components that are known or calculable. Thus, the wetting drag R_{wet} is given by (Elsley and Devereux 1968; Trillo 1971)

$$R_{wet} = R_{tot}(calm\ water) - R_a - R_m - R_w \tag{8.40}$$

where R_{tot} (calm water) is the total drag measured over calm water.

8.2.7 Drag Due to Waves

So far, no analytical method is available for the prediction of the drag due to waves. Its value is obtained from model or full-scale testing (Elsley and Devereux 1968; Trillo 1971). Taking the difference between the total drag over waves R_{tot} (over waves) and that over calm water R_{tot} (calm water), at the same speed, one can obtain the drag due to waves R_{wave}:

$$R_{wave} = R_{tot}(over\ waves) - R_{tot}(calm\ water) \tag{8.41}$$

The wave height, cushion pressure, skirt depth, and vehicle speed have noticeable effects on the drag due to waves.

8.2.8 Total Overwater Drag

For overwater operations, the total drag of an air-cushion vehicle consists of the aerodynamic resistance, momentum drag, wave-making drag, wetting drag, and drag due to waves. Figure 8.14 shows the relative order of magnitude of various drag components as a function of vehicle speed for overwater operations (Trillo 1971). For vehicles with sidewalls, additional sidewall drag, mainly due to skin friction over the immersed surface, should also be taken into account.

The power consumption of an air-cushion vehicle consists of two major parts: power for lift and power for propulsion. In designing an air-cushion vehicle, power for lift and power for propulsion should, therefore, not be considered in isolation. For instance, increasing the clearance height or volume flow would reduce the skirt contact drag, and hence the power for propulsion. However, as discussed previously, the power for lift is proportional to the clearance height. Thus, a compromise must be made in selecting the clearance height so that the total power requirement would be a minimum, and it should also be compatible with other criteria such as skirt wear and ride comfort.

Example 8.2 The air-cushion vehicle described in Example 8.1 is to be used in overland transport. The frontal area of the vehicle is 6.5 m²(70 ft²) and the aerodynamic drag coefficient is 0.35. The coefficient of the skirt contact drag is estimated to be 0.04. Determine the total drag of the vehicle over level ground at a speed of 20 km/h (12.4 mph).

Solution

The total overland drag includes momentum drag, aerodynamic drag, and skirt contact drag.

(a) The momentum drag R_m is given by Eq. (8.31):

$$R_m = \rho Q V_a = \rho \left[h_{c2} l_{c2} D_{c2} \left(\frac{2k_p p_{cu}}{\rho} \right)^{1/2} \right] V_a$$

Figure 8.14 Characteristics of drags of an air-cushion vehicle over water. *Source:* Trillo (1971). Reproduced with permission of R.L. Trillo.

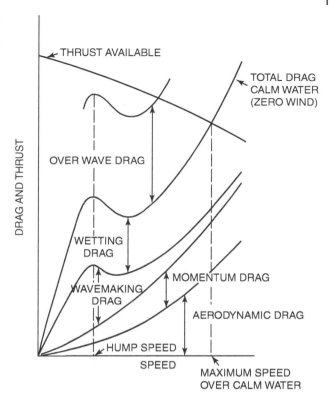

Substituting the appropriate values given into the equation above, one obtains

$$R_m = 110.8 \text{ N } (24.9 \text{ lb})$$

(b) The aerodynamic drag R_a can be determined using Eq. (3.20):

$$R_a = \frac{\rho}{2} C_D A_f V^2 = 43.2 \text{ N}(9.7 \text{ lb})$$

(c) The skirt contact drag R_{sk} can be estimated using Eq. (8.36):

$$R_{sk} = C_{sk} W = 1957 \text{ N } (400 \text{ lb})$$

The total overland drag is the sum of the above three drag components:

$$R_{\text{tot}} = R_m + R_a + R_{sk} = 2.111 \text{ kN } (474.6 \text{ lb})$$

The results indicate that the momentum drag and aerodynamic drag are insignificant at low speeds.

8.3 Suspension Characteristics of Air-Cushion Systems

One of the major functions of an air cushion is to act as a suspension system for the vehicle. To define its characteristics as a suspension, the stiffness and damping in bounce (or heave), roll, and pitch must be determined.

8.3.1 Heave (or Bounce) Stiffness

The heave (or bounce) stiffness can be derived from the relationship between the lift force and vertical displacement. For a simple plenum chamber, this relationship depends, to a great extent, on the fan characteristics. For a practical plenum chamber system, the ducting between the fan and the cushion and the feeding arrangements for the cushion also have a significant influence on its stiffness and damping characteristics. Consider a simple plenum chamber that is in equilibrium at an initial clearance height h_{c0} with initial cushion pressure p_{c0} and volume flow Q_0. Neglecting ducting losses, the fan will operate at point A, as shown in Figure 8.15. Consider that the cushion system is disturbed from its equilibrium position, and that the clearance height decreases by an amount Δh_c. Accordingly, the volume flow will decrease by an amount ΔQ_0, and the pressure will increase from p_{c0} to $p_{c0} + \Delta p_c$. The operating point of the fan shifts from A to A'. Thus, a restoring force that tends to bring the cushion system back to its original equilibrium position is created, and the system is stable in heave.

If the parameters of the cushion system and the fan characteristics are known, the heave stiffness about an equilibrium position can be predicted (Guienne 1964). An approximate method for predicting the stiffness of a simple plenum chamber is described below to illustrate the procedures involved.

The general relationship between the pressure and volume flow of a fan commonly used in air-cushion vehicles is expressed by

$$p = f(Q)$$

and

$$\frac{dp}{dQ} = f'(Q) < 0 \tag{8.42}$$

where $f'(Q)$ is the slope of the pressure–flow characteristic curve of the fan.

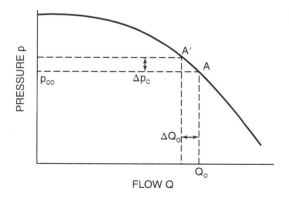

Figure 8.15 Pressure–flow characteristics of a fan.

The volume flow from the cushion for a simple plenum chamber is governed by Eq. (8.3) and the relationship between the cushion pressure and the air escaping velocity is given by Eq. (8.2). By differentiating Eqs. (8.2) and (8.3), the following relationships can be obtained.

From Eq. (8.2),

$$dV_c = \frac{dp_{cu}}{\rho V_c} \tag{8.43}$$

and from Eq. (8.3),

$$dQ = h_c l_{cu} D_c \ dV_c + V_c l_{cu} D_c \ dh_c \tag{8.44}$$

Substituting Eq. (8.43) into (8.44), one obtains

$$dQ = h_c l_{cu} D_c \frac{dp_{cu}}{\rho V_c} + V_c l_{cu} D_c \ dh_c \tag{8.45}$$

Neglecting pressure losses between the fan and the cushion (i.e., $p = p_{cu}$) and combining Eqs. (8.42) and (8.45), one obtains

$$dp_{cu} \left[\frac{1}{f'(Q)} - \frac{h_c l_{cu} D_c}{\rho V_c} \right] = V_c l_{cu} D_c \ dh_c$$

or

$$\frac{dp_{cu}}{\rho V_c^2/2} \left[\frac{\rho V_c}{h_c l_{cu} D_c f'(Q)} - 1 \right] = \frac{2 \, l_{cu} D_c \ dh_c}{h_c l_{cu} D_c} \tag{8.46}$$

Since $p_{cu} = \rho V_c^2/2$, the above equation can be written as

$$\frac{dp_{cu}}{p_{cu}} = \left[\frac{2}{\rho V_c/h_c l_{cu} D_c f'(Q) - 1} \right] \frac{dh_c}{h_c} \tag{8.47}$$

The lift force generated by the cushion F_{cu} is equal to $p_{cu} A_c$; the above equation, therefore, can be rewritten as

$$\frac{dF_{cu}}{dh_c} = \left[\frac{2}{\rho V_c/h_c l_{cu} D_c f'(Q) - 1} \right] \frac{F_{cu}}{h_c}$$
$$= K_h \frac{F_{cu}}{h_c} \tag{8.48}$$

Equation (8.48) gives the equivalent heave stiffness of a simple plenum chamber about the equilibrium position. The heave stiffness is strongly dependent on the slope of the pressure–flow characteristic curve of the fan $f'(Q)$. Since the value of $f'(Q)$ varies with the operating point of the fan, heave stiffness is a function of operating conditions. The general characteristics of the cushion pressure–displacement relationship of a simple plenum chamber under steady-state conditions are shown in Figure 8.16 (Trillo 1971). It shows that the air cushion is essentially a nonlinear system. However, for motions with small amplitudes about an equilibrium position, the system may be linearized. For a peripheral jet system, the equivalent heave stiffness is essentially proportional to the derivative of the cushion pressure with respect to clearance height, which may be obtained from Eq. (8.22).

The damping characteristics of an air-cushion system may be determined experimentally, for instance, using a dynamic heave table (Trillo 1971). The cushion system being tested is mounted

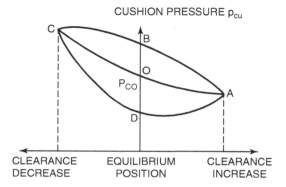

CUSHION PRESSURE p_{cu}

Figure 8.16 Variations of cushion pressure with clearance height of a simple plenum chamber. *Source:* Trillo (1971). Reproduced with permission of R.L. Trillo.

above the heave table, which can move up and down relative to the cushion with various amplitudes and frequencies. At a particular amplitude and frequency, the variation of cushion pressure with displacement in a complete cycle is measured. If the system possesses damping, the variation of cushion pressure with displacement will follow different paths during the upward and downward strokes of the table, as shown in Figure 8.16. The area enclosed by curve *ABCDA* represents the degree of damping the cushion system possesses. The damping of an air cushion is usually not of a simple viscous type. It is asymmetric and dependent on the frequency of motion. However, to simplify the analysis, an equivalent viscous damping coefficient c_{eq} for the air cushion may be derived based on equal energy dissipation:

$$c_{eq} = \frac{U}{\pi \omega Z^2} \tag{8.49}$$

where U is the actual energy dissipated in the air cushion during a cycle that is represented by the area enclosed by curve *ABCDA* in Figure 8.16; and ω and Z are the circular frequency and amplitude of the heave table, respectively.

Figure 8.17 shows schematically an air-cushion system designed for high-speed guided ground vehicles (Morel and Bonnat 1975). One of its unique features is the inclusion of a damper to provide the vehicle with sufficient damping to achieve the required ride quality.

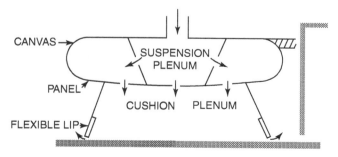

Figure 8.17 An air-cushion system designed for high-speed guided ground vehicles. *Source:* Morel and Bonnat (1975). Reproduced with permission of Swets & Zeitlinger.

8.3.2   Roll and Pitch Stiffness

Stability in the roll and pitch of air-cushion vehicles may be achieved by two methods: differential pressure and differential area. The multiple-cone system developed by Bertin obtains stability in roll and pitch from the pressure differential between the down-going side and the up-going side of the skirt system, as shown in Figure 8.18. When the vehicle rolls, the clearance height on the down-going side decreases. From previous analysis, it is known that the volume flow will decrease, and the cushion pressure will increase. On the up-going side, however, the cushion pressure will decrease because of the increase of the clearance height and volume flow. The increase of the lift force on the down-going side and the decrease of the lift force on the up-going side form a restoring moment that tends to bring the cushion system back to its equilibrium position.

Consider that the simple cushion system shown in Figure 8.18 rolls a small angle $\Delta\theta$ with respect to the equilibrium position. The clearance height on the down-going side will decrease by an average amount Δh_c:

$$\Delta h_c = (B/2)\Delta\theta \tag{8.50}$$

where B is the beam of the cushion.

On the up-going side, the clearance height will increase by the same amount. From Eq. (8.48), the restoring moment ΔM_{ro} corresponding to the angular displacement $\Delta\theta$ is expressed by

$$\Delta M_{ro} = B\Delta F_{cu} = \frac{BF_{cu}K_h}{h_c}\Delta h_c = \frac{B^2 F_{cu}K_h}{2h_c}\Delta\theta \tag{8.51}$$

In the limit, the roll stiffness of the system K_r is given by

$$K_r = \frac{dM_{ro}}{d\theta} = \frac{B^2 F_{cu}K_h}{2h_c} \tag{8.52}$$

Figure 8.19 shows the variation of the restoring moment coefficient C_{ro}, which is equal to $2M_{ro}/WB'$, with roll angle for a 1/5 scale model of the Bertin BC 8 air-cushion vehicle (Bertin 1968). The effects of the difference in clearance height between the cones and the peripheral skirt on the roll characteristics are illustrated in the figure. When the roll angle exceeds a certain range and the

Figure 8.18   Roll stability by differential pressure.

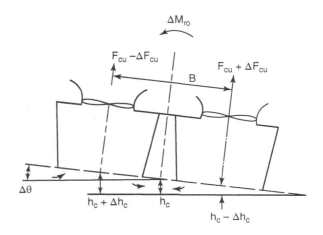

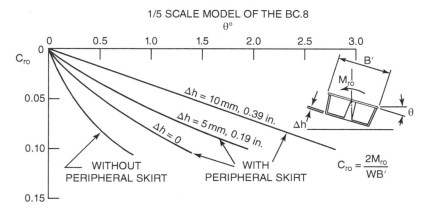

Figure 8.19 Roll characteristics of the Bertin BC 8 cushion system. *Source:* Bertin (1968). Reproduced with permission of the *Canadian Aeronautics and Space Journal.*

down-going side of the skirt comes into contact with the ground, the roll characteristics of the multiple-cone system may change significantly, and considerable hysteresis has been observed (Sullivan et al. 1977).

Roll and pitch stability can also be achieved by using inflated bags (or keels) to divide the cushion into compartments. This method has been used by the British Hovercraft Corporation in their skirt systems, as shown in Figure 8.20(a) (Sullivan 1971). The air pressure in the fan plenum is common to all compartments. However, when the vehicle rolls, on the up-going side the air flow increases and, consequently, the cushion pressure decreases because of increased pressure losses through the cushion feed holes, as shown in Figure 8.5(b). On the down-going side, the air flow decreases, and the cushion pressure increases accordingly. As a result, a restoring moment is generated, which tends to return the system to its original equilibrium position. This method of achieving roll and pitch stability is essentially based on the principle of differential pressure.

The method for obtaining stability in roll and pitch by differential area has been employed by Hovercraft Development Ltd. in the design of their skirt systems. Stability is achieved by the outward movement of the down-going side of the skirt, thus increasing the cushion area of the down-going side, as shown in Figure 8.20(b) (Eggleton and Laframboise 1974). Consequently, the lift force on the down-going side increases and a restoring moment is generated.

8.4 Directional Control of Air-Cushion Vehicles

For vehicles wholly supported on an air cushion, their relative freedom from the surface presents unique problems in directional control. The methods for directional control may be divided into four main categories: aerodynamic control surfaces, differential thrust, thrust vectoring, and control ports. These methods are illustrated in Figure 8.21 (National Research Council of Canada 1969).

Using an aerodynamic control surface, such as a rudder in the slipstream of the air propeller, could provide an effective means for directional control of vehicles wholly supported on an air cushion. However, its effectiveness diminishes with a decrease of the slipstream velocity at low thrust. The control surface may also induce an adverse rolling moment if the center of pressure of the control surface is high relative to the center of gravity of the vehicle.

Figure 8.20 (a) Roll stability by compartmentation. Reprinted with permission from SAE paper 710183 © 1971 Society of Automotive Engineers, Inc. (b) Roll stability by differential area. Reproduced with permission from Eggleton and Laframboise (1974).

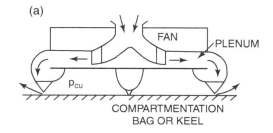

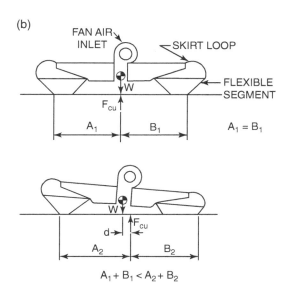

An adequate degree of directional control may be achieved by differential thrust produced by twin propellers fixed side by side, as shown in Figure 8.21. The differential thrust may be obtained by controlling the propeller pitch and/or rotating speed. However, decreasing the thrust on one of the propellers reduces the total forward thrust available, and hence the vehicle speed. In this fixed side-by-side propeller configuration, the thrust is parallel to the longitudinal axis of the vehicle. To provide a lateral force to balance the centrifugal force during a turning maneuver, the vehicle must operate with a certain yaw angle, as shown in Figure 8.22 (National Research Council of Canada 1969).

Using fore and aft swiveling pylon-mounted propellers, the yawing moment and side force required for direction control can be generated. For some designs, the swivel angle is confined to 30° on either side of the longitudinal axis to limit the magnitude of the adverse roll moment. Compared with the fixed side-by-side propeller arrangement, swiveling pylon-mounted propellers can generate a higher yawing moment since the propellers can be mounted further from the center of gravity of the vehicle and less forward thrust is lost for a given yawing moment.

By discharging pressurized air from the so-called "puff-ports" located at each corner of the vehicle or through nozzles mounted at appropriate locations of the vehicle, the yawing moment and side force can be provided. They are usually used as an auxiliary device to supplement other control devices.

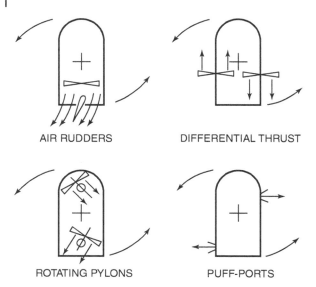

Figure 8.21 Methods for directional control of air-cushion vehicles. *Source:* National Research Council of Canada (1969).

AIR RUDDERS

DIFFERENTIAL THRUST

ROTATING PYLONS

PUFF-PORTS

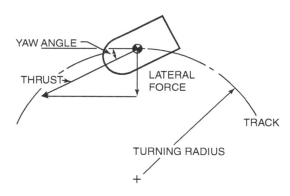

Figure 8.22 Turning of an air-cushion vehicle with a yaw angle. *Source:* National Research Council of Canada (1969).

YAW ANGLE

THRUST

LATERAL FORCE

TRACK

TURNING RADIUS

To further improve the directional control of air-cushion vehicles, surface-contacting devices, such as the wheel for overland operations and the retractable water rod (or the rudder) for overwater operations, have been used. For overland operations, the wheels carry a proportion of the vehicle weight to provide the vehicle with the required cornering force for directional control. The load carried by the wheels ranges from 2 to 30% of the total vehicle weight in existing designs, depending on whether the wheels are also used as a propulsive device. It has been found that using the wheel as a directional control device is quite effective (Wong 1972a).

The cornering force that a wheel can develop for control purposes consists of two major components: the lateral shearing force on the contact area, and the lateral force resulting from the normal pressure exerted on the sidewall of the wheel, which is similar in nature to that acting on a bulldozer blade or a retaining wall, as illustrated in Figure 8.23. The magnitude of this force depends on the sinkage of the wheel and terrain properties, and it may be predicted by the earth pressure theory of soil mechanics discussed in Chapter 2.

Figure 8.23 Development of cornering force by a tire on deformable terrain.

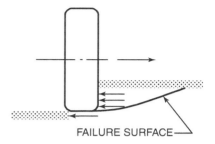

FAILURE SURFACE

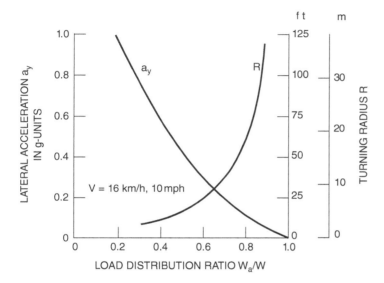

Figure 8.24 Cornering characteristics of an air-cushion vehicle with tires for directional control in clay.

As an example, Figure 8.24 shows the variation of the maximum lateral acceleration a_y that can be sustained under a steady-state turn with load distribution for a particular hybrid vehicle with tires over clay (Wong 1972a). The lateral acceleration shown is calculated from the maximum cornering force that can be developed by the tires of the vehicle. The possible minimum turning radius of the vehicle at a forward speed of 16 km/h (10 mph) is also plotted as a function of load distribution in Figure 8.24.

For overwater air-cushion vehicles, methods like those for controlling the direction of ships may be employed. For instance, rudders immersed in the water have been used in air-cushion vehicles with rigid sidewalls for purposes of directional control.

References

Bertin, J. (1968). French air cushion vehicle developments. *Canadian Aeronautics and Space Journal* **14** (1).

Crewe, P.R. and Egginton, W.J. (1960). The hovercraft: A new concept in maritime transport. *Quarterly Transactions of Royal Institute of Naval Architects* **102** (3).

Eggleton, P.L. and Laframboise, J. (1974). Field evaluation of towed air cushion rafts. Report TDA-500-166. Transportation Development Agency, Ministry of Transport, Ottawa, Canada.

Elsley, G.H. and Devereux, A.J. (1968). *Hovercraft Design and Construction*. Baltimore, MD: Cornell Maritime Press.

Fowler, H.S. (1975). The air cushion vehicle as a load spreading transport device. *Journal of Terramechanics* **12** (2).

Fowler, H.S. (1979). On the lift-air requirement of air cushion vehicles and its relation to the terrain and operational mode. Report 17492 (ME-246). National Research Council of Canada, Ottawa, Canada.

Guienne, P. (1964). Stability of the Terraplane on the ground. *Hovering Craft and Hydrofoil*, July.

Liston, R.A. (1973). Operational evaluation of the SK-5 air cushion vehicle in Alaska. Report TR 413. U.S. Army Cold Regions Research and Engineering Laboratory, Hanover, NH.

Morel, J.P. and Bonnat, C. (1975). Air cushion suspension for Aerotrain: Theoretical schemes for static and dynamic operation. *Proc. IUTAM Symposium on the Dynamics of Vehicles on Roads and Railway Tracks* (ed. H.B. Pacejka). Amsterdam, Netherlands.

National Research Council of Canada. (1969). *Air Cushion Vehicles: Their Potential for Canada*. Ottawa, Canada.

Newman, J.N. and Poole, F.A.P. (1962). The wave resistance of a moving pressure distribution in a canal. *Schiffstechnik* **9** (45).

Silversides, C.R., Tsay, T.B., and Mucha, H.M. (1974). Effect of obstacles and ground clearance upon the movement of an ACV platform. Report FMR-X-62. Forest Management Institute, Department of the Environment, Ottawa, Canada.

Sullivan, P.A. (1971). A review of the status of the technology of the air cushion vehicle. *SAE Transactions* **80**, Society of Automotive Engineers paper: 710183.

Sullivan, P.A., Hinckey, M.J., and Delaney, R.G. (1977). An investigation of the roll stiffness characteristics of three flexible skirted cushion systems. Report 213. Institute for Aerospace Studies, University of Toronto, Toronto, Canada.

Trillo, R.L. (1971). *Marine Hovercraft Technology*. London: Leonard Hill.

Wong, J.Y. (1972a). Performance of the air-cushion–surface-contacting hybrid vehicle for overland operation. *Proceedings of the Institution of Mechanical Engineering* **186** (50/72).

Wong, J.Y. (1972b). On the applications of air cushion technology to overland transport. *High Speed Ground Transportation Journal* **6** (3).

Wong, J.Y. (1973). On the application of air cushion technology to off-road transport. *Canadian Aeronautics and Space Journal* **19** (1).

Problems

8.1 An air-cushion vehicle has a gross weight of 80.06 kN (18 000 lb). Its planform is essentially of rectangular shape, 6.09 m (20 ft) wide and 12.19 m (40 ft) long. The cushion system is of the plenum chamber type. The cushion wall angle is 45° with the horizontal. It operates at an average daylight clearance of 2.54 cm (1 in.). Determine the power required to sustain the air cushion. Also calculate the augmentation factor.

8.2 An air-cushion vehicle has the same weight and planform as those of the vehicle described in Problem 8.1 but is equipped with a multiple-cone system with a peripheral skirt. It has eight cones with a diameter of 2.44 m (8 ft). The average daylight clearance of the cones is 2.54 cm

(1 in.) and that of the peripheral skirt is 1.9 cm (0.75 in.). The wall angles of the cones and the peripheral skirt are 85° with the horizontal. Determine the power required to generate the cushion lift using a suitable peripheral skirt.

8.3 The air-cushion vehicle described in Problem 8.2 is employed for overland transport. The frontal area of the vehicle is 16.26 m^2(175 ft^2) and the aerodynamic drag coefficient is 0.38. The value of the coefficient of skirt contact drag over a particular terrain is 0.03. Determine the total overland drag of the vehicle at a speed of 20 km/h (12.4 mph). Also calculate the total power requirements, including both for lift and for propulsion, at that speed.

8.4 Determine the equivalent coefficient of motion resistance of the air-cushion vehicle described in Problem 8.3 at a speed of 20 km/h (12.4 mph).

8.5 The air-cushion vehicle described in Problem 8.1 is employed for overwater transport. The frontal area of the vehicle is 16.26 m^2(175 ft^2) and the aerodynamic drag coefficient is 0.38. Neglecting the wetting drag, determine the total overwater drag of the vehicle at the hump speed over calm, deep water. Also calculate the total power requirements of the vehicle at the hump speed.

Index

Theory of Ground Vehicles, Fifth Edition. J.Y. Wong.
© 2022 John Wiley & Sons, Inc. Published 2022 by John Wiley & Sons, Inc.
Companion website: www.wiley.com/go/wong/TGV5e